# Math Tools

Georg Glaeser

# Math Tools

## 500+ Applications in Science and Arts

Georg Glaeser
Department of Geometry
University of Applied Arts Vienna
Vienna
Austria

Translation from the German language edition: *Der mathematische Werkzeugkasten: Anwendungen in Natur und Technik* by Georg Glaeser


ISBN 978-3-319-66959-5 ISBN 978-3-319-66960-1 (eBook)
https://doi.org/10.1007/978-3-319-66960-1

Library of Congress Control Number: 2017951186

Mathematics Subject Classification: 00Axx, 53Axx, 62-XX, 97U20



Printed on acid-free paper

This Springer imprint is published by Springer Nature
The registered company is Springer International Publishing AG
The registered company address is: Gewerbestrasse 11, 6330 Cham, Switzerland

# Table of Contents

# 1 Introduction

Mathematics is "the science of numbers, quantities, and shapes and the relations between them".[1] Its history dates back almost three millennia – indeed, many insights found in this book have been known for centuries. The driving force behind the development of mathematics may be found in human nature: As a species, we seem to be uniquely interested in finding rational explanations for regular or repetitive phenomena in order to predict them. If not for this curiosity that appears to be innate to our species, many of the insights in this book may never have been discovered. In cultural terms, it took us a very long time to phrase these explanations in the language of science, and thus, to give them actual predictive power.
Mathematics has been (rather appropriately and beautifully) described as "a self-contained microcosm, with a strong potentiality of mirroring and modelling all the processes of thought and perhaps all of science".[2]

## Mathematical questions, and the involvement of the computer

There was a short period when a substantial amount of people believed that the computer would eventually cause mathematics to be partially replaced. However, mathematics is not merely the calculation of computational steps – its core is to be found in the logical thinking behind those steps.
In other words, the question is not *how*, but *why* certain computational steps have to be made. Once this is determined, the (often tedious) computation itself may actually begin. For this, the computer is truly a blessing, as it allows us to outsource the whole burden of routine operations, and to clear our minds for the comprehension of the relations from which they derive. With a clear mind, it is much easier to tackle more complex problems, and to discover – sometimes – that "one may proceed with the same tried and tested methodologies".

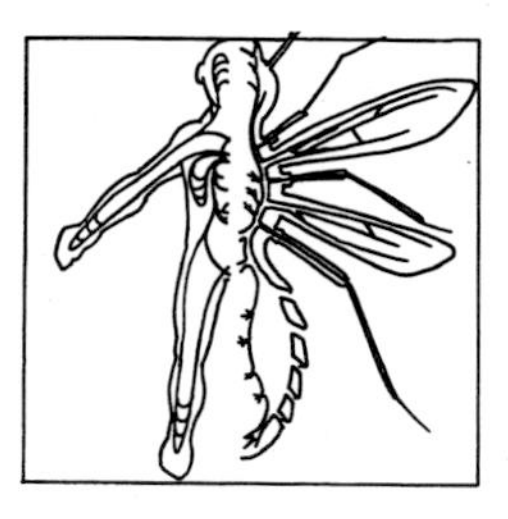

Why is there an upper bound for the size of insects, and a lower bound for the size of warm-blooded animals? The answer follows from an important theorem about similar objects.

[1] `http://www.merriam-webster.com/dictionary/mathematics`
[2] *Marc Kac, Stanislaw Ulam*: *Mathematics and Logic*, Dover Publ., 1992.


G. Glaeser, *Math Tools*, https://doi.org/10.1007/978-3-319-66960-1_1

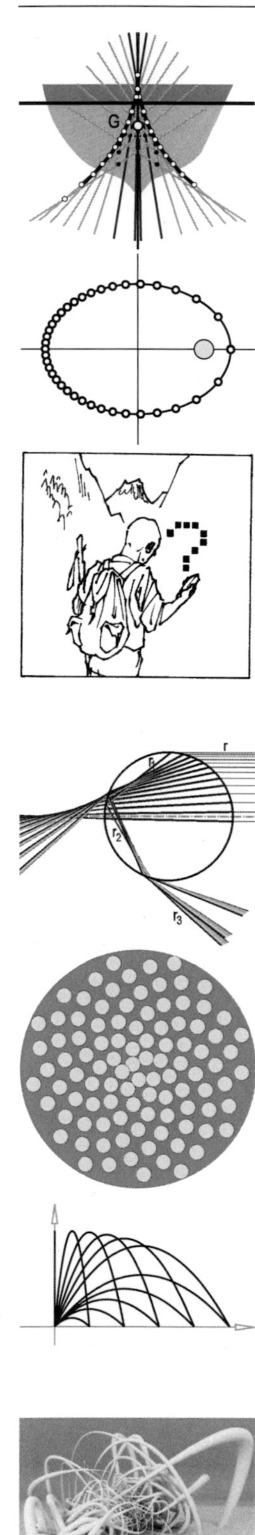


When and why does a ship capsize? This is a typical question for vector mathematics.

What implications does *Kepler*'s Second Law have for the seasons? To answer this question, we need to employ knowledge about trigonometry and ellipses.

How does GPS (="Global Positioning System") work? How many satellites are required to determine a spatial position, and how should their orbital paths be aligned to cover the Earth's surface most efficiently? When should one expect navigational problems? These questions require the techniques of analytic geometry.

Why do rainbows exist? Why does the sun appear as a glowing red fireball even after it has, from a purely physical point of view, already disappeared behind the horizon? In both cases, the answers are to be found in the refraction of light, which may be investigated using differential calculus.

What explains the intriguing spirals in sunflowers, the impressive shapes of antelope horns, or the beautifully geometric forms of snail shells? To answer these questions, we need to employ exponential functions.

How should a lawn sprinkler be moved so that the lawn may be irrigated evenly? How much air is consumed in a multi-level dive? What is the average life expectancy of people that have reached a certain age? Here, we need to employ integral calculus.

Why did natural selection "design" the heartbeat to be slightly irregular, and why can election results be accurately predicted by counting only 10% of the votes cast? The "law of large numbers" and descriptive statistics may be employed here.

$$\sum_{k=0}^{\infty} \frac{f^{(k)}(x_0)}{k!}(x - x_0)^k$$

How were the mathematical proportions of a musical instrument's strings first defined by Pythagoras, and how did musical scales and tonal systems evolve in the history of music? This concerns proportions that do not only make sense mathematically, but that are considered by people to be "harmonic".

Last but not least: How does a computer evaluate function expressions, and under which circumstances should we not rely on its alleged accuracy?

The answers to many of these questions become especially clear when its underlying principles are demonstrated as "animations". Consider a rocking boat, whose motion can be traced from any angle on a computer screen, which makes it easy to see which torsional momenta conspire to return it to its upright state. Consider a visualization of the orbital acceleration of a planet as it avoids being swallowed by its parent star, or a time lapse showing the growth of an antelope's horn. Such animations are easily accomplished through modern computers, but the underlying mathematics has been known since a much earlier time!

## Guiding principles of the book

This book traces its origin to the lectures on *Applied Mathematics* that I have been conducting for many years, for the benefit of students of architecture and industrial design, at the University of Applied Arts Vienna. I have tried to make the topics accessible to a much wider audience – essentially, to all people interested in the relations between mathematics and the most diverse of disciplines. My aim is to help put existing mathematical knowledge into a more structured and practical context.
In the meantime, parts of this book have already been published, mainly in the books
• G. Glaeser: *Der mathematische Werkzeugkasten*, Springer Spektrum, Heidelberg, *4th* edition 2014. Some applications are also to be found in
• G. Glaeser: *Nature and Numbers*, Ambra V/DeGruyter, Vienna, 2014,
• G. Glaeser: *Geometry and Its Applications*, Springer New York, 2012,
• G. Glaeser, K. Polthier: *Bilder der Mathematik*, Springer Spektrum, Heidelberg, $3^{rd}$ edition 2014.
• G. Glaeser, H. Stachel, B. Odehnal: *The Universe of Conics. From the ancient Greeks to 21st century developments.* Springer Spektrum, 2016.
Additionally, some biological applications can be found in
• G. Glaeser, H.F. Paulus: *The Evolution of the Eye.* Springer Nature, Heidelberg/New York 2015,
• G. Glaeser, H.F. Paulus, W. Nachtigall: *The Evolution of Flight.* Springer Nature, Heidelberg/New York 2017.
This book is not a treatise on mathematics in the classical sense, in which definitions, theorems, and proofs are strung together. Many of the practi-

cal examples within encourage interconnections between various, not always strictly technical, physical disciplines, such as biology, geography, archeology, medicine, music, applied arts, etc. Their larger purpose is to heighten awareness of the interconnecting laws underlying art and nature.
Thus, I have attempted to create a *self-contained reference*, aimed not at strict mathematicians, but at more pragmatic readers. The following base principles were followed whenever possible:
• Each chapter starts with a concise introduction to the theory, supported by various practical examples. Very often, a thorough theoretical education is not enough – when solving a particular challenge, one needs to be able to recognize the nature of the mathematical problem at its core.
• I have consciously avoided a strictly mathematical way of phrasing ideas, since it tends to distract more casual readers from the essence of the particular example. Details can always be elucidated once the relevant essence of a theorem has been understood.
• Proofs are conducted with appropriate rigor. However, I try to equip readers with the necessary knowledge to "derive" certain relations purely on the basis of common sense.
• A geometrically insightful sketch is usually preferred to mathematical abstraction, and for this reason, higher-dimensional problems are seldom found in this book.
• The general case – despite being sometimes more elaborate – is usually preferred to the special cases and their various subtleties. Thus, more complex equations of the form $f(x) = 0$, or definite integrals $\int_a^b f(x)dx$, are considered "solved", as the results can always be approximated by a computer to an arbitrary degree of precision.
• References to literature are supplemented with relevant web addresses, for the very simple reason that access to the Internet has become ubiquitous, while access to university libraries is not quite so easy and immediate.

## Structure of the book

The contents have been reduced to several chapters:
• First (Chapter 2), the *fundamentals* of applied mathematics are repeated on the basis of practical examples. A particular focus is laid upon algebraic equations and systems of linear equations.
• The third chapter discusses *proportions* through various examples. Similar bodies are of particular interest, and their analysis leads to simple but nontrivial conclusions that are enormously influential in nature.
• The fourth chapter is dedicated to calculations related to right and oblique triangles – and thus, to *trigonometry*. Here, too, there exists a wide scope of applications in various fields of inquiry.
• In the fifth chapter, the topic of *vector mathematics* is covered – though still with the means of elementary mathematics. Vectors play a fundamental role in physics, and are key to the application of geometrical problems on

the computer. Perhaps for this reason, their prominence and importance has grown enormously in recent times. The ease and elegance of the related calculations are demonstrated through many practical applications.

• The sixth chapter deals with the classical *real functions and their derivatives.* They are fundamentally important for many calculations and have lost much of their terror since the dawn of the computer age.

• The seventh chapter concerns *curves and surfaces.* Locus lines and path curves of so-called constrained motions are discussed in particular. Here, the computer is employed for their visualization. The software used for this purpose has been developed at the University of Applied Arts Vienna. The website of the book contains dozens of executable demo programs. This encourages geometrical creativity, as well as a *sensible* use of modern computing technology. In working with computers, the results should always be evaluated with a healthy dose of common sense.

• The eighth chapter guides the reader through simple and important *applications of calculus.* Complicated evaluations are performed on the computer. This affords enough space to promote the comprehension of general equations for the calculation of surfaces, centers of mass, etc.

• The ninth chapter is dedicated to statistics and probability calculus. The importance of these topics has grown considerably in recent times. A particular emphasis is placed on the analysis of data as a complement to descriptive statistics – in other words, on the art of learning from data.

• Finally, the appendix includes topics that were difficult to place into the preceding chapters, such as *complex numbers*, *Fibonacci numbers*, and the whole subject of *music and mathematics.*

## The accompanying website `www.uni-ak.ac.at/math`

This book is accompanied by its own website, which contains updates, extra examples, additional web addresses to the various topics covered in this book, as well as dozens of executables that allow you to explore mathematical relations interactively. In particular, they make it easy to comprehend complex motions and physical simulations.

The advantages of such a website are self-evident: It can always be kept up to date and augmented without changing the book at its foundation. Readers are encouraged to make frequent use of it. I am always grateful for constructive feedback!

## On the exactness of mathematics

The following "mathematician's joke" shows the discipline's characteristic self-image:

An engineer, a philosopher, and a mathematician are riding through the Scottish Highlands on a train and notice a black sheep within a herd. The engineer says: "I didn't know there were black sheep in Scotland." The philosopher immediately corrects him: "Hold on! This isn't so easy. You should have said: I didn't know there was *at least one* black sheep in Scotland."

The mathematician cannot help but object: "Gentlemen, this is still not precise enough. You should have said: I didn't know there was *at least one* sheep in Scotland with *at least one black side* ..."
Pure mathematics certainly concerns itself with degrees of exactness and descriptive precision that would be absurd in everyday life. In applying mathematics to solve practical problems, an engineer may interpret reality somewhat pragmatically, in order to achieve useful results more quickly. The truth may lie somewhere in between ...

## Acknowledgements

A good book is the result of year-long preparations, based in interaction and cooperation with many individuals.
For their contribution of interesting ideas and/or proof-reading, I would like to thank: Reinhard *Amon* (he contributed substantially to the "musical topics" which form the theoretical foundation of Appendix B), Franz *Gruber* (he came up with many ideas and several illustrations), Stefan *Wirnsperger* and Markus *Roskar* (hand-drawn illustrations), and Wilhelm *Fuhs*. For cooperation and correction, particular thanks are also due to my co-workers Franz *Gruber*, Günter *Wallner*, Eugenie Maria *Theuer* and especially Boris *Odehnal*, and also to the various interested readers that filled my inbox with suggestions for improvements, as well as typo sightings.
Finally, I would like to thank all my students who brought great enthusiasm into the classroom, and who contributed substantially to the book through constructive questions and discussions. They have never failed to give me the impression that mathematics, if taught with excitement and engagement, does not have to be the mere concern of an ivory-tower elite, but has the capacity to address people from all walks of life. The fear of the "dreaded subject" of mathematics, still present in some of us, may one day be replaced by an attitude of "curiosity and fascination".

# 2 Equations, systems of equations

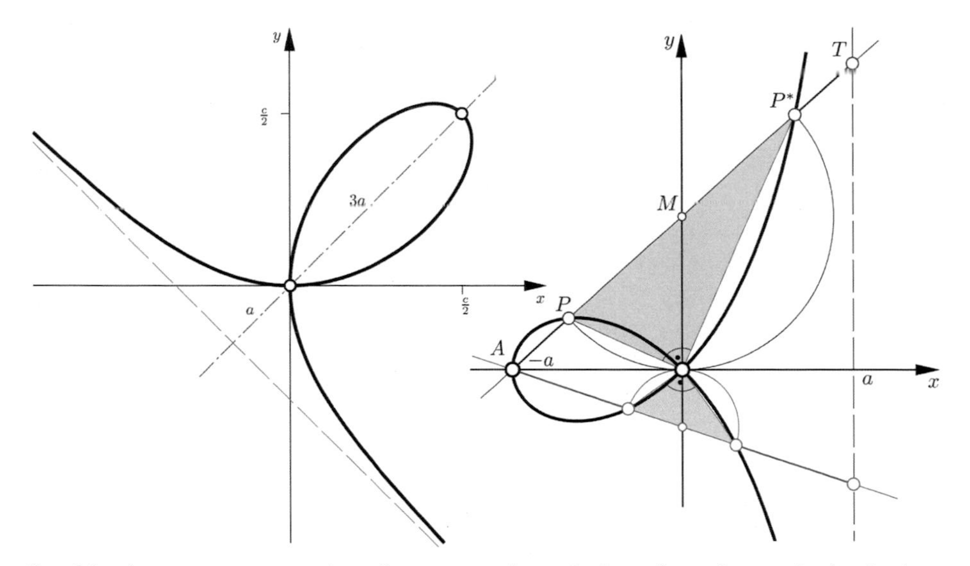


In this chapter, we examine elementary knowledge of mathematical relations from a higher point of view. Having understood the elementary relations thoroughly, we will conclude that many deceptively complex problems are grounded in simpler principles.
We will repeat the most elementary rules of algebra, such as the laws governing power functions, and the rules related to solving linear and quadratic equations.
Linear systems of equations will be particularly relevant to us, and may also be of interest to advanced readers, as they are not always trivial and often exhibit intricate interrelations with other branches of science, such as geography, physics, chemistry, photography, applied arts, and music. We will perform many calculations using real physical units. This will enable us to stay relevant to the solution of practical problems.
Due to the high computational complexity involved, we will only deal with higher-order algebraic equations conceptually. However, we will still find practical applications for their use.
At the end of the chapter, we will discuss further applications that cross over into fields of knowledge that are not exclusively concerned with mathematics.


G. Glaeser, *Math Tools*, https://doi.org/10.1007/978-3-319-66960-1_2

## 2.1 The fundamentals of numbers and equations

### Calculations with floating point numbers and precision

In the following section, we will perform calculations almost exclusively with real numbers. In general, our pocket calculator shows such numbers as decimals. Most calculators perform their task with an accuracy of ≈13–15 digits. We have to keep in mind that the numbers we enter into our calculator might be subject to imprecision (due to estimation, imprecise measurement, or rounding). A definitely *correct digit* (except zero, if placed in order to fix the decimal point) will be called a *valid digit.* Thus, the number 0.000123 possesses *at most* three valid digits. We now have to consider the following theorem:

> After a multiplication or a division, the number of valid digits is equal to the minimum number of valid digits in all operands. The result of an addition or subtraction has no valid digits beyond the last decimal place where *both* operands possess valid digits. One might say that the chain is only as strong as its weakest link!

▸▸▸ **Application**: *average velocity*

Consider a simple equation in physics: $v = s/t$ (velocity = distance over time). A car drives a distance of 88 km in 1 h 6 min . What is the average velocity, and how accurately can it be calculated?

**Fig. 2.1** average velocity

*Solution*:
1 h 6 min is equivalent to $1\,\text{h} + \frac{6}{60}\,\text{h} = 1.1\,\text{h}$. The above equation yields

$$v = \frac{s}{t} = \frac{88\,\text{km}}{1.1\,\text{h}} = 80\frac{\text{km}}{\text{h}}.$$

The distance of the route travelled was taken from a street map, which would indicate 88 km, even if the actual distance is 87.500 km or 88.499 km. In a similar manner, most people might be satisfied with a time measurement that is exact with minute precision. Thus, the actual car might have taken 1 h 5 min 30 s or 1 h 6 min 29 s, and we still would have entered 1 h 6 min into our calculation. However, the time measurement fluctuates between

$$1\,\text{h} + \frac{5}{60}\,\text{h} + \frac{30}{3\,600}\,\text{h} = 1.0917\text{h} \text{ and } 1\,\text{h} + \frac{6}{60}\,\text{h} + \frac{29}{3\,600}\,\text{h} = 1.108\text{h}.$$

We can, therefore, conclude that the "accurate result" might be found anywhere between

$$v = \frac{88.499}{1.0917} = 81.065\,\mathrm{km/h} \quad \text{and} \quad v = \frac{87.500}{1.108} = 78.971\,\mathrm{km/h}.$$

Therefore, it is correct to say that many decimal places of the calculated average velocity are baseless and may lead to misleading and unjustified impressions of accuracy!

⊕ *Remark*: Physicists tend to perform calculations in units of measurement. The velocity – including that of a car – is usually given in m/$s$, and time in seconds. Since $x\frac{1\,\mathrm{km}}{1\,\mathrm{h}} = x\frac{1000\mathrm{m}}{3600\,\mathrm{s}}$, the following frequently employed relation is true:

$$x\,\frac{\mathrm{km}}{\mathrm{h}} = \frac{x}{3.6}\,\frac{\mathrm{m}}{\mathrm{s}} \quad \text{and} \quad y\,\frac{\mathrm{m}}{\mathrm{s}} = 3.6y\,\frac{\mathrm{km}}{\mathrm{h}}.$$

The following example demonstrates the risk of using units of measurement carelessly.
Consider the following – incorrect – chain of equations

$$1€ = 100\,\mathrm{Cent} = 10\,\mathrm{Cent} \cdot 10\,\mathrm{Cent} = 0.1€ \cdot 0.1€ = 0.01€.$$

Can you find the mistake hidden therein? ⊕ ◂◂◂

▸▸▸ **Application**: ***moderate but constant***

Every driver ought to know that it is unwise to make up for lost time by "speeding". A moderate but constant velocity not only consumes less fuel (while also being less stressful), but is also more time-efficient than one might imagine. This provides an answer to the question "when to mount rain tires" in Formula 1 racing: Is it worth it to "put on" rain tires, which allow the vehicle to drive faster on wet surfaces, at the cost of a pit stop?
Let us consider a simple and very idealized example (Fig. 2.2): A race course of $s = 12\,\mathrm{km}$ (where 6 km run across easy and 6 km across difficult terrain) is to be traversed by a runner (at a constant velocity of $12\,\frac{\mathrm{km}}{\mathrm{h}}$) and a mountain biker ($4\,\frac{\mathrm{km}}{\mathrm{h}}$ faster than the runner in simple terrain, and $4\,\frac{\mathrm{km}}{\mathrm{h}}$ slower than the runner in difficult terrain). Which participant takes less time to cross the finish line? How long must the trajectory across simple terrain be for both participants to finish at the same time?
*Solution*:
Time for the runner $t_1 = \frac{12\,\mathrm{km}}{12\,\frac{\mathrm{km}}{\mathrm{h}}} = 1\,\mathrm{h}$.
Time for the mountain biker $t_2 = \frac{6\,\mathrm{km}}{(12{+}4)\,\frac{\mathrm{km}}{\mathrm{h}}} + \frac{6\,\mathrm{km}}{(12{-}4)\,\frac{\mathrm{km}}{\mathrm{h}}} = \frac{9}{8}\,\mathrm{h}$, or $7\frac{1}{2}$ minutes slower!
If the mountain biker is to cross the finish line at the same time as the runner, then the sum of the partial times for the different terrains must be 1 h:

$$\frac{x\,\mathrm{km}}{16\frac{\mathrm{km}}{\mathrm{h}}} + \frac{(12-x)\,\mathrm{km}}{8\frac{\mathrm{km}}{\mathrm{h}}} = 1\mathrm{h} \Rightarrow x = 8\,\mathrm{km}.$$

**Fig. 2.2** the great race

It follows that the easier terrain should cover *two thirds* of the entire course.

⊕ *Remark*: The "accordion effect" is an interesting phenomenon to note at this point: If a line of vehicles travels on the fast lane at a speed of 160 km/h, it may only take a single car, "breaking out" at 130 km/h from behind a truck, to bring the faster vehicles to a full stop: The driver at the front has a certain reaction time, in which the velocity of 160 km/h may still be maintained. To avert a collision with the slower vehicle, the velocity needs to be reduced, until it is significantly lower than that of the obstructing vehicle. However, the same problem arises for the subsequent car in the fast lane. Sufficient distances between the cars help to avoid traffic jam scenarios of this kind. ⊕ ◂◂◂

**▸▸▸ Application**: ***overtaking calculations***

A car (speed $v_1$) drives past another slower moving car (speed $v_2$), travelling in the same direction. How much time does it take for the faster car to pass the slower one and how much distance must the faster car cover?

**Fig. 2.3** overtaking with dog leash as tape measure

*Solution*:

The speed difference is $v_1 - v_2$. If we assume the length of the slower vehicle to be $A$ meters and we should be $B$ meters behind the vehicle before overtaking and cut in $C$ meters after the vehicle, we have to pace across $\Delta s = A + B + C$ meters with a speed of $\Delta v = v_1 - v_2$. The time needed for that is $t = \Delta s / \Delta v$. Clearly, the process depends in an indirectly proportional way on the speed difference. From the beginning of the overtaking process to the end, the faster car must cover a distance of $s_{\text{total}} = v_1 \cdot t$.

*Numerical example*:

$v_1 = 108$ km/h (= 30 m/s), $v_2 = 90$ km/h (= 25 m/s), length of slower vehicle $A = 10$ m.

Good numbers for $A$ and $C$ might be 15 m. Then, we have $\Delta s = 40$ m, $t = 8$ s and $s_{\text{total}} = 30\text{m/s} \cdot 8\text{s} = 240$ m. ◂◂◂

▸▸▸ **Application**: ***making up ground***

Two joggers $A$ and $B$ run together at a speed of 3.5 m/s. Then, $A$ decides to run back, whereas $B$ intends to continue for another 300 m, and then, turn around. He, therefore, immediately speeds up to 4 m/s, and separates with the words "I will join you soon". When is "soon"?

*Solution*:

Jogger $B$ gains 0.5 m every second. Thus, it will take him 1,200 seconds (20 minutes!) to join $A$ again. ◂◂◂

## Calculating with powers of ten

**Fig. 2.4** orders of magnitude in the animal Kingdom I: The *lengths* of the animals are 4,000 mm, 400 mm, 40 mm, 4 mm.

The careful treatment of powers of ten is important whenever one uses mathematics, such as when estimating results or when correctly interpreting computer calculations.

**Fig. 2.5** orders of magnitude in the animal Kingdom II: The *masses* of the animals are $10^9$ mg, $10^6$ mg, $10^3$ mg, 1 mg.

If $n = 1, 2, \ldots$ is any natural number, then the number $10^n$ represents a 1 followed by $n$ zeroes. When applying the rules for general powers that hold for any base, we obtain

$$\begin{aligned} 10^0 &= 1, \quad 10^1 = 10, \\ 10^n \cdot 10^m &= 10^{n+m}, \quad \frac{10^n}{10^m} = 10^{n-m}, \\ 10^{-n} &= \frac{1}{10^n}, \quad (10^n)^m = 10^{nm}, \\ \sqrt[m]{10^n} &= 10^{\frac{n}{m}}. \end{aligned}$$

The numbers $n$ and $m$ do not need to be natural and can each take on any real value.

**▸▸▸ Application**: ***car tire wear per kilometer*** (Fig. 2.6)
How strong is the wear of a car tire's outer surface if a brand new tire loses 1 cm in depth after being driven for about 50,000 kilometers?

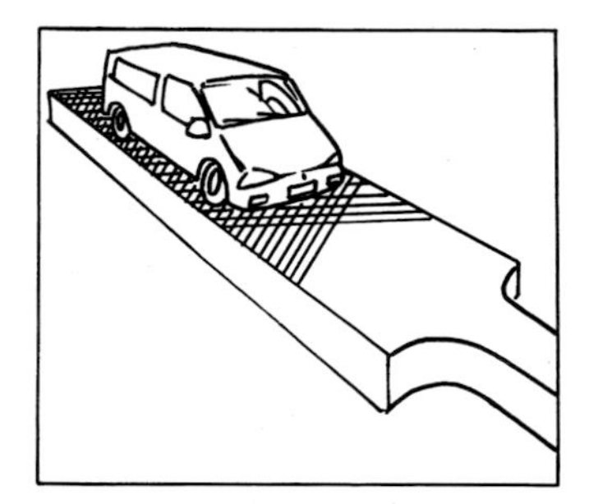
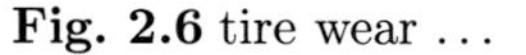

**Fig. 2.6** tire wear …

**Fig. 2.7** … with forward acting turbines

*Solution*:
If the outer surface wears down about $1\,\text{cm} = 10^{-2}\,\text{m}$ after $50{,}000\,\text{km} = 5 \cdot 10^4\,\text{km}$, then its wear per kilometer is

$$\frac{10^{-2}\text{m}}{5 \cdot 10^4} = \frac{10 \cdot 10^{-3}\text{m}}{5 \cdot 10^4} = 2 \cdot 10^{-7}\text{m} = 0.2 \cdot 10^{-6}\,\text{m} = 0.2\mu\text{m (micrometer)}.$$

⊕ *Remark*: A car tire has a diameter of approximately 65 cm (26 inches). Tread wear reduces the tire's diameter by a centimeter, and therefore, the circumference by approximately 2·1.5%. The reduction of speed necessary to prevent being caught by speed cameras at 100 $\frac{\text{km}}{\text{h}}$ is at least 3 $\frac{\text{km}}{\text{h}}$. Tires for aircrafts wear out much faster and have to be replaced more often than those for other types of vehicles. Fig. 2.7 shows a landing on one of the shortest runways in the world (Madeira) and — due to the need to reverse thrust — the shimmering hot air emitted by the turbines. ⊕
◂◂◂

**▸▸▸ Application**: ***hard disk space***
Suppose a hard disk has 100 GB (gigabytes) of storage. How many typed

pages (75 characters per line, 40 lines per page), or alternatively how many color photos (at a resolution of $1,000 \times 1,000$ pixels) can be stored in an uncompressed form?

⊕ *Remark*: Note: 1 typewritten ASCII character takes up 1 byte of space. A color pixel occupies 3 bytes (red, green, and blue components at 1 byte each). 1 gigabyte is equivalent to 1,024 megabytes, 1 megabyte to 1,024 kilobytes, and 1 kilobyte to 1,024 bytes. Nevertheless, we can estimate 1 GB as being equivalent to roughly $10^9$ bytes. Images are rarely stored in uncompressed bitmap format. Modern digital cameras allow taking photos at resolutions of $3,000 \times 2,000$ pixels or more. A 6 megapixel image would require 18 megabytes of space on the medium of storage. Instead, images tend to be stored in compressed form (JPG format) and can thus be fitted into about 2 megabytes – depending on the degree of compression and the type of the image. ⊕

*Solution*:
1 page $= 75 \cdot 40 = 3,000$ characters,
$\frac{100 \cdot 10^9}{3 \cdot 10^3} \approx 30 \cdot 10^6 = 30$ millions of pages,
$1,000 \times 1,000$ occupy $3 \cdot 10^6$ bytes $\Rightarrow \frac{100 \cdot 10^9}{3 \cdot 10^6} \approx 30 \cdot 10^3 = 30,000$ images. ◂◂◂

▸▸▸ **Application**: ***measuring distances through GPS***
Using GPS (Global Positioning System), it is possible to determine one's position at any point on Earth up to a few meters. We will discuss in Application p. 363 how these remarkably precise positions are calculated. In summary, it is necessary to take accurate measurements of one's distance to three satellites. These satellites continuously transmit characteristic signals that propagate at the speed of light. In order to determine the distance of the satellite, one must measure the time that it takes for the signal to reach the receiver, and multiply it by the speed of light. How accurately must such time measurements be taken so that the accuracy of position would fall within 1 m?

*Solution*:
The speed of light $300,000 \, \frac{\text{km}}{\text{s}} = 3 \cdot 10^5 \, \frac{\text{km}}{\text{s}} = 3 \cdot 10^8 \, \frac{\text{m}}{\text{s}}$ implies that the time for the signal to cross a meter is $\frac{1}{3} \cdot 10^{-8} \, \text{s} \approx 3.3 \cdot 10^{-9} \, \text{s}$.
One should be able to measure time to an accuracy of a nanosecond. Even in a microsecond, the GPS signal will already have travelled a distance of 300 m.

⊕ *Remark*: Even atomic clocks (which are used on satellites) can "only" measure microseconds reliably (the magnitude of error of such a clock equates to about 1 second in 6 million years). To overcome this difficulty, the GPS employs a trick: The characteristic signal encodes a description within the wave itself: Since it has a period of 300 m in length, the exact position within the oscillation at the time the wave is received can be observed. Thus, it is possible to say how many nanoseconds have passed since the last microsecond. ⊕ ◂◂◂

▸▸▸ **Application**: ***of billions and trillions***
Comparing the English language with other languages, the big numbers have names that are "false friends". This leads quite often to misunderstandings and wrong translations. $10^9$ is usually a "billion" in English, and a "milliard" in most other languages. $10^{12}$ is called a "trillion" instead of a "billion" elsewhere (`https://en.wikipedia.org/wiki/Names_of_large_numbers`). So, how far is a light year, how much is the gross product of Germany, when you find a German website with the content "Deutschland/Bruttoinlandsprodukt 3,73 Billionen USD (2013)", etc.

*Solution*:
In the case of the light year ($9 \cdot 10^{12}$km): "six trillion miles".
In the case of Germany's gross product: 3.73 billion USD (2013) or $3,73$ Milliarden USD, or, much better: $3.73 \cdot 10^9$ USD (2013). ◂◂◂

## Manipulations of equations

Applied mathematics is constantly working with formulas, which consist of relations between given values (which may be variables or constants) that output new values (variables). Usually, there exists a given "explicit solution" for a preferred variable. If one wants to calculate another variable within such a formula, one must transform the whole equation. We will now briefly remind ourselves of the main rules which may be used during such transformations.

## Elementary operations

Equivalence transformations are elementary (addition and subtraction as well as multiplication and division of equal terms or quantities on both sides of the equation, except division by zero). So, we find

$$a = b \Leftrightarrow a + c = b + c \quad \text{or} \quad a - c = b - c,$$

$$a = b \Leftrightarrow a\,c = b\,c \quad \text{or} \quad \frac{a}{c} = \frac{b}{c} \ (c \neq 0).$$

The product of a sum quantity and a sum is the sum of two products (distributive law):

$$a\,b + a\,c = a(b + c).$$

One often needs to employ

$$(a \pm b)^2 = a^2 \pm 2ab + b^2 \quad \text{and} \quad (a + b)(a - b) = a^2 - b^2.$$

Computing with powers (of the same base):

$$a^n = \underbrace{a \cdot a \cdot a \cdots a}_{n \text{ mal}} \Rightarrow a^n \cdot a^m = a^{n+m}, \text{ and } (a^n)^m = a^{n\,m}.$$

When raising values to a given power, parentheses are important, since

$$(a^n)^m \neq a^{(n^m)}. \tag{2.1}$$

⊕ *Remark*: For example, $(10^{10})^{10} = 10^{100}$ is a number with 100 zeros, and $10^{(10^{10})} = 10^{10000000000}$ is a number with 10 billion zeros. Even $4^{(4^4)} = 4^{256} \approx 10^{154}$ has 154 digits. The largest number that can be written using three digits is obviously $9^{(9^9)}$. This number cannot be represented on any computer (depending on the processor being used, the largest values used by computers fall within the range of $10^{300}$), but can only be processed with software for algebraic computation, such as *Derive*. ⊕

## Representing fractions

"Equating fractions":

$$\frac{a}{b} = \frac{c}{d} \Leftrightarrow a\,d = b\,c.$$

It follows from $\frac{1}{a} = \frac{1}{b}$ that $a = b$ and vice versa. Resolving double fractions means:

$$\frac{\frac{a}{b}}{c} = \frac{a}{b\,c}, \quad \frac{a}{\frac{b}{c}} = \frac{a\,c}{b}, \quad \frac{\frac{a}{b}}{\frac{c}{d}} = \frac{a\,d}{b\,c}.$$

"Common denominator":

$$\frac{a}{b\,c} + \frac{d}{b\,e} = \frac{a\,e + d\,c}{b\,c\,e}.$$

## Proportions, intercept theorem

Let us now address *axioms*, which are simple and apparently intuitive statements that are, nevertheless, unprovable, but that must be assumed in order to deduct logical conclusions from them. An example of this is the *intercept theorem*:

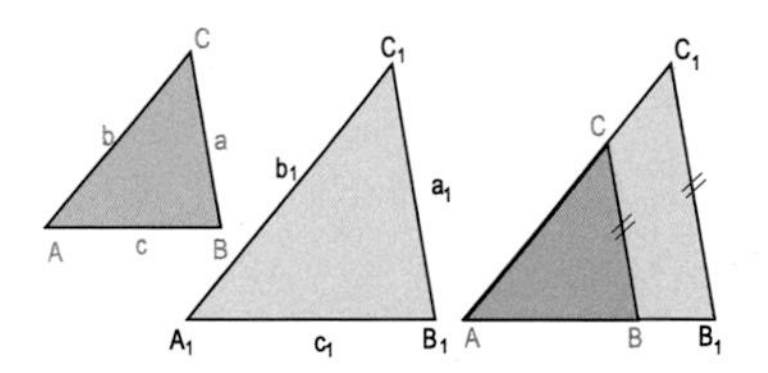


**Fig. 2.8** similar triangles, intercept theorem

Consider a triangle $ABC$, whose side lengths are represented by $a$, $b$, $c$ multiplied by a constant factor $k$ (the similarity factor). Thus, it is possible to describe a similar triangle $A_1B_1C_1$ (Fig. 2.8) with side lengths

$$a_1 = k\,a, \;\; b_1 = k\,b, \;\; c_1 = k\,c.$$

These relations of similarity are not limited to the specific triangles $ABC$ and $A_1B_1C_1$. In fact, they can be applied to any triangle with identical distance proportions.

*Similar triangles have equal angles.*
Furthermore, we have:

$$\overline{AB} : \overline{A_1B_1} = \overline{AC} : \overline{A_1C_1} \quad \text{and} \quad \overline{AB} : \overline{A_1B_1} = \overline{BC} : \overline{B_1C_1}. \tag{2.2}$$

If we move the triangles inside each other so that the legs of an angle join the legs of a corresponding angle of the other triangle, then the remaining two sides are parallel. The two ray sets are then described by formulas (2.2). In other words:

> If you intersect two neighboring sides of a triangle with a pair of parallel lines, then the ratio of the segments on one side equals the ratio of the segments on the other side, and it also equals the ratio of the segments on the parallels.

▸▸▸ **Application**: ***lens equation*** (Fig. 2.9, see also Application p. 54)
Consider a spherical, symmetrical, and convex lens such as a magnifying glass. A "thin lens" (where the spherical radius is much larger than the lens thickness) is a good approximation for light rays near the optical axis adhering to the following three laws (using the notation of Fig. 2.9a):

1. Any ray directed towards the lens center $Z$, "principal ray" $PP^*$, is not refracted.

2. Rays parallel to the optical axis $F\overline{F}$ are refracted so that their refractions are concurrent in $\overline{F}$ on the other side of the lens.

3. The outgoing rays of an object point $P$ are refracted so that their refractions concur in an image point $P^*$.

Let the distance of the object point $P$ to the lens center $Z$ be denoted by $g$ and $\overline{1Z} =: b$ (see Fig. 2.9).

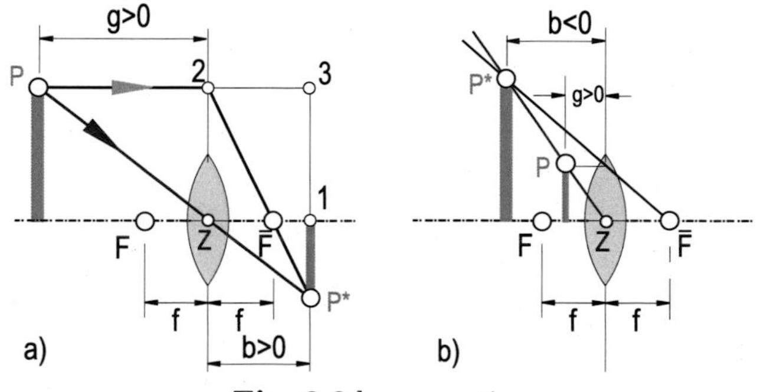


**Fig. 2.9** lens equation

Deduce the *lens equation*, in other words the relation between $g$ and $b$.

*Solution*:
The triangles $P2P^*$ and $Z\overline{F}P^*$ satisfy the conditions of the intercept theorem, which proves our claim, since

$$\overline{P^*P} : \overline{P^*Z} = \overline{P2} : \overline{Z\overline{F}} = g : f.$$

Analogously, for the triangles $P3P^*$ and $Z1P$:

$$\overline{P^*P} : \overline{P^*Z} = \overline{P3} : \overline{Z1} = (g+b) : b.$$

This yields

$$\frac{g}{f} = \frac{g+b}{b} \Rightarrow \frac{1}{f} = \frac{g+b}{bg} = \frac{g}{bg} + \frac{b}{bg}.$$

The simplified version is the easy-to-remember lens equation:

$$\boxed{\frac{1}{f} = \frac{1}{b} + \frac{1}{g}} \tag{2.3}$$

In practice, we often know $f$ and $g$. The distance $b$ of the image can be calculated by Formula (2.3) where it is multiplied by the common denominator $fgb$. This causes the other elements of the equation to move to one side of the equation. After that, $b$ "cancels out":

$$bg = fg + bf \Rightarrow bg - bf = fg \Rightarrow b(g-f) = fg \Rightarrow b = \frac{fg}{g-f}.$$

If we introduce $f$ as a "scalar" and set $g = kf$, we have

$$b = \frac{fkf}{kf - f} = \frac{k}{k-1} f.$$

**Fig. 2.10** convex lens (magnifying glass) with fuel and a magnifying effect

In Fig. 2.9a, the image of the (real) object is reduced and upside-down. It is only magnified if we move closer to the object so that $g < f$. In that case, the image is upright but virtual (see Fig. 2.9b, Fig. 2.10):

$$0 < g < f \Rightarrow 0 < k < 1 \Rightarrow \frac{k}{k-1} < 0.$$

**Fig. 2.11** concave lens (concave mirror) with fuel and a magnifying effect

The value $k = 1$ ($f = g$) is obviously a critical value (division by zero!): Thus, the image is "infinitely large" and is also "infinitely far away". For a very large $k$ ($k \to \infty$), the factor $\frac{k}{k-1}$ converges towards 1, and the value of $b$ converges to $f$ accordingly.

Rays parallel to the optical axis are bundled in the opposite focal point.

⊕ *Remark*: A similar property of "burning mirrors" was allegedly utilized by the Greek mathematician *Archimedes* (298 – 212 BC) to sink Roman ships (Fig. 2.11, left). However, the accuracy of this legend is doubted, because such an undertaking in practice would be quite difficult. After all, cases are known where dried grass ignited due to the internal effect of dewdrops. Anyway, one can still count Archimedes among the greatest mathematicians. He lived in Syracuse and in Egypt, where he discovered the respective laws for the lever and the lift. He also gave the first good approximation of the mathematical constant $\pi$, and even of integral calculus (!). On a side note, he was also the inventor of the Archimedean screw. ⊕ ◀◀◀

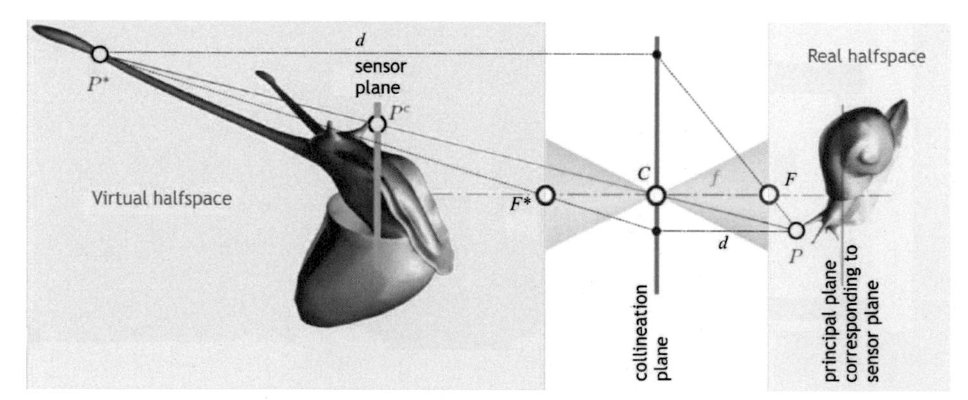


**Fig. 2.12** To each object in space, the lens system creates a collinear virtual object. The sensor plane corresponds to a plane perpendicular to the optical axis. Perfectly sharp points in a photo lie in that plane.

▸▸▸ **Application**: ***spatial images produced by lens systems*** (Fig. 2.12)

The illustrated ideal lens system generates for each point $P$ a virtual point

$P^*$. The pixel on the chip is the – theoretical – section of the ray $PZ$ with the sensor plane.
In reality, there is no "pixel": It is rather a circular "image spot" on the sensor (chip), which is formed by all the light rays emanating from $P$, going through the circular aperture, and forming an oblique circular cone.[1] This circle is called the *blurry circle* in photography. ◂◂◂

▸▸▸ **Application**: ***parallel resistors*** (Fig. 2.13)
The lens equation also occurs in an interesting manner with resistors connected in parallel. Denote by $R_1$ and $R_2$ the resistance of two resistors connected in parallel (so-called partial resistors), then the total resistance $R$ of the system is given by: $1/R = 1/R_1 + 1/R_2$.
The total resistance is always less than each partial resistance. Assuming that the total resistance $R = 50\,\Omega$ and one partial resistance $R_1 = 75\,\Omega$ are given, then we can find the other partial resistance $R_2$:

$$R_2 = \frac{R \cdot R_1}{R_1 - R} = \frac{50\,\Omega \cdot 75\,\Omega}{75\,\Omega - 50\,\Omega} = 150\,\Omega.$$ ◂◂◂

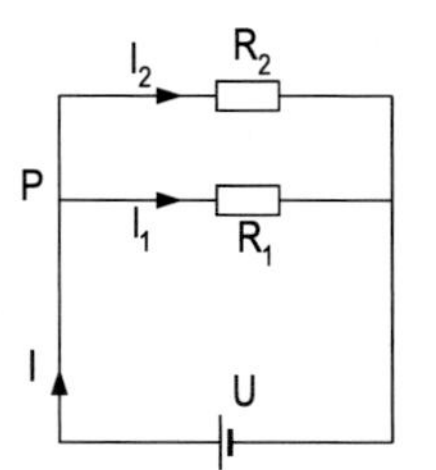


**Fig. 2.13** parallel resistors

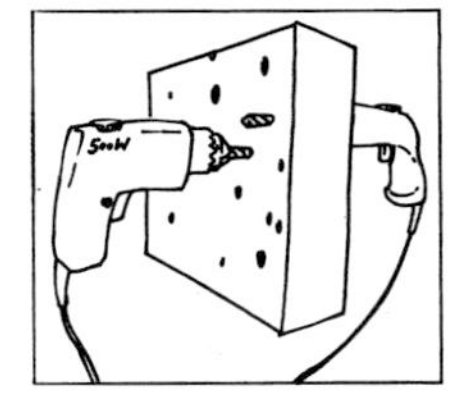

**Fig. 2.14** performing task

▸▸▸ **Application**: ***speeding up a process*** (Fig. 2.14)
A machine completes a task in 8 hours. How much time will it take a second machine to complete the same task, so that both machines can do the job in 4.8 hours?

*Solution*:
Let $W$ denote the total work. From physics we know that *Power* = *Work / Time.*
The performance of the first machine is $P_1 = \frac{W}{8}$, and the performance of the second machine is $P_2 = \frac{W}{x}$. The performance of both machines equals $P = \frac{W}{4.8}$. Since $P = P_1 + P_2$ the equation

$$\frac{W}{8} + \frac{W}{x} = \frac{W}{4.8} \Rightarrow \frac{1}{8} + \frac{1}{x} = \frac{1}{4.8}$$

is a "lens equation" from which $x = 12$ can be calculated as in Application p. 16. ◂◂◂

[1]For further information on this topic, see *G. Glaeser*: Geometry and its Applications. Springer Verlag, New York, 2012.

## Equations involving a single root

▸▸▸ **Application**: ***compound interest*** (Fig. 2.15)
The formula for the return of an output capital $K_0$ over $n$ years, with annual capital (interest rate $p$), reads

$$K = K_0\left(1 + \frac{p}{100}\right)^n.$$

Let the initial deposit be $K_0 = 10,000.00$ €. After five years (with fixed interest rates) the resulting capital $K = 11,000.00$ € (already reduced by the capital gains tax) will be paid out. How much is the net interest rate (without taxes)?

**Fig. 2.15** compound interest

*Solution*:
Firstly, we expect that, in general,

$$\frac{K}{K_0} = \left(1 + \frac{p}{100}\right)^n \Rightarrow 1 + \frac{p}{100} = \sqrt[n]{\frac{K}{K_0}}$$

$$\Rightarrow p = 100\left(\sqrt[n]{\frac{K}{K_0}} - 1\right)$$

and calculate $p$

$$p = 100\left(\sqrt[5]{\frac{11,000}{10,000}} - 1\right) = 1.92.$$

⊕ *Remark*: The recommended strategy is to reshape a formula so that it becomes generalized in such a way that one can only use numerical values. Firstly, calculations may become more accurate, and secondly, this reshaped formula can be used with arbitrary numerical values. ⊕ ◂◂◂

▸▸▸ **Application**: ***mathematical pendulum***
To determine the gravitational acceleration $g$ at a particular place on Earth, one must measure the period of oscillation $T$ of a simple pendulum of length $L$. What is the value obtained for $g$ (in m/s$^2$) if we apply the following formula:

$$T = 2\pi\sqrt{\frac{L}{g}}. \tag{2.4}$$

*Solution*:

$$T = 2\pi\sqrt{\tfrac{L}{g}} \Rightarrow T^2 = 4\pi^2\tfrac{L}{g} \Rightarrow gT^2 = 4\pi^2 L \Rightarrow g = 4\pi^2\tfrac{L}{T^2}.$$

*Numerical example*:
$L = 0.5\,\text{m}$, $T = 1.420\,\text{s} \Rightarrow g = 9.789\text{m/s}^2$ (Beware of the dimensions!)

⊕ *Remark*: Later, we will derive Formula (2.4) with the aid of integral calculus (Application p. 420). Note, incidentally, that the period of oscillation can be measured very accurately since it depends on the angle of deflection and, provided the pendulum is built so that the angle of deflection is sufficiently small, the period is independent of the oscillation amplitude. The pendulum is not, per se, a "perpetuum mobile"; so, energy is to be added constantly – in this case by means of a towing weight. ⊕ ◂◂◂

▸▸▸ **Application**: ***maximum speed in freefall***
Following *Newton*'s Formula (2.5), the air resistance depends on various factors, including the square of the speed $v$. Determine from this the maximum speed in free fall. ◂◂◂

*Solution*:
According to *Newton*, we have

$$F_W = c_W\, A\, \varrho \frac{v^2}{2}. \tag{2.5}$$

Here, $\varrho$ is the density of the flowing fluid (gas or liquid) and $v$ is the velocity of the body relative to the fluid. $A$ is the cross-sectional area of the body and $c_W$ is the drag coefficient specific to the body-type indicating how streamlined a barrier is. It is noteworthy that the force $F_W$ depends only on the mass $m$ (to a certain degree and via the cross-sectional area).
For the freefall we have:

$$m\,a = m\,g - F_W.$$

The current acceleration equals $a$ $(0 \le a \le g)$. The weight $m\,a$ (usually $m\,g$) decreases with increasing resistance. If $a$ reaches the value zero, then the maximum speed is reached. Then, we have

$$m\,g = c_W\, A\, \varrho \frac{v_{\max}^2}{2} \quad \Rightarrow \quad v_{\max} = \sqrt{\frac{2mg}{c_W\, A\, \varrho}}.$$

*Numerical example*: A human body with $A = 0.8\,\mathrm{m}^2$, a drag coefficient $c_W = 0.8$, and a mass of 70 kg achieves at an air density of $1.3\,\mathrm{kg/m}^3$ a terminal velocity of

$$v_{\max} = \sqrt{\frac{2 \cdot 70\mathrm{kg} \cdot 10\frac{\mathrm{m}}{\mathrm{s}^2}}{0.8 \cdot 0.8\mathrm{m}^2 \cdot 1.3\frac{\mathrm{kg}}{\mathrm{m}^3}}} \approx 41\,\mathrm{m/s}$$

(approximately 150 km/h). An ant that is 1 cm long with a mass of 40 mg, with the same $c_W$, and $A = 0.4\,\mathrm{cm}^2$ only achieves

$$v_{\max} = \sqrt{\frac{2 \cdot 40 \cdot 10^{-6}\mathrm{kg} \cdot 10\frac{\mathrm{m}}{\mathrm{s}^2}}{0.8 \cdot 0.4 \cdot 10^{-4}\mathrm{m}^2 \cdot 1.3\frac{\mathrm{kg}}{\mathrm{m}^3}}} \approx 4.5\,\mathrm{m/s}$$

(see Application p. 111).

▸▸▸ **Application**: ***plunge into water*** (Fig. 2.17)
What is the average deceleration experienced by a diver when the diver jumps from an altitude of $H$ m into the water, and he reaches a depth of $T$ m? The following formulas come into play:
Immersion speed is $v_0 = \sqrt{2g\,H}$ (deduction of the formula in Application p. 59, Formula (2.29)), uniform deceleration $a = \frac{v_0^2}{2T}$ (derivation of the formula in Application p. 59, Formula (2.27)).

**Fig. 2.16** different styles of diving ...

*Solution*:
We substitute $v_0 = \sqrt{2g\,H}$ into $a = \dfrac{v_0^2}{2T}$ and obtain

$$a = \frac{(\sqrt{2g\,H})^2}{2T} = \frac{2g\,H}{2T} = \frac{g\,H}{T} = \frac{H}{T}\,g.$$

The average deceleration is directly proportional to the height $H$ from which the diver jumps and "indirectly" proportional to the depth $T$ of immersion.
*Numerical example*: With a dive from a five meter diving board ($H = 5\,\text{m}$) and an immersion depth of $T = 4\,\text{m}$, a negative acceleration of about a quarter of the gravitational acceleration is reached.

**Fig. 2.17** going loco down in Acapulco

Much more extreme, of course, is the practice of the famous "death jumpers" of Acapulco/Mexico. They jump from a height of 40 m and have to roll quickly when submersed in water, because the water at the immersion point is only 3.6 m deep. The deceleration is then $11\,g$ on average.

⊕ *Remark*: The formulas are also valid when applied to the frontal impact of a vehicle crashing against a wall with an airbag of great quality. The deformation of the hood of a car and the relatively small breaking distance by the airbag together produce about $T \approx 0.8\,\text{m}$. Upon impact at $72\,\text{km/h} = 20\text{m/s}$, there is an average deceleration of about $25\,g$. This is almost at the limit ($30\,g$) of what the human organism can survive in the short term. ⊕ ◀◀◀

## 2.2 Linear equations

### General information about algebraic equations

Mathematicians distinguish between algebraic and transcendental equations. For algebraic equations, clear statements about their solutions (number, multiplicity) can be made. One often has to compute the zeros of a function

$$f(x) = a_n x^n + a_{n-1} x^{n-1} + \cdots + a_1 x^1 + a_0. \qquad (2.6)$$

Mathematicians often use the "sum notation"

$$f(x) = \sum_{k=0}^{n} a_k x^k.$$

The natural number $n$ is called the *degree of the equation.* The algebraic equations of degree one are called linear equations; those of degree two are called quadratic equations.
A value $x$ is called a *solution* or a *root* of an equation if $f(x) = 0$ is satisfied.

### The simplest special case of an algebraic equation

For $n = 1$, Formula (2.6) has the form

$$f(x) = a_1 x^1 + a_0$$

which is usually written as

$$y = k\,x + d.$$

If $k \neq 0$, the equation has exactly one solution

$$y = 0 \Rightarrow k\,x + d = 0 \Rightarrow x = -\frac{d}{k}.$$

▸▸▸ **Application**: ***linear profit function ("naive approach")***
A product is to be produced and sold to a wholesaler. The necessary initial investment amount is $K_0$ MU (monetary units). The cost of generating each piece is $E$ MU. Up to what number of pieces is the business "in the red", if one can arrive at a selling price of $V$ MU per piece?

*Solution*:
The *profit function* for $x$ units sold is linear in this simplified case

$$y = V\,x - (K_0 + E\,x) = (V - E)\,x - K_0.$$

The zero is obviously the critical number of pieces. Later on, one starts to make a profit. This is the case for

$$(V - E)\,x - K_0 = 0 \Rightarrow x = \frac{K_0}{V - E}.$$ ◂◂◂

**▸▸▸ Application**: ***conversion between temperature scales***

The unit for measuring temperature in physics is *Kelvin*, in Europe it is *Celsius*, and in the US it is *Fahrenheit*. The conversion from *Celsius* to *Kelvin* is easy: $K = C + 273.15°$. A conversion that is particularly simple is the one from *Celsius* to the *Réaumur* scale (Fig. 2.18): $R = \frac{4}{5}C$. In order to convert from *Celsius* to *Fahrenheit*, observe the following: $0°\,C$ corresponds to $32°\,F$ and $100°\,C$ corresponds to $212°\,F$. Determine conversion formulas between the two scales.

**Fig. 2.18** Rarely seen: On Réaumur's scale, water also freezes at 0°, but it starts boiling at 80°.

*Solution*:

We first describe the temperature in *Celsius* as $c$ and that in *Fahrenheit* as $f$. Then, we can apply the linear ansatz

$$f = k\,c + d.$$

We insert the two nodes, and thus, we obtain two linear equations

$$32 = 0\,k + d \quad \text{and} \quad 212 = 100\,k + d$$

which immediately yield $d = 32$, and thus, $k = \frac{9}{5} = 1.8$. So, we arrive at

$$f = \tfrac{9}{5}\,c + 32 \quad \text{or equivalently} \quad c = \tfrac{5}{9}\,(f - 32).$$

◂◂◂

**▸▸▸ Application**: ***expansion of a body when heated***

If the temperature of a body with volume $V_1$ is raised by $\Delta t$, its volume is increased to $V_2$ according to the following formula

$$V_2 = V_1\,(1 + \gamma\,\Delta t).$$

Determine the material constant $\gamma$ if a given body's volume increases by 3% at $\Delta t = 50°$.

*Solution*:

Reshaping the given formula yields

$$\gamma = \frac{1}{\Delta t}\left(\frac{V_2}{V_1} - 1\right) = \frac{1}{50}(1.03 - 1) = \frac{0.03}{50} = 0.0006 = 6 \cdot 10^{-4}.$$

⊕ *Remark*: The material constant $\gamma$ and $\Delta t$ conform to the same principle regardless of whether we measure in *Celsius* or in *Kelvin* – as is common in physics (see Application p. 24). ⊕ ◂◂◂

## Imperial vs. Metric system

▸▸▸ **Application**: ***conversion bar ↔ psi***
On many pressure gauges (e.g. on diving tanks), one can find values either in bar or in psi (pounds per square inch). A diver knows: A full tank is either $\approx 200$ bar or $\approx 3,000$ psi. Calculate the exact conversion numbers of this linear proportional relation.

*Solution*:
By definition, we have 1 bar $= 10^5\text{kg}/(\text{ms}^2) = 10^5\text{N}/\text{m}^2$.
With 1 N $\approx 0.22481$ pounds and 1 meter $\approx 39.37$ inches (1"=2.54 cm), we then have 1 bar $= 22,481/39.37^2$ psi $= 14.5$ psi or 1 psi $= 0.069$ bar.

⊕ *Remark*: Note that this was not trivial, since we must not confuse masses and weights (forces). A pound is, physically speaking, a weight.
One bar equals approximately the ambient pressure at sea level. Every 10 meters under water, the pressure increases by one bar. ⊕ ◂◂◂

▸▸▸ **Application**: ***conversion miles per gallon ↔ liters per 100 km***
There are some things in daily life that are easier to understand with the use of some simple calculations. US-Americans are more familiar with "miles per gallon", and Europeans only think in "liters per 100 km". What is the formula for this *indirectly proportional* relation?

*Solution*:
1 fluid gallon equals $\approx 3.785$ liters, 1 US mile $\approx 1.61$ kilometers (therefore, e.g., $\approx 62$ miles equal 100 km).
Suppose, we are driving such that this would be equivalent to using $x$ liters per 100 kilometers and result in a distance of $y$ miles per gallon.
Let us assume that we continue driving at this rate until we have used up a gallon. This is the case after $100 \cdot 3.785/x$ km or $100 \cdot 3.785/1.61/x$ miles ($\approx 235/x$ miles). Thus, we have

$$y \approx 235/x.$$

⊕ *Remark*: For example, five liters per 100 km would be equivalent to 235/5=47 miles per gallon. Note that you do not need a computer with Internet access to obtain this result; it may be quickly estimated by means of a mental calculation. By the way, this simple formula was not easy to find anywhere. ⊕ ◂◂◂

The following two questions were part of a written exam taken by the author in order to acquire an American driver's licence. Due to unfamiliar measuring units, together with questions about "turn on red" and special regulations concerning school buses (unusual for Europeans), even seemingly simple tests can become critical ...

▸▸▸ **Application**: ***How long is the braking distance at 55 mph?***
The proposed answers were a) 143 ft, b) 243 ft, c) 443 ft. What to choose?
*Solution*:
Firstly, how many meters per second is 55 mph? We have $v = 55\text{mph} = 55 \cdot 1.61\text{km/h} = 88.5\text{km/h} \approx 25\text{m/s}$.
Secondly, a deceleration of $a \approx 0.6g \approx 6\text{m/s}^2$ might be a good value.
Then we get the time for the braking process: $v = at \Rightarrow t = 25/6\text{s} \approx 4\text{s}$.
This, finally, leads to the braking distance $s = a/2 \cdot t^2 \approx 3 \cdot 4^2 \approx 50\text{m} \approx 165$ feet.
Therefore, the author chose answer a) ...

⊕ *Remark*: The total stopping distance is the sum of the perception distance, the reaction distance, and the actual braking distance (once the brakes are put on). ⊕

◂◂◂

▸▸▸ **Application**: ***How many ounces of 86-proof liquor have the same amount of alcohol as a six-pack of beer?***
This question is less hard to answer once you know what 86-proof liquor means ....
*Solution*:
A six-pack has $6 \cdot 12 = 72$ fluid ounces. The alcohol content by volume of regular beer may be 5 percent, that of 86-proof liquor is 43 percent, that is, almost 9 times as much. Thus, the estimated answer would be $72/9 = 8$ fluid ounces. ◂◂◂

## 2.3 Systems of linear equations

### Linear equations in two variables

The solution of a linear equation is so "trivial" that it needs no further explanation. It is scarcely any more difficult to solve a *system of linear equations.* Let us consider some introductory examples:

▸▸▸ **Application**: ***mixing task***

How many liters of hot water ($95° C$) does one need to raise the temperature of $n$ liters of cold tap water ($15° C$) to that of bathwater ($35° C$)? What is the ratio of cold and hot water?

*Solution*:

Let $x$ and $y$ be the amount of cold water and hot water respectively.

1. Mass comparison: $x + y = n$
2. Energy comparison: $15\,x + 95\,y = 35 \cdot n$

Multiply the first equation by 15 and subtract it from the second equation. One will thus need $\frac{n}{4}$ liters of hot water. The ratio of *cold : hot* is, therefore,

$$x : y = \tfrac{3n}{4} : \tfrac{n}{4} = 3 : 1.$$

⊕ *Remark*: One could also have solved the above example as follows: The temperature difference between bathwater and cold water is $-20°$, and $60°$ for hot water. In this case, we use the energy comparison equation $-20\,x + 60\,y = 0$, from which it also follows that $x : y = 3 : 1$. ⊕ ◂◂◂

▸▸▸ **Application**: ***total resistance and partial resistance***

Two resistors $R_1$ and $R_2$ connected in (Fig. 2.13) behave like $1 : n$ and have a total resistance of $R$. What are the individual resistances?

*Solution*:

For resistors connected in parallel, it holds that (see Application p. 19)

$$\frac{1}{R_1} + \frac{1}{R_2} = \frac{1}{R}.$$

By assumption, we have

$$R_1 : R_2 = 1 : n \Rightarrow \frac{1}{R_1} : \frac{1}{R_2} = n : 1.$$

We now set $x = \dfrac{1}{R_1}$ and $y = \dfrac{1}{R_2}$. Then, we have two linear equations

$$x + y = \frac{1}{R}, \quad x : y = n.$$

Following the second equation $x = n\,y$, and thus, from the first equation,

$$n\,y + y = \frac{1}{R} \Rightarrow y = \frac{1}{n+1}\frac{1}{R} \text{ respectively } x = \frac{n}{n+1}\frac{1}{R}.$$

◂◂◂

Since we had two unknowns in the above examples, we needed two equations. By skilfully adding, subtracting or using "substitution", we were able to solve such a "$(2,2)$-System" quickly.
We now derive formulas for the solution $(x/y)$ of a general system of linear equations in two variables, so we do not always have to worry about using such "tricks".
A $(2,2)$-system has the general form

$$\begin{aligned} &(\mathrm{I}) \quad a_{11}\,x + a_{12}\,y = b_1, \\ &(\mathrm{II}) \quad a_{21}\,x + a_{22}\,y = b_2. \end{aligned} \tag{2.7}$$

Both of the equations are linear, for we have $y = -\dfrac{a_{i1}}{a_{i2}}x + \dfrac{b_i}{a_{i2}}$ $(i = 1, 2)$. We want to find a pair of values $(x/y)$ that satisfies both equations. In order to "eliminate" one of the unknowns, we multiply (I) by $a_{21}$ and (II) by $a_{11}$ and get:

$$\begin{aligned} &(\mathrm{I}) \quad a_{21}\,a_{11}\,x + a_{21}a_{12}\;y = a_{21}\,b_1, \\ &(\mathrm{II}) \quad a_{11}\,a_{21}\,x + a_{11}\,a_{22}\,y = a_{11}\,b_2. \end{aligned} \tag{2.8}$$

Now, we subtract the upper from the lower equation, and thus, we get

$$(\mathrm{II}) - (\mathrm{I}) \quad 0 \cdot x + (a_{11}\,a_{22} - a_{21}a_{12})\,y = a_{11}\,b_2 - a_{21}\,b_1 \tag{2.9}$$

or

$$y = \frac{a_{11}\,b_2 - a_{21}\,b_1}{a_{11}\,a_{22} - a_{21}\,a_{12}} = \frac{D_y}{D}.$$

Now, $x$ can be calculated from $(I)$ or $(II)$. The result is

$$x = \frac{a_{22}\,b_1 - a_{12}\,b_2}{a_{11}\,a_{22} - a_{21}\,a_{12}} = \frac{D_x}{D}.$$

The result in this form is ideally is verified by computers; it is easily programmable. It must be mentioned that there exist more efficient methods for solving linear equations. This applies, in particular, to systems of linear equations with three or more variables, but they require – at least programmatically – much greater effort, because different cases have to be considered. We gave formulas for $x$ and $y$ that always work as long as the denominator $D$ is not 0. If the denominator "disappears", then the system has no solution. The only exception being: The equations are (scalar) multiples of each other.

Then, of course all (infinitely many) pairs $(x/y)$ satisfying equation $(I)$ are solutions of the system.
It is difficult to learn these formulas by heart, because they involve confusing indices of the coefficients. Here we are aided by *Cramer*'s Rule (Gabriel *Cramer*, 1704–1752, a Swedish mathematician and philosopher) – a service that will subsequently be important. It is based on the common "matrix notation", in which system (2.7) – omitting variables $x$ and $y$ for short – is written as follows:

$$\left(\begin{array}{cc|c} a_{11} & a_{12} & b_1 \\ a_{21} & a_{22} & b_2 \end{array}\right)$$

The right column can be described as a "spare column". From this matrix, one builds the three "determinants" (square number schemes) as follows:

$$D = \begin{vmatrix} a_{11} & a_{12} \\ a_{21} & a_{22} \end{vmatrix}, \quad D_x = \begin{vmatrix} b_1 & a_{12} \\ b_2 & a_{22} \end{vmatrix}, \quad D_y = \begin{vmatrix} a_{11} & b_1 \\ a_{21} & b_2 \end{vmatrix}. \tag{2.10}$$

The sub-determinants $D_x$ and $D_y$ arising from the "major determinant" $D$ are formed by respectively replacing the $x$ or $y$ column by the spare column. The "value" of such a determinant can now be defined as the difference between the product of elements in the *principal diagonal* (from top left to bottom right) and the product of the members in the *anti-diagonal*, for example,

$$D = \begin{vmatrix} a_{11} & a_{12} \\ a_{21} & a_{22} \end{vmatrix} = a_{11}\,a_{22} - a_{21}\,a_{12}. \tag{2.11}$$

▸▸▸ **Application**: ***intersecting two lines***
Where do the lines $2\,x + 4\,y = 5$ and $3\,x - 5\,y = 0$ meet?
*Solution*:
The intersection of two lines in the plane leads to a $(2,2)$-system. The determinants that are sought in this case are

$$D = \begin{vmatrix} 2 & 4 \\ 3 & -5 \end{vmatrix} = -22\ (\neq 0),\ D_x = \begin{vmatrix} 5 & 4 \\ 0 & -5 \end{vmatrix} = -25,\ D_y = \begin{vmatrix} 2 & 5 \\ 3 & 0 \end{vmatrix} = -15.$$

According to *Cramer*'s Rule, we have the intersection $S\left(\frac{25}{22}\Big/\frac{15}{22}\right)$. See also Application p. 57.

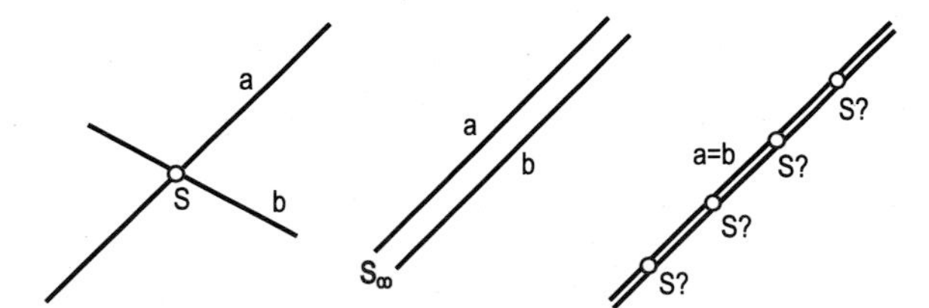


**Fig. 2.19** three cases when intersecting two lines

⊕ *Remark*: If the principal determinant vanishes, the lines are either parallel – so that the "point at infinity" $S_\infty$ is the solution – or they are identical – so that every point is a solution (Fig. 2.19, the situations depicted in the center and right).
If one uses the "substitution method", i.e. expressing either variable by the other and substituting them into one equation, one recognizes the exceptional case of $D = 0$ as follows: When the lines are identical, one obtains a "true statement" without further information about $x$ and $y$; if they are different, then it is a "false statement".

When calculating with computers, it may be quite advantageous to work with "points at infinity" so that, instead of using $D = 0$, we use a very small value that is about $D = 10^{-10}$. This saves us from addressing a large number of different cases since the result will be "true" for values of precise decimals. ⊕ ◂◂◂

## Linear equations in three or more variables

The advantage of *Cramer*'s Rule becomes even more apparent when we solve three linear equations containing three variables, that is, a $(3,3)$-system. We can then follow the exact same rules that apply for a $(2,2)$-system (though this can be a bit tedious to prove by recalculation). We will just need to know how to calculate a "three-row" determinant: Perform this calculation by calculating three of the "two-row" sub-determinants; this is called "expanding by complementary minors". This technique will be explained by means of several examples (Application p. 30, Application p. 32). Expanding by the minors complementary to the first column gives:

$$\begin{vmatrix} a_{11} & a_{12} & a_{13} \\ a_{21} & a_{22} & a_{23} \\ a_{31} & a_{32} & a_{33} \end{vmatrix} = a_{11} \begin{vmatrix} a_{22} & a_{23} \\ a_{32} & a_{33} \end{vmatrix} - a_{21} \begin{vmatrix} a_{12} & a_{13} \\ a_{32} & a_{33} \end{vmatrix} + a_{31} \begin{vmatrix} a_{12} & a_{13} \\ a_{22} & a_{23} \end{vmatrix}. \tag{2.12}$$

We can obviously associate $a_{ik}$ to an element in the sub-determinant created by deleting the $i$-th row and $k$-th column. It should be noted that each summand is given a sign according to the following scheme

$$\begin{vmatrix} + & - & + \\ - & + & - \\ + & - & + \end{vmatrix}. \tag{2.13}$$

▸▸▸ **Application**: ***intersection of three planes*** (Fig. 2.20)
We would like to find the common point $S$ of the three planes

$$2x + y = 3, \quad y + 2z = 0, \quad y - z = 1$$

(see also the chapter on vector calculus).

*Solution*:

We write the three equations in matrix form. It must not be forgotten that zeros have to be placed if a variable does not show up:

$$\left( \begin{array}{ccc|c} 2 & 1 & 0 & 3 \\ 0 & 1 & 2 & 0 \\ 0 & 1 & -1 & 1 \end{array} \right).$$

The principal determinant is now

$$D = \begin{vmatrix} 2 & 1 & 0 \\ 0 & 1 & 2 \\ 0 & 1 & -1 \end{vmatrix} = 2\begin{vmatrix} 1 & 2 \\ 1 & -1 \end{vmatrix} - 0\begin{vmatrix} 1 & 0 \\ 1 & -1 \end{vmatrix} + 0\begin{vmatrix} 1 & 0 \\ 1 & 2 \end{vmatrix} = -6.$$

With the same sub-determinants, $D_x$ assumes the following value

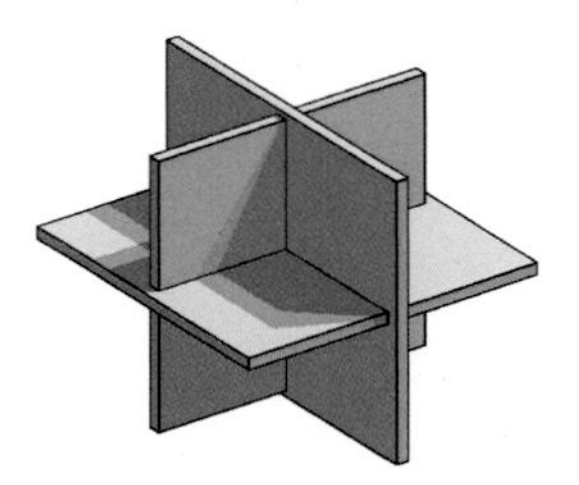

**Fig. 2.20** the intersection of three planes ...

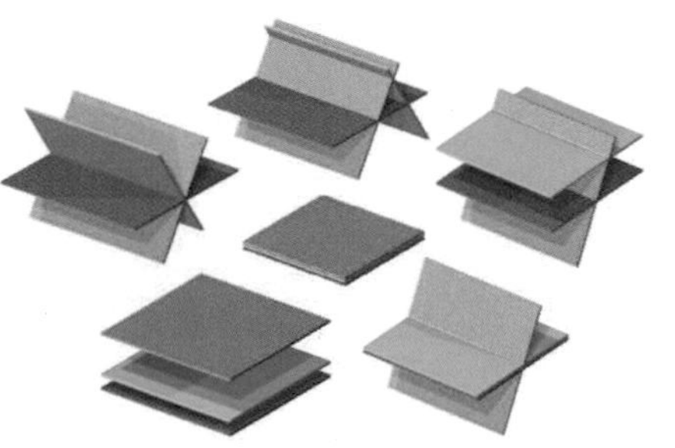

**Fig. 2.21** ... including special cases.

$$D_x = \begin{vmatrix} 3 & 1 & 0 \\ 0 & 1 & 2 \\ 1 & 1 & -1 \end{vmatrix} = 3\begin{vmatrix} 1 & 2 \\ 1 & -1 \end{vmatrix} - 0\begin{vmatrix} 1 & 0 \\ 1 & -1 \end{vmatrix} + 1\begin{vmatrix} 1 & 0 \\ 1 & 2 \end{vmatrix} = -7.$$

Finally, we have

$$D_y = \begin{vmatrix} 2 & 3 & 0 \\ 0 & 0 & 2 \\ 0 & 1 & -1 \end{vmatrix} = 2\begin{vmatrix} 0 & 2 \\ 1 & -1 \end{vmatrix} = -4, \quad D_z = \begin{vmatrix} 2 & 1 & 3 \\ 0 & 1 & 0 \\ 0 & 1 & 1 \end{vmatrix} = 2\begin{vmatrix} 1 & 0 \\ 1 & 1 \end{vmatrix} = 2.$$

According to *Cramer*'s Rule, now $x = \frac{D_x}{D}$, $y = \frac{D_y}{D}$, $z = \frac{D_z}{D}$, and the intersection point has the coordinates

$$S\left(\frac{7}{6}\Big/\frac{2}{3}\Big/-\frac{1}{3}\right).$$

⊕ *Remark*: If the principal determinant vanishes, the lines of intersection of any pair of planes are either parallel or identical (Fig. 2.21). If the planes are identical, then each point of the plane(s) is a solution. The planes can also form a "pencil" when they have a common line. All of the points on this line are then a solution. This common line can also be the common "line at infinity" of three parallel planes. The planes can intersect each other even along three parallel lines (one of which may also be a line at infinity). Then, the common point at infinity of this line is a solution. In computer calculations – particularly in animations or simulations where such special cases can sometimes occur, but not always – one can continue in the cases with a zero determinant by setting the determinant to a very small value (e.g., $D = 10^{-10}$) in order to circumvent nasty case distinctions (see Application p. 29). ⊕

◀◀◀

▸▸▸ **Application**: *Δ-connection, Y-connection* (Fig. 2.22)
When transforming a Δ-connection into a Y-connection, the following equations occur (see Application p. 27):

$$r_k + r_i = \frac{R_j(R_k + R_i)}{R_1 + R_2 + R_3} = A_i.$$

The indices $i$, $j$, $k$ run "cyclically", starting with $i = 1$, $j = 2$, and $k = 3$. The values $r_i$ are to be expressed in terms of the $R_i$.

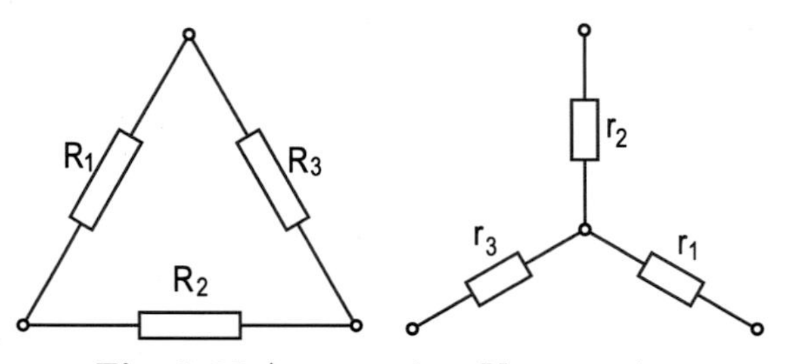


**Fig. 2.22** Δ-connection, Y-connection

*Solution*:
Once again, we write down the three equations explicitly:

$$\begin{aligned} r_3 + r_1 &= s\,R_2(R_3 + R_1) = A_1, \\ r_1 + r_2 &= s\,R_3(R_1 + R_2) = A_2, \\ r_2 + r_3 &= s\,R_1(R_2 + R_3) = A_3 \end{aligned}$$

(with $s = 1/(R_1 + R_2 + R_3)$). In matrix form, the scheme is as follows:

$$\left(\begin{array}{ccc|c} 1 & 0 & 1 & A_1 \\ 1 & 1 & 0 & A_2 \\ 0 & 1 & 1 & A_3 \end{array}\right).$$

The principal determinant is now

$$D = \begin{vmatrix} 1 & 0 & 1 \\ 1 & 1 & 0 \\ 0 & 1 & 1 \end{vmatrix} = 1\begin{vmatrix} 1 & 0 \\ 1 & 1 \end{vmatrix} - 1\begin{vmatrix} 0 & 1 \\ 1 & 1 \end{vmatrix} + 0\begin{vmatrix} 0 & 1 \\ 1 & 0 \end{vmatrix} = 2.$$

Furthermore, we have

$$D_1 = \begin{vmatrix} A_1 & 0 & 1 \\ A_2 & 1 & 0 \\ A_3 & 1 & 1 \end{vmatrix} = A_1\begin{vmatrix} 1 & 0 \\ 1 & 1 \end{vmatrix} - A_2\begin{vmatrix} 0 & 1 \\ 1 & 1 \end{vmatrix} + A_3\begin{vmatrix} 0 & 1 \\ 1 & 0 \end{vmatrix} = A_1 + A_2 - A_3.$$

We expand the remaining determinants for calculating $r_2$ and $r_3$ by expanding along the middle and right columns respectively:

$$D_2 = \begin{vmatrix} 1 & A_1 & 1 \\ 1 & A_2 & 0 \\ 0 & A_3 & 1 \end{vmatrix} = -A_1\begin{vmatrix} 1 & 0 \\ 0 & 1 \end{vmatrix} + A_2\begin{vmatrix} 1 & 1 \\ 0 & 1 \end{vmatrix} - A_3\begin{vmatrix} 1 & 1 \\ 1 & 0 \end{vmatrix} = A_2 + A_3 - A_1,$$

$$D_3 = \begin{vmatrix} 1 & 0 & A_1 \\ 1 & 1 & A_2 \\ 0 & 1 & A_3 \end{vmatrix} = A_1 \begin{vmatrix} 1 & 1 \\ 0 & 1 \end{vmatrix} - A_2 \begin{vmatrix} 1 & 0 \\ 0 & 1 \end{vmatrix} + A_3 \begin{vmatrix} 1 & 0 \\ 1 & 1 \end{vmatrix} = A_3 + A_1 - A_2.$$

The new resistors are then

$$r_i = \frac{D_i}{D} = \frac{A_i + A_j - A_k}{2}.$$

The indices $i$, $j$, $k$ again run in a cyclical order, starting with $i = 1$, $j = 2$, and $k = 3$.

⊕ *Remark*: One remembers formulas better when they display regularities. Often such regularities in output are an indication that one has calculated correctly. *Einstein* once said that a formula is only correct if it is "beautiful". ⊕ ◂◂◂

▸▸▸ **Application**: ***parabola on three points*** (Fig. 2.23)
There are very common techniques for the approximation of complicated curves by simpler ones such as parabolas. A famous example of such a technique is the *Kepler*'s Fass-Rule (Fass = German word for barrel, see Section 8, page 413). For this rule, one derives the coefficients $a$, $b$, $c$ of the equation of a parabola $y = ax^2 + bx + c$ passing through three points $P(u_i/v_i)$.

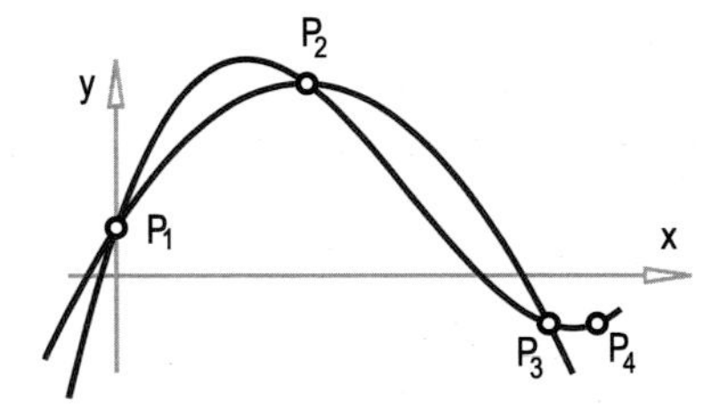


**Fig. 2.23** quadratic parabola (3 points) vs. cubic parabola (4 points)

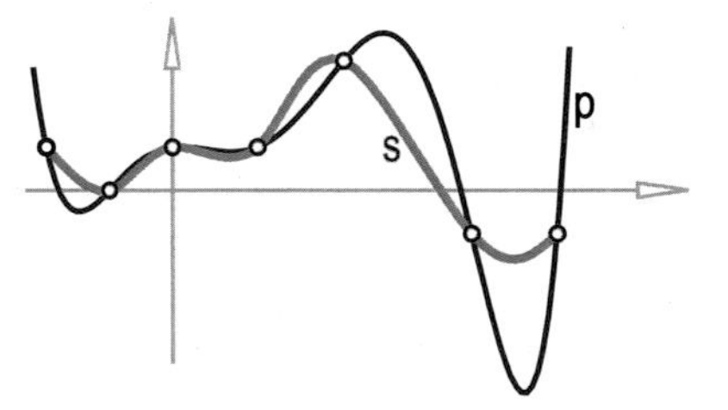


**Fig. 2.24** parabola of degree 6 (7 points) vs. cubic spline

*Solution*:
The coordinates of the three points have to "fulfil" the equation of the parabola, so we immediately get three equations with three unknowns $a$, $b$, and $c$:

$$\begin{aligned} u_1^2\, a + u_1\, b + c &= v_1, \\ u_2^2\, a + u_2\, b + c &= v_2, \\ u_3^2\, a + u_3\, b + c &= v_3. \end{aligned}$$

We solve this by means of *Cramer*'s Rule. The principal determinant is:

$$D = \begin{vmatrix} u_1^2 & u_1 & 1 \\ u_2^2 & u_2 & 1 \\ u_3^2 & u_3 & 1 \end{vmatrix}.$$

⊕ *Remark*: Generally, one can interpolate $n + 1$ points by a "parabola of degree $n$" (Fig. 2.23 and Fig. 2.24). However, complicated curves in computer graphics are rarely approximated by parabolas of degrees higher than 3 – such curves tend to oscillate (especially at their end points, Fig. 2.24). Instead of choosing a higher degree curve, it is better to use a sequence of "cubic parabolas" which are joined as smoothly as possible. The overall approximating curve is a cubic spline. Such splines also require the solution of systems of linear equations. For more details, read the chapter on differential calculus (Application p. 290). ⊕ ◀◀◀

Systems of linear equations having higher degrees are not uncommon in applied mathematics. Often one can "linearize" complex problems by introducing additional variables. This allows us to obtain a simple system of linear equations with many unknowns, rather than a complicated system with a few unknowns.
Such $(n, n)$-systems can also be solved using an adjusted version of *Cramer*'s Rule. The determinants of degree $n$ develop gradually by a summation of determinants of degree $n - 1$ until they reach a degree of three. The higher the level, the more complex *Cramer*'s Rule becomes, and one often uses other methods with computers leading to a more efficient way to obtain solutions. Here, we have a simple example of using a $(4, 4)$-system, this can be solved by skilfully inserting determinants (the "substitution method"):

▶▶▶ **Application**: ***combustion of alcohol in carbon dioxide and water***
Consider the "chemical reaction formula" when one has sufficient oxygen to burn alcohols – such as the ethanol found in sugarcane liquor, which decomposes $C_2H_5OH$ – to $CO_2$ and $H_2O$.
*Solution*:
The following holds true

$$n_1 \cdot C_2H_5OH + n_2 \cdot O_2 = n_3 \cdot CO_2 + n_4 \cdot H_2O.$$

The four unknowns $n_i$ can be calculated incrementally if we compare the quantity of the atoms:

$$\begin{array}{ll} \text{Carbon } (C) & 2\,n_1 = n_3, \\ \text{Hydrogen } (H) & (5+1)\,n_1 = 2\,n_4, \\ \text{Oxygen } (O) & n_1 + 2\,n_2 = 2\,n_3 + n_4. \end{array}$$

The first striking observation: We have three equations but four unknowns. However, there is an additional condition: The $n_i$ must be integers. First, we substitute $n_1 = 1$ and hope to obtain integer solutions for the remaining $n_i$S (If we do not obtain such solutions: Chemists also reckon with fractions!). Now we only need to solve a $(3, 3)$-system:

$$2 = n_3, \quad 6 = 2\,n_4, \quad 1 + 2\,n_2 = 2\,n_3 + n_4.$$

The solutions can immediately be seen: The first equation evaluates to $n_3 = 2$, the second one evaluates to $n_4 = 3$, and thus, the third evaluates to $n_2 = 3$. In fact, all solutions are integers (otherwise you will have fractions or will have to try $n_1 = 2, \ldots$). The chemical reaction formula is thus

$$C_2H_5OH + 3 \cdot O_2 = 2 \cdot CO_2 + 3 \cdot H_2O.$$

◂◂◂

▸▸▸ **Application**: ***explosive combustion of gasoline vapor***

Gasoline, especially, contains alkanes that have 6 to 9 carbon atoms in their molecules such as $C_8H_{18}$. This is used as a fuel due to the fact that the mixture of gasoline vapor and air combusts explosively into water and carbon dioxide given a certain composition. Calculate the coefficients of the chemical reaction equation

$$n_1 \cdot C_8H_{18} + n_2 \cdot O_2 = n_3 \cdot CO_2 + n_4 \cdot H_2O.$$

*Solution*:

We calculate the four unknowns $n_i$ gradually by comparing the quantities of the atoms:

| | |
|---|---|
| Carbon ($C$) | $8\,n_1 = n_3$ |
| Hydrogen ($H$) | $18\,n_1 = 2\,n_4$ |
| Oxygen ($O$) | $2\,n_2 = 2\,n_3 + n_4$ |

We have three equations but four unknowns. The $n_i$ must be integers. First let us put again $n_1 = 1$ and hope to obtain integer solutions for the remaining $n_i$.

The $(3,3)$-system is now

$$8 = n_3, \quad 18 = 2\,n_4, \quad 2\,n_2 = 2\,n_3 + n_4.$$

The first equation gives $n_3 = 8$, the second gives $n_4 = 9$, and thus, the third gives $n_2 = (16+9)/2 = 12.5$, which is not an integer. Thus, we try $n_1 = 2$. Therefore, the appropriate $(3,3)$-system is

$$16 = n_3, \quad 36 = 2\,n_4, \quad 2\,n_2 = 2\,n_3 + n_4,$$

which gives the chemical reaction formula

$$2 \cdot C_8H_{18} + 25 \cdot O_2 = 16 \cdot CO_2 + 18 \cdot H_2O.$$

◂◂◂

## 2.4 Quadratic equations

### The pure quadratic equation

The equation

$$x^2 = D$$

has two real solutions if $D > 0$:

$$x_1 = \sqrt{D}, \ x_2 = -\sqrt{D}.$$

The solutions for $D = 0$ are: $x_1 = x_2 = 0$. The solutions for $D < 0$ are *complex conjugates*. More on this in Section B.

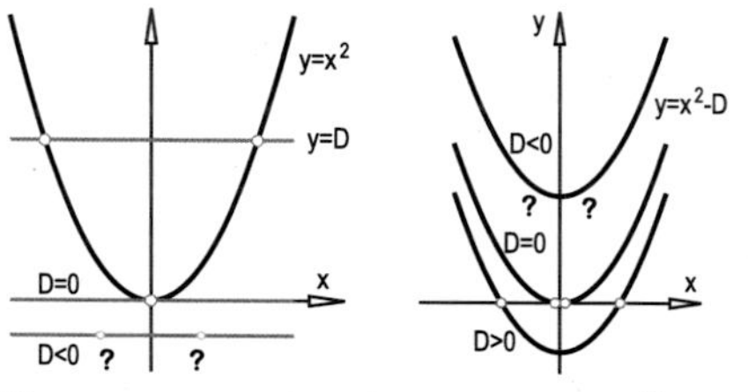


**Fig. 2.25** graphical solution of $x^2 = D$

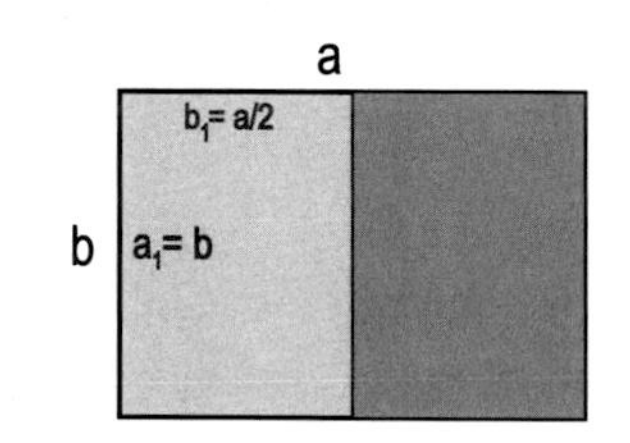


**Fig. 2.26** convenient paper format

▸▸▸ **Application**: ***convenient paper format*** (Fig. 2.26)
Let the side lengths of a rectangular sheet of paper be $a$ and $b$ with a ratio $a : b$ such that two similar rectangles arise when folding the sheet. How should one read the aspect ratio? How long are the sides of the paper if the surface of the sheet is $1\,\mathrm{m}^2$?
*Solution*:

$$\frac{a_1}{b_1} = \frac{b}{\frac{a}{2}} = \frac{a}{b} \ \Rightarrow \ b^2 = \frac{a^2}{2} \ \Rightarrow \ a = b\sqrt{2}$$

(the solution $a = -b\sqrt{2}$ is not relevant).
If the area of the rectangle ($1\mathrm{m}^2$) is the only information given, $b$ can be calculated, and thus, $a$ as well:

$$a\,b = b\sqrt{2}\,b = b^2\sqrt{2} = 1$$

$$\Rightarrow \ b = \sqrt{\frac{1}{\sqrt{2}}} = 0.841\mathrm{m} \ \Rightarrow \ a = 1.189\mathrm{m}.$$

This format is called $A0$. Repeatedly folding the sheet gives the formats $A1 = 0.841 \times 0.595$, $A2 = 0.595 \times 0.420$, $A3 = 0.420 \times 0.297$, $A4 = 0.297 \times 0.210$ etc. (with areas $\frac{1}{2}\mathrm{m}^2$, $\frac{1}{4}\mathrm{m}^2$, $\frac{1}{8}\mathrm{m}^2$, $\frac{1}{16}\mathrm{m}^2$, $\ldots$). ◂◂◂

## The general quadratic equation

In the general form of the quadratic equation,

$$Ax^2 + Bx + C = 0,$$

the sign of the so-called *discriminant*

$$D = B^2 - 4AC$$

determines the number of real solutions: For $D > 0$, it gives two real solutions; for $D = 0$, it gives one real solution; and for $D < 0$, it gives two complex conjugate solutions.
The solutions can be given directly

$$x_{1,2} = \frac{-B \pm \sqrt{D}}{2A}. \tag{2.14}$$

***Proof***: We first divide equation $Ax^2 + Bx + C = 0$ by $A$, and we get

$$x^2 + \frac{B}{A}x + \frac{C}{A} = 0.$$

Now, we complete to a full square

$$\left(x + \frac{B}{2A}\right)^2 + \frac{C}{A} - \left(\frac{B}{2A}\right)^2 = 0.$$

Hence, we have a purely quadratic equation:

$$\left(x + \frac{B}{2A}\right)^2 = \left(\frac{B}{2A}\right)^2 - \frac{C}{A} = \frac{B^2 - 4AC}{4A^2}.$$

The solutions are given by Formula (2.14). ⊙

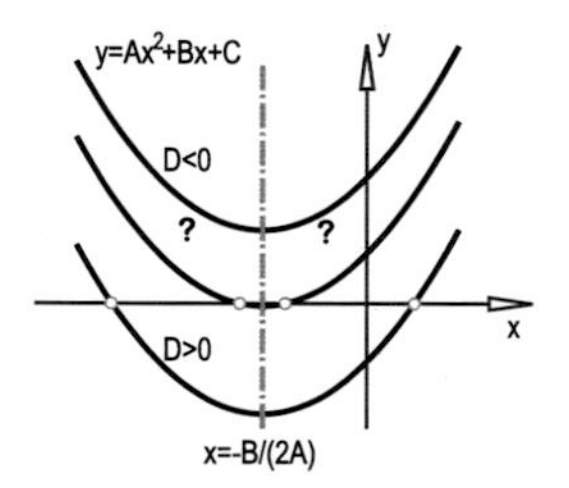


**Fig. 2.27** graphical solution

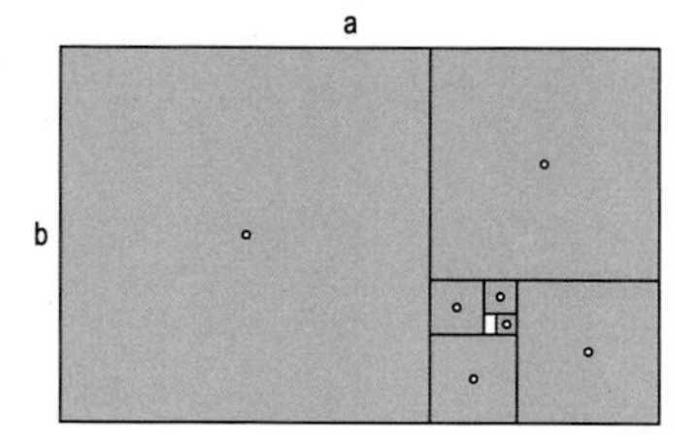


**Fig. 2.28** harmonic rectangle

▸▸▸ **Application**: ***Golden Ratio, harmonic rectangles*** (Fig. 2.28)
The side lengths $a$, $b$ of a rectangle are chosen such that $b < a$ where

$$b : a = a : (a + b) \Rightarrow a^2 = ab + b^2. \tag{2.15}$$

We then call this rectangle *harmonic*. Calculate the ratio of the side lengths (the *Golden Number*).
In addition, prove the following: One rectangle divides into a square and a smaller rectangle, and this smaller rectangle is again considered to be harmonic (this ratio is called the *Golden Section*).

*Solution*:
We first set $b = 1$ and then,

$$\frac{1}{a} = \frac{a}{a+1} \Rightarrow a^2 - a - 1 = 0 \Rightarrow$$
$$a_{1,2} = \frac{-(-1) \pm \sqrt{(-1)^2 - 4 \cdot 1 \cdot (-1)}}{2} = \frac{1 \pm \sqrt{5}}{2}.$$

We do not consider the negative solution (however, see Application p. 528). Thus

$$\boxed{a : b = \frac{1+\sqrt{5}}{2} : 1 \approx 1.62 : 1.} \tag{2.16}$$

Assuming that Formula (2.15) applies, we also get

$$a^2 = ab + b^2 \Rightarrow a^2 - ab = b^2 \Rightarrow a(a-b) = b^2 \Rightarrow (a-b) : b = b : a.$$

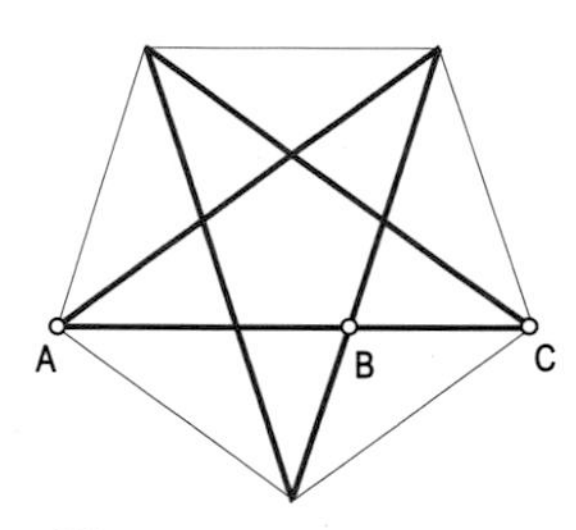


**Fig. 2.29** pentagram

**Fig. 2.30** Modulor, Parthenon

⊕ *Remark*: Harmonic rectangles or the golden ratio often occur in art.[2] Examples include the "Modulor" by *Corbusier* (Fig. 2.30, left),[3] the "magic pentagon" ("pentagram", Fig. 2.29: $\overline{AC} : \overline{AB} = \frac{1+\sqrt{5}}{2} : 1$), and the Parthenon (Greek temple, Fig. 2.30). In nature, the remarkable number $\frac{1+\sqrt{5}}{2}$ plays an important role. More on this can be found in Appendix B. ⊕ ◂◂◂

▸▸▸ **Application**: ***upward vertical throw*** (Fig. 2.31)
Let $v_0$ be the initial velocity and let $g$ be the gravitational acceleration. The air resistance in this example is negligible.
Then, for an elapsed time of $t$ seconds, the height $h$ is given by

$$h = G(t) - B(t) = v_0 t - \frac{g}{2} t^2. \tag{2.17}$$

$G(t)$ is the component of the upwards uniform motion; $B(t)$ is the component of the downwards uniformly accelerated motion.
Compute $t$ at given values of $h$ and $v_0$.

[2]See `http://www.uni-hildesheim.de/~stegmann/goldschn.pdf` for an interesting article on the historical development of the Golden Section.

[3]Le Corbusier (Charles Edouard Jeanneret): *The Modulor: A Harmonious Measure to the Human Scale Universally Applicable to Architecture and Mechanics and Modulor 2 (Let the User Speak Next)*. 2 Volumes. Birkhäuser, Basel, 2000.

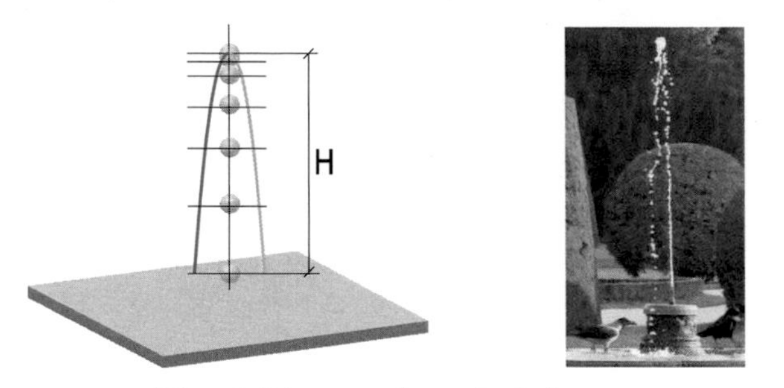


**Fig. 2.31** upward vertical throw

*Solution*:
We collect together powers of $t$ as follows: $\frac{g}{2}t^2 - v_0\,t + h = 0$ and find

$$A = \frac{g}{2},\ B = -v_0,\ C = h \Rightarrow D = (-v_0)^2 - 4\frac{g}{2}h = v_0^2 - 2gh.$$

This results in

$$t_{1,2} = \frac{v_0 \pm \sqrt{v_0^2 - 2gh}}{g}.$$

The values $t_1$ and $t_2$ are the times at which the thrown object reaches height $h$, the difference $t_2 - t_1$ being the time it takes to rise and fall back down to this height. For the highest point (the current height $H$), there must be a double solution:

$$D = 0 \Rightarrow v_0^2 - 2gH = 0 \Rightarrow H = \frac{v_0^2}{2g}. \tag{2.18}$$

Conversely, measuring from the current height $H$, the initial velocity $v_0$ can be determined as:

$$v_0 = \sqrt{2gH} \tag{2.19}$$

⊕ *Remark*: This formula unexpectedly provides a good service in Application p. 56. ⊕

⊕ *Remark*: If we *obliquely throw upwards* (with an angle tilted at $\alpha$ to the horizontal), the value of $v_0 \sin\alpha$ is used instead of $v_0$ (Application p. 142). ⊕ ◂◂◂

▸▸▸ **Application**: ***balance of the attraction forces*** (Fig. 2.32)
Let $m_1$ and $m_2$ be two masses with centers of gravity separated by a distance $d$. The center of gravity of a body $P$ is located at a distance $x$ from the center of gravity of $m_1$, on the line joining the centers of gravity of $m_1$ and $m_2$. For which $x$ will the attractions between $m_1$ and $m_2$ compensate?
*Solution*:
According to *Newton*, the attraction is proportional to the mass, but it is inversely proportional to the square of the distance. For the body to be in equilibrium, one must have:

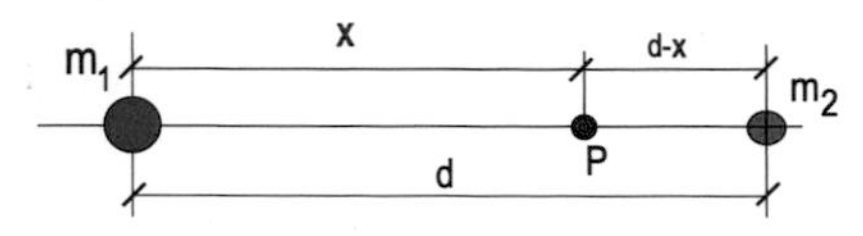


**Fig. 2.32** Where does $P$ have to lie in order to be attracted equally by $m_1$ and $m_2$?

$$\frac{m_1}{x^2} = \frac{m_2}{(d-x)^2} \quad (0 < x < d)$$

$$\Rightarrow m_1(d^2 - 2dx + x^2) = m_2 x^2 \Rightarrow \underbrace{(m_1 - m_2)}_{A} x^2 \underbrace{-2m_1 d}_{B} x + \underbrace{m_1 d^2}_{C} = 0.$$

Now we turn to Formula (2.14):

$$x_{1,2} = \frac{2m_1 d \pm \sqrt{4m_1^2 d^2 - 4(m_1 - m_2)m_1 d^2}}{2(m_1 - m_2)} = \frac{m_1 \pm \sqrt{m_1 m_2}}{m_1 - m_2} d.$$

Only one of the two solutions (namely the one associated with the "–") is relevant (since the forces must have opposite signs). This implies $x < d$ and

$$x = \frac{m_1 - \sqrt{m_1 m_2}}{m_1 - m_2} d. \tag{2.20}$$

⊕ *Remark*: It often proves to be convenient to consider only the *ratio* of the two masses. This could provide a better understanding of the relations. We can rewrite Formula (2.20) to show this by dividing the right term numerator and denominator by $m_1$

$$\Rightarrow x = \frac{1 - \sqrt{\frac{m_2}{m_1}}}{1 - \frac{m_2}{m_1}} d. \tag{2.21}$$

Now we employ a small trick: we have

$$\frac{a-b}{a^2 - b^2} = \frac{a-b}{(a+b)(a-b)} = \frac{1}{a+b}.$$

In our case, let $a = 1$ and $b = \sqrt{\frac{m_2}{m_1}}$. Thus, we can simplify Formula (2.21) to get

$$x = \frac{1}{1 + \sqrt{\frac{m_2}{m_1}}} d.$$

*Numerical example*: For a spacecraft between the Earth and the Moon (which have a mass ratio of $1:81$, $d \approx 384{,}000$km), this means

$$x = \frac{d}{1 + 1/9} = \frac{9}{10} d.$$

So, the spacecraft must be about 346,000 km from Earth, or rather 38,000km from Earth (measured from the centers of both the Earth and the Moon). ⊕

⊕ *Remark*: While the attraction of the Sun is quite strong, it is completely offset by the centrifugal force (the Earth, the Moon and the spacecraft rotate around the Sun at about the same speed). ⊕ ◂◂◂

## 2.5 Algebraic equations of higher degree

In applied mathematics, algebraic equations of higher degree(s) such as

$$f(x) = \sum_{k=0}^{n} a_k x^k = 0$$

are not uncommon. The solutions of such equations are called the "roots" or "zeros" of the equation. Equations of degree three and four can be solved by means of rather complicated formulas without too much effort and can even be solved exactly with a detour via the *complex numbers.*
Without going into too much detail or even using any "complex numbers" (see B), we only want to collect the most important facts on algebraic equations. The theory of these equations kept mathematicians busy over many centuries. The most important theorem was given and proved by C.F. *Gauß* in his PhD thesis:

> *The Fundamental Theorem of Algebra*: An algebraic equation of degree $n$ has $n$ solutions when so-called "multiple solutions" and also when so-called "complex solutions" are counted.

⊕ *Remark*: The German mathematician Carl Friedrich *Gauß* (1777 – 1855) is one of the most important mathematicians of all time. He was involved in many fields of mathematics and its applications, especially in astronomy. Even up to the present, many of his elegant methods have remained unsurpassed. ⊕
For us, an important consequence of the fundamental theorem of algebra (see Section 2, page 41) is the following: *The number of real solutions is at most $n$, but it may reduce by multiples of* 2. *Multiplicities also count.*

**Fig. 2.33** images formed by refraction

▸▸▸ **Application**: ***refraction at a plane*** (Fig. 2.33)
To determine the "break point" of a light ray at the planar interface between two "media" (a water surface, a surface of a thick glass plate, etc.), we must

solve an algebraic equation of degree 4, like $f(x) = a_4\,x^4 + a_3\,x^3 + a_2\,x^2 + a_1\,x + a_0 = 0$. Two or four solutions are real. However, only one of these is relevant. It can, above all, be calculated accurately and efficiently by a computer.

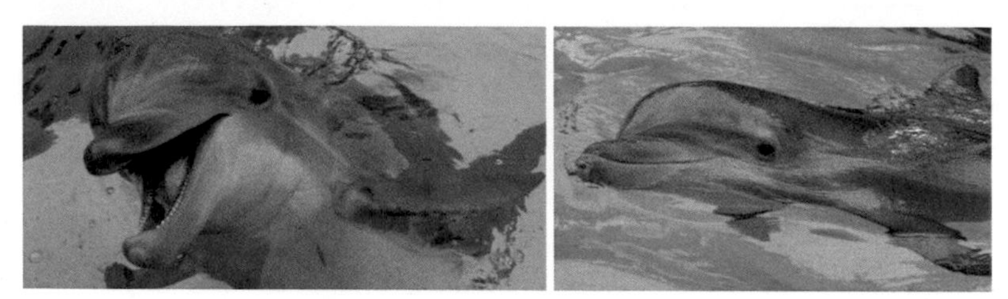

**Fig. 2.34** right: heavily skewed beneath the surface

⊕ *Remark*: Consider the "elevation" of the pool bottom in Fig. 2.33. Both the photo on the left and the computer calculation on the right have a constant value as their depth! The right image shows four (!) dolphins that swim right above each other. The top dolphin is barely recognizable in the image. Also if we look at Fig. 2.34, the animals are "undistorted" above, but heavily skewed beneath the surface. ⊕ ◂◂◂

▸▸▸ **Application**: ***ray tracing***

Realistic computer images, to which we are now accustomed, are often generated by ray tracing programs. The "scene" is composed of "primitive algebraic building blocks", such as flat polygons (triangles), spheres, cones, cylinders, or annular surfaces (tori). Lines of sight are considered by the individual *pixels* of the screen ("picture elements"). They either disappear "into nothing", or else they meet such blocks. The identification of all possible intersections of the line of sight is done by means of the rigorous and efficient method of solving algebraic equations up to degree 4. If several intersections exist, only the foremost – thus seen as the visible – is considered. ◂◂◂

In general, when dealing with algebraic equations of a higher degree, we will have to make do with approximate solutions, as will be the case with non-algebraic expressions of the form $f(x) = 0$. We deal with this issue in the differential calculus section (*Newton*'s method, Chapter 6).
Sometimes the degree of an equation decreases. This occurs especially when we can use a substitution of the form $u = x^2$, $u = x^3$, etc.

▸▸▸ **Application**: ***reducible equation of degree six*** (Fig. 2.35)

Find all real solutions of the equation

$$x^6 - 2\,x^3 + 1 = 0.$$

*Solution*:
By means of the substitution $u = x^3$, we obtain the easily solvable quadratic equation:

$$u^2 - 2\,u + 1 = 0 \Rightarrow u_{1,2} = 1 \text{ (double solution!)}$$

Since $u = x^3 = 1$, a cubic equation arises. The latter has one real solution. Overall, our equation of degree six has, therefore, a – doubly counted – real zero. This solution might not be found by a program that works numerically. It will only estimate zeros (up to a certain precision) within intervals where the function changes sign (or vanishes). Thus, in this special case, it *may* find the only zero, but it will do so only by chance. ◂◂◂

Sometimes trivial solutions of a higher degree equation split off so that the degree of the equation can be reduced. Under these circumstances, the remaining solutions can be calculated exactly.

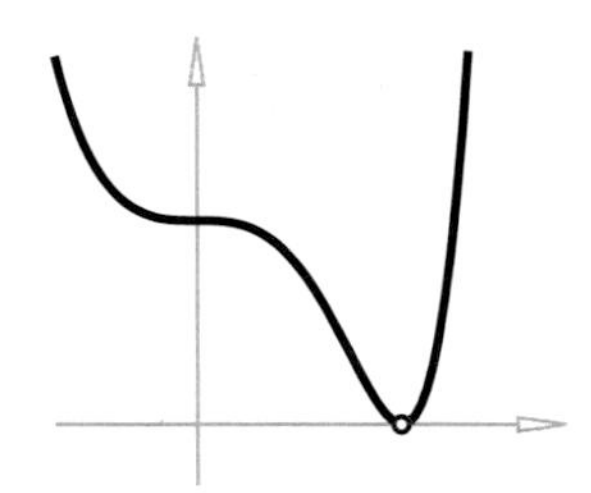

**Fig. 2.35** $f(x) = x^6 - 2x^3 + 1$

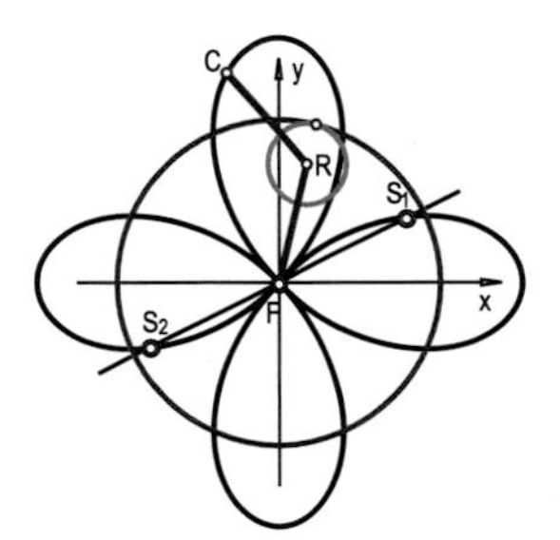


**Fig. 2.36** Cartesian quadrifolium

▸▸▸ **Application**: ***algebraic curve of degree six (quadrifolium)*** (Fig. 2.36)
*Descartes*'s *quadrifolium* is given by the equation

$$(x^2 + y^2)^3 = 27\,x^2y^2. \tag{2.22}$$

(An equation in $x$, $y$ where neither variable is isolated like $y = f(x)$ is called an *implicit equation.*) Find the up to six points of intersection with a straight line through the origin.

*Solution*:
In the general linear equation $y = kx + d$, we set $d = 0$ since the line passes through the origin (0/0). Then we insert into this Formula (2.22) and get

$$[x^2(1 + k^2)]^3 = 27\,k^2\,x^4 \Rightarrow (1 + k^2)^3\,x^6 = 27\,k^2\,x^4.$$

Now, we can cut out $x^4$ if we assume $x \neq 0$. Then, we have only a purely quadratic equation whose solutions can be directly written as

$$(1 + k^2)^3\,x^2 = 27\,k^2 \Rightarrow x_{1,2} = \pm\sqrt{\frac{27\,k^2}{(1 + k^2)^3}} = \pm\frac{3k\sqrt{3}}{\sqrt{(1 + k^2)^3}}.$$

The value $x = 0$ counts in the algebraic sense as a *four-fold solution.* In fact, one can see in Fig. 2.36 that the curve meets a general line through the origin at two points $S_1$ and $S_2$ (with the $x$-coordinates $x_1$ and $x_2$) and it additionally intersects at the origin four times.

**Fig. 2.37** generalization of the quadrifolium by varying the kinematic generation

⊕ *Remark*: The quadrifolium can be generated *kinematically* by a so-called *planetary gear* in which a rod $FR$ uniformly rotates about a fixed origin $F$, while a second rod $RC$ – of the same length – (Fig. 2.36) rotates about $R$ with triple angular velocity. The same motion is generated by rolling a circle $R$ on a fixed circle $F$. The ratio of the radii is $1:4$. This generation allows a generalization by varying the ratio of the radii. This produces a trefoil, quintifolium, etc. The trefoil is a curve of degree four. The degrees of the other curves are significantly higher. See also page 542. ⊕

◂◂◂

▸▸▸ **Application**: ***cubic curves*** (Fig. 2.38)

The folium of René Descartes (1596–1650) is best described in the coordinate system that bears his name (Cartesian). Its cubic equation is $x^3 + y^3 = c\,xy$. The intersection with rays $y = kx$ through the origin lead to simple linear equations: $x^3(1+k^3) = c\,kx^2 \Rightarrow x = \frac{c\,k}{1+k^3}$.

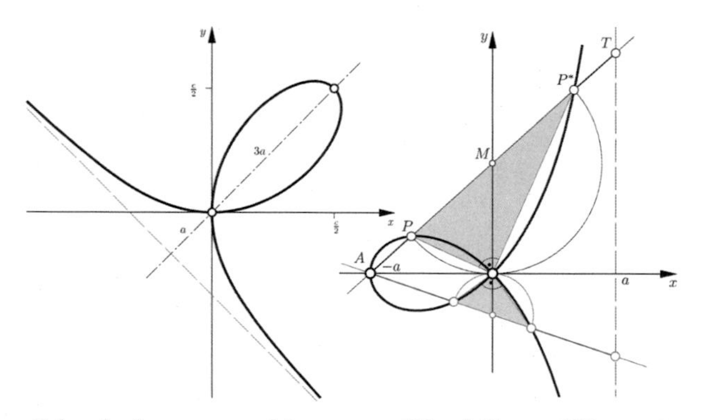


**Fig. 2.38** Two "classics" among cubic curves: The folium of Descartes and the right strophoid. Both curves are described by algebraic equations of degree 3.

The strophoid is easily constructed: On each ray through the point $A(-a/0)$, one draws those points which are at distance $d$ from the ray's intersection $M = (0/d)$ with the $y$-axis. This yields two points $P$ and $P^*$ on the curve. The curve has a double point at the origin, and its tangents there are the two angle bisectors of the coordinate axes. The algebraic equation of the curve is $(a-y)y^2 = x^2(a+x)$. Again, points on rays can be found via a linear equation.

◂◂◂

▸▸▸ **Application**: ***the Delian cube duplication problem***[4] (Fig. 2.39)
Legend has it that the Delphic oracle once suggested that the volume of the cube shaped altar in the temple to Apollo should be doubled. It was, therefore, necessary to determine the side lengths of two cubes whose volumes formed the ratio 1 : 2. The solution is, of course, the cubic root of 2:

$$x^3 = 2a^3 \Rightarrow x = a \cdot \sqrt[3]{2}.$$

A remarkable approach is attributed to *Menaechmus*, who presented the problem to Plato's school. In his approach to the problem, he used curves that would later be classified by *Apollonius of Perga* (262–190 BC) as parabolas. We consider four cubes altogether with side lengths $a$, $x$, $y$, and $2a$. Let the volume of the second cube be twice that of the first, the volume of the third be twice that of the second, and that of the fourth be twice the volume of the third.

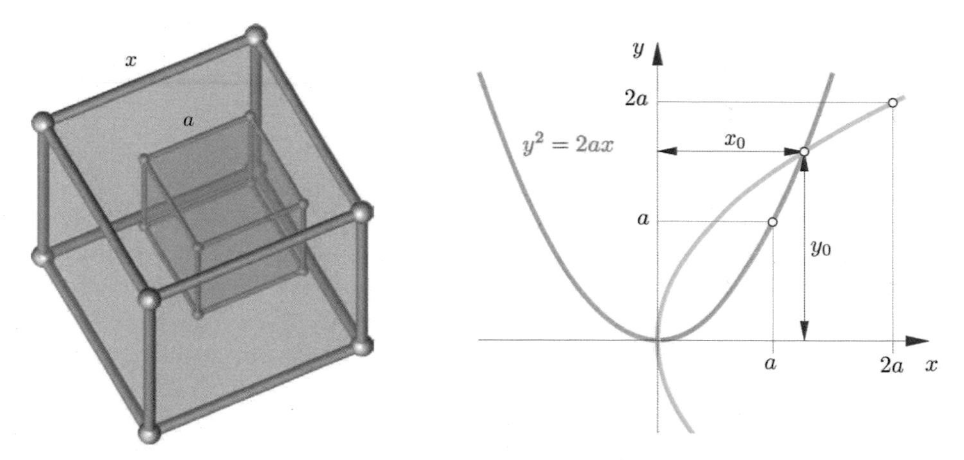


**Fig. 2.39** left: *Menaechmus*'s method, right: intersection of quadratic parabolas

Then, from $a : x = x : y = y : 2a$, we obtain the system of equations

$$x^2 = ay, \quad y^2 = 2ax.$$

These two equations can be represented graphically as parabolas through the origin whose remaining point of intersection yields the side length of the desired "intermediate cube." By manipulating the equations, the system can be interpreted as the intersection of a circle with an equilateral hyperbola:

$$x^2 + y^2 = 2ax + ay, \quad x^2 - y^2 = ay - 2ax.$$

In each case, it amounts to a solution that the classical Greeks might have deemed inelegant. ◂◂◂

[4] G. Glaeser, K. Polthier: *Bilder der Mathematik*, Springer Spektrum, Heidelberg, $3^{rd}$ edition 2014.

## 2.6 Further applications

This section contains exercises related to the previous sections. They are mostly independent of one another and can be partially skipped without affecting one's comprehension of the concepts. The examples are somewhat more complex than the previous ones. The reader is encouraged to consider at least the comments on the results in this section.

▸▸▸ **Application**: ***How often do the hands of a clock overlap?***

*Solution*:

This problem can be solved quickly (otherwise you have to give a lavish argument): We start at exactly 12 noon, where both hands overlap. After a full turn of the minute hand, the hour hand has performed 1/12 (30°) of a full rotation. When the minute hand has rotated 12/11 of the full angle of rotation, then the hour hand has rotated 1/11 (almost 33°) of the full revolution, so that there is coverage again. Then, over a period of 12 hours, the hands will have overlapped 12/(12/11) times, that is, exactly eleven times.

**Fig. 2.40** clocks that drive each other

⊕ *Remark*: The same question is significantly more challenging to solve when two clocks that move at different speeds are linked via hinges on their minute hands (Fig. 2.40) and the whole installation is left to move on its own ... ⊕ ◂◂◂

▸▸▸ **Application**: ***counter-directed motions***

A train with an average speed of $c_1$ drives from $A$ in the direction of $B$ (distance $\overline{AB} = d$). At $B$, a train starts $\Delta t$ hours later and drives with an average speed $c_2$ towards $A$. When do the trains meet?

*Solution*:

Let $x$ and $y$ be the respective distances that the two trains have travelled when they reach their meeting point. Thus $x + y = d$. Furthermore, let $t$ be the time that it takes for the first train to reach the meeting point. Then,

$x = t\,c_1$ and $y = (t - \Delta t)\,c_2$. This results in $t\,c_1 + (t - \Delta t)\,c_2 = d$, and thus, $t = \dfrac{d + \Delta t\,c_2}{c_1 + c_2}$. ◂◂◂

▸▸▸ **Application**: ***rock erosion*** (Fig. 2.41)
The striking Table Mountain with an altitude of 1,087 m is a landmark of Cape Town (South Africa). It is one of the oldest mountain ranges on Earth. Its age is estimated to be an impressive 600 million years (so, it is almost 10 times as old as the Alps). It is believed that the peak was over 5 000 m at its birth. What was the average erosion per year? How high was the mountain 65 million years ago, i.e. in the period when the Alps emerged and the dinosaurs became extinct?

**Fig. 2.41** Table Mountain in Cape Town with its typical "tablecloth"

*Solution*:

| | |
|---|---|
| height difference | $5{,}000\,\text{m} - 1{,}000\,\text{m} = 4{,}000\,\text{m} = 4 \cdot 10^3\,\text{m}$ |
| time difference | $600$ million years $= 6 \cdot 10^8\,a$ |
| erosion per year | $\frac{4 \cdot 10^3\,\text{m}}{6 \cdot 10^8\,a} = \frac{40 \cdot 10^2\,\text{m}}{6 \cdot 10^8\,a} \approx 6.7 \cdot 10^{-6}\,\frac{m}{a} = 6.7 \cdot 10^{-3}\,\frac{\text{mm}}{a}$ |

The erosion in 65 million years $(6.5 \cdot 10^7 a)$ is then

$$6.7 \cdot 10^{-6}\,\frac{m}{a} \cdot 6.5 \cdot 10^7 a \approx 44 \cdot 10^1\,\text{m} = 440\,\text{m}.$$

The mountain was 440 m higher 65 million years ago, so a little over 1,500 m high. ◂◂◂

▸▸▸ **Application**: ***if the Antarctic ice melts ...***
The Antarctic (the mainland has a surface area of about 12 million km$^2$) has only relatively recently become frozen – but now it is all the more so! The major part of the fresh water on Earth is bound in its ice sheet, which is about 2 km thick on average. By how much will the sea level rise if the ice melts completely?

*Solution*:
We will set the unit to kilometers. The volume of ice on the mainland is $24 \cdot 10^6\,\text{km}^3$. When ice melts, it loses about 10% of its volume, so that about

**Fig. 2.42** the Antarctic and one of its inhabitants – the emperor penguin

$21 \cdot 10^6\,\text{km}^3$ remains.

The Earth has a surface area of about $500 \cdot 10^6\,\text{km}^2$, of which $70\% \approx 350 \cdot 10^6\,\text{km}^2$ is water. Let $\Delta$ be the difference in the heights of the sea level, then

$$350 \cdot 10^6\,\text{km}^2 \cdot \Delta = 21 \cdot 10^6\,\text{km}^3 \Rightarrow \Delta \approx \frac{21}{350} = 0.06\,\text{km} = 60\,\text{m}.$$

⊕ *Remark*: 60 m is quite a lot. With this rise of sea level, whole groups of islands would disappear, large parts of Florida would be flooded, etc. The emperor penguins probably would die out – not because the temperature would become too high, but because once again – as before – mammals and reptiles would start to live on the Antarctic. These mammals and reptiles would then steal the eggs of emperor penguins and eat their defenceless pups (see Application p. 421).

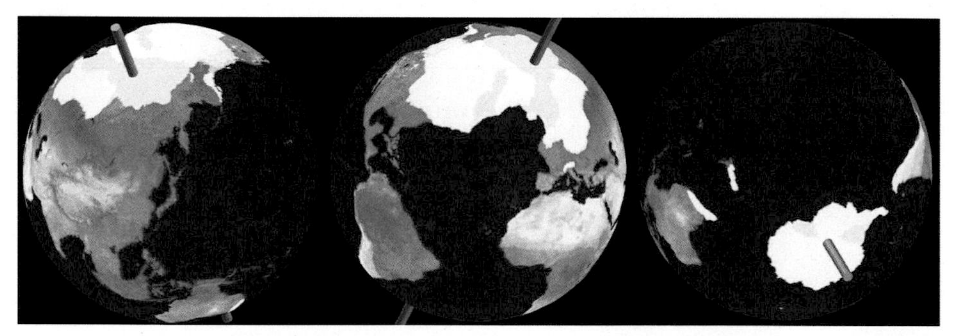

**Fig. 2.43** The last glacial maximum ($\approx 24,500$ BC): Vast ice sheets covered large parts of northern Europe, North America, Siberia, and also parts of the southern hemisphere.

Conversely, the sea level rose to this level of 60 meters about 15 million years ago (because the Antarctic was free of ice). During the ice ages that have occurred several times in the last 100,000 years (where, for example, thick ice was superimposed over central and northern Europe), the sea level was 125 m (!) lower than today. Today divers in Southern France have found underground entrances of caves where Stone Age people lived! The low sea allowed a small group of people to walk across the Bering Strait from Siberia to Alaska. During that time, they had more than 22,000 years to migrate to the two American continents (they joined together only 3 million years ago). ⊕ ◂◂◂

**▸▸▸ Application**: ***The sun dies out!***

Our sun is a giant nuclear reactor where hydrogen fuses into helium. Here 4.5 million tons of mass per second are converted into energy. The sun has a mass that is 332,000 times the mass of Earth (Application p. 91). How long would it theoretically take to use up the total mass of the Sun?

*Solution*:

We try to get an idea of the given magnitudes by relating them to more familiar objects: The loss of mass per second would be – "downscaled to earthly conditions" – $1/332,000$ of $4,500,000$ tons, i.e. 13.6 tons per second. Let us reduce further down to a ball with a diameter of only 1 m. The Earth has a diameter of about $13,000\,\text{km} = 13 \cdot 10^6\,\text{m}$. Our nickel-iron sphere with a diameter of 1 m is about $1/(13 \cdot 10^6)^3$ of the mass of Earth, and per second, it would lose $13.6 \cdot 10^3\text{kg}/(13 \cdot 10^6)^3$ of its mass, which is in milligrams

$$\frac{13.6 \cdot 10^9\,\text{mg}}{13^3 \cdot 10^{18}} \approx 0.006 \cdot 10^{-9}\,\text{mg}.$$

A year has approximately 30 million seconds. Thus, the unit sphere loses

$$0.006 \cdot 10^{-9} \cdot 30 \cdot 10^6\,\text{mg} \approx 0.2 \cdot 10^{-3}\,\text{mg}$$

every year or five milligrams every 5,000 years, or 1 kg every five billion years – that is, about three tons of deadweight.

⊕ *Remark*: Studies have shown that the Sun has a 10 billion year lifespan of which it has already elapsed half of its time. Regarding the loss of mass via nuclear fusion, the limited lifespan of the Sun is not obvious: This proportion is practically negligible! Firstly, the limited lifespan is given by the uranium component of solar hydrogen, of which it only makes up about 70%, and secondly, only in the innermost part of the Sun are the temperatures high enough to propel nuclear fusion. Thus, only 10 to 20% of the solar mass is available for fusion of hydrogen helium. ⊕

⊕ *Remark*: The "fading" of the Sun as the source of all life is an old phobia of many indigenous people. Mayans and Aztecs in the historical parts of Central America made human sacrifices as a result of such fears! ⊕ ◂◂◂

**▸▸▸ Application**: ***diameter of a molecule***

An amount of $12\,g$ of carbon contains

$$N_A = 6.022 \cdot 10^{23} \tag{2.23}$$

of ${}^{12}C$ molecules. This number is called the "Avogadro" constant (formerly *Loschmidt*'s number).

**Fig. 2.44** the Sun dies out ...

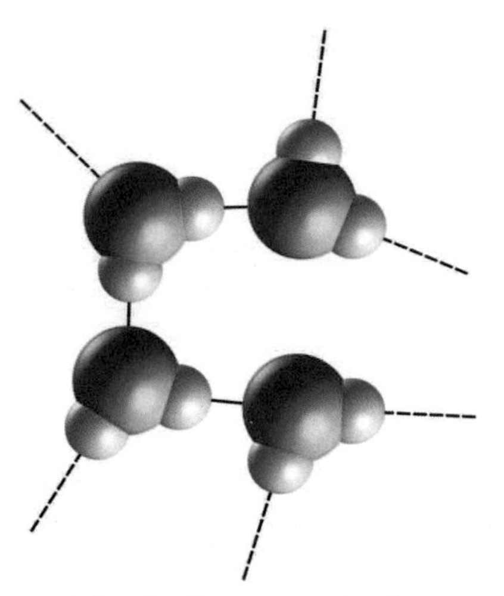

**Fig. 2.45** water molecules

Water molecules consist of 1 oxygen atom and 2 hydrogen atoms (angle $\approx 105°$). The same number of molecules is contained in one mol of any substance (*mol*= molecular weight in grams).
Water (chemical formula $H_2O$) has a molecular weight of 18 since each water molecule contains two hydrogen atoms (each with an atomic weight of 1) and an oxygen atom (an atomic weight of 16).
So, in 18 g of water, there are $N_A$ water molecules.
One can estimate the diameter of a water molecule.

*Solution*:

18 g of water has a volume of $18\,\text{cm}^3$. These fit into a cube with a side length of $\sqrt[3]{18}\,\text{cm} \approx 2.6\,\text{cm}$.
Suppose we pack every single water molecule into an "elementary cube" of side length $d$. Then, $N_A$ of these cubes would fit into said cube with a side length of 2.6 cm. Thus, we have

$$d^3\,N_A = 18\,\text{cm}^3 \Rightarrow d = \sqrt[3]{\frac{18\,\text{cm}^3}{N_A}} = \sqrt[3]{\frac{18}{6\cdot 10^{23}}}\text{cm} = \sqrt[3]{\frac{180}{6\cdot 10^{24}}}\text{cm}$$

$$\Rightarrow\ d = \sqrt[3]{\frac{30}{10^{24}}}\text{cm} \approx 3\cdot 10^{-8}\,\text{cm} = 3\cdot 10^{-10}\text{m} = 0.3\cdot 10^{-9}\,\text{m}.$$

We have shown that the diameter of a molecule is less than one nanometer.

⊕ *Remark*: According to the above result, we can say that, in 2 g of hydrogen, there are $N_A$ hydrogen molecules $H_2$. From this, one can determine the mass of a hydrogen atom, $H$, which used to be the so-called *atomic mass* unit until 1961 (since then, $\frac{1}{12}$ of the atomic mass of $^{12}C$ has been used as the legal unit): ⊕

⊕ *Remark*:

$$u = \frac{1}{2}\cdot\frac{2}{6.022\cdot 10^{23}}\,\text{g} = 0.166\cdot 10^{-23}\,\text{g} = 1.66\cdot 10^{-27}\,\text{kg}.$$

Accordingly, a helium atom has a mass of $\approx 4\,u$, one carbon atom has $12\,u$ (since exactly 1961!). An oxygen atom contains $\approx 16\,u$, a gold atom $\approx 197\,u$ etc. – each of the corresponding values are located in the "Periodic Table of Elements". The mass of the electron is (almost) negligible: An electron has a mass of $\frac{1}{1,824}u \approx 9\cdot 10^{-31}$ kg (note that this is residual mass). ⊕ ◂◂◂

**▸▸▸ Application**: ***continental drift***

*Pangea* broke up into two roughly equal continents called *Laurasia* and *Gondwana* roughly 280 million years ago. About 150 million years ago, *Gondwana* further broke and has been drifting apart since then – as seen, for example, in the movement of South America and Africa away from each other. What used to be a daring theory (Alfred *Wegener*, 1912) can be confirmed through measurement today. Assume that the drift velocity $v$ was reasonably constant and calculate how far Africa and South America drift apart every year or how far they drift apart every second when their current distance is about 5,000 km. ◂◂◂

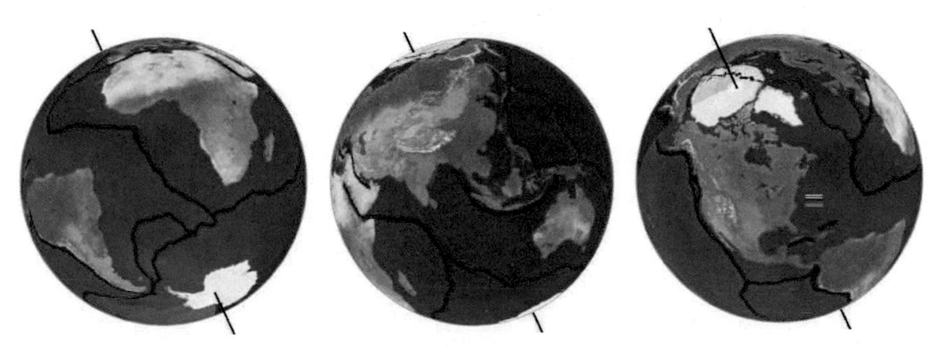

**Fig. 2.46** drifting continents, tectonic plates

*Solution*:

$$v = \frac{5{,}000\,\mathrm{km}}{150\cdot 10^6\,a} = \frac{5\cdot 10^6\,\mathrm{m}}{150\cdot 10^6\,a} \approx \frac{3\cdot 10^{-2}\,\mathrm{m}}{a} = \frac{3\cdot 10^{-2}\,\mathrm{m}}{365\cdot 24\cdot 3{,}600\,\mathrm{s}} = \frac{3\cdot 10^{-2}\,\mathrm{m}}{3\cdot 10^7\,\mathrm{s}} = \frac{1\cdot 10^{-9}\,\mathrm{m}}{\mathrm{s}}.$$

We have a drift velocity per year of 3 cm, and as a result of Application p. 49, we know this involves about *three water molecules per second*, given that we know their diameters.

⊕ *Remark*: We can measure the original distance using the drift velocity of 3 cm per year by looking at the current positions (currently at 1 m accuracy) of many places in Africa or South America over large time intervals by means of GPS (Application p. 13, Application p. 363). Then, the drift velocity is obtained by calculating the average, and this becomes more accurate given the more measurements one has taken and the longer the time interval is (for example, 5 years). ⊕

**▸▸▸ Application**: ***a global conveyor belt as an air motor***

The enormous amount of 20 million cubic meters of salt water (which is almost half the amount of freshwater on Earth) flows every second, most of it at great depths. It flows in streams at a rate of 1 to 3 km per day repeatedly around the world (Fig. 2.47). In certain places, for example, in the Gulf of Mexico, this stream is "caught" and rises so that it heats up, quickly reaching the polar regions (the Gulf Stream!), where it cools down rapidly and descends again. Looking at the six projections in Fig. 2.47, the question arises: How long will a full cycle take?

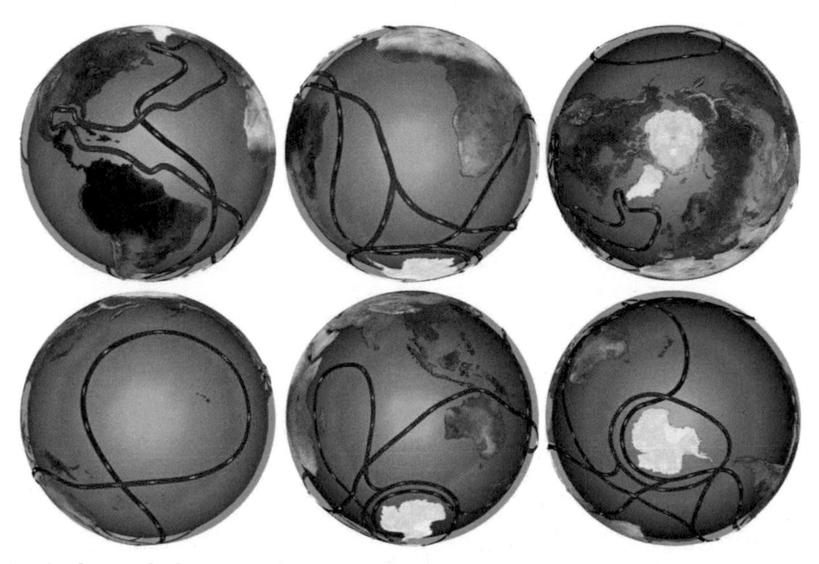

**Fig. 2.47** thermohaline circulation, colloquially shown as a global conveyor belt in different views of the globe. Images using a single view of the globe can easily lead to misinterpretations of the currents (particularly around the Antarctic).

*Solution*:
An important preliminary remark: The circulation belt is usually depicted on a "rectangular projection" where each estimation of length – especially near the poles – leads to incorrect results.The sphere is doubly curved and cannot be unfolded distortion-free into the plane (Application p. 321). The multiple belts encircling the South Pole (bottom right) is clearly visible in the image. It is in total no more than a huge loop in the Pacific (bottom left). If we estimate the length of all flows thoroughly, then the length is roughly three times the Earth's circumference (120,000 km). Assuming the flow covers a distance of 1 to 3 km daily, then assuming a yearly distance of 600 km – just to have a "nice number" – the cycle lasts for 200 years. ◂◂◂

▸▸▸ **Application**: ***lens power***
One can use Formula (4.29) to obtain the formula for the focal length of a thin biospheric lens (Fig. 2.48).

*Solution*:
Formula (4.29) applies to both spherical surfaces (radii $r_1$ and $r_2$, with refractive indices $n_1 = n$ and $n_2 = 1/n_1 = 1/n$) where Formula (4.29) is:
$\frac{1}{g} + \frac{n}{b} = \frac{n-1}{r}$, thus,

$$(1) \quad \frac{1}{g_1} + \frac{n}{b_1} = \frac{n-1}{r_1} \quad \text{and} \quad (2) \quad \frac{1}{g_2} + \frac{1/n}{b_2} = \frac{1/n-1}{r_2}.$$

Now, we can connect "in series": The image width $b_1$ of the first refraction is the object's distance, $g_2$, and the result for the second refraction is the image distance $b = b_2$ (negative!):

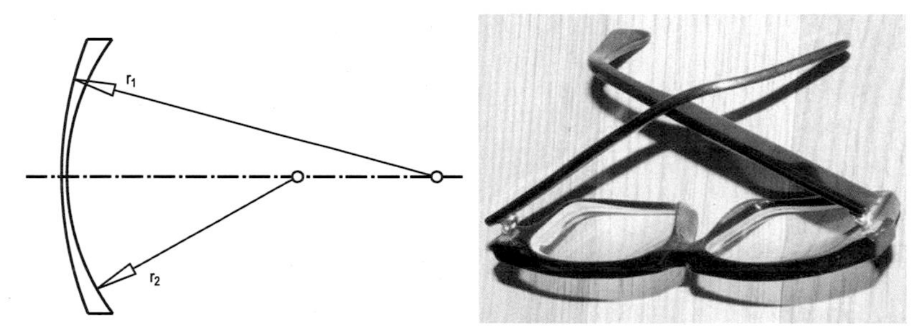


**Fig. 2.48** aspherical lens: two slightly different curved spherical surfaces

$$\frac{1}{-b_1} + \frac{1/n}{b} = \frac{1/n-1}{r_2} \quad \Rightarrow \quad (3) \quad -\frac{n}{b_1} + \frac{1}{b} = \frac{1-n}{r_2}.$$

If we add (1) and (3) and set $g_1 = g$, then we obtain

$$\frac{1}{g} + \frac{1}{b} = (n-1)\left(\frac{1}{r_1} - \frac{1}{r_2}\right).$$

Objects at infinity ($g = \infty$) are mapped to the focal point (focal length $f$). With $1/g = 0$, we get

$$\frac{1}{f} = (n-1)\left(\frac{1}{r_1} - \frac{1}{r_2}\right).$$

*Numerical example*: Setting $r_1 = 16\,\text{cm}$, $r_2 = 8\,\text{cm}$, and $n = 1.4$ results in

$$\frac{1}{f} = (1.4-1)\cdot\left(\frac{1}{16\,\text{cm}} - \frac{1}{8\,\text{cm}}\right) = \frac{-0.4}{16\,\text{cm}} = -\frac{0.025}{\text{cm}} = -\frac{2.5}{\text{m}} \quad \Rightarrow \quad f = -0.4\,\text{m}.$$

The value denoted by $1/f$ (where $f$ is measured in meters!) is called the lens power and it is expressed in diopters. Our lens has $-2.5$ dpt (a suitable power for a spectacle lens made for near-sighted people, Fig. 2.48 on the right).

⊕ *Remark*: The refractive power of the cornea is normally about 43 diopters (dpt) (focal length $f \approx 2.33\,\text{cm}$), the refractive index of the lens is about 19 diopters. Visual defects are expressed in positive or negative deviations. The emmetropic eye has a total of 65 diopters (focal length $f \approx 1.54\,\mathrm{cm}$). Its value is not determined by combining the power of the lens and the cornea as discussed. With hyperopia, the rays meet behind the fovea centralis (with a positive deviation); with myopia, they meet just before (with a negative deviation). The cornea has a refractive index that can be compared with that of water. Therefore, underwater light from the cornea is hardly broken. But because the lens in these changing conditions cannot compensate enough, our vision becomes blurry when underwater. ⊕ ◂◂◂

▸▸▸ **Application**: ***depth of field (DOF) for a photographic lens*** (Fig. 2.51)
Derived from the lens equation in Application p. 16, $\frac{1}{f} = \frac{1}{b} + \frac{1}{g}$ (Formula (2.3)) follows with $b = n\,f$

$$g = \frac{n}{n-1} f. \tag{2.24}$$

Let us imagine the lens from Fig. 2.9a built into a camera lens. Behind the lens, the photosensitive layer will be perpendicular to the optical axis. Because of the lens equation, the only points that appear in sharp focus are those that lie in the plane $\gamma$ parallel to the photo plane. Assume that for $b$, a tolerance of $t\,f$ is allowed. (In general, $t$ is considered to be very small, e.g. $t = 0.01$). What is the "depth of field" in this case? How far is the object allowed to deviate from $\gamma$?

*Solution*:
We have $n\,f - t\,f \le b \le n\,f + t\,f$. The extreme image distances

$$b_1 = f(n+t) \text{ and } b_2 = f(n-t)$$

correspond, according to Formula (2.24), to extreme object distances

$$g_1 = \frac{n+t}{n+t-1} f, \quad g_2 = \frac{n-t}{n-t-1} f.$$

Then, the difference $g_2 - g_1$ equals the depth of field

$$s = \left(\frac{n-t}{n-t-1} - \frac{n+t}{n+t-1}\right) f.$$

We try to simplify the expression in parentheses by finding the common denominator

$$N = (n-t-1)(n+t-1) = [(n-1)-t][(n-1)+t] = (n-1)^2 - t^2.$$

The numerator $Z$ is then $Z = (n-t)(n+t-1) - (n+t)(n-t-1) =$

$$= (n^2 - t^2) - (n-t) - [(n^2 - t^2) - (n+t)]n^2 - t^2 - n + t - [n^2 - t^2 - n - t] = 2t$$

and we have $s = \frac{Z}{N} f = \frac{2t}{(n-1)^2 - t^2} f$. If $t$ is small, then $t^2$ is much smaller (for example, $t = 0.01 \Rightarrow t^2 = 0.0001$). Thus, we can write

$$s \approx \frac{2t\,f}{(n-1)^2}.$$

The result should not be interpreted rashly by assuming that a better depth of field is a larger one, i.e. the larger $f$, the larger $s$. Both $t$ and $n$ depend on $f$. Let us replace $tf$ by $t_0$ (where $t_0$ is a constant value representing time). From the lens equation, we calculate

$$\frac{1}{b} = \frac{1}{f} - \frac{1}{g} = \frac{g-f}{fg} \Rightarrow n = \frac{b}{f} = \frac{g}{g-f} \Rightarrow n - 1 = \frac{g}{g-f} - 1 = \frac{f}{g-f}.$$

Then, we get

$$s \approx \frac{2\,t_0}{\left(\frac{f}{g-f}\right)^2} = 2\,t_0\left(\frac{g-f}{f}\right)^2 = 2\,t_0\left(\frac{g}{f} - 1\right)^2$$

and recognize:

| The smaller the focal length $f$, or the further the object, the larger is the DOF. |
|---|

**Fig. 2.49** wide-close shooting vs. telephoto shooting

*Numerical example*: We want to take a picture of an object that has a distance of 1.50 m, once with a wide-angle lens ($f_1 = 30$ mm) and a second time with a telephoto lens of ($f_2 = 150$ mm).
Let $g = 1,500$ mm, i.e.

$$g/f_1 = 1,500\,\text{mm}/30\,\text{mm} = 50, \quad g/f_2 = 1,500\,\text{mm}/150\,\text{mm} = 10.$$

For the depth of fields, we have

$$s_1 = 2\,t_0 49^2, \quad s_2 = 2\,t_0 9^2 \Rightarrow s_1 : s_2 \approx 30 : 1.$$

Thus, we see that the telephoto lens can have a sharp image only with a *much* smaller depth range. Yet, you have to remember that such a lens allows us to capture details that might be displayed in sharp focus but they only appear tiny. Enlarging a section of the corresponding image will mostly lead to a loss of image resolution.

**Fig. 2.50** depth I **Fig. 2.51** depth II

⊕ *Remark*: The lens equation is only accurate when the incident light rays do not deviate too much from the optical axis. This environment is called Gaussian space. The depth of this field can be increased very effectively by choosing a large aperture. Pictures in the macro area are very critical because they are mapped very close to the simple focal length. Fig. 2.51 shows two fighting male stag beetles (*Lucanus cervus*). Here, the depth could be achieved only through the widest aperture – this one must either use an ultra-light-sensitive film or a flash. Nevertheless, the foot joints in the front and rear area are blurred. In contrast, the blur of the background in Fig. 2.50 (garter snakes) – due to the small aperture of 2.8 and the short distance as a result of the short exposure time 1/250 s (Application p. 70) – is quite desirable. The background should be completely neutralized.
Intentional blur is an important design element in photography. In this specific case, it was important to capture the snakes' eyes and tongues in sharp focus, so that the viewer's attention is drawn to these elements of the photograph.
Even the human brain works in this manner, with the eyes acting like an "external branch". One could say that we have the advantage of a "selective perception". ⊕

◂◂◂

▸▸▸ **Application**: ***How long does an electron flash expose?***
The built-in flash of a DSLR cannot be arbitrarily "synchronized". One can illuminate only with relatively long exposure times (about 1/250 sec). Nevertheless, images that should actually be blurry may be engraved and appear sharper than preferred. Guess the actual exposure time by the flash of the pictures shown in Fig. 2.52.

*Solution*:
A ball (depicted in the middle of Fig. 2.52) is launched by a spring to a height of approximately 1.2m. At a height of 40 cm, the ball has just enough speed to make the remaining 80 cm.
With Formula 2.19 (Application p. 38), we deduce the instantaneous velocity of the ball $v = \sqrt{2gh} \approx 4\,\mathrm{m/sec} = 4{,}000\,\mathrm{mm/sec}$. During the time the picture was taken, the ball barely moves – perhaps 1 mm. This means that the flash exposes approximately $1/4{,}000$ of a second. So, under certain circumstances,

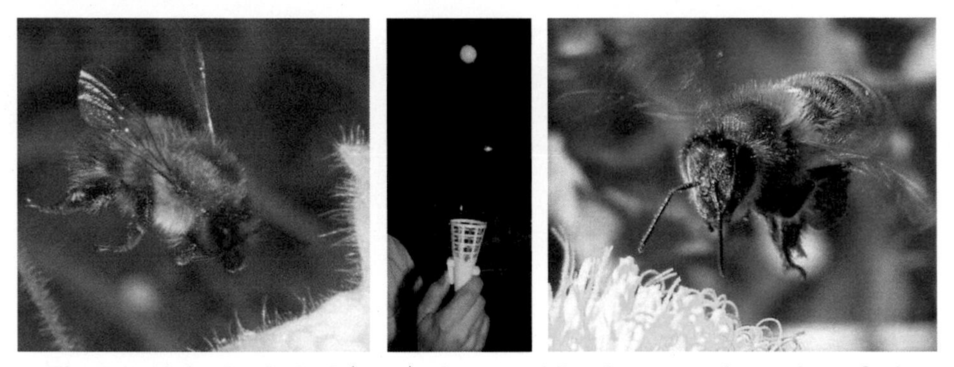

**Fig. 2.52** left: the flashed (crisp) photo; middle: the test; right: without flash

one can "freeze" the blazing fast moving wings of a bumblebee (Fig. 2.52, left).

⊕ *Remark*: For comparison: 1/250 seconds (in sunlight) is definitely not sufficient to depict the wings of a bee in sharp focus (picture right). In direct sunlight, flash cannot be used to "freeze": The residual light during the synchronization time leads to the exposure of the film. Here, you have to work with an external and relatively expensive flash.
Digital cameras have no problem with flash synchronization, because they require no mechanical motion of a mirror. ⊕ ◂◂◂

▸▸▸ **Application**: ***the product of two pencils of rays*** (Fig. 2.53)
A straight line $a$ rotates about a fixed point $A(0/0)$ with a constant angular velocity of 1 while a straight line $b$ rotates about a fixed point $B(4/0)$ with a proportional angular velocity of a) $-1$, b) $+1$, c) 2. Consider the locus of all points of intersection.

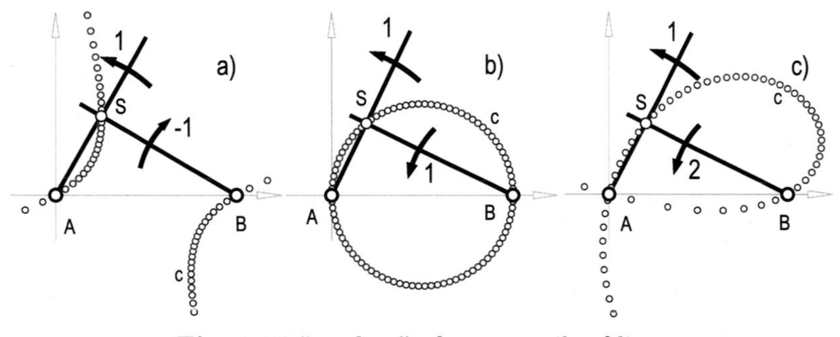


**Fig. 2.53** "product" of two pencils of lines

*Solution*:
The overall calculation of all points is, of course, done by computer. Here, I will just say how to "draw up" the line equations: A straight line through the origin $O$ with an inclination angle $\alpha$ to the $x$-axis has the equation $y = k_1\,x$ with $k_1 = \tan\alpha$ (or in the usual notation, of $-k_1\,x + y = 0$). A straight line

through the point $B(4/0)$ with the inclination angle $\beta$ and $k_2 = \tan\beta$ has the equation $-k_2\,x + y = -4\,k_2$. The principal determinant of our $(2,2)$-system is, therefore, $D = \begin{vmatrix} -k_1 & 1 \\ -k_2 & 1 \end{vmatrix} = k_2 - k_1$. It vanishes if $k_1 = k_2$, a case which shall be excluded for the moment (since the lines are always parallel in this case). The other two determinants are

$$D_x = \begin{vmatrix} 0 & 1 \\ -4\,k_2 & 1 \end{vmatrix} = 4\,k_2 \quad \text{and} \quad D_y = \begin{vmatrix} -k_1 & 0 \\ -k_2 & -4\,k_2 \end{vmatrix} = 4\,k_1\,k_2.$$

Thus, the intersection of the lines is $S\left(\frac{4\,k_2}{k_2-k_1} / \frac{4\,k_1\,k_2}{k_2-k_1}\right)$.
Fig. 2.53 shows some specific solutions. In the general case, an equilateral hyperbola is defined as being generated by two indirectly congruent pencils of lines (Fig. 2.53a), a circle by two directly congruent pencils (Fig. 2.53b). If the ratio of the angular velocities equals $1:2$, the locus of intersection points is a cubic curve (Fig. 2.53c). If the two lines start at the same position with the ratio of the angular velocities equalling $-1:1$, then the hyperbola collapses to the bisector of $AB$. If the ratio of angular velocities is $1:1$ and the lines are parallel in the beginning, then they remain parallel. If the ratio equals $2:1$, then the point of intersection traces a circle (the central angle is twice the angle of circumference – see Fig. 4.52). ◂◂◂

▸▸▸ **Application**: ***supersonic speed in free-fall***

Felix *Baumgartner*'s stratosphere dive in October 2012 helps us find out how long one must fall within a (near-)vacuum in order to break the sound barrier, and it helps us to figure out the length of the distance travelled.

*Solution*:

Properly formulated, the problem is easily solved: The acceleration of gravity (1 $g$) is approximately 10 meters per second squared (at an altitude of 40 km it is still almost the same). This simply means that the instantaneous velocity increases every second by 10 meters per second. After 32 seconds, the diver is, therefore, able to reach 320 m/s, which corresponds to the speed of sound at about $-20°$ Celsius (the speed of sound is dependent on temperature and increases by about 6 m/s given each $10°$ temperature increase.) Since the uniform acceleration is the average speed – equal to half the maximum speed, i.e. 160 m/s, the distance travelled is thus $32 \cdot 160 = 5{,}120$ meters.

⊕ *Remark*: If one theoretically accelerated (by means of rocket-power) for one year (that is 30 million seconds) with 1 $g$, one would reach 300,000 km/s – this is the speed of light. However, this is purely theoretical, because if you spin this idea further, you will reach superluminal speed ...
Realistically, however, just about 10 $g$ (100 m/s speed increase per second) is bearable! At this rate, in just under two minutes, one could reach the necessary 11.2 km/s, in order to be able to escape the Earth's gravitational field (Application p. 403). ⊕
◂◂◂

▸▸▸ **Application**: ***A body brakes.***

A body (for example, an automobile) moves with velocity $v_0$ and should be brought to a standstill within $d$ meters. What is the average deceleration $a$?

*Solution*:

For the instantaneous velocity $v$, the following applies in the case of a uniform deceleration $a$:

$$v = v_0 - a\,t. \tag{2.25}$$

For $v = 0$, this results in $t = \frac{v_0}{a}$. Inserting this variable into the formula for the distance

$$s = v_0\,t - \frac{a}{2}t^2 \tag{2.26}$$

one obtains

$$d = v_0\frac{v_0}{a} - \frac{a}{2}\left(\frac{v_0}{a}\right)^2 = \frac{v_0^2}{2a}.$$

From this, one can gather that the length of the braking distance depends on the square of the output speed. Thus, we obtain for the average deceleration

$$a = \frac{v_0^2}{2d}. \tag{2.27}$$

◂◂◂

▸▸▸ **Application**: ***bungy jumping*** (Fig. 2.54)

Bungy jumping (also "bungee jumping" or "bungi jumping") is an "invention" of the inhabitants of New Zealand. In this daring activity, a person jumps from a fixed spot, at a great hight, to which they have been tied by an elastic rope. Until the rope length $L_0$ is reached, the person will fall freely. After that, the elastic rope will produce an increasing braking effect until the maximum cable length of $L_{max}$ is reached. How long does the bungy jumper dive in absolute free fall?

To give a specific example: A jump from the Bloukrans River Bridge, South Africa (216 m height) is currently the longest possible dive: $L_0 = 90\,\text{m}$, $L_{max} = 170\,\text{m}$.

*Solution*:

The modified Formulas (2.25) and (2.26) for the vertical fall (initial velocity $v_0$, current velocity $v$, acceleration due to gravity $g \approx 9.81 m/s^2$) are:

$$v = v_0 + g\,t, \quad s = v_0\,t + \frac{g}{2}t^2. \tag{2.28}$$

When horizontal bounce applies, then $v_0 = 0$ (since there is no downward velocity component). The formula is valid until the time $T_0$, when the cable is tensioned. Then, we have

$$L_0 = 0 \cdot t + \frac{g}{2}T_0^2 \Rightarrow T_0 = \sqrt{\frac{2L_0}{g}}.$$

**Fig. 2.54** bungy jumping

The corresponding velocity is

$$V_0 = 0 + gT_0 = \sqrt{2gL_0}. \tag{2.29}$$

In our special case ($L_0 = 90\,\text{m}$, $L_{max} = 170\,\text{m}$):
$T_0 \approx 4.28\,\text{s}$, $V_0 \approx 42\text{m/s} \approx 150\text{km/h}$

⊕ *Remark*: From the time $T_0$, $g$ is reduced by the (ever-increasing) cable delay. The speed continues to increase until the rope force exceeds the bungee jumper's weight. This occurs when the rope reaches its maximum length and the jumper hangs at rest (i.e. when the rope force and gravity are in equilibrium).
We will examine the rather complex conditions in the final phase of the flight in more detail by means of differential calculus in (Application p. 274). Here, we only determine the (initially not very telling) *average* deceleration $a$ and use it in the result of Formula (2.27) of Application p. 59. Using $V_0$ instead of $v_0$, the braking distance is the rope expansion $d = L_{max} - L_0$

$$a = \frac{V_0^2}{2(L_{max} - L_0)} = \frac{gL_0}{L_{max} - L_0}.$$

The average deceleration of the rope $b$ is naturally greater than $g$, because the total acceleration consists of still-operating gravity and the counteracting deceleration of the rope.
For example, let $L_0 = 90\,\text{m}$, $L_{max} = 170\,\text{m}$:

$$a \approx 11\frac{m}{s^2} \Rightarrow b \approx 21\frac{m}{s^2} \approx 2.1g.$$

The average deceleration $b$ of the rope is naturally not constant but it is, according to *Hooke*'s law, proportional to the current cable strain $\varepsilon = \dfrac{L - L_0}{L_0}$. If $b$ is $2.1\,g$ on average and increases linearly from 0 to a maximum value $b_{max}$, then $b_{max}$ will be approximately equal to twice the average acceleration, i.e., about $4\,g$. ⊕ ◂◂◂

## ▸▸▸ Application: *gravity through freefall*

During free fall, you are – at least in a vacuum – weightless. A person who jumps out of an airplane barely reaches speeds above 50 m/s, because air drag and weight are balanced. A plane can overcome the barrier of air resistance by motor force. How long can one thus create weightlessness with an airplane?

*Solution*:

A plane cannot rise arbitrarily high because it takes a certain air density to fly. At an altitude of about 10 km, it could plunge towards the ground "like an eagle". Through proper use of the engines, constant acceleration can be achieved just as for a body in vacuum. At a certain instant, the pilot has to end the plunge and must not accelerate any further. Team and materials should not be overtaxed. From Application p. 59, we know that acceleration and deceleration have to take the same amount of time in order to reach a deceleration of 1 g. To stay on the safe side, you could thus accelerate from a height of 10 km to a height of 6 km and then be located in level flight again at a height of 2 kilometers. This would correspond to a free fall of 4,000 m. With

$$s = \frac{g}{2}t^2 \Rightarrow t = \sqrt{\frac{2s}{g}} \approx \sqrt{800}\,\mathrm{s} \approx 28\,\mathrm{s}$$

we compute 28 seconds of weightlessness. The maximum speed would be $v = g\,t \approx 280\,\mathrm{m/s}$, thus, below the speed of sound.

⊕ *Remark*: In fact, aircraft are used to perform physical experiments in weightlessness. An Airbus 300 is used in practice. Here weightlessness means achieving the ascent phase of a "parabolic flight": the machine faces straight up with an increasing speed and is then forced into an inverted flight parabola – this is where weightlessness occurs (the acceleration of gravity acts by reversing the parabola of the achieved speed, and so we get the same path acceleration along the opposite curve of the parabola). After the culmination point of the parabolic flight, the conditions in free fall are shown as described. ⊕ ◀◀◀

### ▸▸▸ Application: *squaring the circle*

A square (in blue) is to be increasingly "rounded" so that the circumference or the area remain the same, as with the square shown on the left of Fig. 2.55 and even more with the one shown in the center of Fig. 2.55. In particular, the radius of the limit circle is to be determined so that the circumference or the area are equal to those of the initial square.

*Solution*:

Both in the circumference, as well as in the area of the circle, the circle constant $\pi$ is utilized. One can, therefore, deliver no exact constructive solution to the problem, but only one through calculation:

Given the side $a$ of the square, its circumference is $U = 4a$ and its area is $A = a^2$. With $U = 2\pi\,r$ or $A = \pi\,r^2$, we immediately have the radius of the limit circle:

$$2\pi\,r = 4a \Rightarrow r = \frac{2a}{\pi} \quad \text{or} \quad a^2 = \pi\,r^2 \Rightarrow r = \frac{a}{\sqrt{\pi}}.$$

For the intermediate circles (tangent distances $y$ and the radius $x$ of the rounding circles), we have:

$$4a = 4y + 4 \cdot \frac{2\pi x}{4} \quad \text{or} \quad y^2 + 4xy + 4\frac{\pi x^2}{4} = a^2.$$

With equal circumference, we have the linear condition $y = a - \frac{\pi}{2}x$; with equal area, we have the quadratic condition $y = -2x \pm \sqrt{a^2 + (4-\pi)x^2}$. The limit positions are reached at $y = 0$.

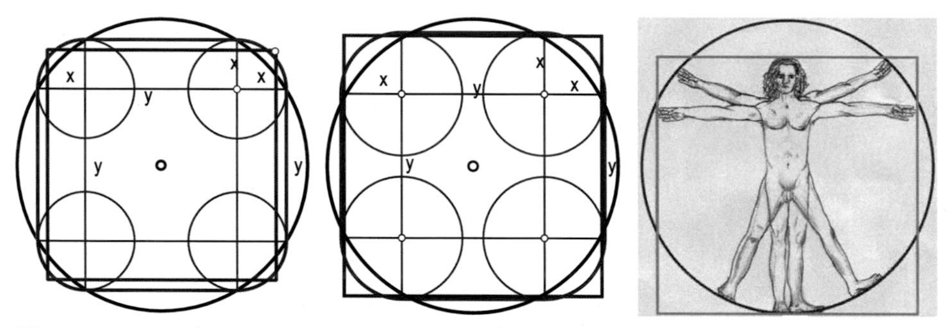


**Fig. 2.55** converting a square into a circle (left: equal circumference; middle: equal area; right: an interpretation of *Leonardo*'s Vitruvian Man)

⊕ *Remark*: Leonardo da Vinci is supposedly believed to have found a graphical solution for land conversion (squaring the circle). He suggested that the radius of the circle should enlarge a thousand times (thus increasing the area to a million times larger) and this giant circle would be composed of a million equal sectors. Such a sector would be virtually indistinguishable from an isosceles triangle whose surface can be transformed into a rectangle, and then easily into a square. For a practitioner who is satisfied with a finite number of decimal places, this is a perfectly workable solution! In any case, his drawing, which is one of the most famous sketches in the world, depicts man inside a square or circle (Fig. 2.55, right). ⊕ ◂◂◂

### ▸▸▸ Application: *quickly estimating a large area*

In Kolontar (near Ajkai in Hungary), the dam of a lake broke in October 2010. This lake was filled with toxic aluminium sludge. As the news spread across the world, it was rumoured (and this information is still available online on some websites) that an area of $40{,}000\ \text{km}^2$ was flooded. In fact, the flooded area covered a surface of only $40\ \text{km}^2$ (which is still a considerable amount). As an ordinary citizen, how do you quickly assess the maximum amount of sludge brought by the flood in cubic meters or the realistic surface area of the flooded region at a given time?

*Solution*:

A Google Earth image (Fig. 2.56) with the plotted scale shows the dam (depicted in orange because at that time it was still filled with sewage sludge). The dam can be rapidly approximated by a polygon. Now you can convert

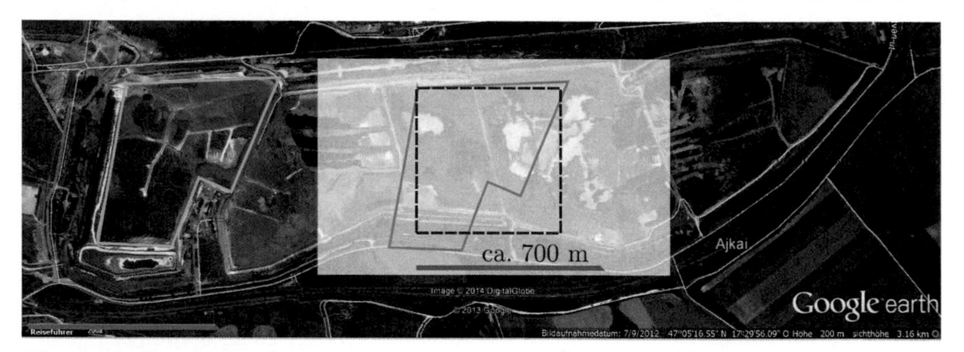


**Fig. 2.56** Google Earth image with plotted scale (bottom left). The dam is in orange and approximated by a polygon. Middle: Estimated conversion of the polygon to a square (indicated by a red dashed line) of the same area.

the polygon "instinctively" and generously to a square, with no claim to great accuracy. Taking the scale of the image into account, the following can be said: If the square has 500 to 600 meters as a side length, then the area of the polygon will be approximately $300,000\ \mathrm{m}^2$. If the lake of sludge has a depth of 3 meters (a number that was bandied about and appeared realistic), the resulting area will be significantly larger as it will be covered by 1 million cubic meters of mud. With a sludge depth of 3 cm (instead of 3 m) the surface is reduced a hundredfold to $100 \cdot 300,000\ \mathrm{m}^2 \approx 30\ \mathrm{km}^2$, which is close to the actual result. ◂◂◂

**Fig. 2.57** put all together ...

▸▸▸ **Application**: ***all the gold in the world*** (Fig. 2.57)
An annual amount of $M_1 = 2,500$ tons of gold are produced worldwide. The American United States Geological Survey estimates that more than 80 percent of the current gold production of humanity was produced after 1900 and amounts to a total of $M_2 = 160,000$ tons of gold produced by human hands. This is "nothing" in relation to the estimated $M_3 = 30 \cdot 10^9$ tons produced throughout the Earth's crust (`http://www.zeit.de/2008/16/Stimmts-Gold`). How big would the respective cubes made up of the entire gold mass of $M_1$, $M_2$ and $M_3$be? Gold (aurum) has a density of $\varrho_{Au} = 19.2$ per $\mathrm{dm}^3$.

*Solution*:
We compute with meters and tons. Then 19.2 tons have a volume of $1\ \mathrm{m}^3$. The

corresponding volumes are thus $V_1 = 2,500/19.2 \approx 130\text{m}^3$, $V_1 = 160,000/19.2 \approx 8,300\text{m}^3$, $V_3 = 30 \cdot 10^9/19.2 \approx 1.56 \cdot 10^9\text{m}^3$. By extracting the cubic root, we get the edge lengths of the cubes: 5 m, 20 m, and $\approx 1,160$ m.

⊕ *Remark*: Remarkably, there is a seemingly insignificant difference between the first two cubes. But the volume increases dramatically with the edge lengths of the cubes. This will be dealt with in more detail in the second chapter. ⊕ ◂◂◂

▸▸▸ **Application**: ***a crown of pure gold***

Legend says that King Hiero confronted the great all-round scientist Archimedes with a tricky question, and the answer was a matter of life or death for his goldsmith: He wanted to know if his crown was made of pure gold, but this should be determined without destroying the crown. Archimedes's answer: He brought a balance beam (Fig. 2.58) to balance between a piece of pure gold on one side and the crown on the other side. It changed upon immersion within water and was not at equilibrium, which spelt bad news for the goldsmith ...

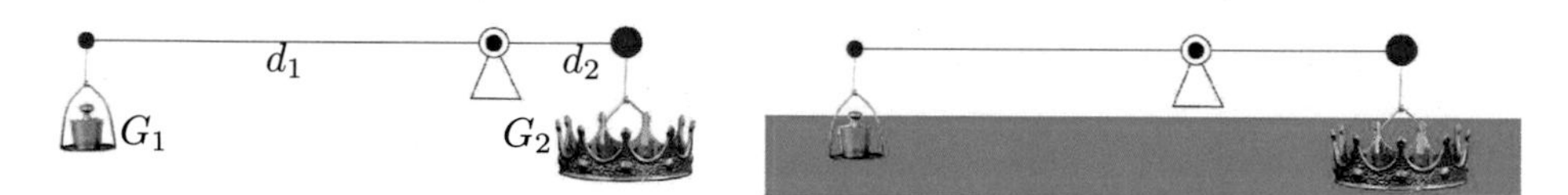


**Fig. 2.58** the gold test of *Archimedes*: $G_1 : G_2 = d_2 : d_1$

*Solution*:

Here Archimedes combined two of his most important inventions: the law of the lever and Archimedes's principle concerning buoyancy. Suppose a calibrated weight of pure gold (weight $G_1$, mass $M_1 = G_1/\varrho_1$; the density $\varrho_1 = \varrho_{Au}$ being as given in the previous example). With a volume of $V_2$ and a density of $\varrho_2$, the crown has the weight $G_2 = V_2 \cdot \varrho_2 \cdot g$. If the beam balance (Fig. 2.58, left) is in equilibrium, then $G_1 : G_2 = d_2 : d_1$ from which $G_2$ can be deduced. If we now submerse the two weights completely in water, their buoyancy forces reduce by about the weight of either displaced amount of water. Thus, the respective density decreases indirectly by about the density of water, i.e., about 1 kg per dm$^3$:

$G_1^* = M_1 \cdot (\varrho_1 - 1) = G_1 \cdot (\varrho_1 - 1)/\varrho_1$ and $G_2^* = M_2 \cdot (\varrho_2 - 1) = G_2 \cdot (\varrho_2 - 1)/\varrho_2$.

With the same density $\varrho_1 = \varrho_2$, the balance remains in equilibrium. However, if $\varrho_2 < \varrho_1$, then the buoyancy of the crown is slightly higher, and as a result, the crown rises. Equilibrium is reached again only when the bearing point is moved to the left by a (measurable) distance of $\Delta$ (Fig. 2.58, right). With a little skill, you can calculate $\varrho_2$ from all known values, and it subsequently determines a possible silver content in the crown. This works well when using a weight $G_1$ that is not made of pure gold. ◂◂◂

►►► **Application**: ***swarm rules*** (Fig. 2.59)
At times, shoaling fish or flocking birds seem to "dance" in the water or in the sky. One might be excused for thinking that there is a complicated and deliberate choreography behind it. Animals often gather or travel together in large numbers. In many cases, this swarming behavior serves as a defense against predation. Yet, how can we explain the complex and intriguing motions of swarms, as these collections of animals change directions within a fraction of a second, split up into groups and then reunite?

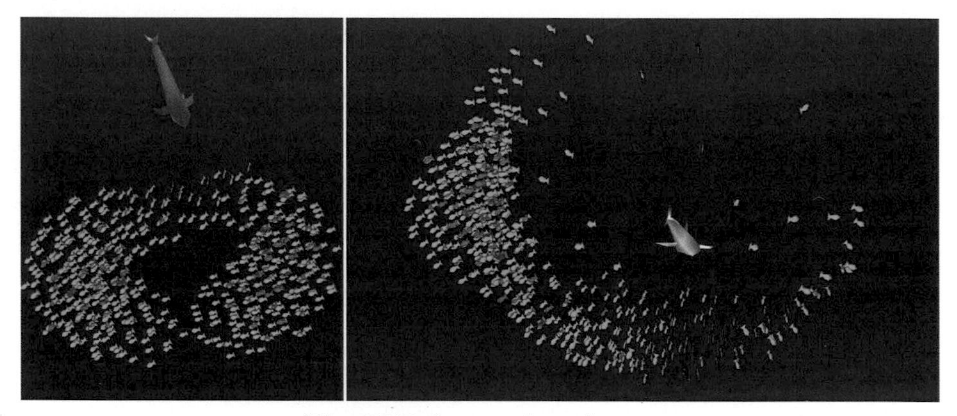

**Fig. 2.59** three major rules ...

*Solution*:
One might assume the existence of an "alpha specimen" that determines the motion of the swarm. However, how is it possible that this individual always stays at the front of the pack? In fact, there is no such leader of the pack. All members of the swarm are equal while they are in motion and merely follow three very simple rules:
1. Move in a common direction.
2. Always keep a certain distance to your neighbours.
3. If a predator is approaching, escape.

In moments of danger, the distances between neighbours increase due to reaction time. The swarm becomes wider and may even "tear apart", but as soon as the predator has left the scene, the remaining group usually reunites. As a virtual shark attacks a swarm in a computer simulation, all individuals obey the abovementioned rules. This simulation yields extremely realistic behaviour and may, thus, be taken as heuristic "proof" that the swarm rules actually exist. What is more, predators are usually distracted by swarming behaviour, and the chances of survival are larger for the individual. ◄◄◄

►►► **Application**: ***sum of cross-sections*** (Fig. 2.60)
Thinkers as far back as Leonardo da Vinci have suspected that, as trees branch out into ever more intricate formations, their total cross-section, nevertheless, stays roughly the same. This apparent rule has been explored and

**Fig. 2.60** Computer simulations based on different parameters of trunk thickness, iteration count, and branching angle. The sum of all cross-sections is constant.

refined by computer graphics engineers. Let us attempt an analysis based

**Fig. 2.61** African Aloe tree

on the African Aloe tree (Fig. 2.61). It may not be exact, but it is sufficient to imagine the branches as having a locally circular cross-section. If this is true, then the following reasoning works relatively well: Wherever a first branching occurs, we select the center $M$ of a sphere, which then includes the cross-section circles of the branches as small-circles. We can measure the radii of these circles. If the largest (lower) circle has a diameter of 1 unit (= 100%), then the smaller circles have radii of 0.57 (57%), 0.38 (38%), and 0.7 (70%). We know that the circle area increases quadratically as the radius grows linearly. In fact, our result of $0.57^2 + 0.38^2 + 0.7^2 = 0.96$ is relatively close to our expectation of $1^2 = 1$. It would seem that Leonardo's rule works well in the case of the smallest sphere. Now, we increase the sphere radius. The number of branches has grown to 25, which would suggest an average cross-section surface of 1/25 (1/5 of the maximum diameter). Leonardo's rule does not seem to apply this time, as the total cross-section seems to have grown smaller. This irregularity is even more pronounced when considering the third sphere, where the total cross-section is even smaller. ◄◄◄

# 3 Proportions and similar objects

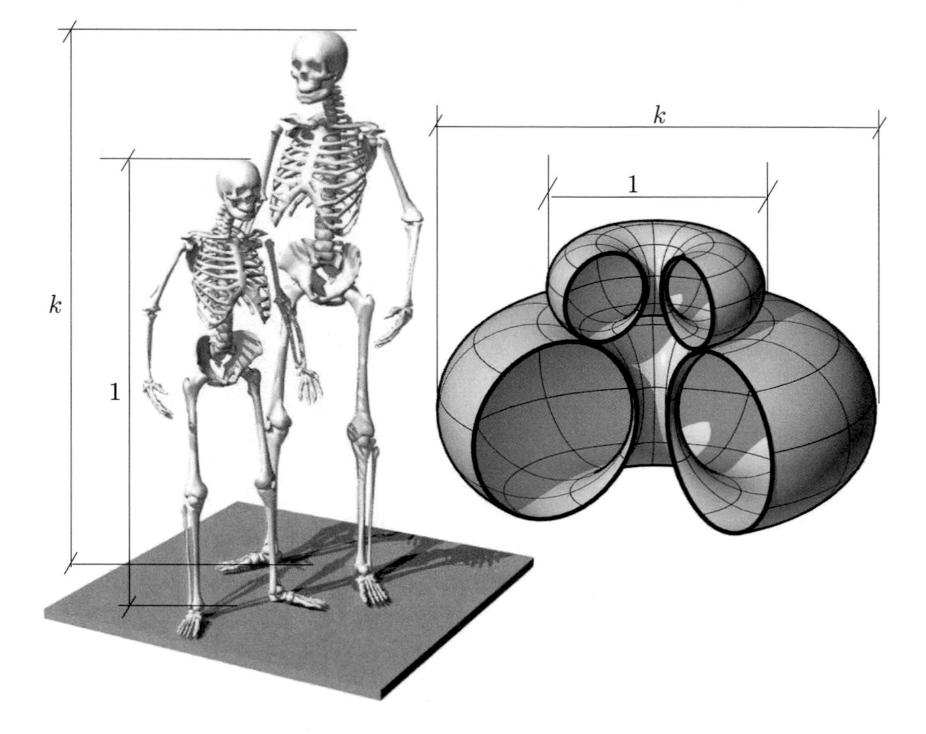


In this chapter, we deal with similar objects and other proportions. The insights thus won have many practical applications and benefit the understanding of many natural phenomena.

In particular, the surface area and cross-section area of a body do not change at the same rate as its volume or mass. Once these "relations of scale" have been understood, a multitude of interesting conclusions may be drawn. One such conclusion is, for instance, that one should be careful when drawing comparisons between large and small objects in nature, even if they share many visual characteristics.

It becomes clear that it is advantageous for large animals to be warm-blooded, while it is largely impossible for small animals to be so. In relative terms, however, smaller animals are significantly stronger. An enlargement of agile, mostly flying insects immediately changes their properties. The gigantic monsters that roam through fictional worlds would quickly fold under their own weight. Similarly, superstructures like skyscrapers, bridges, and ocean liners have to follow different rules than their miniature models.

Gravity and attraction play an increasingly significant role as the sizes and masses of objects increase, until these forces begin to dominate shapes and behaviours on the cosmic scale of stars and planets. A clever use of proportions allows us to quickly and efficiently derive astronomical phenomena, such as the calculation of orbital times and velocities of planets and satellites.


G. Glaeser, *Math Tools*, https://doi.org/10.1007/978-3-319-66960-1_3

## 3.1 Similarity of planar figures

Let us start by defining what we mean by similarity:

Objects are considered to be *similar* if they differ only in scale but not in form. Angles are identical, and pairs of lengths are at a constant proportion $k$ with respect to each other.

In the plane, all circles are similar to each other – as are all squares, equilateral triangles, etc. In three-dimensional space, spheres, cubes, equilateral tetrahedra, etc., are similar to other objects within their respective group.
*In the plane*, the following important theorem applies:

If the lengths of a two-dimensional object are enlarged by a factor $k$, then its area is enlarged by the factor $k^2$.

***Proof***: 1. This theorem evidently applies to each triangle where the following holds (Fig. 3.1):

$$\text{area} = \tfrac{1}{2}\cdot\text{base length} \times \text{height}.$$

If the base line and its height are enlarged by a factor $k$, then the area is enlarged by the factor $k^2$.

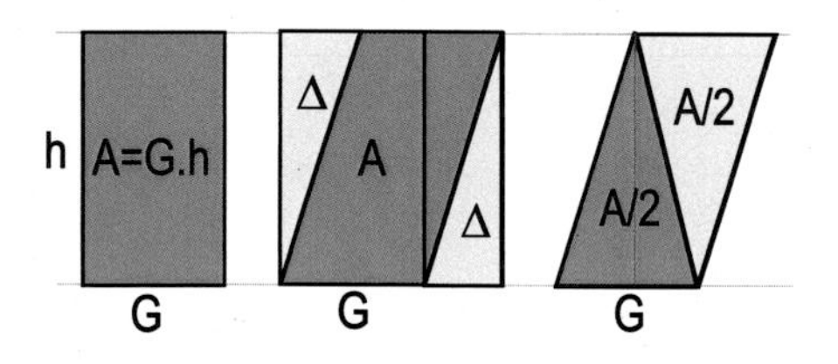


**Fig. 3.1** on the area of a triangle

**Fig. 3.2** triangulated 2D object

2. Each closed, planar polygon – even with "holes" as in Fig. 3.2 – can be "triangulated", i.e. split into triangles. Curvilinear objects can be approximated to an arbitrary precision by polygons. ⊙

▸▸▸ **Application**: ***Einstein's proof of the Pythagorean theorem***
Almost everyone knows the formula $a^2 + b^2 = c^2$ for the lengths of the sides of a right triangle. There are hundreds of different proofs for it. Particularly noteworthy is the proof that the eleven year-old *Albert Einstein* (1879–1955) discovered.

*Solution*:
Einstein imagined a triangle $ABC$ as being composed of the similar (since

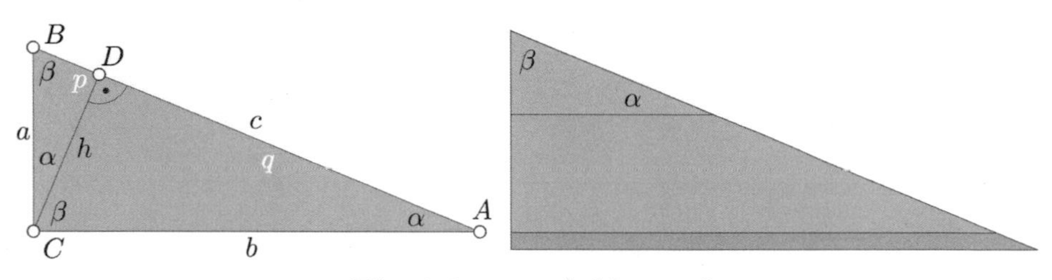


**Fig. 3.3** a remarkable proof

the angles are the same) component triangles CBD and ACD (Fig. 3.3). All three similar triangles (they have hypotenuses $a$, $b$, $c$) can be obtained from a prototype with a hypotenuse of length 1 by multiplying its sides by the factors $a$, $b$, $c$. Let $F$ be the area of this prototype. Since the area of a triangle grows as the square of the scaling factor, we have $F \cdot c^2 = F \cdot a^2 + F \cdot b^2$, and one has only to cancel $F$ to obtain the Pythagorean theorem. ◂◂◂

The *circumference* of an object increases *linearly* (multiplied by the factor $k$). Thus, in all similar objects, we may find relations that are dependent on the scale of the object: *area* : *circumference*.
Three examples shall be given:
1. square: area $a^2$, circumference $4a \Rightarrow$ *area*: *circumference* $= \frac{a}{4}$,
2. equilateral triangle: area $\frac{\sqrt{3}}{4}a^2$, circumference $3a \Rightarrow$
 *area*: *circumference* $= \frac{a}{4\sqrt{3}}$,
3. circle: area $\pi a^2$, circumference $2\pi a \Rightarrow$ *area*: *circumference* $= \frac{a}{2}$.

Thus, we conclude:

When a planar figure is enlarged, the ratio *area*: *circumference* increases proportionally to the scaling factor.

▸▸▸ **Application**: ***a strange thread around the equator***
Imagine the Earth as a completely smooth sphere (radius 6,370 km). We now wind a thread (tightly) around the equator. Then we extend this very long thread by only 10 m and lift the thread uniformly from the surface so that it is stretched again. How high is the thread?
*Solution*:
The solution is initially surprising: The radius $R$ of the circle (circumference $U_0 = 2\pi R$) lifts the thread up to around $\frac{10\,\text{m}}{2\pi} = 1.59\,\text{m}$. A circle with radius $R + \frac{10\,\text{m}}{2\pi}$ has the circumference:

$$U = 2\pi\left(R + \frac{10\,\text{m}}{2\pi}\right) = U_0 + 10\,\text{m}.$$

If we work with a *factor* of $k$, this is at least plausible: We increase the circumference of $U_0 = 40{,}000{,}000\,\text{m}$ at $U_r = 40{,}000{,}010\,\text{m}$. Thus, it is $k =$

$\frac{U_r}{U_0} = 1.000\,000\,25$. The radius $R$ of the circular tensioned thread increases from $6,370,000\,\text{m}$ to $6,370,000\,\text{m} \cdot k = 6,370,001.59\,\text{m}$. This *in proportion* is very little, but yet it is 1.59 m.

**Fig. 3.4** How much land can you cover with a bull hide?

⊕ *Remark*: A thread that became famous in history is the one that, as legend has it, led to the founding of Carthage (Fig. 3.4):
The Numidian king *Jarbas* unkindly granted the Phoenician Princess *Elyssa* (also known as *Dido*, see Fig. 3.4) only as much land as could be encompassed by a bull hide. *Jarbas* could not know that *Elyssa* would cut the skin into thin strips so that she could fence off a vast territory ... ⊕ ◂◂◂

### ▸▸▸ Application: *enlargement or diminishment with a photocopier*

When a photocopier shows an enlargement by 141%, the surface area doubles (for instance, from A4 to A3): $141\% \approx \sqrt{2}\cdot\%$. At $71\% \approx 1/\sqrt{2}\cdot\%$, the surface area is halved. ◂◂◂

### ▸▸▸ Application: *side length of the parachute by Leonardo da Vinci*

Five-hundred years ago, Leonardo da Vinci sketched the first parachute in the form of a four-sided pyramid (Fig. 3.5, left). It might have worked, as there is a hole at the top that had a stabilizing effect. The man on the left of Fig. 3.5 must still have had a rough landing. If a square base with 7 meters on each side brings a body of 70 kg mass to the floor in a reasonably safe manner, how big is the base if twice the mass has to be landed safely at the same rate of fall?

*Solution*:
The air resistance increases at low speeds linearly with the cross-sectional area, i.e. with the square of the side length. We need twice the resistance, so the $\sqrt{2}$-fold side length (10 m). ◂◂◂

### ▸▸▸ Application: *shutter speed, apertures, and sensor sensitivity*

In manual exposure photography, the shutter speed (= exposure time) and

**Fig. 3.5** parachute options à la Leonardo

the aperture need to be configured. The shorter the exposure time, the less light reaches the light-sensitive sensor. Thus, if the light conditions are to be kept constant, and the aperture number doubled (thus halving the aperture) in order to achieve a greater depth of field (see Fig. 2.51), a four-fold exposure time must be configured!
How long, for instance, must one expose with an aperture of 16 if the exposure time at an aperture of 8 takes 1/500 seconds?

*Solution*:
An aperture setting of 16 only lets in one quarter of the light onto the light-sensitive surface in comparison to the aperture setting 8. It is, therefore, necessary to configure an exposure that takes four times as long (1/125 seconds). In digital photography, it is possible to compensate by adjusting the sensitivity of the sensor (see the next example). ◂◂◂

### ▸▸▸ Application: *inferences about brightness*

If the exposure time, aperture setting, and sensor sensitivity are known, then the brightness of a scene can be estimated with good accuracy. The object in Fig. 3.6 was photographed three times based on the following camera configurations: the left photo with 1/200 s exposure, an aperture setting of 11, and ISO 100; the center photo with 1/160 s exposure, an aperture setting of 10, and ISO 400; and the right photo with 1/80 s exposure, an aperture setting of 7.1, and ISO 1600. What can be inferred about the brightness of the three scenes?

*Solution*:
If we take the interior scene as the reference, then the picture in the shade, due to half the exposure time, an aperture setting 1.4 times greater (decrease of incident light by a factor of $1.4^2 \approx 2$), and a fourfold sensor sensitivity, must have been more strongly lit by a factor of approximately $2 \cdot 2 \cdot 4 = 16$.
The scene in the sun – when compared to the scene in the shade – underwent an exposure that was $200/160 = 5/4$ shorter still, with an aperture setting 1.1 times greater ($1.1^2 \approx 6/5$ times more incident light), and with light-sensitivity

**Fig. 3.6** in the sunlight, in the shade, and inside a building

reduced by a factor of four. Thus, it can be estimated that the scene reflected about $5/4 \cdot 6/5 \cdot 4 = 6$ times as much light as the scene in the shade. It can thus be estimated that the leftmost scene was illuminated $6 \cdot 16 \approx 100$ as brightly as the building interior.

⊕ *Remark*: If the objective was to picture the toy animal as well as possible, then the center scene is clearly preferable: Strong shadows, as in the left picture, may distract some viewers from the subject. The sensor sensitivity ISO 400 does not yet produce strong noise artefacts, but allows for a higher aperture number, and thus, for greater depth of field. ISO 1600 (as in the interior) distorts the colors somewhat more – on the other hand, one is often left with no choice when photographing interior scenes, and when no photo flash is available. The quality of a photo sensor depends significantly on how strongly it produces noise artefacts in low-light scenarios.

What is true for the light-sensitive sensor also applies, to some degree, to us people: We need a "daily quantity of light" for our well-being. Whether in the baking sun, or beneath a cloudy sky: We take in at least ten times more sunlight under the open sky than in interior spaces. Not even daylight lamps can replace the outdoors! What is more, glass surfaces change the structure of light to a significant degree – for instance, by filtering out large amounts of UV light. For this reason, it takes much longer to get sunburnt in a vehicle where all windows are fully shut. ⊕ ◂◂◂

### ▸▸▸ Application: *zooming in as much as possible*

The *optical* zoom factor is an important technical specification while shooting (with digital zoom one obtains no information). Compare the images in Fig. 3.7 in terms of resolution and the zoom factor (left: 10-megapixel photo with a resolution of $2,592 \times 3,888$ pixels, center: 3-megapixel shot with a high-quality HD camcorder together with a bolted teleconverter with a resolution of $1,440 \times 1,920$ pixels).

*Solution*:

By measuring, we note: In the left image, a third of the image width is covered by the tower. On the image in the middle, the tower covers half of the width. On the image in the middle, the optical zoom factor is 50% larger than on the left image (although on the right, the image format 4:3 shows less height). For the tower width on the left, $2,592/3 = 864$ pixels are available, while, in the middle, there are $1,440/2 = 720$ pixels. The *optical resolution* on the

**Fig. 3.7** three extreme zoom shots. The left figure is a 10-megapixel photo. The middle one has only 3 megapixels, but has a 50% stronger optical zoom than the left photo. Right: An almost "spacey" morning mood, but it is not manipulated!

left is thus only slightly better, despite the 10-megapixel camera. A stronger zoom would already provide more details in a 3-megapixel photo. *Optical zooming* increases the resolution quadratically more than an increase of the number of pixels. Assume we had the same number of pixels with the same ratio of side lengths, then the right image would provide more than twice as much ($1.5^2 = 2.25$) information.

⊕ *Remark*: The left image is – however this comparison may end – better than the image in the middle, since there is no drop in brightness at the edges (this is called "vignetting" and it is primarily driven by the teleconverter. In the center image of Fig. 3.7, the vignetting was amplified by software as it was not clearly visible in the original). Of course, the right image is the more interesting one because it represents an interesting mood, almost resembling a rocket launch. ⊕ ◂◂◂

## 3.2 Similarity of spatial objects

In *space*, the analogous and important result on scales reads:

Let an object and a similar copy of it be given. Corresponding measures $L$ in lengths may be in a ratio of $1 : k$ (scaling factor $k$), then corresponding areas (surfaces) are in a ratio $1 : k^2$ and corresponding volumes are in ratio $1 : k^3$:

$$L_1 : L_2 = 1 : k, \quad S_1 : S_2 = 1 : k^2, \quad V_1 : V_2 = 1 : k^3. \tag{3.1}$$

***Proof***: 1. Each surface can be *triangulated* (Fig. 3.8). Each triangle's area on the surface changes with the square of the scaling factor. So this rate also applies to the sum of all triangular faces, i.e. going into more detail by getting closer to the surface.

**Fig. 3.8** surfaces ...

**Fig. 3.9** ... and volumes

2. Each volume can be approximated in an arbitrarily accurate way by using *voxels* ("voxels" – short for "volume elements" – analogous to how pixels refer to "picture elements"). Consider Fig. 3.9, which shows a Lion family built from *Lego* bricks at the Expo 2000 in Hannover. For each given cuboid, the volume increases by the cube of the scaling factor. So the rate applies to all bodies. ⊙

**▸▸▸ Application**: ***gold masks***

The only non-plundered Egyptian grave chamber is that belonging to Tutanchamun, who lived around 1300 BC and 1,300 years after the time of the Great Pyramids. The death masks were placed over each other using onion skins and we see they are quite similar to each other. They were made of hammered gold and decorated with semi-precious stones, giving the mask a 50% larger diameter, $1.5^2 = 2.25$ times more than the surface area. ◂◂◂

**▸▸▸ Application**: ***construction of the pyramids of Giza*** (Fig. 3.11)

What percentage of the mass of the Great Pyramid of Giza (pyramid of Chephren, Menkaure) was installed at the time when the pyramid reached

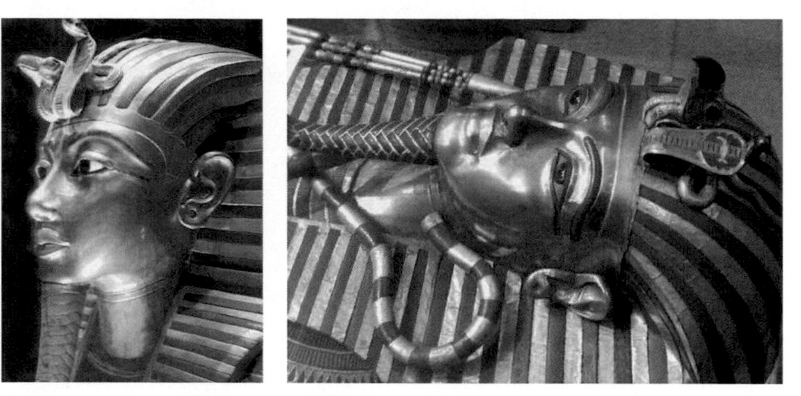

**Fig. 3.10** different sized death masks

one-third (one-half, three-fourths) of its final height? What percentage of the surface was done at this time?

**Fig. 3.11** the pyramids of Giza at the outskirts of Cairo

*Solution*:
The following statement applies to each of the pyramids:
We expect the missing upper part of the pyramid to be similar to the totally built pyramid. This residual pyramid had 2/3 (1/2, 1/4) of the final height. The missing mass was, therefore, $(2/3)^3 = 8/27$ ($(1/2)^3 = 1/8$, $(1/4)^3 = 1/64$) of the total mass. Thus, $1 - 8/27 = 19/27$ ($1 - 1/8 = 7/8$,$1 - 1/64 = 63/64$) of the mass has already been installed, or about 70% (87.5%, 98.4%). For the surfaces, the square of the scaling factor matters. Accordingly, for the time in question, there is already $1-(2/3)^2 = 5/9$ ($1-(1/2)^2 = 3/4$, $1-(1/4)^2 = 15/16$) of the surface finished, this is about 56% (75%, 94%).

⊕ *Remark*: The surface of the great pyramids was originally smoothly polished and reflected sunlight. Thus, the pyramids offered a completely different impression than today. Only about 500 years ago, the precious surface material was almost completely removed and used for the construction of buildings in Cairo. Only the uppermost part of Khafre's Pyramid is still reasonably intact. ⊕ ◂◂◂

▸▸▸ **Application**: ***weight comparison***

A man of 1.60 m height has a mass of 50 kg. What is the mass of another man who is 2 m tall and has a similar figure?
Because of the same density, the masses are in ratio $M_1 : M_2$ and so are the volumes $V_1 : V_2$,

$$k = 2 : 1.6 = 1.25 \Rightarrow k^3 \approx 2 \Rightarrow M_2 \approx 2 \cdot M_1 = 100\,\text{kg}.$$

In practice, however, larger people often have different proportions. In fact, they tend to have muscles of a relatively smaller gauge.[1] ◂◂◂

▸▸▸ **Application**: ***the surface of the Moon and Mars***

The Moon has a fourth of the diameter of the Earth, and therefore, a sixteenth of the surface, i.e. approximately 31 million square kilometers. This is almost exactly the area of Africa. Mars has a diameter that amounts to a little more than half the diameter of the Earth (0.532). Thus, the surface is little more than a quarter of the Earth's surface ($0.532^2 = 0.283$), which is the area of all continents together. ◂◂◂

▸▸▸ **Application**: ***subjective size of the Moon***

The distance of the Moon from the Earth varies between 356,410 km and 406,760 km. How many times greater in these extreme cases do the following appear: a) the surface of the crescent, b) the supposed volume c) the brightness of the Moon?

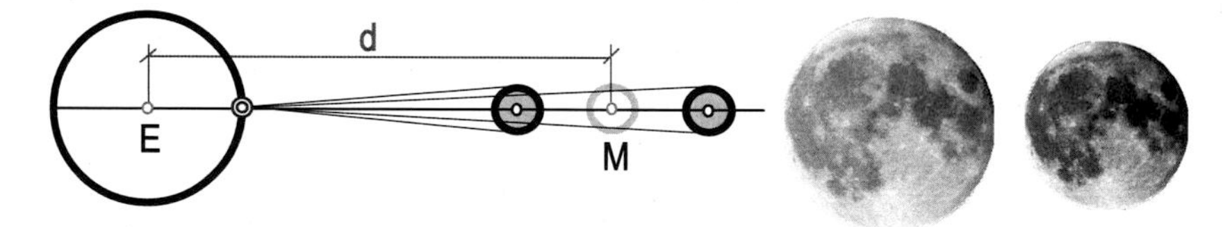


**Fig. 3.12** subjective Moon size at varying distances

*Solution*:
The given circumstances relate the respective centers of the Earth and the Moon. In fact, we have to subtract the Earth's radius $R = 6,370\,\text{km}$, though the result is, of course, hardly affected (Fig. 3.12). We have

$$1 : k = (356,410\,\text{km} - R) : (406,760\,\text{km} - R) \approx 1 : 1.144$$

$$\Rightarrow A_1 : A_2 = 1 : 1.31, \quad V_1 : V_2 = 1 : 1.50.$$

The area of Moon's disc in the firmament fluctuates by almost a third, and with it, the radiation strength of the Earth fluctuates through the Moon as well. The Moon's volume subjectively varies by as much as 50%. However, these extreme values are only achieved at intervals of several months. ◂◂◂

[1]The exact mathematical result, therefore, does not necessarily coincide with the various "BMI tables" (*body mass index*), which can be found on the Internet.

▸▸▸ **Application**: ***diver's fears*** (Fig. 3.13)
Viewed through a diving mask, submerged objects appear $\frac{4}{3}$ times their actual size (based on the length scale). So, a 1.3 m long reef shark appears to have a length of 1.7 m, but with an alleged mass increasing by a factor of $(\frac{4}{3})^3 \approx 2.4$ (240%). This may be seen as a possible contributing reason behind reported sightings of gigantic sharks. ◂◂◂

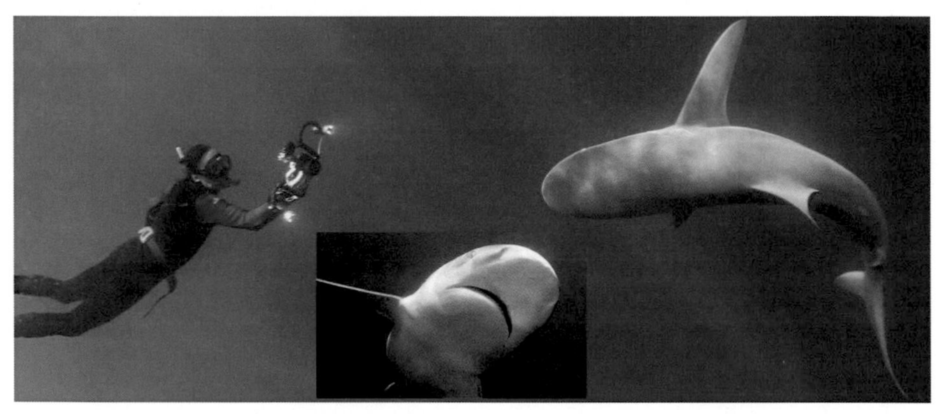

**Fig. 3.13** Diver's fears: The author faces a 2.6 m dusky shark (*Carcharhinus obscurus*), photo from above: Sean Hedger.

▸▸▸ **Application**: ***diver's reality***
The greatest danger to divers is not sharks. It is the so-called decompression sickness ("caisson disease" or "the bends"). It occurs when a diver stays at a great depth for a long time and then emerges too fast. Due to the increased external pressure, more nitrogen is accumulated in the blood and in the tissues, while the oxygen in the respiratory air is exhausted. During ascent, the reduction of external pressure (*Henry*'s law) causes nitrogen bubbles to form in the blood and tissues. If they are not "exhaled" in time by a very slow ascent, they expand and are "caught" in the joints and tissues. This initially results in extreme strong joint pain ("bends") and can be fatal if the diver is not brought up in time for recompression in a hyperbaric chamber (decompression chamber).
*Numerical example*: For each 10 m depth increase in water, the external pressure increases by 1 bar. In a water depth of 30 m, there is a pressure of 1 bar+3 bar = 4 bar. The diver now makes a rapid ascent (> 20 m per minute), for example because the diver's air supply has run out. The resulting rise in nitrogen bubbles are then not fully exhaled and they increase in volume due to the pressure reduction. The surface volume is now a quarter of the external pressure; hence, it has (4 bar → 1 bar) quadrupled. The radius – and thus the diameter – of each bubble has increased by a factor of $\sqrt[3]{4} \approx 1.6$. Diagnosis: The "bends" get worse the longer the diver has been exposed to high pressure. ◂◂◂

## 3.3 On small scales not as on large scales

In this section, we will see that bodies with increasing size must exhibit differences even though they share certain external properties. Supposed similarities often lead to fallacies.

Let us first consider the following example:

▸▸▸ **Application**: ***optimizing the size of catalyst elements***

Which scale factor $k$ has to be applied to a catalyst object (Fig. 3.15: catalyst elements for contact lens storage) so that the active surface doubles? How many times heavier is the enlarged object?

*Solution*:

$$S_1 : S_2 = 1 : 2 = 1 : k^2 \Rightarrow k = \sqrt{2}$$
$$\Rightarrow V_1 : V_2 = M_1 : M_2 = 1 : k^3 = 1 : (\sqrt{2})^3 = 1 : 2.83.$$

Thus, the volume increases by almost three times. So, two identical catalyst objects have a greater surface area than a similar one with twice the mass.

**Fig. 3.14** surface enlargement through fanning

⊕ *Remark*: In order to increase the surface area of a catalyst element without increasing the mass, the element is fanned out (Fig. 3.15). Nature benefits from this: The antennae of male butterflies (Fig. 3.14, right) are fanned out, and the surface of the olfactory organ is extended. ⊕ ◂◂◂

In general, the following important theorem is valid:

> When you magnify an object, the volume increases faster than the surface. Specifically: The ratio $V : S$ = *Volume*:*Surface* is proportional to the scaling factor $k$.

***Proof***: The volume increases by the factor $k^3$, and the surface increases by the factor $k^2$. Therefore, the quotient $V : S$ increases by the factor $k^3 : k^2 = k$. ⊙

**Fig. 3.15** catalyst elements

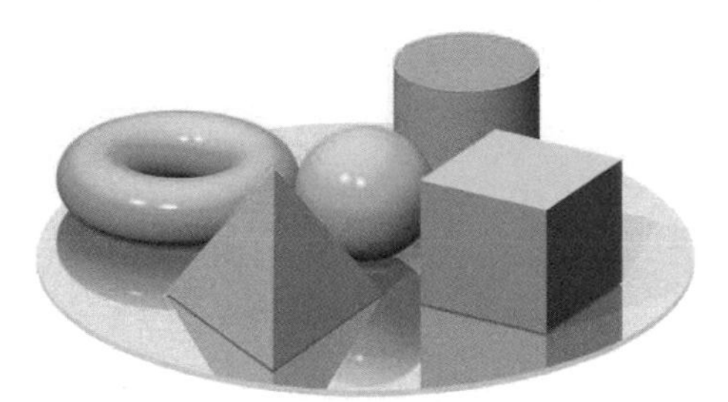

**Fig. 3.16** basic building blocks

Examples (Fig. 3.16):

1. A cube (edge length $k \cdot a$) has volume $(k \cdot a)^3$ and surface area $6(k \cdot a)^2$, which yields $\Rightarrow \dfrac{V}{S} = \dfrac{a}{6}\, k$.

2. A regular tetrahedron (4 equilateral triangles with side length $k{\cdot}a$ and area $A$) has volume $\frac{1}{3}\sqrt{\frac{2}{3}}\, k\, a{\cdot}A$ and surface area $4A$, which gives $\Rightarrow \dfrac{V}{S} = \dfrac{a}{6\sqrt{6}}\, k$.

3. A sphere (radius $k\, r$) has volume $\frac{4\pi}{3}(k\, r)^3$ and surface area $4\pi(k\, r)^2$, which results in $\dfrac{V}{S} = \dfrac{r}{3}\, k$.

4. A cylinder of revolution (radius $a$, height $b$) has volume $k^3\, \pi a^2 \cdot b$ and surface area $k^2(2\pi a \cdot b + 2\pi\, a^2)$, which implies $\dfrac{V}{S} = \dfrac{ab}{2(a+b)}\, k$.

5. A torus (major radius $a$, minor radius $b$, $b < a$) has volume $k^3\, 2\pi a \cdot \pi b^2$ and surface area $k^2\, 2\pi a \cdot 2\pi b$, which yields $\dfrac{V}{S} = \dfrac{b}{2}\, k$.

**Fig. 3.17** guttation drops as minimal surfaces

⊕ *Remark*: Of all the possible bodies of the same volume, the sphere delivers the maximum ratio $\frac{V}{S}$. The sphere belongs to the so-called *minimal surfaces*, which differ from slightly deviating surfaces in having less (minimal) surface area (Fig. 3.17). ⊕

**▸▸▸ Application**: ***floating needles*** (Fig. 3.18)
It is well-known that a water strider is supported by the surface tension of water. One can hardly imagine that this is also true for a steel needle. Although steel has almost eight times the density of water ($\varrho = 7.8\,\mathrm{g/cm^3}$), a sufficiently small steel needle – if it is very carefully placed on the surface of water – "swims". Find the reason for this. How much bigger/heavier could an aluminium needle be ($\varrho = 2.7\,\mathrm{g/cm^3}$)? Is a gold needle ($\varrho = 19.4\,\mathrm{g/cm^3}$) able to float?

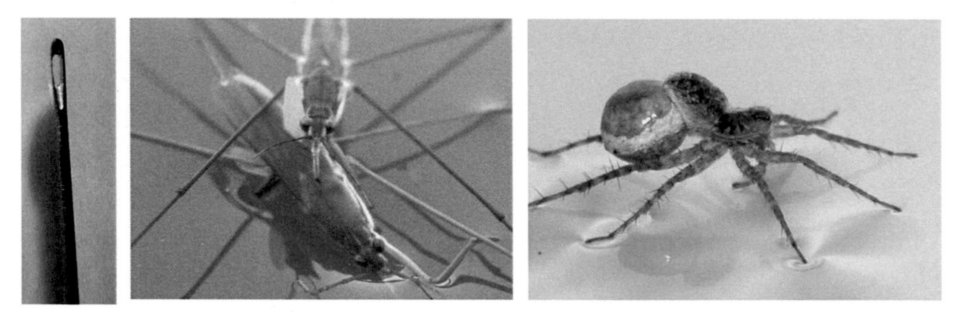

**Fig. 3.18** floating steel needle, water striders, spider (not a water spider!)

*Solution*:
The smaller the steel needle, the greater its surface area relative to its volume or weight. From a certain critical length (depending on the needle's shape), the needle's surface, and thus the surface tension of the water, is large enough to carry the steel needle.
A similarly shaped aluminium needle can be $k = 7.86/2.7 \approx 3$ times as long. The volume is then $k^3$ times as large, but the weight is only $k^2$ times as large, and thus, the surface has increased as well.
Without much ado, the needle can be made from pure gold if it is $k = 7.86/19.4 \approx 1/2.5$ shorter. ◂◂◂

**▸▸▸ Application**: ***less heat loss through a larger body*** (Fig. 3.20).
Explain why large whales are much more likely to live in Arctic waters than small whales (lengths: dolphin ... 2m, blue whale ... 20m).
*Solution*:
Since the body shapes are fairly similar, the ratio $\frac{V}{S}$ of the whale is $k = 10$ times as large as the dolphin's ratio. The surface of the blue whale can be kept warm by its circulation of blood. Therefore, it is 10 times easier to keep it warm.

⊕ *Remark*: The Arctic waters have lots of oxygen, and therefore, they are more nutritious. The largest whales are, thus, more likely to be found there. They only

**Fig. 3.19** Arctic fox and desert fox (ear size)

**Fig. 3.20** blue whale and dolphin (larger bodies have comparatively less surface)

swim in warmer waters so that their newborn calves do not freeze in the colder waters. ⊕

⊕ *Remark*: African elephants have cooling problems because of their body size. That is why they have big ears (surface area and of course "fan function").
The "cold-blooded" great white shark, which is constantly in motion (Application p. 217), loses so little warmth to the outside that its body temperature can be up to ten degrees more than the water temperature. ⊕

⊕ *Remark*: Arctic foxes and desert foxes almost have the same shape. It is striking that desert foxes have much larger ears (Fig. 3.19). These are used for listening to the often almost noiseless and odorless preys (scorpions, etc.), as well as for cooling. Naturally, the Arctic fox has to forgo both. ⊕ ◂◂◂

**Fig. 3.21** The swan is one of the largest flying birds (up to 14 kg mass).

▸▸▸ **Application**: ***weight limit for volant animals***

Why do volant animals (especially birds) have a weight limit (10 – 15kg)?

*Solution*:

Buoyancy while flying is related to the wing-span ("tearing edge") and wing surface. The change of both will be slower compared to the change of the volume when increasing the length scale. The larger the bird, the more difficult it is for it to take off. Large raptors launch themselves from rocks to reach the initial velocity required. The non-volant ostrich (50 – 100 kg) would need a start-up speed of several hundred kilometers per hour in order to take off.

**Fig. 3.22** bee-eaters, wasp with prey at departure, a stowaway (1 mm)

⊕ *Remark*: *Pteranodon* (the biggest flying dinosaur) probably had a wingspan of about seven meters, but only a mass of approximately 15 kg. It would never have been able – like in *Jurassic Park III* – to drag an object of human size through the air. A similar performance can only be achieved by much smaller creatures, like the wasp on the right of Fig. 3.22, which has paralyzed a spider and is transporting the spider in flight. ⊕ ◂◂◂

### ▸▸▸ Application: *cooling of the planet*

After a huge explosion in space, two planets are produced whose diameter have a ratio of $1:2$. Which planet will cool faster?
The volumes have a ratio of $1:8$; the surfaces have a ratio of only $1:4$. That is, the larger planet has half as much surface area in proportion to its volume, and therefore, cools down more slowly.

**Fig. 3.23** odd cocktail "on the rocks" and ice transportation at 35° in the shade

⊕ *Remark*: Our Moon is probably 4.5 billion years old[2] and is believed to have merged from 20 smaller "moonlets" that spalled off following an asteroid strike on the Earth. The fact that the density of the Moon is less than that of the Earth may be taken as an indication supporting this theory. The larger density of the Earth is mainly due to the nickel-iron core. The Moon, with a diameter of only about a quarter of the Earth's diameter, has, therefore, already cooled by the above considerations, while the Earth is still liquid inside (the frozen continents float like

[2]In National Geographic, issue September 2001, there is a clear explanation for the relatively precise dating of the formation of the Earth.

a milk skin on the surface of coffee). Due to their own gravity, all comets or moons with a diameter of more than 500 km attain a spherical shape as long as they are liquid inside. For example, the two quite small moons of Mars are not spherical.

What applies to cooling generally also applies to temperature compensation. Fig. 3.23 illustrates how much the size of a frozen block influences its melting time. Towards the end of the austral winter, the giant icebergs on the sea ice of the South Atlantic cover an area which is twice as large as Canada. The enormous amounts of ice have to be thawed only once (Application p. 226)! ⊕ ◂◂◂

The fact that during enlargement or reduction, the ratio *Volume* : *Surface* changes, had an enormous impact on the respective development of insects and warm-blooded animals:

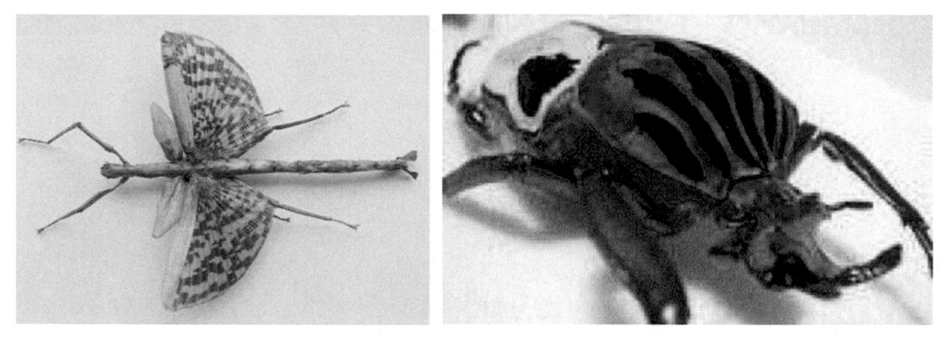

**Fig. 3.24** an elongated and a "rotund" insect

▸▸▸ **Application**: ***maximum size of insects*** (Fig. 3.24)

In general, insects are shorter than 10–15 cm (the heaviest, though not the largest, insect is the up to 12 cm-long Goliath beetle, see Fig. 3.24, right). Their oxygen supply is the so-called trachea, which is a system of tubes in their exoskeleton. The oxygen exchange takes place via the surface of this system. For a length over 15 cm, this oxygenation is no longer enough: There is a mismatch in the ratio *volume (weight) : tracheal surface* which is getting too large. The ideal size of an insect is clearly in the range of 1 cm or less. Smaller insects are much more resistant than bigger ones. This is also due to the deteriorating ratio between body weight and muscle strength (compare Application p. 85).

⊕ *Remark*: As always, exceptions prove the rule: There is an up to 30 cm-long stick insect (*Palophus titan) (Fig. 3.24, left), and the giant dragonflies in carbon rocks have a wing-span of up to 75 cm! For an extremely elongated body shape, only a very short trachea is required, and such a trachea may then have a relatively large diameter.* ⊕ ◂◂◂

▸▸▸ **Application**: ***surface enlargement for a slenderly built body***

The following table is intended to show how much the surface increases for a cuboid – starting from a cube – with a consistent volume of $1\,\mathrm{m}^3$ (and thus a constant mass) which is getting slimmer. The same also applies for more complex shapes. The dragonfly in Fig. 3.25 needs a small cross-section in order to achieve high flight speeds. Surprisingly enough, the beetle can still fly pretty well!

| type | proportions | measures [m] | surface factor | increase |
|---|---|---|---|---|
| T0 | 1 : 1 : 1 | $1^2 \times 1$ | 6.0/m | |
| T1 | 1 : 1 : 2 | $0.79^2 \times 1.59$ | 6.3/m | 5% |
| T2 | 1 : 1 : 4 | $0.63^2 \times 2.52$ | 7.1/m | 19% |
| T3 | 1 : 1 : 8 | $0.5^2 \times 4$ | 8.5/m | 42% |
| T4 | 1 : 1 : 16 | $0.4^2 \times 6.35$ | 10.4/m | 73% |
| T5 | 1 : 1 : 32 | $0.31^2 \times 10.08$ | 12.9/m | 115% |
| T6 | 1 : 1 : 64 | $0.25^2 \times 16$ | 16.1/m | 169% |

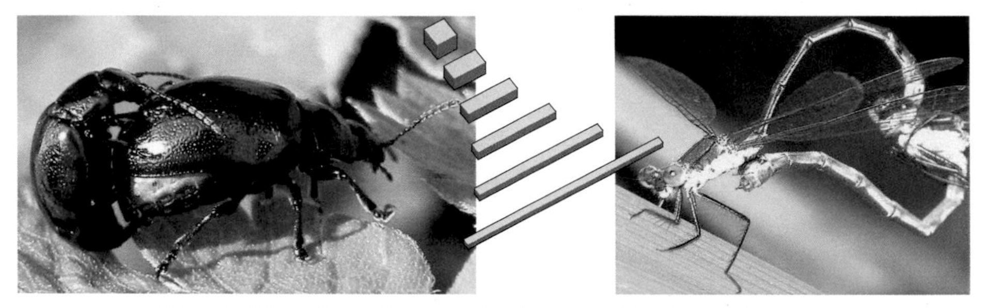

**Fig. 3.25** various degrees of thinness with the same mass

◂◂◂

▸▸▸ **Application**: ***optimal oxygen supply*** (Fig. 3.26)

In contrast to insects, the bodies of higher developed animals are supplied with oxygen via their blood (Fig. 3.26). More blood allows more oxygen to be transported. The amount of blood increases in proportion to the volume and there is no mismatch. The oxygenation by blood thus limits the size of the living being but it is not the only factor – there are other factors at play as well (see Application p. 85).

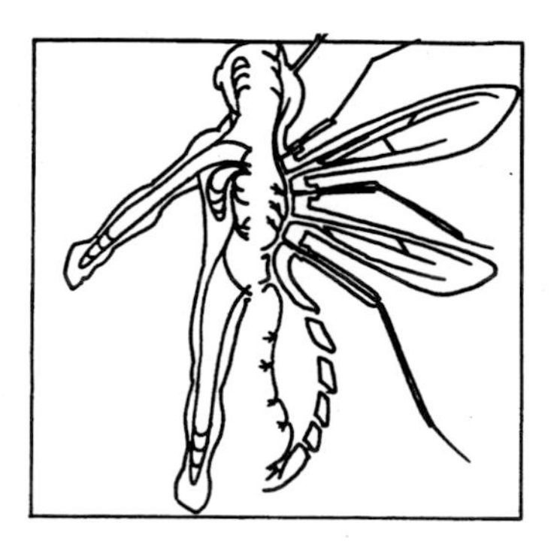

**Fig. 3.26** veins and trachea

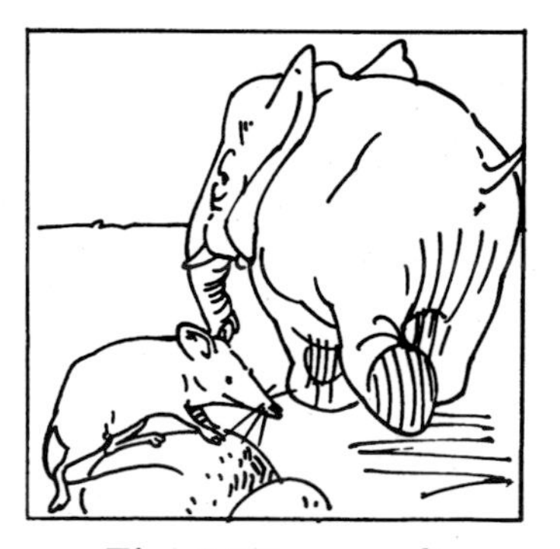

**Fig. 3.27** mammals

⊕ *Remark*: The only problem is the accumulation of blood in the lungs: The surface of the lungs has to increase disproportionately, which is made possible by a rapid increase in the number of alveoli in large animals. Infants initially have fewer alveoli, and they disproportionately increase their number until the age of eight, to be ready by "maturity". ⊕ ◂◂◂

▸▸▸ **Application**: ***minimum size of warm-blooded animals*** (Fig. 3.27)
For animals with constant body temperature, there is now a clear lower limit for the height, namely, about the size of a pygmy shrew (Fig. 3.27) in mammals and the bee hummingbird (a hummingbird species with only 2.5 $g$ of body weight) in birds. For both, there is an extremely small ratio of weight (and thus warming amount of blood) to surface area over which these animals constantly lose heat (energy). To make up for this loss, the shrew has to eat constantly. Smaller shrews from earlier geological periods, therefore, subsisted partially on nectar, extremely high-energy food. The same holds true for the hummingbird. ◂◂◂

For similar reasons, one can explain why large animals in relation to their mass are much weaker than smaller animals: The strength of a muscle is not dependent on its volume but on the cross-section:

**Fig. 3.28** an ant and an elephant at work

▸▸▸ **Application**: ***relative body strength*** (Fig. 3.28, Fig. 3.29)
An ant can carry many times its weight. Although an elephant is objectively incomparably stronger, it can only carry much smaller loads relative to its weight. In addition, an elephant already needs very thick legs (due to a disproportional cross-sectional enlargement of the muscles), in order to bear the enormous weight.

⊕ *Remark*: This is even true when comparing relative strengths of small and large articulate animals: 5 mm long jump spiders can jump much further (in relation to their size) than the quite tremendous (and 10,000 times heavier) tarantulas. ⊕
*Numerical example*: Pygmy Shrew: 4.3 – 6.6 cm (without the tail) with 2.5 – 7.5 g, elephant: up to 3.5 m and 4,000 kg. If the elephant were as slim as the shrew, it would weigh less than 2,000 kg. ◂◂◂

**Fig. 3.29** Left: It took minutes to overtake the millipede – the short legs of the victim are strong levers. Right: An ant tries to escape a meat-eating sticky plant (sundew, *Drosera capensis*).

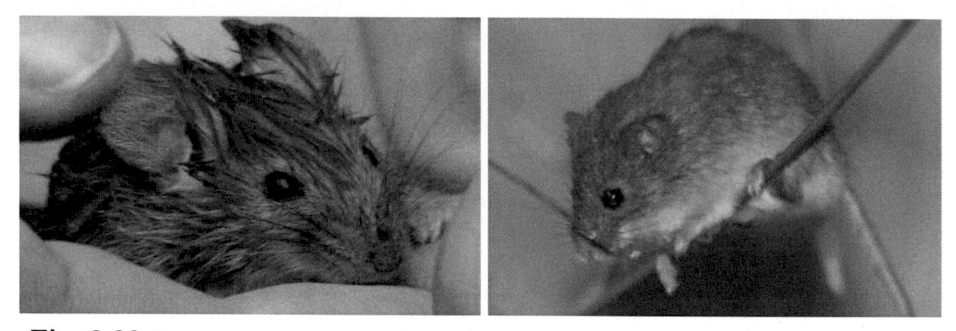

**Fig. 3.30** One cannot get much smaller. Left: Escaped after being trapped in a cat's mouth!

If the ratio $V_2 : V_1$ of the volumes is given, then the similarity factor is calculated by taking the cube root:

$$k = \sqrt[3]{V_2/V_1}. \tag{3.2}$$

Then, one can again use the formulas (3.1).

►►► **Application**: ***allometries in the animal kingdom***

Biologists are understandably cautious when using the term "similarity" and instead use the term "allometry" in this context, when certain organs are not in the same proportion as most others. Nevertheless, a domestic cat (5 kg) and a leopard (50 kg) are not too allometric in spite of their similar shape (Fig. 3.31). How do their shoulder heights respectively relate to their coat surface or their tread surface (i.e. the surface of their paw prints) ?

*Solution*:

Since their bodies have the same density, the ratio of the masses equals that of the volumes.

$$k^3 = 10 \Rightarrow k = \sqrt[3]{10} = 2.15, \; k^2 \approx 4.6.$$

The shoulder height has a ratio of $1 : 2.15$. The coat surfaces or the treading surfaces have a ratio of $1 : 4.6$. ◄◄◄

**Fig. 3.31** similar cats

**Fig. 3.32** similar white rhinos (low allometry)

▸▸▸ **Application**: ***similarity between a pup and a dam***
Relatively little allometry can be seen in the two rhinos in Fig. 3.32 and the tigers in Fig. 3.33. In the middle image of Fig. 3.33, it can be seen only at second glance that a "baby tiger" slurps water. The difference in size to the mother can only be estimated from the image on the right. How can the masses of the young animals be guessed?

*Solution*:
Fig. 3.32: Since the rhinos in both images are standing in parallel, length measurements (for example, the length of the spine) can be compared quite well. The scaling factor is about $k \approx 0.75$. The mass of the baby animal is $k^3$ of the mother's mass, so at a good 40%. Since adult rhinos have a mass of $1,500 - 2,000$ kg it is likely that the cub would weigh around $600 - 650$ kg. On the right, the shoulder height of the baby animal may be $k = 0.4$ of his mother's shoulder height, resulting in a mass of about $k^3 = 0.4^3 = 0.064 \approx 1/16$, which equals about 100 kg.
Fig. 3.33: Here, the length dimensions are not directly comparable. If the shoulder height of the young tiger were 45% of the shoulder height, the masses would be in the ratio $0.45^3 : 1^3 \approx 1 : 10$. Siberian tigers are the largest

**Fig. 3.33** similar Siberian tigers (low allometry)

living carnivores. Even the females have a mass of 100 – 170 kg. So, young tigers must have a mass of about 15 kg.

**Fig. 3.34** two ants of the same species (worker, guardian)

⊕ *Remark*: The ants in Fig. 3.34 are directly comparable. The larger guardian ant is about 1.5 times as long as the worker ant, and in absolute terms it is $1.5^2 = 2.25$ times stronger. Relatively speaking, it has a mass that is $1.5^3 \approx 3.4$ times as large as the mass of the worker ants. It is, thus, weaker than the smaller ant. ⊕ ◂◂◂

**▸▸▸ Application**: ***wild comparisons***

Compare the animals in Fig. 3.35 and Fig. 3.36 and consider the differences in the functioning of their various organs.

*Solution*:

Fig. 3.35: The whirligig beetle is a predator at the water surface with a rigid and extremely resistant exoskeleton. It has compound eyes and floats – "translating" its body length to that of a human – at a rate of 500 km/h with 60 beats per second. Like almost all insects, the beetle can even fly. The seal is warm-blooded and has ordinary eyes like humans (only with a specific focal length for underwater vision, see Application p. 52). It swims through serpentine body motion (flexible inner skeleton)!

Fig. 3.36: The flower chafer uses its antennae to smell, chitin hair for pollinating plants, and can fly excellently. Water buffaloes have their horns as

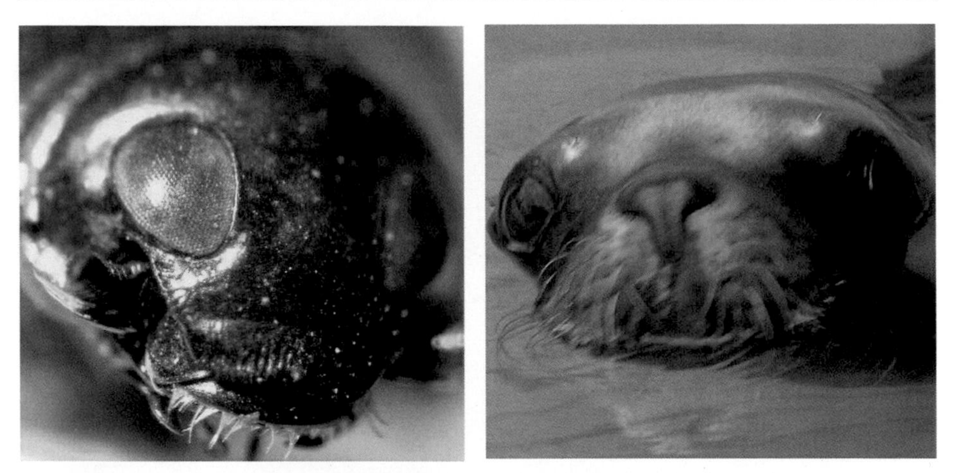

**Fig. 3.35** a whirligig beetle (5 mm) and a seal (1.5 m)

**Fig. 3.36** a flower chafer (2 cm) and a water buffalo (2.5 m)

a defence mechanism, hair as refrigeration, and safety features – and quite obviously, they cannot fly. ◂◂◂

▸▸▸ **Application**: ***more wild comparisons*** (Fig. 3.37)
A giraffe with a height of 6 m can have a mass of up to 1,600 kg. How much mass does a 2 m high baby giraffe have? What would one estimate the mass of a 12 cm long sea horse to be?

*Solution*:
Young giraffes have comparatively shorter necks. Thus, we should compare the calf with a giraffe of height 5 m. With 2/5 of the height, one can estimate $(2/5)^3 \approx 1/16$ of the mass, i.e., 100 kg.
Now the comparison with the sea horse. Here we really just get a rough estimation. The length proportion is 1 : 50, thus the volume (and mass)

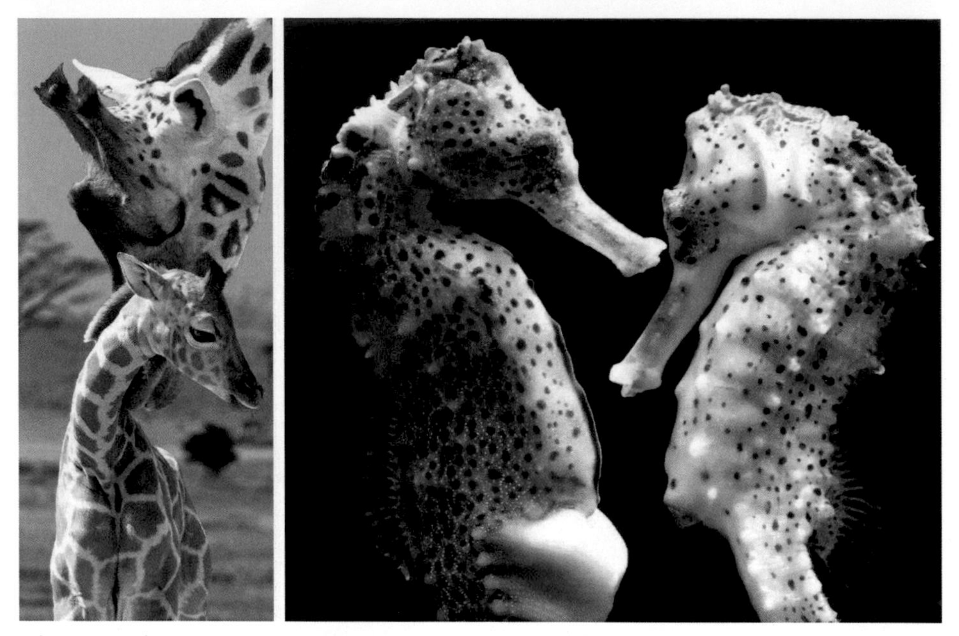

**Fig. 3.37** A geometrical resemblance is certainly there – including the mane or the dorsal fin – but that is about it ...

comparison should be $1 : 50^3 \Rightarrow 1,600/125,000$ kg or 13 grams. (Sea horses have masses between 1 g and 1 kg, depending on the species.)

⊕ *Remark*: As a species, we rely disproportionately on our eyesight – a predilection that makes us prone to judging by appearance. Sea horses and giraffes may be extremely different animals, and yet, one would be hard-pressed to deny a superficial resemblance beyond the mere difference in size.[3] Some even conclude, on the basis of appearance alone, that humans have fish-like ancestors – but looks may be deceptive, and our naive intuitions are often mistaken. Incidentally, evolutionary biology has already settled the question of what caused the giraffe's famous neck to grow so long: it did not evolve, as it is often believed, because the animals fed on trees that grew their foliage on increasingly higher branches in a desperate arms race to survive in the African savanna. Rather, their necks seem to have been molded by the ubiquitous mechanism of sexual selection. ⊕ ◂◂◂

[3] G. Glaeser, H. Paulus: *The Evolution of The Eye*, Springer Nature, 2015.

## 3.4 Centrifugal and gravitational forces

In this section, we want to derive some rather interesting findings about astronomy and space travel by means of proportionalities. *Newton* had already realized that there is a mutual force of attraction between two bodies that increases linearly with respect to their masses, but decreases with respect to the square of the distance. Moreover, we know that the centrifugal force that acts on a rotating body increases quadratically with its track speed while the distance to the center of rotation enters the formula for the attraction force only linearly. The entire universe is based on the interplay between attraction and centrifugal force: Planets, moons, and satellites pursue their course through space and arrive precisely to the second at pre-calculatable positions.

⊕ *Remark*: The gravitational force is the decisive force *only* "on a large scale". This is because the force is a product of mass and acceleration. The smaller the objects are, the smaller is the influence of the gravitational force. Insects can already blithely crawl on vertical or even overhanging walls against gravitational forces. Mini-mosquitoes or mini-spiders can be airborne in the wind for kilometers.
In the molecular region, there are very different forces that are of significance. There are enormous forces that hold molecules or even just atomic nuclei together, these dominant forces are at the nanoscale. Their effect ends abruptly "at the edge" of the molecule or atomic nucleus. ⊕

▸▸▸ **Application**: ***astronomical calculations***
Given the following data:

| | diameter [km] | distance to Earth's center [km] | density [kg/dm$^3$] |
|---|---|---|---|
| Earth E | 12,740 | 6,370 | 5.516 |
| Sun S | 1,390,000 | 149,500,000 | 1.409 |
| Moon M | 3,470 | 384,000 | 3.341 |

How do the volumes, the masses, the forces of attraction act (relative to the Earth's surface)?
We relativize these absolute values: The lengths are given in terms of the Earth's radius. Density, volume $V$, and mass $M$ of the Moon and the Sun are given as multiples of the respective measures of the Earth, and the attraction of the Sun and the Moon are in relation to the gravity of the Earth to a certain point on the Earth's surface:

| | radius | distance $d$ | density $\varrho$ | $V$ | $M = \varrho V$ | $F \widehat{=} M/d^2$ |
|---|---|---|---|---|---|---|
| E | 1 | 1 | 1 | $1^3 = 1$ | 1 | 1 |
| S | 109 | 23,500 | 0.26 | $109^3 = 1.3 \cdot 10^6$ | 332,000 | 0.0006 |
| M | 0.27 | 60 | 0.61 | $0.27^3 = 0.02$ | 0.012 | 0.000003 |

An astonishing result concerning the force of attraction: The Sun has about 200 times the pull of the moon! Nevertheless, the tide of the Earth is determined by the Moon, rather than the Sun! The reason for this is that the attraction by the Sun is largely "eliminated" by the centrifugal force of the Earth's orbit around the Sun. ◂◂◂

▸▸▸ **Application**: ***the gravitation on the surface of a celestial body***
On Earth, we have a gravitational acceleration of 1 g. What is the acceleration on another planet or on the Sun with $k$ times the diameter and $d$ times the density?

*Solution*:
The mass of the body increases with $d\,k^3$. Gravitation decreases at the same time with the square of the distance from the center (factor $1/k^2$). Altogether, the increase is, thus, $d\,k$.
Since Mars, for instance, has a lower density than the Earth ($d \approx 0.7$) and its diameter is about half ($k = 0.532$), we can expect little less than 40% of the gravitational force on its surface. On our Moon ($d \approx 0.8$, $k = 0.25$), we have 20%. For both Jupiter ($k = 10$) and the Sun ($k = 100$), we have $d \approx 1/4$. Thus, we can expect 2.5 g or ≈ 25 g, respectively. ◂◂◂

▸▸▸ **Application**: ***track speed of a satellite***
Satellites are orbiting the Earth at altitudes of 150 km. To be in "equilibrium of forces", the force of gravity (weight $G$) and the centrifugal force $F$ must "cancel" each other out. Now, we want to derive a relation between linear speed, orbital period, and altitude.
Let $R$ be the radius of the Earth ($R = 6{,}370$ km), and $r$ the distance between a satellite (mass $m$) and the center of the Earth. With respect to the Earth's surface, the satellite has weight $G_R = mg$. The attraction decreases with the square of the distance from the Earth's midpoint (*Newton*). At a distance $r = k\,R$, the attraction is thus reduced to the weight $G_r = G_R \cdot \left(\frac{R}{r}\right)^2 = \frac{G_R}{k^2}$.
For the centrifugal force, we apply the well-known formula $F = mv^2/r$. With $F = G_r$, the airspeed at a distance $r$ equals

$$\frac{m\,v_r^2}{r} = m\,g\left(\frac{R}{r}\right)^2 \Rightarrow v_r^2 = g\,\frac{R}{r}\,R = \frac{g\,R}{k} \Rightarrow k\,v_r^2 = g\,R = \text{constant}.$$

So, we have

$$k\,v_r^2 = 9.81 \cdot 6.37 \cdot 10^6 \frac{\text{m}^2}{\text{s}^2} \Rightarrow v_r\sqrt{k} \approx 7{,}905\frac{\text{m}}{\text{s}} = 7.90\,\frac{\text{km}}{\text{s}} = 28{,}460\frac{\text{km}}{\text{h}}$$

or

$$v_r = \frac{7.90}{\sqrt{k}}\frac{\text{km}}{\text{s}} = \frac{28{,}460}{\sqrt{k}}\frac{\text{km}}{\text{h}}. \tag{3.3}$$

The factor $k$ is calculated from the altitude $H$ by

$$k = \frac{r}{R} = \frac{R+H}{R}.$$

The circumference of the Earth is $U_R = 40,000\,\text{km}$. With increasing height, the circumference – in this case, the trajectory of the journey – increases linearly in $k$:

$$U_r = k\,U_R.$$

Thus, the so-called "sidereal period of rotation" $T$ (relative to the fixed stars) equals

$$T = \frac{U_r}{v_r} = \frac{k \cdot 40,000\,\text{km}}{\frac{7.9}{\sqrt{k}}\,\frac{\text{km}}{\text{s}}} \approx k^{3/2} \cdot 5,060\,\text{s} \approx k^{3/2} \cdot 1.41\,\text{h}. \tag{3.4}$$

The results will be compiled in a clearly arranged table.

| $H$ [km] | 150 | 500 | 2,000 | R=6,370 | 31,850 | 377,000 |
|---|---|---|---|---|---|---|
| $k = (R+H)/R$ | 1.024 | 1.078 | 1.314 | 2 | 6 | 60.3 |
| $v_r$ [km/h] | 28,140 | 27,300 | 24,830 | 20,120 | 11,620 | 3,665 |
| period [h] | 1.46 | 1.57 | 2.12 | 3.99 | 20.7 | 660 |

We recognize that the greater the altitude, the lower the airspeed will be and the longer the orbital period will be as well. If we compare two satellite orbits ($k_1$ and $k_2$), then, by Formula (3.4), we have

$$T_1 : T_2 = k_1^{3/2} : k_2^{3/2} \Rightarrow T_1^2 : T_2^2 = k_1^3 : k_2^3 = r_1^3 : r_2^3.$$

Thus, we have verified for a circular orbit:

The squares of the revolution periods are proportional to the cubes of the radii (*Kepler*'s Third Law).

◂◂◂

▸▸▸ **Application**: ***sidereal and synodic orbital period of the Moon***

The last column of the table above relates to the orbit of the Moon, which can also be seen as a satellite of the Earth with an average distance of $384,400\,\text{km}$. You have to remember that the common center of gravity of the Moon and the Earth is not the center of the Earth, so that you should rely only on a few decimals. The Moon takes about $660\,\text{h} \approx 27.3\,d$ in *sidereal* time to complete an orbital period around the Earth. The period from a new moon to the next (a "synodic month") takes longer because the Moon must also compensate the Earth's rotation during this rather long time. It sums up to $\approx 29.5$ days.

⊕ *Remark*: The Moon increases its distance annually by a few centimetres, thus it is getting slower. Millions of years ago, the Moon appeared bigger in the Firmament and solar eclipses were more pronounced. ⊕ ◂◂◂

►►► **Application**: ***relative position of a satellite***[4] (Fig. 3.38)
When plotting the position of a satellite on a global map, the result becomes a wavy trajectory whose shape depends on the imaging method used for the map (yet, in no case, would it look like a sine curve, as one might suspect). The trajectory leaves the map somewhere on an edge and enters on the opposite edge at an appropriate place. This is, of course, due to the fact that the world map must be "cut to pieces" somewhere.
Until now, we have not taken into account that during the orbital period of the satellite, the Earth rotates. If the satellite rotates above the equator in the same direction as the Earth, then it cannot be seen from an observation station $B$ after precisely one rotation, because $B$ has already turned further by this time (per hour $360°/24 = 15°$). Therefore, the satellite may require additional time. Consequently, the trajectory of the satellite is "out of phase" with the next orbit.

⊕ *Remark*: To be more precise: The Earth rotates exactly once around its axis every 23 h 56 min = 23.933 h (a sidereal day), so the Earth turns 15.04° per hour. The remaining 4 min is needed for the Earth to turn back to the same position relative to the Sun: Within one day, the Earth has made ≈ 1° of its orbit about the Sun. (This value varies slightly in the course of the calendar year.) ⊕
Now, if we place Fig. 3.38 "under the microscope", we see that the parallel shift of the wave train is in increments of 23.1° longitude. This corresponds to 1.54 h of the orbital period. Using Formula (3.4), $k$ is calculated and the flying height $H$ is consequently deduced as well

$$k^{3/2} \cdot 1.41\,\text{h} = 1.54\,\text{h} \Rightarrow k = 1.058 \Rightarrow H = R(k-1) \approx 373\,\text{km}.$$

Furthermore, the track speed equals $v_r \approx 27,600\,\text{km/h}$ according to Formula (3.3). The values of $H$ or $v_r$ closely match those instantaneous values given in Fig. 3.38 (in fact, these values vary slightly). ◄◄◄

►►► **Application**: ***geostationary satellites*** (Fig. 3.39)
In Application p. 92, we saw that high-flying satellites fly more slowly in order to overcome the already small gravitational force by means of centrifugal force. When a satellite flying above the equator (in the same direction as the globe's rotation) has an orbital period of exactly 23.93 h (see above), it will remain above the same spot on the equator. In fact, there are now almost 300 (!) satellites "positioned" above the equator: weather satellites and especially communication satellites (mostly TV-satellites). This means that one satellite follows the next at an approximate distance of only one degree of the central angle. For such a "geostationary" satellite, we derive the following condition with Formula (3.4):

---

[4]See `http://spaceflight.nasa.gov/realdata/tracking/index.html` – with kind permission of NASA.

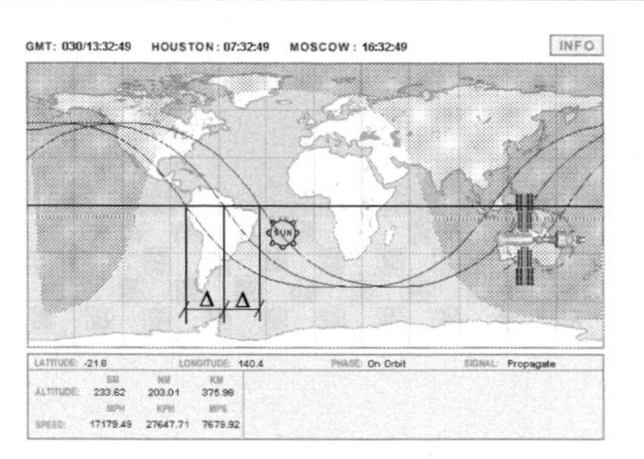


**Fig. 3.38** In a rectangular map, the classical trajectory resembles a collection of repeatedly shifted sinusoids.

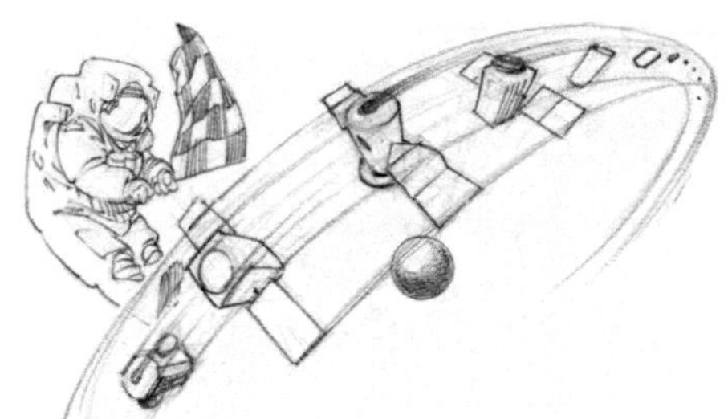

**Fig. 3.39** More than 300 geostationary satellites and only one orbit for them makes it a little bit crowded up there ...

$$T = k^{3/2}\,1.41\,\text{h} = 23.93\,\text{h} \Rightarrow k = 16.97^{2/3} \approx 6.60. \qquad (3.5)$$

This corresponds to a distance of $6.60 \cdot R$ from the center of the Earth or an altitude of $5.60 \cdot R \approx 35,700\,\text{km}$. The associated track speed is, in accordance with Formula (3.3), still $11,000\,\text{km/h}$. The word "positioned" must, therefore, be seen as applying only relatively!

A geostationary satellite has to fly above the equator: Each satellite's orbit does indeed have the center of the Earth as its center. However, a "fixed point" on a non-available equator point rotates in a plane which does not contain the center of the Earth.[5] ◂◂◂

▸▸▸ **Application**: ***mental maths: the Moon vs. geostationary satellites***

The well-known orbit of the Moon (orbital period ≈ 27 days) (Application p. 93) allows us to make a quick estimation of the distance of the geostationary satellites (orbital period 1 day).

*Solution*:

$(r_{\text{geostat}} : r_{\text{Moon}})^3 = (1:27)^2 \Rightarrow r_{\text{geostat}} : r_{\text{Moon}} = (1:3)^2 \Rightarrow r_{\text{geostat}} \approx \frac{1}{9} \cdot r_{\text{Moon}}.$

◂◂◂

▸▸▸ **Application**: ***airborne weight loss***

If we were to fly a plane at a speed of $28,460\,\text{km/h}$ (altitude would be very small, speed sidereal and not relative to the ground), we would – like the astronauts in the space station – become weightless.

Of course, this is not possible because our plane would burn up in the dense atmosphere. Therefore, we suppose that we fly at cruising speed, that is, at about $830\,\text{km/h}$ "ground speed". If we do this above the equator "in line with the Earth's rotation" (i.e. flying to the east), we arrive, together with the speed of a point above the equator of $\dfrac{40,000\,\text{km}}{24\,\text{h}} \approx 1,670\,\dfrac{\text{km}}{\text{h}}$, at a sidereal

[5] An interesting website is `http://www.sat.dundee.ac.uk`, where you can find new daily recordings from geostationary satellites from all over the world.

speed of 2,500 km/h. This corresponds to about 1/12 of 28,460 km/h. The centrifugal force increases or decreases at the same radius with the square of the speed. Therefore, the force on the plane equals $1/12^2 = 1/144$ of the force that would annul our weight. Our weight in the aircraft is, thus, almost one per cent lower than in the hibernation state at the North Pole (sidereal velocity 0). With a mass of 70 kg (weight force approximately 700N), this is about 5N.

⊕ *Remark*: Leonardo *da Vinci* (1452–1519) still believed that the weight of a body is dependent on its speed. As "proof", he argued that a horse with a rider moving in full gallop can stand shortly on one leg. His theory was refuted a century later by *Galileo Galilei* (1564–1642). After all: Leonardo's daring theory contained a (very small) grain of truth if we stick with the example of the riding horse: It is at least true that the rider is lighter when he rides toward the east ... ⊕ ◂◂◂

### ▸▸▸ Application: *when the centrifugal force is dominant*

Fig. 3.40 shows the rapid rotation of a human performing multiple somersaults. Three revolutions per second with a radius of perhaps 30 centimetres is very fast for us. After all, the centrifugal force is then already so strong that one's hair almost exclusively obeys the centrifugal acceleration $a$. Find out what the size of $a$ is.

**Fig. 3.40** The centrifugal force can easily outdo other forces.

*Solution*:
We use the well-known formula $a = r\,\omega^2$, where $r$ is the radius in meters and $\omega$ is the angular velocity, and get

$$a = 0.3\,\mathrm{m} \cdot (6\pi/\mathrm{s})^2 \approx 120\,\mathrm{m/s}^2 \approx 12\,g.$$

Humans can stand 12-fold gravity only for a short time!
Twice as many rotations per second produce four times the acceleration, which is more than a person could cope with: The person will at least have internal bleeding due to blood emerging from the veins.
A technique in which this is utilized: Dryers that tumble with 900 revolutions per minute (or 15 per second) and have a drum radius of 30 cm produce accelerations of $300\,g$, extracting every drop of water from the laundry. ◂◂◂

## 3.5 Further applications

►►► **Application**: ***The king's chamber in the Great Pyramid*** (Fig. 3.41)
The Cheops pyramid was originally 146 m high. The height of the King's Chamber was chosen so that the horizontal plane through the sarcophagus produces a cross-section of the pyramid whose area is exactly half the surface of the pyramid's base. How high is the horizontal plane?

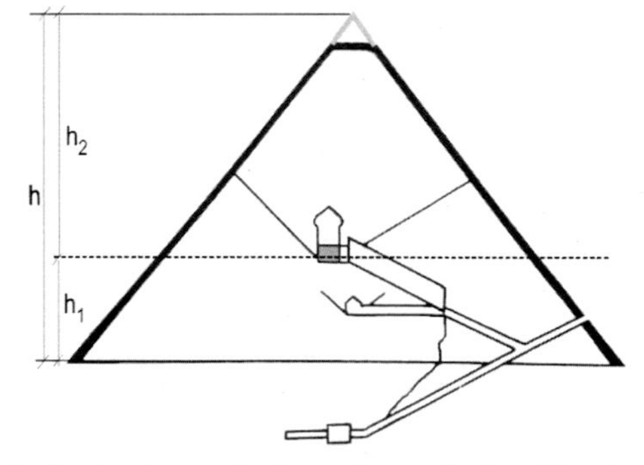


**Fig. 3.41** the horizontal plane through the King's Chamber

*Solution*:
Let $h = 146\,\text{m}$ be the total height of the pyramid and $A$ be the area of the base. Furthermore, let $h_2$ be the height of the horizontal plane above which there is a similar pyramid whose base is $A_2 = A/2$. We find

$$h_2 : h = \sqrt{A_2 : A} \Rightarrow h_2 = h/\sqrt{2} \approx 103.25\,\text{m}.$$

Thus, the horizontal plane is at an altitude of $h_1 = h - h_2 = 42.75\,\text{m}$.

⊕ *Remark*: The subsoil of the pyramid – a square whose 230 m long sides are precisely aligned with the cardinal directions – was leveled up to a precision of centimetres before the beginning of the construction process. This was before the invention of the spirit level or the compass ... ⊕ ◄◄◄

►►► **Application**: ***misleading diagrams*** (Fig. 3.42)
In the "statistics" in Fig. 3.42a, there are three similar barrels, and the ratio of their heights and radii can be defined as

$$1 : 1.3 : 1.3^2.$$

They intend to illustrate that an oil company has increased its production in the last two years by 30% ($k = 1.3$) in every sector. However, the volumes of the barrels have ratios of

$$1 : 1.3^3 : (1.3^2)^3 \approx 1 : 2.2 : 4.8.$$

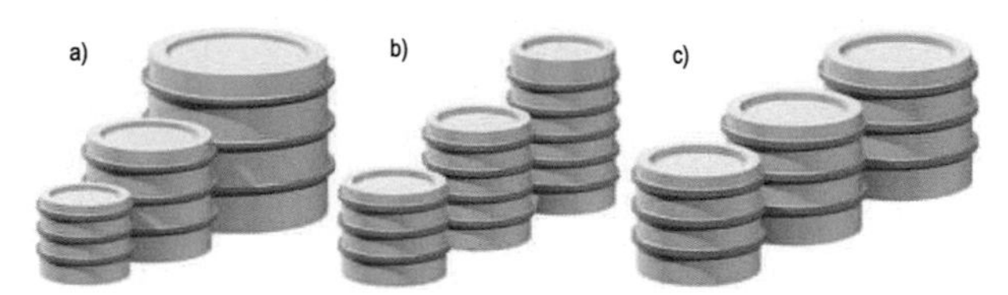


**Fig. 3.42** a) falsifying chart, b) and c) correct charts

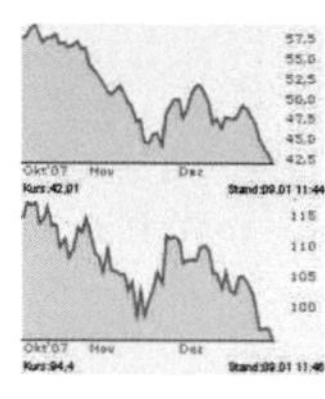

**Fig. 3.43** stock prices: distortion in the ordinate suggests comparable losses

Thus, the diagram in Fig. 3.42a pretends an increase by 120% each year. The diagram in Fig. 3.42b, however, is correct because the radii of the barrels were left unchanged.
Even the information conveyed by Fig. 3.42c is correct, because both the radius and height are multiplied by the factor $k = \sqrt[3]{1.3}$.

⊕ *Remark*: Fig. 3.43 shows the price development of the shares of two major banks during the real estate crisis in late 2007 (as of 9. 1. 2008). At first glance, it seems "caught" between the same two banks. A closer look at the scale shows: The upper bank varies between the values 42 and 58 (which is almost 40%, based on the lower value); the lower one is between 116 and 94 (which is "only" a little more than half). The "fever chart" looks similar, because of the assignment of equal fluctuation ranges (in the mathematical sense), but they are stretched along the ordinate. ⊕ ◂◂◂

▸▸▸ **Application**: ***ostrich and chicken eggs compared*** (Fig. 3.44, left)
An ostrich's egg looks like a large chicken egg, but has about 24-times the mass of a chicken egg. What is the ratio of the diameters and what is the ratio of the surface areas? Why do ostrich eggs need extreme heat for hatching?

**Fig. 3.44** ostrich and chicken eggs in comparison

*Solution*:

$$M_1 : M_2 = 1 : 24 \Rightarrow V_1 : V_2 = 1 : 24 \text{ (equal consistency)}$$
$$\Rightarrow d_1 : d_2 = 1 : \sqrt[3]{24} \approx 1 : 2.9 \Rightarrow S_1 : S_2 = 1 : (\sqrt[3]{24})^2 \approx 1 : 8.3.$$

The ratio of the surfaces: The volume of the chicken egg is much smaller than that of the ostrich egg (using the factor $k = 2.9$, see Application p. 80). In order to heat the entire interior of the egg, a long exposure to a higher outdoor temperature is required. Therefore, ostriches live mainly in the hot semi-deserts of southern Africa.

⊕ *Remark*: How long do you have to cook an egg? This question depends not only on whether you want to have the egg soft-boiled or hard-boiled, but also on the size of the egg. Large eggs take longer! ⊕ ◂◂◂

▸▸▸ **Application**: ***ground pressure*** (Fig. 3.45)
Another comparison between ostriches and chickens: Both birds prefer to run, rather than take to the air (the ostrich uses its wings only to maneuver at rapid speeds). Which bird has the greater ground pressure?

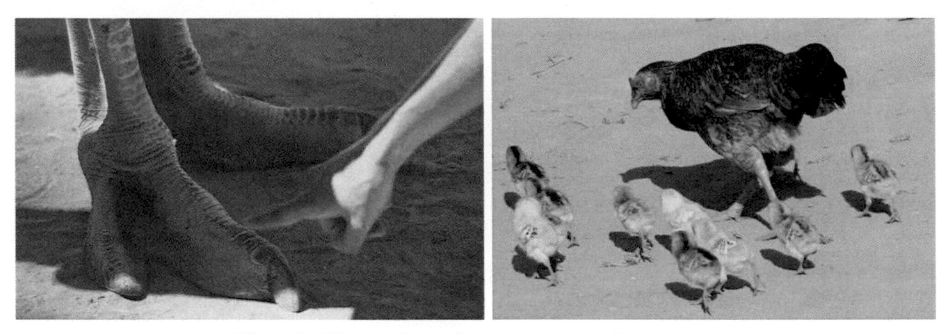

**Fig. 3.45** tread surfaces or ground pressure

*Solution*:
A statement can be made immediately: The young domestic chickens in Fig. 3.45 (right) have a lower ground pressure than their mother (see Application p. 86): They are smaller, and if they are of similar shape, their tread surface is larger in relation to their weight (*pressure* = *force* / *area*). One would be tempted to think now that an ostrich weighing 50–100 kilograms would produce significantly greater ground pressure. After all, the ostrich might weigh twenty or thirty times the weight of an adult domestic fowl. On the other hand, the foot of an ostrich has oversized dimensions (Fig. 3.45, left) and it is expected to offset maybe twenty to thirty times the tread weight. Thus, the ground pressure of both birds is approximately the same.

⊕ *Remark*: A bunch of ostriches look like certain dinosaurs (for example, the raptors), because they are quite similar with regard to their shape. These protozoa were probably also able to run quickly at a pace that they could possibly sustain. However, the latter property requires the animal to be warm-blooded. ⊕

⊕ *Remark*: A man has a relatively large contact surface (Fig. 3.44, right, Fig. 3.46, left). Things change once he sits on a bike (Fig. 3.46, top left). ⊕

⊕ *Remark*: Having stated the above, an elephant has a smaller ground pressure than a cow. A giraffe weighs a little more than a cow, and has slightly larger hooves. So, their ground pressure is comparable. The Anopheles mosquito in Fig. 3.46 on the right stands on four legs. However, the ground pressure produced by the mosquito given with comparable physique and tread surface would be merely 1/500 of the ground pressure produced by a giraffe (1 cm versus 500 cm). At this scale, adhesion prevails, and the mosquito can, therefore, dangle from the ceiling. ⊕ ◂◂◂

**Fig. 3.46** Left: man, bicycle, ox, horse. Center and right: two extremes

▸▸▸ **Application**: ***camera sizes*** (Fig. 3.47)
The chip size of a digital camera is a criterion for the image quality. Many professional cameras have a "full-frame sensor". Meanwhile professional cameras with a "crop-factor" are also being offered on the market. A sensor size of 50%, for instance, does not necessarily mean a camera has only half the body size. The cameras pictured below have a width of 15.5 cm and 12.5 cm. What weight difference is to be expected for the bodies and the lenses respectively?

**Fig. 3.47** Camera sizes and weights of two professional cameras and two comparable lens configurations (left: telezoom up to 300 mm, right: macro lenses).

*Solution*:
The factor is $k = 15.5/12.5 \approx 1.24$. If all measurements (including the thickness of the metal) are scaled equally, this should result in a weight difference of $k^3 \approx 1.9$. Due to the fact that the larger camera has a comparatively larger battery, the weight factor is about 2.2. When it comes to the lenses, the difference should be bigger: A lens with a focal length of $f = 300$ mm, for instance, needs to have only $f = 150$ mm for the smaller sensor. This would result in an extreme weight difference ($2^3 = 8!$). Telelenses, however, are "mainly air", and therefore, the weight difference is usually 2 to 3. ◂◂◂

►►► **Application**: ***Tetra Pak in various sizes***
A drink is to be launched on the market in "Tetra-Pak" packaging, respectively containing 0.25 liters and 0.5 liters. How many times more packaging material does a "Tetra-Pak" of 0.5 liters need? By what percentage must the edge length of the tetrahedron (Fig. 3.48) or the cuboid (Fig. 3.49) increase?

**Fig. 3.48** Tetra Pak in various ...

**Fig. 3.49** ... sizes and shapes

*Solution*:

$$V_1 : V_2 = 1 : 2 \Rightarrow k = \sqrt[3]{2} = 1.26 \Rightarrow S_1 : S_2 = 1 : k^2 = 1 : \sqrt[3]{4} = 1.59$$

So you need about 59% more packing material for the half-liter drink. The edge length increases by 26%. ◄◄◄

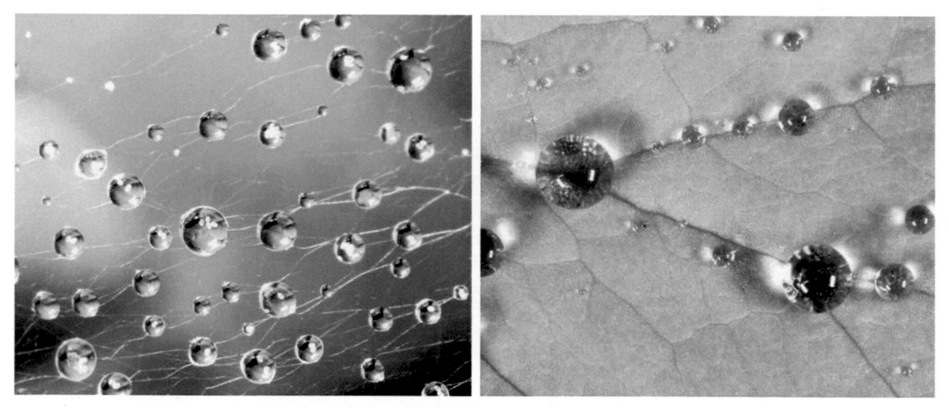

**Fig. 3.50** Droplets distribution. Right: 64 small droplets must unite to create a droplet of fourfold diameter.

►►► **Application**: ***spherical dew*** (Fig. 3.50)
If left to cool and condense, the morning dew that attaches itself to leaves (right) and cobwebs (left) produces wonderful constellations of approximately equally sized water droplets. These droplets are spherical due to surface tension, which aims to enclose the highest possible volume. Their individual sizes are so similar because of cohesiveness, which causes molecules to be attracted towards "strategic centers" such as equidistant points. The droplets

on the cobweb have a radius of about 1–2 mm, just like the smallest droplets on the leaf. If 64 droplets join together, they produce a larger drop of fourfold diameter (right). ◂◂◂

### ▸▸▸ Application: *dissolving a powder in a liquid*

The same quantity of an effervescent powder is to be dissolved in two glasses of water. For the first glass, the powder is coarse; for the other glass, the particles of the powder are half the diameter. How many times faster will the finer powder dissolve than the coarser one?

**Fig. 3.51** dissolving a powder in theory ...

**Fig. 3.52** ... and in practice

*Solution*:
The radii have a ratio of 2 : 1, the volumes are 8 : 1, and the surface areas are 4 : 1. In the finer powder there are 8 times as many spherules available, each with $\frac{1}{4}$ of the surface of a larger spherule (Fig. 3.51). Overall, all the small spherules have, thus, twice the surface area, and the liquid will *initially* dissolve twice as fast. Here, the radii of the large and the small spherules changes, and other conditions occur: The small spherules become smaller in proportion. The above example is only correct if the radii of the spherules do not change. ◂◂◂

### ▸▸▸ Application: *dipping a cylinder* (Fig. 3.53)

An arbitrary axially symmetric cylinder (such as in Fig. 3.53 on the left) is immersed vertically into water. How deep does it subside?
Note: The cross-sections are not only similar, but even congruent. The volume of a cylinder equals

$$volume = cross\text{-}section \times height.$$

*Solution*:
Let $A$ be the cross-sectional area, $h$ be the cylinder height, $\varrho$ be the density, and $t$ be the depth of immersion. Then the mass of the cylinder $M_Z = \varrho \cdot A \cdot h$ and the mass of water displaced is $M_W = 1 \cdot A \cdot t$. According to *Archimedes*'s *principle*, we have

$$M_z = M_W \Rightarrow \varrho \cdot A \cdot h = A\,t \Rightarrow t = h \cdot \varrho.$$

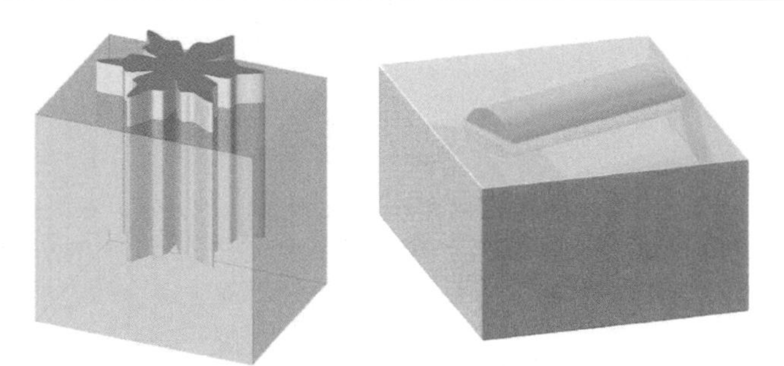

**Fig. 3.53** dipping a general cylinder or a cylinder of revolution

⊕ *Remark*: If the cylinder is not fixed, it will eventually – depending on its shape and the specific weight – even at small disturbances (wave motion) assume a stable equilibrium position, such as on the right in Fig. 3.53. We will examine this physical process more accurately in Application p. 218. ⊕ ◂◂◂

**▸▸▸ Application**: ***dipping a conical object in water***

A fully axially symmetric cone (density $\varrho < 1$) floats
a) with the top down or
b) with its base down
in the water. How deep does it plunge? What percentage of the lateral surface is wet?

*Solution*:

Fixing some unit of measurement, let the volume of the cone be $V$ and its mass be $M = \varrho \cdot V$. We apply *Archimedes*'s principle: *The weight of the body is equal to the weight of the displaced fluid* (water: $\varrho_w = 1$). Mass and weight are proportional ($G = M \cdot g$); hence, the law also applies to the masses.

First case a) (Fig. 3.54a):

The penetration depth is $t_1$ units ($0 < t_1 < 1$). The volume $V_m$ of the ousted water is the volume of a similar cone with $V_w = V \cdot t_1^3$. Again, according to *Archimedes*'s principle, the displaced weight before and the total weight afterwards are equal:

$$M_w = M \;\Rightarrow\; V \cdot t_1^3 = \varrho \cdot V \;\Rightarrow\; t_1 = \sqrt[3]{\varrho}.$$

For the lateral surface, we find

$$total : moistened = 1 : t_1^2 = 1 : \sqrt[3]{\varrho^2}.$$

In Fig. 3.54a, we have $\varrho = 0.5$. This results in $t_1 \approx 0.79$ and $t_1^2 \approx 0.63$, that is, 79% of the body's height is immersed in the water and 63% of the surface area is wet.

Shifting our attention to the case b) (Fig. 3.54b):

The cone stands $h$ units out of the water. The volume (and hence the mass) of the displaced water is then $V_w = V(1 - h^3)$, and thus, following *Archimedes*'s

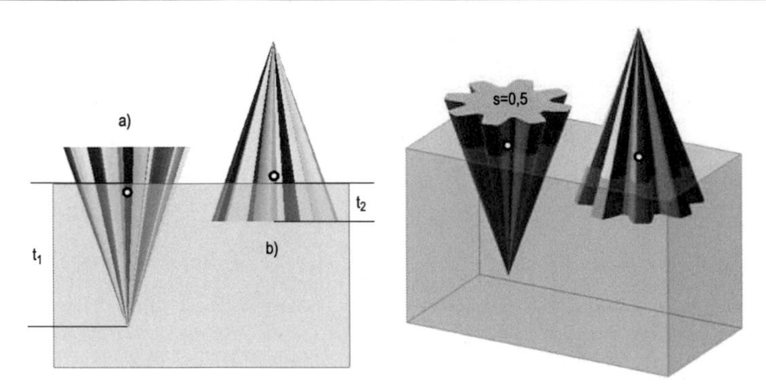


**Fig. 3.54** immersing a cone in water

principle, we have

$$V(1-h^3) = \varrho \cdot V \;\Rightarrow\; h = \sqrt[3]{1-\varrho}.$$

The penetration depth is

$$t_2 = 1 - h = 1 - \sqrt[3]{1-\varrho}.$$

For the surface area, we get

$$total : moistened = 1 : (1-h^2) = 1 : (1 - \sqrt[3]{(1-\varrho)^2}).$$

Setting $\varrho = 0.5$, Fig. 3.54b results in $t_2 \approx 0.21$ and $1 : (1-h^2) \approx 1 : 0.37$. That is, 79% of the body height is above the water and only 37% of the surface area is wet. Compare these results with the case of a).

⊕ *Remark*: Upon immersion of other bodies, such as a ball, the displaced volume is not similar to the immersed body, and so you have to use other, more complex methods – see Application p. 283. Generally, one must always remember that a body tries to adapt a stable equilibrium position during immersion. The cone will tilt under certain conditions! This reminds us of an iceberg, which protrudes by about one-tenth of its volume out of the water. It is often observed that an iceberg begins to rotate unexpectedly in order to take a different, more stable position. ⊕

◂◂◂

▸▸▸ **Application**: ***Why do big air bubbles ascend faster?*** (Fig. 3.55)
Every diver knows: Big air bubbles ascend faster than small ones (Fig. 3.55). The rule of thumb is the following: Ascending faster than 20 m per minute should be avoided – and this also applies somehow to small air bubbles.

*Solution*:
We first simplify the situation and assume that bubbles are spherical. Then, larger bubbles paradoxically have – in relation to their volume – less surface and less cross-section. Due to Archimedes's principle (Application p. 64), the buoyancy force equals the weight of the displaced water bubble, whereas the

**Fig. 3.55** One major, but simplified, rule for divers: Do not ascend faster than the *small* bubbles – no matter what happens.

cross-section is mainly responsible for the friction force. One could compare the situation with the free fall of small and big spheres of the same material with *extreme* air-resistance (Application p. 21): The big spheres, then, are faster than the small ones.

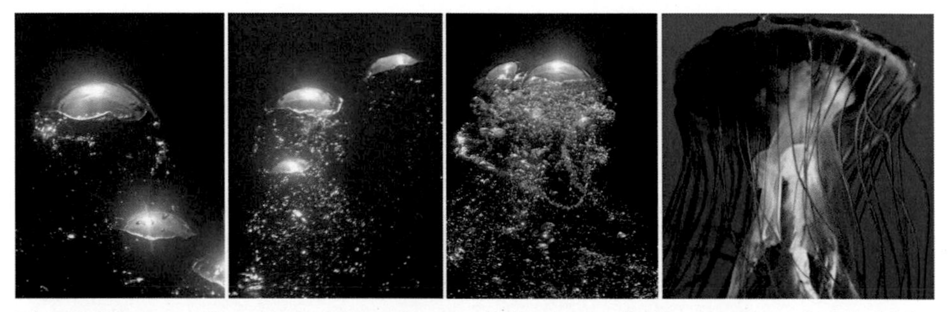

**Fig. 3.56** Ascending air bubbles change their shape all the time. The big bubbles roughly develop the shape of spherical caps and resemble the shape of jellyfish.

When looking at photo series or videos of ascending water bubbles (Fig. 3.56), one notices two things: Firstly, big bubbles look like spherical caps (Application p. 411), and secondly they "explode" and form new smaller spheres every once in a while as they ascend. This might have the following explanation: Small bubbles have a spherical shape due to their surface tension. Since they ascend at a lower speed, they are barely deformed. When the bubbles get bigger due to decreasing pressure, the streaming dents the sphere on the back.

⊕ *Remark*: Application p. 77 explains that air (80% Nitrogen) is dissolved under pressure into the blood system of the diver. If the diver ascends too quickly, the air cannot be fully exhaled by the lungs and nitrogen embolism can cause serious damage. ⊕ ◂◂◂

**▸▸▸ Application**: ***large and small people***

Two similarly constructed human beings have a mass of 50 kg and 75 kg, respectively. Since their masses (and hence their volumes) have a proportion of 1 : 1.5, we can say the following according to the above statements (see Application p. 497):

- Their body sizes form the ratios $\sqrt[3]{1} : \sqrt[3]{1.5} = 1 : 1.14$. Thus, if the lighter body is, for instance, 160 cm tall, the heavier body will have a height of about 183 cm.

  ⊕ *Remark*: A newspaper article from March 27, 2007: "The 2.36 m tall Chinese *Bao Xishun* has found the love of his life. The happy woman is a whole 68 cm shorter than him." The next photo shows two reasonably equal slim people. The husband was about $(2.36/1.68)^3 \approx 2.75$-times as heavy at the time of their wedding. ⊕

- Your blood levels and liver capacities behave like volumes. So, if the lighter person is drinking 2 glasses of wine, the heavier person can drink 3 glasses and feels about the same effect. In practice, of course, other components play a role.

- A smaller person gets cold faster than a heavier person (bad ratio of *Volume* : *Surface*). One should always bear this in mind when going for a walk with kids in winter. ◂◂◂

**▸▸▸ Application**: ***How many ants weigh as much as an elephant?***

**Fig. 3.57** threatening gestures in different scales

*Solution*:

With the best of intentions, these two cannot be considered similar (Fig. 3.57)! The following calculation is, therefore, initially wrong: The elephant and the ant have – like most living creatures – the specific weight of water.

Their lengths, therefore, behave like $k \approx 3.5\,\mathrm{m} : 7\,\mathrm{mm} \approx 500$. The ratio of their volumes (and masses) is $k^3 \approx 500^3 \approx 125$ million. However, the ant has a much more fragile build. The ratio is, therefore, greater. So, we adjust upwards and estimate: An elephant weighs as much as 1 billion ants ...

⊕ *Remark*: Now, let us continue with our estimations: The surface of the Earth is about 150 billion $\mathrm{km}^2$. Of these, 120 million $\mathrm{km}^2$ are colonized by small animals like ants. Just eight ants per $\mathrm{km}^2$ are sufficient to counterbalance an elephant. In the world, there are about 100,000 elephants. They are "counterbalanced" by 800,000 ants per $\mathrm{km}^2$ or 0.8 ants per $\mathrm{m}^2$. The total mass of all insects and invertebrates (worms, etc.) currently living on Earth is greater than the total mass of all other currently living animals! Among the mammals, bovine lifestock, i.e., cows, are the absolute leader. Cattle biomass is several times greater than that of humans. Cows even weigh more than all the ants on this planet taken together. Animal biomass accounts for only a tiny fraction of plant biomass though (less than 1%). This amounts to an estimated 2 trillion tons,[6] i.e. $2 \cdot 10^{15}\,\mathrm{kg}$. With an average density of water ($1\,\mathrm{kg/dm}^3$), this corresponds to a "water cuboid" with dimensions $22\,\mathrm{km} \times 22\,\mathrm{km} \times 4\,\mathrm{km}$. (Our oceans are on average 4 km deep.) Floods of this magnitude may be caused by an underground earthquake which results in a tsunami. ⊕ ◂◂◂

### ▸▸▸ Application: *bound carbon*

Biomass can be measured in different ways, but in strict scientific applicati-

**Fig. 3.58** By far the most bound carbon can be found in plants (left: 15 m high trees, right: 30 cm high bushes).

ons, it is defined as the mass of organically bound carbon (C) that is present in a given environment. The total live biomass of bacteria probably exceeds that of plants and animals.[7] The further up animals are in the food chain, the more its total biomass tends to decrease. In oceans, biomass usually follows the food chain phytoplankton ⟶ zooplankton ⟶ predatory zooplankton

[6] Maxeiner, Mirsch: *Life counts*, Berlin Verlag, Berlin 2000.
[7] G. Glaeser: *Nature and Numbers*. Birkhäuser/deGruyter, 2013.

⟶ filter feeders ⟶ predatory fish. Terrestrial biomass generally decreases significantly at each higher level (plants ⟶ herbivores ⟶ carnivores). Trees, for instance, possess a vastly larger amount of biomass than all animals together. Fig. 3.58 on the left shows trees that are 15 meters tall. The right photo depicts bushes with a height of 30 cm. Thus, one can conclude that they should have roughly 100,000 times more biomass ($50^3 = 125,000$), as the trees are 50 times taller. ◂◂◂

**Fig. 3.59** Stag beetles fight for a female. The air supply plays an important role.

▸▸▸ **Application**: ***Are larger animals preferred by evolution?*** (Fig. 2.51, Fig. 3.59)

Male specimens of the stag beetle resemble each other, even if their size varies between 35 mm and 80 mm. Two males (50 mm and 60 mm length) fight for a female. How many times heavier or stronger is the larger one? Why might the stronger beetle still face a problem?

*Solution*:

The scaling factor equals $k = 60/50 = 1.2$. So, the larger males are $k^3 \approx 1.73$ times heavier and as much as $1.2^2 = 1.44$ times stronger (see Application p. 85). Yet, their air supply via the trachea may be 1.2 times worse. In addition, the stronger beetle will require more force to control its extremities (longer lever).

⊕ *Remark*: While you can let yourself be bitten in the finger by a male – since the lever of its antlerpliers is large, the same experiment is not advisable for much smaller females. These antlers must be short, because they act as effective pliers in rotten trees "where they eat their way" in order to lay eggs. ⊕

⊕ *Remark*: The playful fight (in Fig. 3.60) between two elephants is typically a "showdown" lasting for many hours in which body mass plays a crucial role. The young bull Abu from the Schönbrunn Zoo in Vienna has a mass of about 1800 kg. His mother is – as can be estimated from the picture to the left – 25% taller, and thus, twice as heavy. Abu's jostling ended fatally for a zookeeper in February 2005! ⊕ ◂◂◂

**Fig. 3.60** battle of giants: For non-heavy-weights, this is very dangerous.

**Fig. 3.61** Collective mimicry: Larvae of a beetle form a bee.

▸▸▸ **Application**: ***collective mimicry*** (Fig. 3.61)

In the Mojave Desert in California, there is a beetle (*Meloe franciscanus*) whose 2 mm long larvae congregate in hundreds on top of a straw to collectively mimic the shape of a female bee and, thereby, develop additional fragrances to attract male bees. Before the bee discovers the mistake, many larvae will already cling onto its body, and they thus manage to get transported to the bee's nest, where they will feed.[8] How many larvae can mimic the body of a 14 mm long bee?

Supplementary question: If the first "transport" takes 25% of larvae and the rest of the larvae again try to form a female bee, how long will this bee be?

*Solution*:

The aspect ratio is about 7 : 1. The larvae appear somewhat more elongated than a bee, so that it is difficult to set their shapes in relation. In order to achieve the volume of the bee, the ratio is $7^3 : 1 = 343 : 1$. Thus, there are several hundred larvae involved in this. If there are now 25% fewer larvae, the shrunken volume leads to a scaling of the figure by a factor of $k = 3/4$, and thus, the length changes by a factor of $\sqrt[3]{k} \approx 0.91$.

The newly formed shape is now 12.7 mm long instead of the initial 14 mm. ◂◂◂

▸▸▸ **Application**: ***When and how did the universe originate?***

With super-telescopes, one can make out stars from the Earth even if they are some 3 billion light years away (practically, this means that one is looking into the past). The light of more distant stars is no longer able to travel through

[8] See `https://www.nps.gov/moja/learn/nature/upload/201204MOJAscience.pdf`

the atmosphere. With space telescopes (e.g., the Hubble Telescope), one can make out stars that have only 1/50 of the limit light intensity. Since brightness decreases with the square of the distance, you can thus, theoretically, see stars that are $\sqrt{50} \approx 7$ times further away (more than 20 billion light years).

⊕ *Remark*: Since the universe originated 12–15 billion years ago by the "Big Bang", it should be possible to observe this! Indeed, scientists presume that some images show only 300,000 years old parts of the universe. During the period, so it is believed, light was able to escape the "primordial soup" for the first time. ⊕ ◂◂◂

▸▸▸ **Application**: ***linear expansion of bodies when heated***
(Application p. 24, Application p. 384)

A body is heated by the temperature $\Delta t$ with volume $V_1$, so that its volume increases to $V_2$ according to the formula

$$V_2 = V_1\,(1 + \gamma\,\Delta t).$$

How big is the linear expansion?

*Solution*:
We have $k^3 = \dfrac{V_2}{V_1} = 1 + \gamma\,\Delta t$. The scaling factor for lengths (widths, heights) is thus $k = \dfrac{L_2}{L_1} = \sqrt[3]{\dfrac{V_2}{V_1}}$. The "expansion formula" is, therefore,

$$L_2 = L_1\,\sqrt[3]{1 + \gamma\,\Delta t}.$$

◂◂◂

▸▸▸ **Application**: ***"magnification" for macro lenses***

Before the advent of digital cameras, one thing was clear: A macro lens with magnification 1 : 1 could take a rectangle with a frame of size $36 \times 24\,\mathrm{mm}^2$. What are the rules for digital cameras now?

The chips of digital cameras have different sizes today. The extremely expensive chips of professional cameras still have the "classic" format of $36 \times 24\,\mathrm{mm}^2$. These are followed by mid-priced chips with about $22 \times 15\,\mathrm{mm}^2$. This allows us to fill rectangles of the same size using the same lens.

Specifically: If I take a picture of a 24 mm long beetle using a professional camera with a large (expensive) chip of 8 megapixels and another one using a smaller chip (but also of 8 megapixels) then – assuming that the same lens is used – the image recorded by the smaller chip has a higher resolution (factor $(36/22)^2 \approx 2.6$)! However, one has to consider two things: Firstly, the quality of the picture does not only depend on the number of pixels, but is also strongly affected by the quality of the chip. Larger chips deliver (even) better results. Secondly, smaller chips indirectly extend the lens's focal length (which is in fact linked with the classic size $36 \times 24\,\mathrm{mm}^2$), in our case by a factor of $36/22 \approx 1.6$. This, in turn, causes a significant (quadratic) decrease in the depth of field (Application p. 54). The professional camera will still emerge as the winner ... ◂◂◂

▸▸▸ **Application**: ***sucker cups?*** (Fig. 3.62)
Surface tension (Application p. 80) is similar to adhesion – which is the mutual attraction of particles to the surfaces of various substances.

**Fig. 3.62** adhesion makes various substances possible

Geckos have small extensions on each toe. These are composed of small, closely spaced suction pads, which, in turn, carry up to a thousand branched hairs that are microscopically small (half a million per foot approximately). Thanks to this ingenious system, these animals can wait for prey on overhanging walls. Yet, the exploitation of adhesion will soon reach its limits, because the increased contact area cannot keep up with weight gain. A larger and heavier animal cannot have the incredibly large number of hairs that would be necessary on each foot to enable it to adhere to a wall. Besides, animals of increasing size lack the relative force that is needed to consistently overcome of gravity. Nonetheless, there have been experiments with tape trying to imitate the "Gecko-principle".[9] ◂◂◂

▸▸▸ **Application**: ***Can large animals jump higher?***
Explain the following paradox: In principle, all animals with good jumping conditions (long jumps correspond to leg proportions and muscle distribution) have the potential to jump about two meters high. This is true for insects as well as large mammals. For small animals (with a body length of less than 5 cm) the limit height is strongly reduced by air resistance, and the resistance increases as the animal gets smaller (Application p. 21).
*Solution*:
First of all, when dealing with a high jump, we always look at the center of gravity, which is higher for a larger animal. We will treat the "standing height" separately in our calculations. Furthermore, in dealing with accelerations, we are always concerned with *how long* a force can act.
We need the following rules:
(1) The jump height depends *on the square* of the initial velocity. When the body has taken off, only the force of gravity acts. Duplicating the speed at the top means quadrupling the jump height!

[9]See `www.welt.de/data/2003/06/03/106227.html`

**Fig. 3.63** bounce in small and large animals

(2) The initial rate is greater, the longer the acceleration through the muscles takes place. The duration of the acceleration depends on the length of the legs.
(3) The acceleration results from the relation Force = Mass × Acceleration. We first substitute (3) into (2) and subsequently (2) into (1), so we find:
(4) The jump height depends on the square of the expression
$(Force \times Time)/Mass$.
The force, as we know, increases quadratically with increasing body length, the time depends linearly on the leg length, and therefore, linearly on the body length. The terms Force × Time and mass increase cubically as the length increases linearly. This leads to an amazing intermediate result (which we must subsequently adapt a little bit):
*The absolute jump height does not depend on the height of the body.*

⊕ *Remark*: In fact, the skip of the small Galagos (small lemurs that have a mass of only one tenth of a kilogram and are also known as "bush babies") is almost as high as a leopard's jump.
The "prehistoric horse" (*Propalaeotherium*) was the predecessor of today's horse and had a head-body length of 55 cm (with a shoulder height of 30 cm). If a present-day horse can lift its center of gravity by 120 cm and thus jump over a hurdle of 220 cm, then the Propalaeotherium could lift its center to the same extent (but it could only jump over 140 cm high hurdles).
From this example, it is also clear why the Cuban Javier *Sotomayor*'s legendary world record high jump of over 2.45 m has remained intact since 1993. He must credit his success to his enormous size of 2.08 m, amongst other factors: It is not because taller people can jump higher, but because his center of gravity lies higher! In addition, he has very long and accordingly trained legs and a very slight torso.
Now as regards small animals (of less than 5 cm): Even the widest jumping grasshopper can reach only a quarter of the jumping height of larger animals despite the fact that these insects are true professionals in jumping. This is due to air resistance:

**Fig. 3.64** Jumping force does not count here, but the exit speed does!

A small beetle can achieve a falling speed of 3 meters per second. It cannot be more than that, simply because the air resistance slows them down immediately. Now going back to physics: If I throw a stone up at 3 meters per second, it flies as high as 45 cm! To throw the stone 2 m high, I have to throw it up at almost 7 meters per second!

Even if a small insect were able to jump at 7 meters per second (theoretically, it could), it would be immediately slowed down by air resistance and its altitude would thus be drastically reduced. Therefore, it simply makes no sense for an insect to jump away with so much energy.

Rat fleas, which are only 2 to 3 millimeters in size and were once feared as carriers of the plague, are truly gifted jumpers that can reach heights 50 times greater than their body size. Their top speed is less than two meters per second, though this is already their top speed in free fall.

Dolphins and larger whales may skip several meters high out of the water. The maximum jumping height is reached when the animals shoot out of the water vertically. Here it is all about the maximum speed under water, but not the bounce! Thus, if an arbitrarily large cetacean (or even a penguin) can make a jump perpendicular to the water surface at 10 m/s, its center of gravity rises about 5 feet from the water according to Formula (2.18). ⊕ ◂◂◂

## ▸▸▸ Application: *Do large aircraft need relatively less fuel?*

*Solution*:

Assuming that both large and small commercial aircraft seem fairly similar, the question is quickly answered: Yes, they need a relatively smaller amount of fuel. A plane that can carry 320 passengers needs to be only twice as large as one that carries 40 people ($40 \cdot 2^3 = 320$). Then, it has only four times the air resistance, and the per capita consumption is half as much for the small model.

⊕ *Remark*: Large commercial aircraft can fly even higher than small ones and they have a lower air resistance. On the other hand, the ascent requires an enormous effort, which is why the situation is clearly only suitable for long haul flights.

Numerical example: Lufthansa has indicated a fuel consumption of an average of 4.4 liters per person on a 100 km route. The wide-bodied Airbus A380-800, however, consumes 3.4 liters. The planned new Boeing 787 is supposed to get through with as little as 2.5 liters. ⊕ ◂◂◂

**Fig. 3.65** length runs ...

▸▸▸ **Application**: ***in shipping: "length runs"*** (Fig. 3.65)

*Solution*:

One criterion for water resistance is the width of the hull, and it increases in similar constructions only with the root of the submerged hull.
Another criterion is the area of the surface that is in contact with the water, and this decreases proportionally with increasing body length $L$. Occasionally, one even finds the formula for the maximum velocity $v_{max} \approx 2.5\sqrt{L}$ nautical miles/hour (= knot).

⊕ *Remark*: In a downhill race, much longer skis are used than in the technical disciplines, again because "length runs". For the interview with the winner, the skis are replaced by shorter ones so that the brand is clearly visible inside the frame of TV screens. ⊕ ◂◂◂

▸▸▸ **Application**: ***transmission matters*** (Fig. 3.66)

Explain how the depicted electric screwdriver works and determine its angular velocity if the electric motor makes 3,500 revolutions per minute.

*Solution*:

Let us consider the inside of the screwdriver. A 3.6 volt battery (Fig. 3.66, left) provides the necessary electricity to turn the small electric motor with a speed of about 3,500 revolutions per minute. The motor shaft propels a small gear with only six teeth (Fig. 3.66, middle). This so-called sun gear transfers its rotational momentum to three rotationally-symmetric positioned planetary gears. The planetary gears are positioned so that they are connected to a fixed outer gear with 48 teeth. (Their number of teeth – in this case 19 each – depends on the sizes of the sun gear and outer gear and has no direct impact.) This causes the centers of the planetary gears to rotate with one eighth $(6:48)$ of the drive shaft's angular velocity.
The centers of the planetary gears form a fixed equilateral triangle. On its reverse, in a second step, a further 6-tooth sun gear propels three other

**Fig. 3.66** left: electric screwdriver, middle: the small sun gear transfers its rotational momentum to three planetary gears, right: the axes are fixed, and the outer gear is allowed to rotate.

planetary gears with the same transmission ratio as before. The connecting triangle between the planetary gears in the second layer is finally connected to the axis of the screwdriver. The $8 \cdot 8 = 64$ revolutions of the electric motor thus cause a single turning of the screw (roughly $3,500/64 = 55$ revolutions per minute).

⊕ *Remark*: This explains the large angular momentum $M$. Given the motor's constant power $P$, we can apply the equation $P = M \cdot \omega$, where $\omega$ is the angular velocity. Planetary gears have many applications in technology, such as in gearboxes, cable winches, and bicycle hub gears.

Another variation on the same principle is also notable: If the axes of the planetary gears are fixed and the outer gear is allowed to rotate, then it rotates significantly slower than the drive shaft (Fig. 3.66, right). ⊕ ◂◂◂

### ▸▸▸ Application: *Where do the similarities lie?* (Fig. 3.67)

The fern on the left exhibits remarkable similarities to the right image of tire tracks in the snow. But where do the actual similarities lie?

*Solution*:

There is a middle line from which two sets of branches extrude at roughly constant angles. The similarity is further improved by the extreme perspective of the photo on the right, as the distances between the branches appear to shrink in the direction of the tip. The fern had to be photographed from the front in order to appear sharp in the picture. If we now compare the branching distances, we will come to the conclusion that they follow an arithmetic series. By and large, each distance follows from the previous one by the addition of a constant, and after a finite number of points, this distance converges towards zero. The tire tracks display a typical projective scale. The angle and composition is comparable to a picture of train tracks running towards the horizon. This means that, in theory, an infinite number of branching points would fit in the image.

**Fig. 3.67** Live or dead matter (1)?

**Fig. 3.68** Live or dead matter (2)?

⊕ *Remark*: Fig. 3.68 represents an extreme comparison: On the left, you can see how plants sprout from gravel. On the right, you can see patina on a copper plate at a magnification of a hundred times (micrograph, Institute of Chemistry, University of Applied Arts Vienna): Dead matter might also be found on Mars – only at a small scale though, certainly *not* in any greater quantity! ⊕ ◂◂◂

▸▸▸ **Application**: ***the Reynolds number*** (Fig. 3.69)
The correlation via the dimensionless Reynolds number $Re = \frac{l \cdot v}{\nu}$ describes the quotient of inertial force to viscous force acting on a body in a fluid (e.g., air or water).[10] This number increases in proportion to both the object's size $l$

[10]G. Glaeser, H.F. Paulus, W. Nachtigall: *The Evolution of Animal Flight.* Springer Nature, Heidelberg/New York 2017.

and the flow velocity $v$, and it decreases (indirectly proportionally) with the kinematic viscosity $\nu$ of the fluid. The latter depends to a large extent on the temperature. At room temperature, for instance, it is 15 times greater in air than in water. Therefore, objects moving at a certain flow velocity in water have a *Re* number 15 times greater than objects moving at the same velocity in air.

**Fig. 3.69** Is there any chance to compare this mathematically? Same *Re*-numbers $\Rightarrow$ comparable conditions! Left: Landing of a blue tit, right: water flea.

If two bodies have the same *Re* numbers, they are comparable in terms of fluid mechanics. Let us say a water flea moves at a Reynolds number of 300. Then, its inertial force is 300 times greater than its viscous force. This number is derived from a body length of $l = 3$ mm, a "jump" velocity of $v = 10$ cm/s, and the corresponding value for the kinematic viscosity of water.

⊕ *Remark*: If one wishes to study a model of a water flea (*Daphnia pulex*) that is ten times larger, the surrounding water must flow at one-tenth of the velocity in order to arrive at the same *Re* number and the same balance of forces. Alternatively, a wind tunnel could be used: Then, the model must be surrounded by air moving at a flow velocity that is about 15 times higher (1.5 m/s), which is easier to achieve. If the experiment succeeds, one could combine tests in water with tests carried out in air. One medium allows us to determine one parameter, while the other medium lends itself to determining another parameter. Such procedures are commonly used in fluid-mechanical measurements. ⊕ ◂◂◂

▸▸▸ **Application**: ***form follows function*** (Fig. 3.70, Fig. 3.71)
The smallest hummingbirds and hummingbird moths may be compared with horizontal propellers, at least in terms of size, weight, wing shape (no curvature), wing surface and also wing load, wing motion, and the principle of lift to rise in the air.

**Fig. 3.70** form follows function: bird vs. insect

It could also be said that they fly in the same range of Reynolds numbers. Conversely: Within the specified Reynolds number range, certain physical conditions need to be provided for an optimal hovering flight. These conditions, which match the ones listed in the first sentence, have been met by two very distinct biological lineages. The biologist would consider this as a case of convergent evolutions and analogous structures. In technology, this would qualify as an example of the phrase "form follows function".

**Fig. 3.71** form follows function: fish vs. mammal

⊕ *Remark*: Fig. 3.71 shows how mammals (dolphins) converged to shark design within millions of years. Sharks are, evolutionary speaking, *much* older than whales. *Mesonix*, the ancestor of dolphins, was a terrestrial animal that went into the water to feed. ⊕ ◂◂◂

# 4 Angles and trigonometry

In this chapter, we deepen our understanding of right and oblique triangles. We will also discuss various applications of angular or trigonometric functions.

The first part is the connecting tissue that binds the subsequent sections together. The Pythagorean theorem has been proven for $2,500$ years – its underlying relation has been known for much longer than that, for it illuminates the relations between angles and trigonometric functions. The inverse of the theorem – namely that a triangle is right if the Pythagorean formula can be applied to it – is important for some proofs.

In the second section, we will discover that mathematicians prefer to measure angles in radians rather than degrees. We will further show that our eyes tend to measure angles rather than distances, which has consequences for the subjective impressions of photographs.

The following section deals with trigonometric functions and their inherent relations, and also answers the question of how these functions were used to calculate in antiquity. We will then discuss calculations in oblique triangles, which usually derive from applications of the Law of Sines and the Law of Cosines. This will allow us to solve various problems directly.

The last section includes a vast array of practical applications of the principles discussed in the preceding material.


G. Glaeser, *Math Tools*, https://doi.org/10.1007/978-3-319-66960-1_4

## 4.1 The family of Pythagorean theorems

The *Pythagorean theorem* in the right triangle (catheti $a$, $b$, hypotenuse $c$) reads

$$\boxed{a^2 + b^2 = c^2.} \tag{4.1}$$

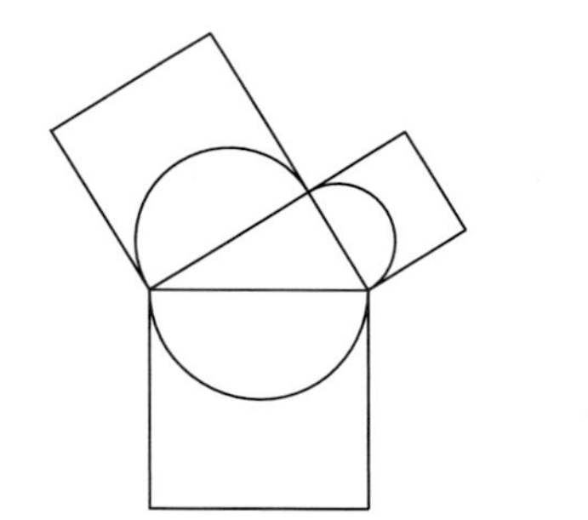

**Fig. 4.1** Pythagorean theorem

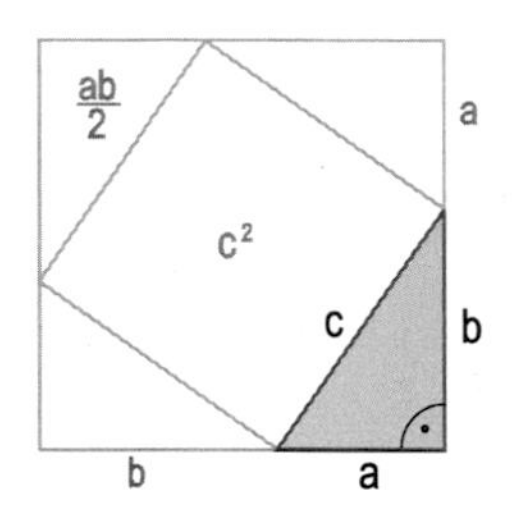


**Fig. 4.2** figure for the proof

***Proof***: The proof behind the theorem is of such great importance that it has been conducted throughout history in over 200 (!) different ways. The following proof is among the easiest to understand.
In Fig. 4.2, the area of the square can be written in two different ways:

$$A = (a + b)^2 = a^2 + 2ab + b^2 \quad \text{and} \quad A = c^2 + 4\frac{ab}{2},$$

from which, through equalization, the formula (4.1) follows directly. $\odot$
Fig. 4.1 illustrates that the sum of the areas of the squares above the legs is equal to the area of the square above the hypotenuse. The same applies to the three similar figures (such as half-circles) that are placed above the triangle sides.
The converse statement of the Pythagorean theorem reads as follows:

| If $a^2 + b^2 = c^2$ applies to a given triangle, then this triangle is right. |
|---|

***Proof***: Let us conduct the proof “indirectly”.
We assume that $a^2 + b^2 = c^2$ is true and the triangle is not right (Fig. 4.3). This leads to a contradiction. (In mathematics, if a statement is not false, then it is true. There exists a form of “black-white thinking” that is grounded in the principle of consistency, or the absence of contradictions. Conversely, a proof is not valid if at least one of its inferences is not true under all circumstances.)
By inscribing the height $h$ onto $b$, a triangle may be interpreted as the sum or difference of two right triangles $AHB$ and $BCH$. In both, the (already proven) Pythagorean theorem applies, or

$$(1) \;\; c^2 = (b \pm d)^2 + h^2 \quad \text{and} \quad (2) \;\; a^2 = d^2 + h^2.$$

We subtract (2) from (1) and get

$$c^2 - a^2 = (b \pm d)^2 - d^2 = b^2 \pm 2bd.$$

According to our initial assumption, however, $c^2 - a^2 = b^2$ holds. Thus, $2bd = 0$ and $d = 0$ must be true. This means that $H = C$ and that the triangle is, indeed, right. ⊙

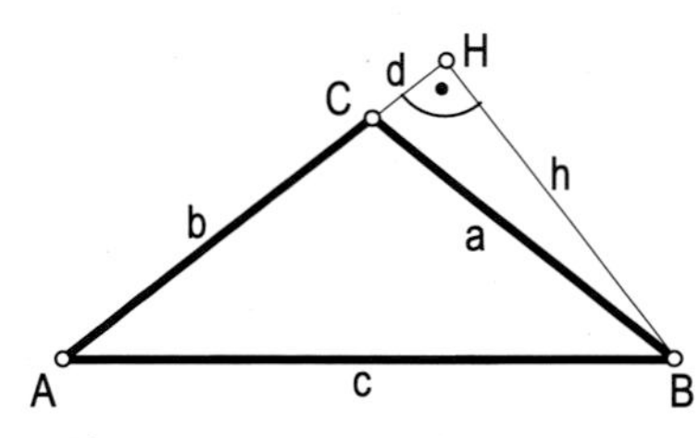


Fig. 4.3 proving the converse

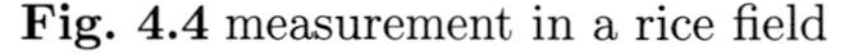

Fig. 4.4 measurement in a rice field

▸▸▸ **Application**: ***Egyptian triangle*** (Fig. 4.4)

The ancient Egyptians were aware that a triangle with sides 3, 4, and 5 is right. They took a rope, made 12 knots at equal distances, banded them together into a noose, and stretched it across three corners (at the knots $0 = 12$, 3, and 7). This enabled them to split their rice fields into right-angled sections. The front view of the Chephren pyramid actually consists of two Egyptian triangles!

⊕ *Remark*: The ancient Indians also knew of a right triangle with the side lengths 5, 12, and 13. The search for right triangles with integer sides was of great interest to the Pythagorean School (Pythagoras 580 – 496 BC). Their solutions are commonly known as "Pythagorean triples":

$$[k(m^2 - n^2),\ 2kmn,\ k(m^2 + n^2)], \quad k, m, n \text{ are integers}$$

$k = 1,\ m = 2,\ n = 1$ produces the Egyptian and $k = 1,\ m = 3,\ n = 2$ the Indian triangle. ⊕ ◂◂◂

▸▸▸ **Application**: ***King's chamber in the Cheops pyramid*** (Fig. 4.5)

The Cheops pyramid was built more than 4,500 years ago, that is, 2,000 years before Pythagoras. We find within it several examples of Pythagorean triples, one of which is found in the granite King's chamber (Application p. 97). The measurements of the right triangle inscribed in the figure correspond to 15, 20, 25 Egyptian ells. The height of the chamber is not rational, but apparently corresponds to $\sqrt{15^2 - 10^2} \approx 11.2$ ells. ◂◂◂

The triangles $ABC$, $ACH$, and $CBH$ (Fig. 4.6) are similar (equal angles). From the proportions $a_c : h = h : b_c$, $c : a = a : a_c$, and $c : b = b : b_c$, we infer the *altitude theorem* and the *cathetus theorem*

$$\boxed{a_c\, b_c = h^2} \quad \text{and} \quad \boxed{c\, a_c = a^2,\ c\, b_c = b^2.} \tag{4.2}$$

**Fig. 4.5** burial chamber, the only Cheops (Khufu) statue (12 cm), granite quarry Syene

In addition, the two catheti theorems incidentally yield

$$a^2 + b^2 = c\,(a_c + b_c) = c^2.$$

Thus, the "Pythagorean theorem" is shown in a second way.

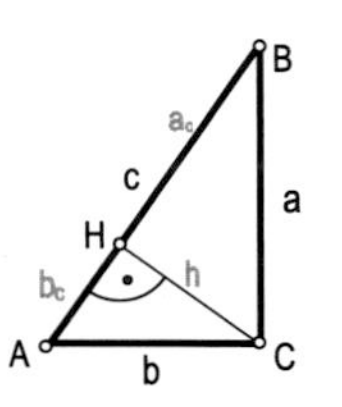


**Fig. 4.6** altitude and cathetus theorems

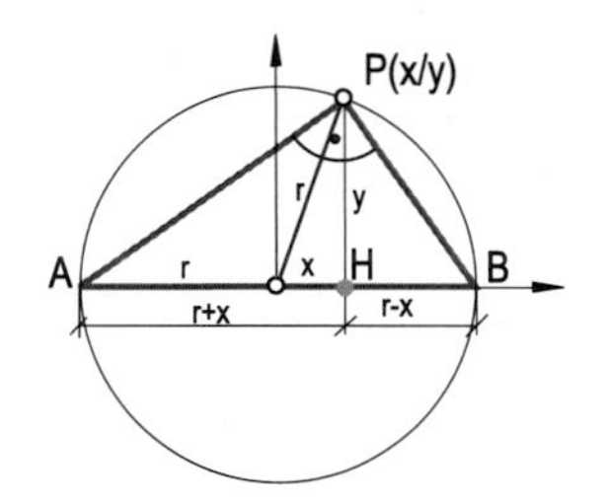

**Fig. 4.7** *Thales*'s theorem

**▸▸▸ Application**: ***construction of the golden ratio and approximate construction of the number*** $\pi$ (Fig. 4.8)

Explain the two specified constructions of $\frac{1+\sqrt{5}}{2}$ in *Euclid*'s figure (see Application p. 37) or the approximate construction of $\pi$ up to two decimal places using the method of *d'Ocagne*.

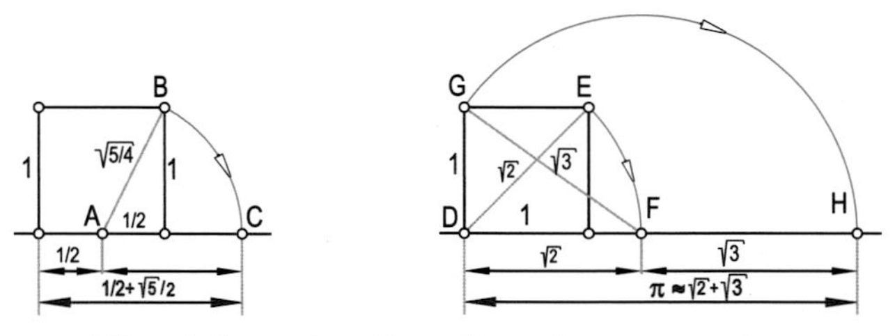


**Fig. 4.8** construction of two famous numbers

*Solution*:

For the construction of the golden ratio, we have

$$\overline{AB}^2 = (1/2)^2 + 1^2 = 5/4 \Rightarrow \overline{AC} = \overline{AB} = \sqrt{5}/2.$$

In order to verify the approximate construction of $\pi$, we calculate

$$\overline{DE}^2 = 1^2 + 1^2 = 2 \Rightarrow \overline{DF} = \overline{DE} = \sqrt{2},$$

$$\overline{FG}^2 = \overline{DF}^2 + 1^2 = 3 \Rightarrow \overline{FH} = \overline{FG} = \sqrt{3},$$

$$\Rightarrow \overline{DH} = \overline{DF} + \overline{FH} = \sqrt{2} + \sqrt{3} \approx 3.146 \approx \pi \quad (\text{Error} \approx 0.15\%).$$

◂◂◂

Furthermore, the fundamental theorem holds:

*Thales's theorem*: Each angle in a semicircle is a right angle.

***Proof***: Even this significant theorem – Thales (ca. 630 – 550 BC) was one of the first Greek mathematicians – has already been proved in numerous ways. One of these proofs indicated the equation as a special case of the "angle of circumference theorem" (Fig. 4.52). Here is another proof (Fig. 4.7):

The triangles $AHP$ and $BHP$ are right: Thus, the Pythagorean theorem applies twice:

$$\overline{AP}^2 = \overline{AH}^2 + \overline{HP}^2 = (r+x)^2 + y^2,$$

$$\overline{BP}^2 = \overline{BH}^2 + \overline{HP}^2 = (r-x)^2 + y^2.$$

Consequently,

$$\overline{AP}^2 + \overline{BP}^2 = (r^2 + 2rx + x^2 + y^2) + (r^2 - 2rx + x^2 + y^2) =$$

$$= 2\underbrace{(x^2 + y^2)}_{r^2} + 2r^2 = 4r^2 = \overline{AB}^2.$$

The Pythagorean theorem also applies to the triangle $ABP$. Thus, the triangle has a right angle. ⊙

▸▸▸ **Application**: ***road bridge*** (Fig. 4.9)

Bridges often have arched structures. These arcs may be circular, parabolic, or "catenary shaped" (according to statics, the catenary below forms a "support line arc" – see Application p. 430). In our case, let the arc be circular (which has the advantage that it includes congruent components). Given the span $s$ and a maximum height $h$, we want to find the circle radius $r$ and the column height $t$ at a distance $x$.

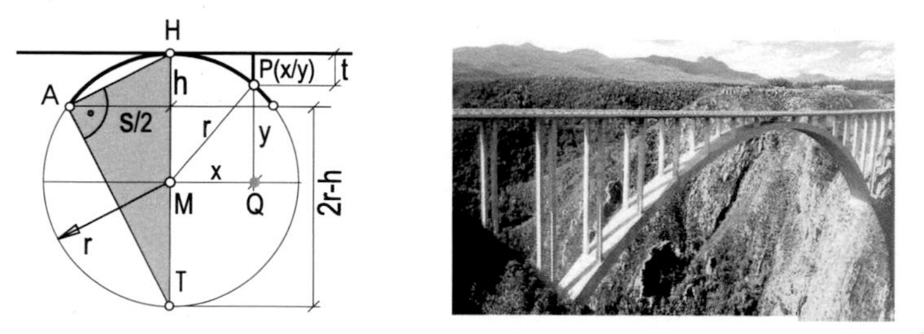


**Fig. 4.9** Bloukrans River Bridge, South Africa

*Solution*:
Following *Thales*'s theorem, the triangle $AHT$ is right, so the altitude theorem (geometric mean theorem) is applicable $(s/2)^2 = h(2r - h)$. By rearranging the theorem, we can now calculate $r$

$$\frac{s^2}{4} = 2rh - h^2 \Rightarrow r = \frac{s^2 + 4h^2}{8h}. \tag{4.3}$$

The pillar height at a distance $x$ results from the auxiliary rectangular triangle $PQM$ according to the Pythagorean theorem $t = r - y = r - \sqrt{r^2 - x^2}$. ◂◂◂

### ▸▸▸ Application: *tree of Pythagoras*

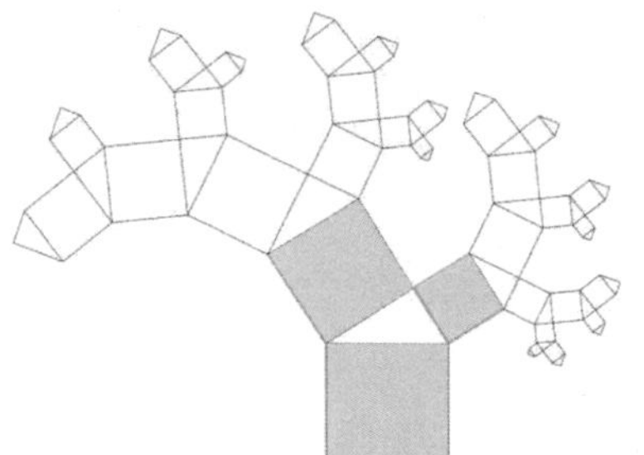

**Fig. 4.10** the tree of Pythagoras

Even in antiquity, repeating the Pythagorean figure in Fig. 4.1 with recursive scaling, i.e. fractal-like, as shown in Fig. 4.10, seemed like an obvious thing to do. The resulting figure shows no sharp contour line as the depth of the recursion grows. Zooming in does not change this. A nice generalization can be found in the model below (Fig. 4.11) designed by F. *Gruber* and G. *Wallner*.

**Fig. 4.11** a generalization of the tree of Pythagoras ◂◂◂

## 4.2 Radian measure

In mathematics, one rarely computes with the familiar units of degrees (the subdivision of the circle into 360 parts comes from the Babylonians), but rather with radians (*Arcus*):

> The radian $\varphi$ of an angle $\varphi^\circ$ is defined as the length of the arc on the unit circle corresponding to the central angle $\varphi^\circ$.

For $\varphi^\circ = 180^\circ$, we get the half circle with the length $\pi$. Thus,

$$\boxed{\varphi = \varphi^\circ \frac{\pi}{180^\circ}} \quad \text{or} \quad \boxed{\varphi^\circ = \varphi \frac{180^\circ}{\pi}.} \tag{4.4}$$

Specifically, the radian $\varphi = 1$ corresponds to the angle $\varphi^\circ = 180^\circ/\pi \approx 57^\circ$. For the corresponding circular arc on a circle of radius $r$, we have

$$\boxed{b = r\,\varphi.} \tag{4.5}$$

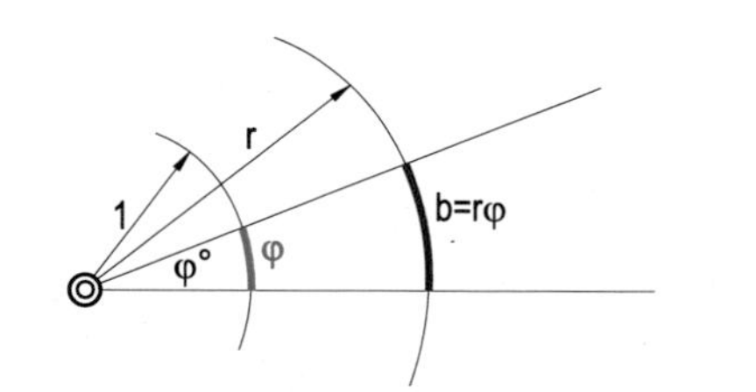


**Fig. 4.12** radians

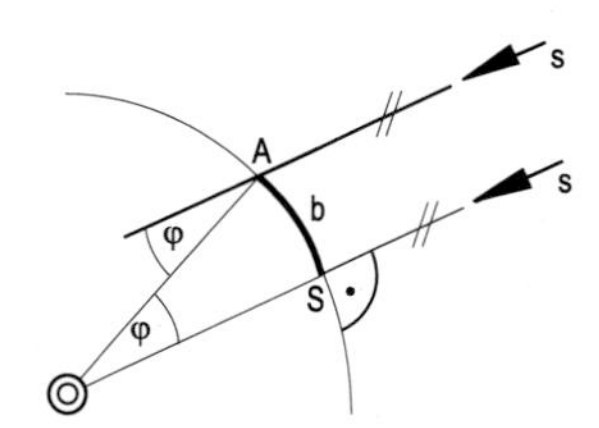


**Fig. 4.13** calculation of the circumference of the Earth

▸▸▸ **Application**: ***calculation of the circumference of the Earth*** (Fig. 4.13)
Even the ancient Greeks knew that the Earth must have a spherical shape (this knowledge was then lost for a long time). Already in the 3rd century BC, *Eratosthenes* was able to make an astonishingly accurate calculation of the Earth's circumference (!), and he did so as follows: The ancient city of Syene was located almost exactly on the Tropic of Cancer (near the present-day Aswan). It was probably well-known that there was a well in which, only at noon on the day of the summer solstice, one could see the Sun directly above one's head. That is, precisely on that day, at that time, and at that particular place, the Sun's rays hit the ground of the well vertically. By means of amazingly accurate measurements, *Eratosthenes* found out that, at the same time in Alexandria (800 km north of Syene), the Sun's rays had an angle of incidence of 7.2° (measured from the vertical). Assuming that the surface of the Earth is spherical, the following simple proportion can be

determined:
$b = r\varphi \Rightarrow 800\,\text{km} = r\left(7.2° \frac{\pi}{180°}\right) \Rightarrow r \approx 6,370\,\text{km} \Rightarrow U \approx 40,000\,\text{km}.$

⊕ *Remark*: There had been indications that the Earth has a spherical shape since ancient times: Navigators who sailed toward the equator along the West African coast gave accounts of an ever-changing starry sky. The calculation of *Eratosthenes* was obviously not a "proof" of the spherical shape, but demonstrated that the surface must be curved. Empirical evidence can be provided as follows: The first assumption was that the Sun's rays are always "parallel".
In reality, the Sun is not a point light source and the various light rays form an angle of up to half a degree. If an objects casts a shadow, it is framed by a "penumbra aura" which becomes easier to see the farther the object is away. For the sake of simplicity in our calculations, we want to assume that the Sun's rays are "parallel".
If we draw comparisons now with many other places, we will, of course, get the same result. Even then, the Earth could theoretically still have the shape of a "spindle torus" which arises when a circle rotates about one of its chords (the sphere is formed by rotating a circle about one of its diameters). You ought to consider that photographs of the Earth from space have only recently been available! ⊕ ◂◂◂

▸▸▸ **Application**: ***What is a thumb width?*** (Fig. 4.14)
To communicate the positions of objects that are far apart, we use terms like "thumb widths", as in: "Do you see the black spot two thumb widths next to the striking group of trees". Knowing the approximate distance $d$ of the object, we can estimate its size. What is the size of an object corresponding to "one thumb width" at $d$ m distance?

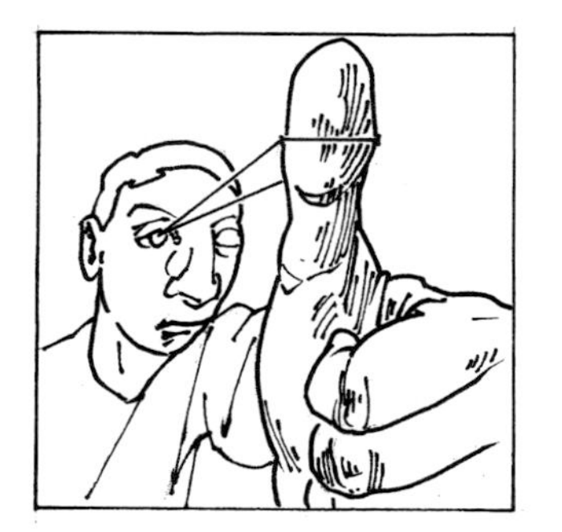

**Fig. 4.14** a "thumb's width"

**Fig. 4.15** How far is it to the plane?

*Solution*:
With an outstretched arm, our thumb is $r = 60\,\text{cm}$ away from our non-squinting eye. The thumb by itself is about $b = 2\,\text{cm}$ wide. The radian of the visual angle $\varphi$ is, therefore, $\varphi = \dfrac{b}{r} \approx \dfrac{1}{30}$ (the viewing angle does not change much for bigger or smaller people, because even arm length and width remain roughly proportional to the thumb). This applies to the width of the "thumb-wide" object at a distance $d$ meters, and thus, we have

$$D = \frac{d}{30}\,\text{m}.$$

Conversely, we can estimate the size of an object whose distance is known: A house may have one side of length 12 m and may be "half of a thumb's width" high. From this, it follows that the house is $d = 720\,\text{m}$ away, since $\frac{12\,\text{m}}{d} = \frac{1}{2} \cdot \frac{1}{30}$.

⊕ *Remark*: One can also work with "palms": One palm (the width of the hand at the transition of the metacarpus to the fingers) has a viewing angle of 8°, regardless of age and gender.

The Sun and the Moon appear about less than half a degree in the firmament. If the aircraft in Fig. 4.15 is 100 m long, then since $100\,\text{m} = \frac{0.4°\pi}{180°}\,r$, the aircraft must be about $r = 14\,\text{km}$ away from the observer. ⊕ ◂◂◂

### ▸▸▸ Application: *How long is a mile?*

*Solution*:

The length of a nautical mile is defined as the length of one arc minute (1/60 of a degree) on the equator. Therefore, 40,000 km corresponds to $360° = 360 \cdot 60'$ and 1 nautical mile corresponds to $40{,}000\,\text{km}/360/60 = 1.852\,\text{km}$.

⊕ *Remark*: This unit at sea is actually very useful: Angles to stars can easily be measured on the high seas. Ships move, and when they go "straight", the trajectory is seen locally as a great circle on the globe, that is, a circle with the same circumference as the Earth.

One kilometer (1,000 m) was previously defined in terms of the circumference of the Earth, namely as 1/40,000 of it. Since 1983, the definition of the unit of length has depended on the definition of the second: A meter is the distance that light traverses in 1/299,792,458 of a second within vacuum. ⊕ ◂◂◂

### ▸▸▸ Application: *curved perspective* (Fig. 4.16)

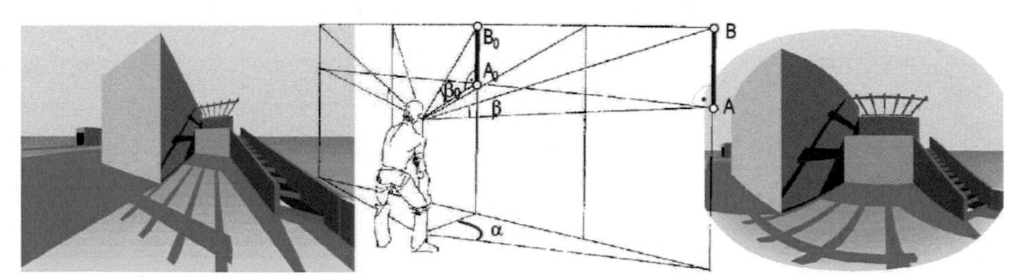


**Fig. 4.16** left: classic view; right: curved perspective

If we place ourselves close to a "big" object, for example, in front of a wall (Fig. 4.16, middle), then the visual angles (in the picture denoted by $\beta$) change for the equally long segments $AB$ and $A_0B_0$ depending on our distance to the object. If we interpret the visual angles $\alpha$ and $\beta$ as new units of length, we get a completely different perspective image of a scene (Fig. 4.16, right) than we would obtain by classical photography (Fig. 4.16, left). It corresponds to the cylindrical projection of the image projected onto the spherically curved retinal image – or the subjective perception when the object is "scanned" by a rolling of the eyeball. Viewed mathematically, instead of plotting metric lengths, we simply lay off the arc length of the azimuth angle $\alpha$ and the elevation angle $\beta$. The vertical rod $AB$ as shown in Fig. 4.16 (left) may be located at $\alpha = 60°$ to the right of the principal direction. The lower end point $A$ lies in the horizontal plane through the eye (horizon). The bar appears at an elevation angle of $\beta$. Which coordinates do the endpoints of the perspective image of the rod have?

**Fig. 4.17** *Degas* painted a curved perspective.

*Solution*:
The radians of $\alpha$ and $\beta$ are

$$\alpha = 60 \cdot \frac{\pi}{180} = \frac{\pi}{3} \approx 1.047 \quad \text{and} \quad \beta = 15 \cdot \frac{\pi}{180} = \frac{\pi}{12} \approx 0.262.$$

Thus, $A$ and $B$ have the coordinates $A(1.047/0)$, $B(1.047/0.262)$. The image can of course be enlarged by scaling all of its lengths. Applying such a coordinate transformation to all image points, one obtains a curved perspective as in Fig. 4.16 (bottom right) or in Fig. 4.17 (right).

⊕ *Remark*: Some famous artists used this technology in order to produce curved perspectives (Fig. 4.17). At the expense of linearity, this would allow one to avoid an extreme perspective distortion (see also Fig. 4.16, right).
Insects can see by conical facets. Therefore, they preferably measure objects in angles. The insect's visual impression of a scene seen with compound eyes (Fig. 4.18) is expected to correlate to the curved perspective just described.
Other curved perspectives are produces by so-called "fisheye lenses" (Fig. 4.19, left; Fig. 2.2; Fig. 4.42), or by reflections in curved surfaces, especially in spheres (Fig. 4.19, right). ⊕ ◂◂◂

▸▸▸ **Application**: ***area of a spherical triangle*** (Fig. 4.20)
Let $ABC$ be a spherical triangle, bounded by three great circles (center =

**Fig. 4.18** Compound eyes produce curved perspectives.

**Fig. 4.19** fisheye perspective, perspective in the eye of the beholder

center of the sphere). The sum of the three angles $\alpha$, $\beta$, and $\gamma$ is always greater than $180^\circ$. Measured in radians, the "spherical excess" $\alpha+\beta+\gamma-\pi$ is then exactly the area of the doubly curved triangle if the sphere has radius 1 (otherwise, the area increases with the square of the radius). Verify this formula for the blue area in the right image of Fig. 4.20.

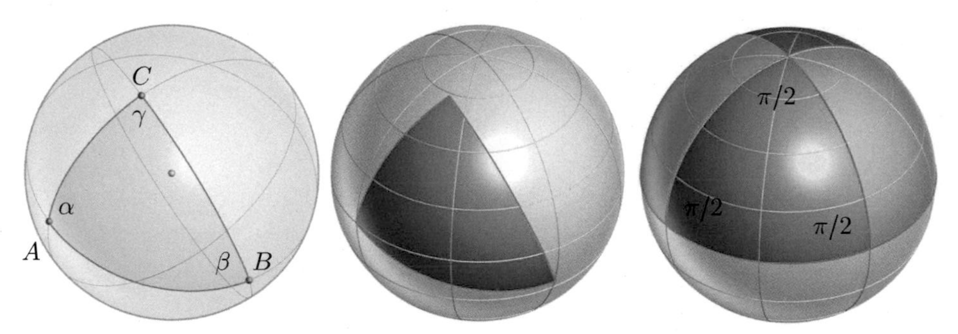


**Fig. 4.20** The area of a spherical triangle equals the "spherical excess".

*Solution*:
In the special case of the right image, all three angles are $90^\circ$, i.e. $\pi/2$. The

spherical excess is then $3\pi/2 - \pi = \pi/2$. The triangle covers 1/8 of the sphere's surface, which is (with $r = 1$) $4\pi/8 = \pi/2$.

⊕ *Remark*: Areas on the sphere are quite essential, because our planet is spherical. In order to determine the area of a continent, one has to work with "spherical triangulations". ⊕ ◂◂◂

▸▸▸ **Application**: ***How often does Greenland fit into Africa?*** (Fig. 4.21)
If you ask people to estimate estimate, the solutions offered to this question tend to be of the order of "four to six times" or so. The correct answer, however, is 15 times! This has to do with the fact that Africa is "around the equator" and Greenland is far North. The rectangular maps of the Earth distort Greenland much more than Africa.

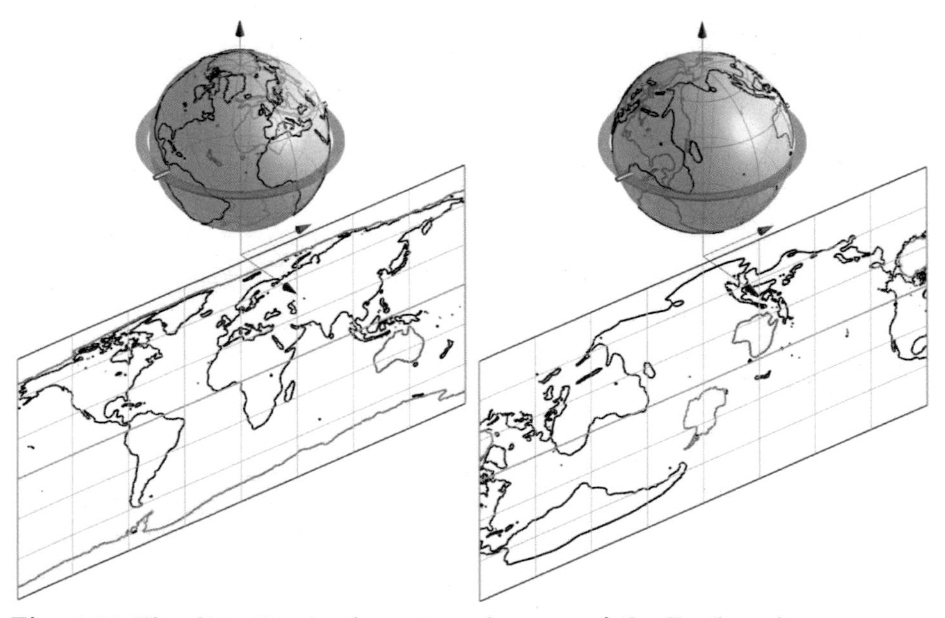

**Fig. 4.21** The distortion in the rectangular map of the Earth makes area comparisons hard. In the right image, another great circle takes over the role of the equator. Both Africa and Greenland are close to the equator, and the size comparison is more realistic. Compare also Australia and Antarctica in the left and the right rectangular map.

⊕ *Remark*: Even on the "usual" rectangular map, one can see that Africa and South America were once joined together. This, however, is only possible since both continents are close to the equator. Australia and Antarctica used to be one continent as well, but this cannot be seen on such extremely distorted maps. ⊕ ◂◂◂

## 4.3 Sine, cosine, tangent

The trigonometric functions sine, cosine, and tangent play major roles in mathematics. They are the basis of all calculations based on angle measurement.

Let us consider the unit circle: A point $P$ on the circle can be described in the Cartesian coordinate system with the coordinates $(x, y)$. The corresponding polar angle is $\varphi$ (Fig. 4.22). Then, we define:

$$\boxed{x = \cos\varphi, \quad y = \sin\varphi, \quad \tan\varphi = \frac{y}{x} = \frac{\sin\varphi}{\cos\varphi}.} \tag{4.6}$$

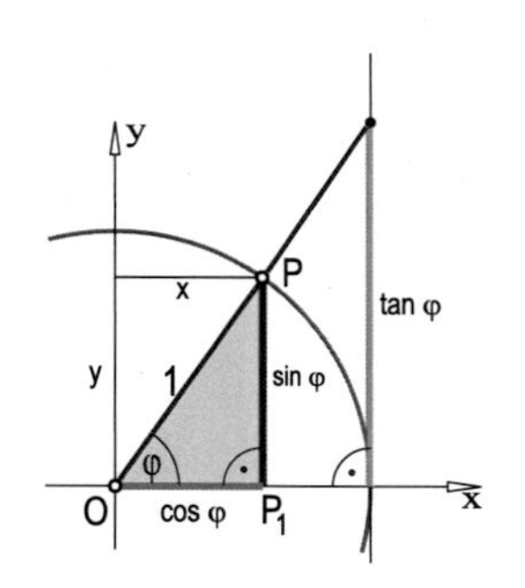


**Fig. 4.22** sine, cosine, tangent

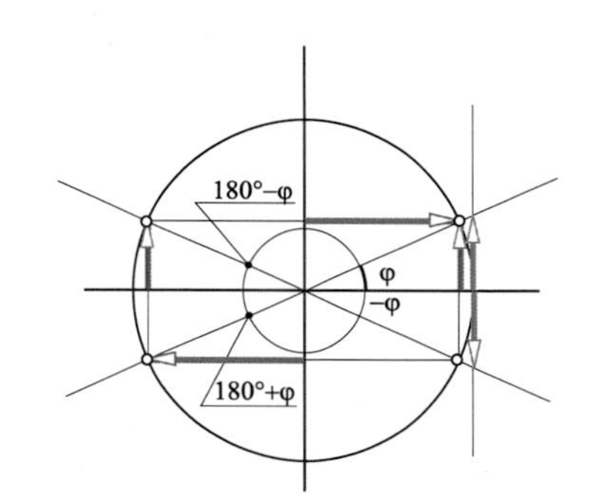


**Fig. 4.23** $\sin(\pm\varphi)$, $\sin(180° \pm \varphi)$

The Pythagorean theorem implies the following important relation:

$$\boxed{\sin^2\varphi + \cos^2\varphi = 1.} \tag{4.7}$$

The ratio $\tan\varphi : 1 = \sin\varphi : \cos\varphi$ shows that, on the plane, the line parallel to the $y$-axis (vertical) on the right is tangent to the circle at $x = 1$ (see Fig. 4.22).

The cotangent function is rarely used (because there is no such button available on pocket calculators). However, it is only the reciprocal of the tangent:

$$\cot\varphi = \frac{1}{\tan\varphi} = \frac{\cos\varphi}{\sin\varphi}.$$

From Fig. 4.23, the following relations can be read off:

$$\begin{aligned} \sin(180° - \varphi) &= \phantom{-}\sin\varphi, & \sin(-\varphi) &= -\sin\varphi, \\ \cos(180° - \varphi) &= -\cos\varphi, & \cos(-\varphi) &= \phantom{-}\cos\varphi, \\ \tan(180° - \varphi) &= -\tan\varphi, & \tan(-\varphi) &= -\tan\varphi. \end{aligned} \tag{4.8}$$

We now consider the right triangle $OP_1P$ in Fig. 4.22. It is built

- by the cathetus $A$ with length $x$ adjacent to the angle $\varphi$,
- by the cathetus $G$ of length $y$ opposite to the angle $\varphi$, and
- by the hypotenuse $H$ with length 1.

The following holds true:

$$\boxed{\cos\varphi = \frac{A}{H}, \quad \sin\varphi = \frac{G}{H}, \quad \tan\varphi = \frac{G}{A}.} \tag{4.9}$$

These equations are valid in any right triangle with a leg angle $\varphi$. Consider similar triangles!

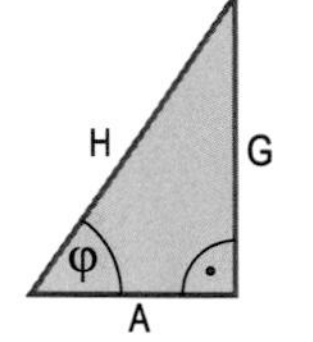


Thus, in *right triangles*, we have

$$\boxed{A = H\cos\varphi, \quad G = H\sin\varphi.} \tag{4.10}$$

Furthermore, we can apply $\alpha = 90° - \beta$. By swapping the notations, we obtain the important relations

$$\boxed{\cos\alpha = \sin(90° - \alpha) \quad \text{and} \quad \sin\alpha = \cos(90° - \alpha).} \tag{4.11}$$

In particular, since $\cos 45° = \sin 45°$ and $\cos^2 45° + \sin^2 45° = 1$, we have

$$\cos 45° = \sin 45° = \frac{1}{\sqrt{2}}. \tag{4.12}$$

By virtue of half of the equilateral triangle, we immediately recognize

$$\cos 60° = \sin 30° = \frac{1}{2},$$

and thus, since $\cos^2 60° + \sin^2 60° = 1$, we infer

$$\sin 60° = \cos 30° = \frac{\sqrt{3}}{2}.$$

Overall, we have now obtained the following important values that should be memorized quickly:

| $x$ | $0°$ | $30°$ | $45°$ | $60°$ | $90°$ |
|---|---|---|---|---|---|
| $\cos x$ | 1 | $\sqrt{3}/2$ | $\sqrt{2}/2$ | $1/2$ | 0 |
| $\sin x$ | 0 | $1/2$ | $\sqrt{2}/2$ | $\sqrt{3}/2$ | 1 |
| $\tan x$ | 0 | $1/\sqrt{3}$ | 1 | $\sqrt{3}$ | $\infty$ |

(4.13)

A "memory hook" for the sign-values is the following table (the cosines are obtained by reading the table backwards):

| $x$ | $0°$ | $30°$ | $45°$ | $60°$ | $90°$ |
|---|---|---|---|---|---|
| $\sin x$ | $\sqrt{0}/2$ | $\sqrt{1}/2$ | $\sqrt{2}/2$ | $\sqrt{3}/2$ | $\sqrt{4}/2$ |

We do not have to know the following *(harmonic) addition theorems* (the Prosthaphaeresis Formulas) by hcart, but they will be of importance when we deal with rotations in Section 7:

$$\begin{aligned} \sin(\alpha+\beta) &= \sin\alpha\cos\beta+\cos\alpha\sin\beta, \\ \cos(\alpha+\beta) &= \cos\alpha\cos\beta-\sin\alpha\sin\beta. \end{aligned} \tag{4.14}$$

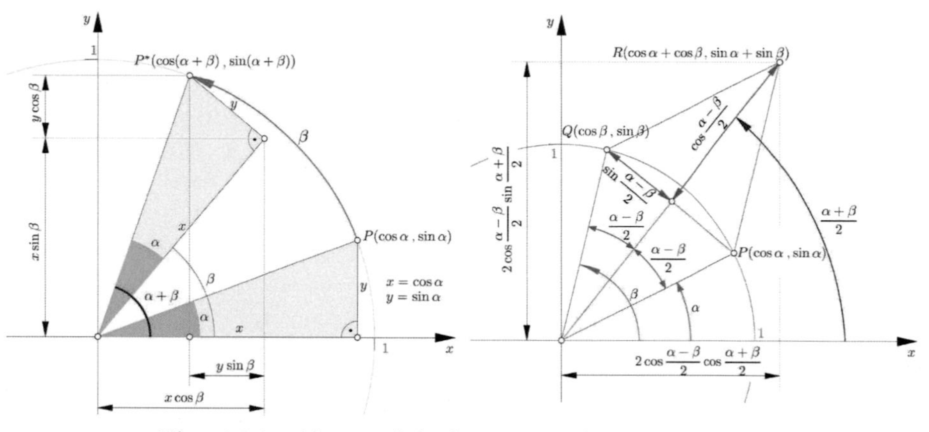


**Fig. 4.24** evidence of the harmonic addition theorems

***Proof***: Let $P_\alpha(\cos\alpha/\sin\alpha)$ and $P_\beta(\cos\beta/\sin\beta)$ be points on the unit circle whose radii subtend angles $\alpha$ and $\beta$ with the $x$-axis. Now we turn to $P_\beta$ through the angle $\alpha$ and get $P_{\alpha+\beta}(\cos(\alpha+\beta)/\sin(\alpha+\beta))$. According to Fig. 4.24 (left), its coordinates are

$$P_{\alpha+\beta}(\cos\alpha\cos\beta-\sin\alpha\sin\beta/\sin\alpha\cos\beta+\cos\alpha\sin\beta). \qquad \odot$$

▸▸▸ **Application**: ***trigonometric formulas***

Use the addition theorems in order to show

$$\cos 2x = 1-2\sin^2 x, \quad \text{and thus,} \quad \sin\frac{x}{2} = \pm\sqrt{\frac{1}{2}(1-\cos x)}. \tag{4.15}$$

***Proof***:

$$\cos 2x = \cos(x+x) = \cos^2 x - \sin^2 x = (1-\sin^2 x) - \sin^2 x = 1-2\sin^2 x$$

$$\Rightarrow \sin x = \sqrt{\frac{1}{2}(1-\cos 2x)} \Rightarrow \sin\frac{\alpha}{2} = \sqrt{\frac{1}{2}(1-\cos\alpha)}.$$

The actual names of the variables do not matter. $\odot$ ◂◂◂

▸▸▸ **Application**: ***How were the tables for trigonometric functions created in early times?***

It is from a historical point of view that this question is of particular interest. Without these tables, only a few decades ago, we would not have been able to make exact calculations! The ancient Greeks used to create tables for the trigonometric functions which were afterwards taken over by Arabic mathematicians who refined them. Even the Vikings are said to have determined the latitude at sea with tangent tables.
We will discuss how calculators and computers calculate values in the section on differential calculus and series expansion (Application p. 382).

*Solution*:
From the important values given in Table 4.13, we compute intermediate values using the addition theorems (4.14):

$$\sin 15^\circ = \cos 75^\circ = \sin(60^\circ - 45^\circ) \quad \text{and} \quad \sin 75^\circ = \cos 15^\circ = \sin(90^\circ - 15^\circ).$$

Through transition to the half angle formula (4.15), we get new values, which, in turn, means that the addition theorems lead to new values, etc. This was a lot of work, but one could determine arbitrarily many intermediate values in this way. If you wanted a more precise value, you could *interpolate* linearly between two neighboring values (the functions can only be replaced by a straight line in very small intervals). The tangent values are derived as the quotient of the sine and cosine. ◂◂◂

Sometimes, you need the *Simpson formulas (second addition theorems)*

$$\begin{aligned}\sin\alpha + \sin\beta &= 2\sin\frac{\alpha+\beta}{2}\cos\frac{\alpha-\beta}{2},\\ \cos\alpha + \cos\beta &= 2\cos\frac{\alpha+\beta}{2}\cos\frac{\alpha-\beta}{2}.\end{aligned} \tag{4.16}$$

***Proof***: By virtue of Fig. 4.24, right: The points $O$, $P$, $R$, and $Q$ form a rhombus with side length 1. $R$ initially has the coordinates $R(\cos\alpha+\cos\beta/\sin\alpha+\sin\beta)$. Half the diagonal $OM$ has the length $\cos\frac{\alpha-\beta}{2}$. This also allows the coordinates of $R$ to be specified by

$$R\left(2\cos\frac{\alpha-\beta}{2}\cos\frac{\alpha+\beta}{2}\Big/2\cos\frac{\alpha-\beta}{2}\sin\frac{\alpha+\beta}{2}\right)$$ ⊙

▸▸▸ **Application**: ***three-phase current*** (Fig. 4.25)

In the generation of three-phase currents, a magnetic rotor rotates within three induction coils and induces three dephased alternating currents. These may be used individually or jointly. Calculate the sum of all voltages $y = \sin x + \sin(x + \frac{2\pi}{3}) + \sin(x + \frac{4\pi}{3})$ and the voltage difference between two consecutive phases $y = \sin x - \sin(x + \frac{2\pi}{3})$ (see Application p. 177).

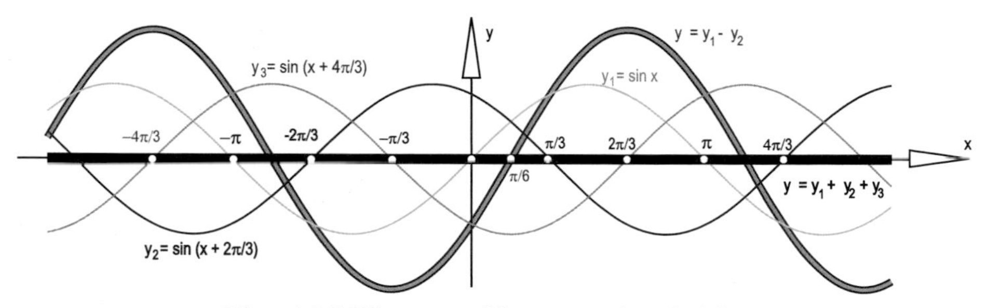


**Fig. 4.25** Waves amplify or cancel each other.

*Solution*:
We show that the sum of two phase-shifted sinusoids (displacement $x_0$, equal amplitude and frequency) is again a sinusoid of the same frequency but of a different amplitude:
With the second addition theorems, we get

$$\sin x + \sin(x + x_0) = 2 \sin \tfrac{x+(x+x_0)}{2} \cos \tfrac{x-(x+x_0)}{2} = 2 \cos \tfrac{-x_0}{2} \sin(x + \tfrac{x_0}{2}).$$

For $x_0 = \frac{2\pi}{3}$ ($\Rightarrow 2\cos\frac{-\pi}{3} = 1$), we find the relation

$$\sin x + \sin(x + \tfrac{2\pi}{3}) = \sin(x + \tfrac{\pi}{3}).$$

If $\sin x = -\sin(x + \pi)$, then $\sin(x + \frac{\pi}{3}) = -\sin(x + \frac{4\pi}{3})$ and the sum is equal and opposite to the third function. The total sum will vanish:

$$\sin x + \sin\left(x + \frac{2\pi}{3}\right) + \sin\left(x + \frac{4\pi}{3}\right) = 0. \tag{4.17}$$

The voltage difference between two neighboring phases is calculated by

$$\sin x - \sin(x + x_0) = \sin x + \sin(-x - x_0) =$$
$$= 2 \sin \tfrac{x-(x+x_0)}{2} \cos \tfrac{x+(x+x_0)}{2} = 2 \sin \tfrac{-x_0}{2} \cos(x + \tfrac{x_0}{2}).$$

For $x_0 = \frac{2\pi}{3}$, we have an increase in the amplitude by a factor of $2\sin(-\frac{\pi}{3}) = -\sqrt{3}$. The voltage difference is described by

$$y = -\sqrt{3} \cos\left(x + \frac{\pi}{3}\right) = -\sqrt{3} \sin\left(\frac{\pi}{2} - \left(x + \frac{\pi}{3}\right)\right) = \sqrt{3} \sin\left(x - \frac{\pi}{6}\right).$$

In individual voltages with 230 volts, this results in a maximum voltage difference of about 400 volts. ◂◂◂

▸▸▸ **Application**: ***inclined plane*** (Fig. 4.26)
In order to avoid picking up a barrel of 100 kg one can roll it on a ramp, that is, roll it on a board with an upward inclination of $\alpha = 30°$. What force $F = |\vec{F}|$ is necessary (friction ignored) to compensate for the driving component $T = |\vec{T}|$ of weight $G$?
*Solution*:

$$T = F = G \sin\alpha, \; G = m\,g = 100\text{kg} \cdot 9.81\text{m/s}^2 = 981\,\text{N},$$

$$\Rightarrow F = 981\text{N} \cdot \frac{1}{2} = 490.5\,\text{N}.$$ ◂◂◂

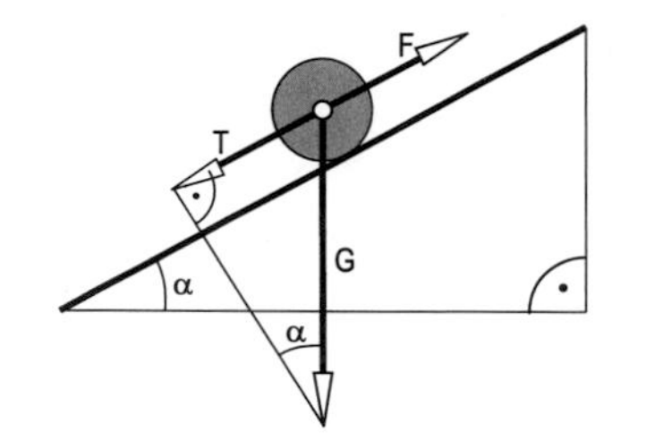

**Fig. 4.26** inclined plane in theory ...

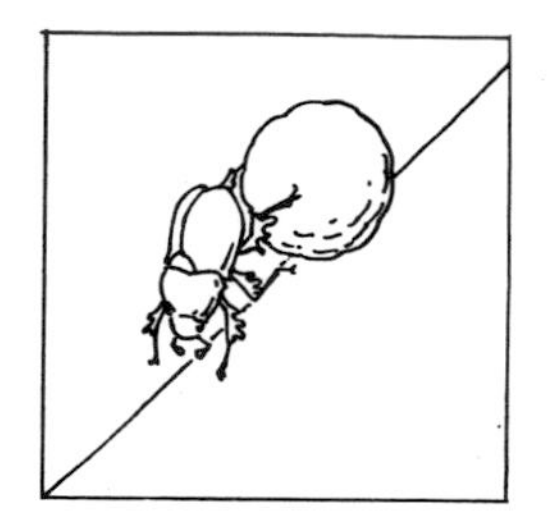
**Fig. 4.27** ... and in practice

▸▸▸ **Application**: ***pyramid ramps***
The ancient Egyptians certainly used inclined planes ("ramps") to lift heavy stone blocks (i.e. to draw them on sledges). In Application p. 74, we have seen that by far the greatest amount of cuboid stone blocks was used to build the the lower layers of the pyramids. For these lower layers, a ramp was obviously the best choice in order to move the blocks to the (still growing) height of the truncated pyramid. Ramps for greater altitudes were practically feasible. Obviously, the older architects worked with much more sophisticated methods (a "tipper" amongst other things).
The quarry for the Cheops pyramid was about 400 m south of the pyramid. What inclination angle should this ramp have had in order to raise limestone blocks to an 8% higher height? How long would such a ramp be to transport the last block of the pyramid, the "pyramidion", to the top of the Great Pyramid?
*Solution*:
The slope of the ramp (inclination angle $\alpha$) is $\tan\alpha = 8 : 100$. For the 400 m long baseline, this results in a lift of $400 \cdot \frac{8}{100}\,\text{m} = 32\,\text{m}$. After all, this is a good fifth of the height, and about half of all blocks are positioned at that height. However, to transport even the pyramidion on the ramp, one would need a five times longer baseline (2,000 m). Just the dams of the ramp would have required a multiple of the volume of the whole pyramid! ◂◂◂

▸▸▸ **Application**: ***cable forces*** (Fig. 4.28)
A symmetric wedge with the tip angle $\alpha$ is pressed with the force $F$ into a padding. How big are the normal forces $F_1$ and $F_2$ against the flanks of the wedge?
*Solution*:

$$F_1 = F_2 = \frac{\frac{F}{2}}{\sin\varphi} = \frac{F}{2\sin\frac{\alpha}{2}}.$$

Therefore, $F_1$ increases as $\sin\varphi$, i.e. $\alpha$ decreases. ◂◂◂

▸▸▸ **Application**: ***V-belt*** (Fig. 4.30)
Let the radii be $R$ and $r$ and the central distance $d$ of two circles be given. How long is the belt?

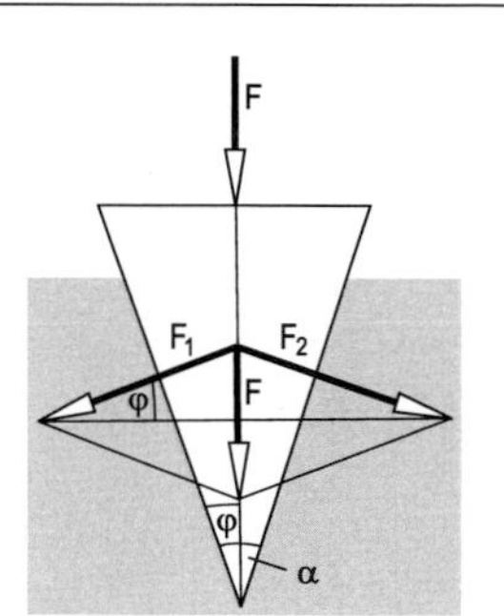


**Fig. 4.28** cable forces in theory ...

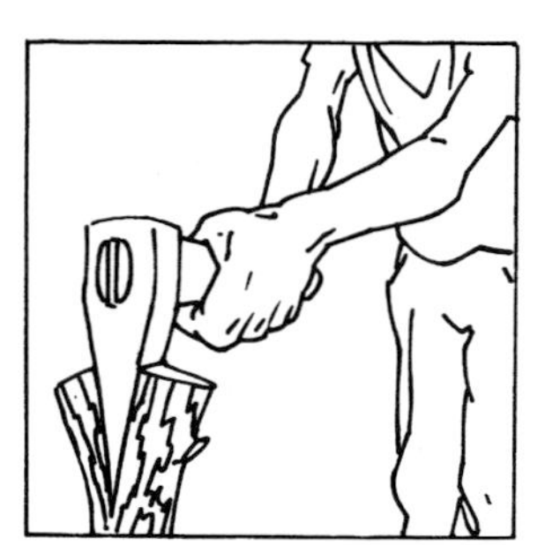

**Fig. 4.29** ... and in practice

*Solution*:
Though a seemingly simple example, we will need a little trick that must be "seen": By a parallel displacement of the common tangent segment of the two circles, we obtain a right triangle, to which the Pythagorean theorem applies: $t = \sqrt{d^2 - (R-r)^2}$. Furthermore, we have $\sin\varphi = (R-r)/d \approx \varphi$ (in radians!). Thus, the belt length is determined as

$$L = 2\left[R\left(\frac{\pi}{2} + \varphi\right) + t + r\left(\frac{\pi}{2} - \varphi\right)\right]. \quad ◂◂◂$$

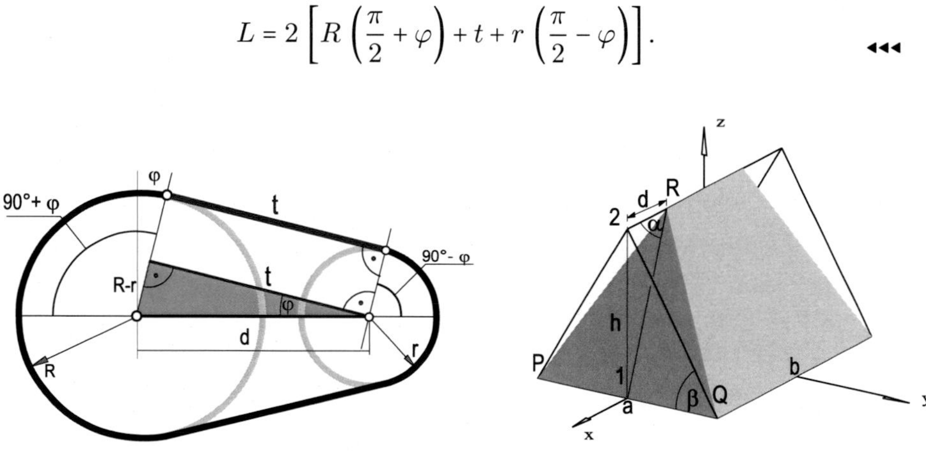


**Fig. 4.30** V-belt

**Fig. 4.31** hipped roof

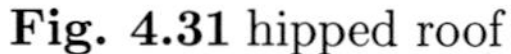

▸▸▸ **Application**: ***hipped roof*** (Fig. 4.31)
Calculate the volume of the space bounded by the hipped roof with eaves of given lengths $a$, $b$ and given roof slope angles $\alpha$, $\beta$.
*Solution*:
The altitude of the triangle $Q12$ equals

$$h = \frac{a}{2}\tan\beta.$$

Thus, the area $A$ of the triangle $PQ2$ is

$$A = \frac{ah}{2}.$$

The volume equals the difference between, on the one hand, the volume of the three-sided prism with cross-sectional area $A$ and length $b$ and, on the other hand, twice the volume of the tetrahedron $PQ2R$ whose base triangle $PQ2$ also has the area $A$, and whose altitude $R2$ has the length $d$. The altitude $d$ arises from the right triangle $12R$ with

$$d = h/\tan\alpha$$

$$\Rightarrow V = A\,b - 2\,\frac{A\,d}{3} = A\left(b - \frac{2d}{3}\right).$$ ◂◂◂

▸▸▸ **Application**: ***cornering a motorcycle*** (Fig. 4.32)
If a motorcycle (weight $G = m\,g$) with velocity $v$ travels around a curve (radius $r$), it must be tilted by an angle $\alpha$ to avoid being offset by the centrifugal force ($F = mv^2/r$). Show that $\alpha$ is independent of the mass?

**Fig. 4.32** cornering in theory **Fig. 4.33** and practice

*Solution*:
We have

$$\tan\alpha = \frac{F}{G} = \frac{v^2}{gr}$$

independent of the mass! In addition, the frictional force must be greater than $F$.

*Numerical examples*:
1) From the picture (Fig. 4.33, left), the inclination angle $\alpha$ and the curve radius $r$ are known. How fast was the moped traveling on the path of the garden way?
$\alpha = 15°,\ r = 10\,\mathrm{m} \Rightarrow v = \sqrt{g\,r\,\tan\alpha} \approx 5\,\mathrm{m/s} \approx 18\,\mathrm{km/h}$.
2) How big must the angle be if $v$ and $r$ are given as such?
$r = 70\,\mathrm{m},\ v = 72\,\mathrm{km/h} = 20\,\mathrm{m/s} \Rightarrow \tan\alpha \approx 0.58 \Rightarrow \alpha \approx 30°$.
3) The pictured extreme carver (Fig. 4.35) tacks a curve with a radius of $r = 5$ m at a speed of $v = 12$ m/s. This results in $\tan\alpha = 2.88$, which means that the board has to be inclined at an angle of 71° in order for the carver to maintain his balance. More than three times the Earth's gravity is exerted on the carver. "Touching the snow" as in the photo is a physical necessity, given a standard slope inclination of 20°. The athlete is forced to literally drag himself across the snow if he is to master the curve. ◂◂◂

**Fig. 4.34** The angle of inclination depends on the speed, but not on the mass!

**Fig. 4.35** extreme snowboarding

▸▸▸ **Application**: ***development of a cone of revolution*** (Fig. 4.36)
A cone of revolution (radius $r$, height $h$) transforms by development ("flattening") into a circular sector of radius $s$ with the central angle $\omega$. Compute $\omega$ in relation to half of the cone's angle $\alpha$ of aperture.

*Solution*:
According to Fig. 4.36, the arc length of the cone's base circle must match the arc of the circular sector. Therefore,

$$2\pi\, r = s\,\omega \Rightarrow \omega = 2\pi\,\frac{r}{s}.$$

With $\frac{r}{s} = \sin\alpha$, we obtain the formula

$$\boxed{\omega = 2\pi\,\sin\alpha \quad \text{or} \quad \omega^\circ = 360^\circ\,\sin\alpha.} \tag{4.18}$$

The second variant of the formula arises from the first through multiplication by $\frac{180}{\pi}$. ◂◂◂

▸▸▸ **Application**: ***crank slider (Otto-cycle engine)*** (Fig. 4.37, left)
The piston drives the crank pin $A$ via the pin $B$ and the push rod $AB$, and this causes the rotation of the crank shaft $L$. Calculate the distance $\overline{LB}$ as a

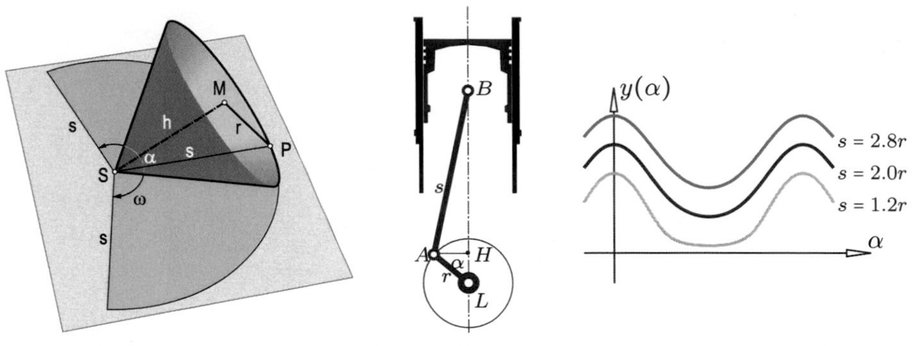


**Fig. 4.36** development of the cone of revolution

**Fig. 4.37** petrol engine, time-distance diagrams

function of the rotation angle $\alpha$. This distance is the key to the entire kinematic analysis of the transmission (velocity and acceleration distribution).

*Solution*:

The distance $y = \overline{LB}$ is, of course, dependent on the angle of rotation $\alpha$, and this, in turn, is proportional to the time $t$. Thus, we obtain the path-time diagram of $B$ via the equation $y = y\big(\alpha(t)\big) = y(t)$:

$$\overline{LB} = \overline{LH} + \overline{HB},$$

$$\overline{LH} = r\,\cos\alpha,\ \overline{AH} = r\,\sin\alpha,\ \overline{HB} = \sqrt{s^2 - \overline{AH}^2},$$

$$y = \overline{LB} = r\,\cos\alpha + \sqrt{s^2 - r^2\sin^2\alpha}.$$

Fig. 4.37 on the right shows such diagrams $y = y(\alpha)$ for different ratios $r : s$. For $s >> r$ (say: $s$ much bigger than $r$), we get

$$\sqrt{s^2 - r^2\sin^2\alpha} = s\sqrt{1 - \left(\frac{r}{s}\right)^2\sin^2\alpha} \approx s$$

and the path-time diagram approaches an ordinary sinusoid. The graphs of $y = \sin x$ and $y = \cos x$ differ only by parallel displacement along the $x$-axis by $\frac{\pi}{2}$. Therefore, in both cases, we are dealing with a sine curve

$$y = r\,\cos\alpha + s.$$

The demo program `otto_engine.exe` on this book's website was created using $y(\alpha)$. For mechanical engineers, an additional velocity diagram is still important. This is obtained, as we will see in the chapter on *differential calculus*, by means of the derivative function (Application p. 276). ◂◂◂

Quite often, we need the following formula:

$$\boxed{\sin 2\alpha = 2\sin\alpha\cos\alpha.} \qquad (4.19)$$

***Proof***: We consider the right triangle $ABC$ in Fig. 4.38 (with $\overline{AC} = 1$). It has the area

$$A_1 = \frac{1}{2} \sin\alpha \cos\alpha.$$

We reflect it in $AB$ and obtain an (isosceles) triangle $ACC^*$ of twice the area (area $A_2 = \sin\alpha\cos\alpha$). The height $CH$ in this triangle is $\sin 2\alpha$ and this allows us to calculate the area $A_2$ in the following way:

$$A_2 = \frac{1 \cdot \sin 2\alpha}{2} = \sin\alpha\cos\alpha = 2\,A_1.$$

⊙

▸▸▸ **Application**: ***area of a regular $n$-gon*** (Fig. 4.39)
Calculate the area of a regular $n$-gon with circumradius $r$.

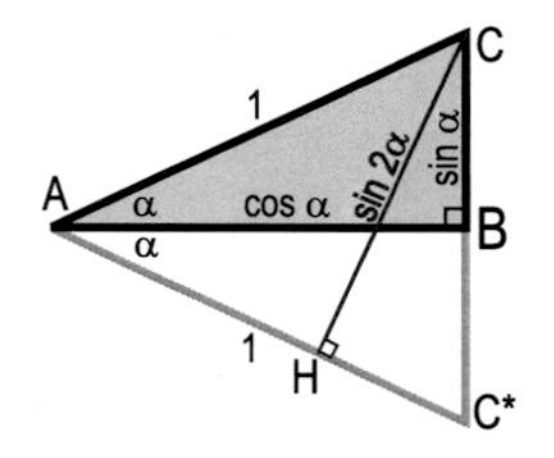


**Fig. 4.38** $\sin 2\alpha = 2\sin\alpha\cos\alpha$

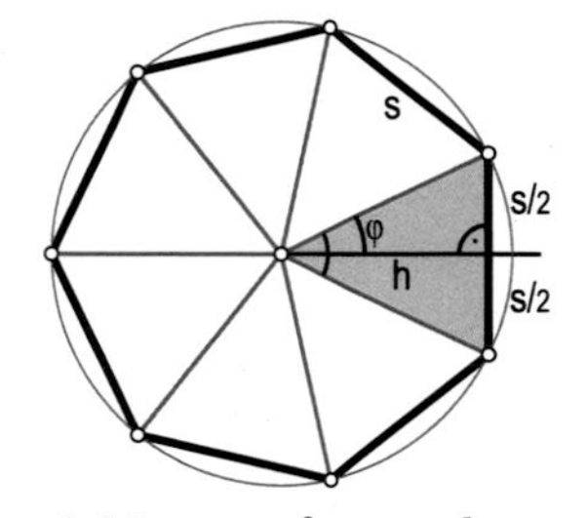


**Fig. 4.39** area of a regular $n$-gon

*Solution*:
The following holds:

$$\frac{s}{2} = r\,\sin\varphi,\ h = r\,\cos\varphi \quad \text{with} \quad \varphi = \frac{360°}{2n}.$$

From this, the area is computed

$$A = n \cdot \frac{s}{2} \cdot h = n\,r^2\,\sin\varphi\cos\varphi = n\,r^2 \frac{\sin 2\varphi}{2} \quad \Rightarrow$$

$$A = \frac{n\,r^2}{2} \sin\frac{360°}{n}. \tag{4.20}$$

⊕ *Remark*: We let $n$ "grow" to a large value. Then the area of the polygon "converges" to $\pi\,r^2$, that is, the area of the circle of radius $r$. Even *Archimedes*, more than 2,000 years ago, proceeded in a similar manner – namely, via the *circumference* of the regular $n$-gon – and approximated the number $\pi$ quite well.
If we use the "limit notation", then we have

$$A_\circ = r^2 \lim_{n\to\infty} \frac{n}{2} \sin\frac{360°}{n} = r^2\,\pi.$$

After changing to radians ($360° = 2\pi$), we arrive at the following by cutting out $r^2$ and dividing by $\pi$

$$\lim_{n\to\infty} \frac{n}{2\pi} \sin\frac{2\pi}{n} = 1.$$

If we write $x$ instead of $\frac{2\pi}{n}$ (so $x$ converges to 0, as $n$ tends to infinity), then we get an important relation that we need in differential calculus:

$$\lim_{x\to 0} \frac{1}{x} \cdot \sin x = \lim_{x\to 0} \frac{\sin x}{x} = 1. \tag{4.21}$$

In words:

For very small angles, the radians equal the corresponding sines. ⊕ ◂◂◂

▸▸▸ **Application**: ***upward inclined throw and maximum throw distance***
Calculate the height $h$ after $t$ seconds by modifying Formula (2.17) in Application p. 38:

$$h = G(t) - B(t) = (v_0 \sin\alpha)t - \frac{g}{2}t^2. \tag{4.22}$$

Herein, $v_0$ and $g$ are again top speed and acceleration due to gravity. The initial angle to the horizontal direction shall be denoted by $\alpha$. The air resistance is neglected. Calculate the throw with a given $v_0$ and $\alpha$. For which $\alpha$ can the best throw be achieved?

**Fig. 4.40** inclined upward throw

*Solution:*
By solving the quadratic equation (4.22), we obtain

$$t_{1,2} = \frac{1}{g}\left(v_0 \sin\alpha \pm \sqrt{v_0^2 \sin^2\alpha - 2gh}\right).$$

The highest point is reached for $D = v_0^2 \sin^2\alpha - 2gh = 0$. Thus,

$$\text{at time } t_0 = \frac{v_0 \sin\alpha}{g} \text{ at the height } H = \frac{v_0^2 \sin^2\alpha}{2g}.$$

The total time of the throw until the impact in the plane $h = 0$ equals $2t_0$. For the horizontal component (the $x$-coordinate of the thrown object), the following applies: $x(t) = t\, v_0 \cos\alpha$. From the total time of the throw, the throw distance is computed as

$$w = 2t_0 v_0 \cos\alpha = \frac{2\sin\alpha\cos\alpha\, v_0^2}{g} = \frac{1}{g} v_0^2 \sin 2\alpha.$$

**Fig. 4.41** *Galileo*'s experiment **Fig. 4.42** A high long jump or a long high jump?

The theoretical maximum throwing distance is obtained from the maximum of $\sin 2\alpha$. The sine curve has a maximum value of 1 at 90°. From $\sin 2\alpha = 1$, we deduce $2\alpha = 90°$, and therefore, $\alpha = 45°$, in order to obtain the maximum throwing distance.

⊕ *Remark*: The air resistance has to be taken into account, at least since *Galileo*'s experiment with balls of iron and wood on the Tower of Pisa (Fig. 4.41). The trajectory is then a cubic parabola. When throwing a rounder's ball, a shot put, a javelin, or when attempting a long jump (Fig. 4.42), it is not possible at an initial angle of $\alpha = 45°$ to obtain the greatest initial velocity $v_0$, mostly for physiological reasons. Yet, the initial velocity is even more important than the initial angle: The throw (displacement) grows *quadratically* with $v_0$. Therefore, the best sprinters are often the best long jumpers (*Carl Lewis*). ⊕ ◂◂◂

## Equations of type $P \sin x + Q \cos x + R = 0$

This type of equation is quite common in practice (Application p. 144, Application p. 335) and leads to a quadratic equation:
First, we rearrange

$$P \sin x = -Q \cos x - R \Rightarrow P^2 \sin^2 x = (Q \cos x + R)^2$$

$$\Rightarrow P^2 (1 - \cos^2 x) = Q^2 \cos^2 x + R^2 + 2QR \cos x.$$

If we use the abbreviation $u = \cos x$, then we have the quadratic equation

$$(P^2 + Q^2)\, u^2 + 2QR\, u + R^2 - P^2 = 0$$

with the two (sometimes coincident and sometimes not real) solutions for $u = \cos x$:

$$\begin{aligned} u_{1,2} &= \frac{-2QR \pm 2\sqrt{Q^2R^2 + (P^2+Q^2)(P^2-R^2)}}{2(P^2+Q^2)} = \\ &= \frac{-QR \pm \sqrt{P^4 + P^2Q^2 - P^2R^2}}{P^2+Q^2} = \frac{-QR \pm P\sqrt{P^2+Q^2-R^2}}{P^2+Q^2}. \end{aligned}$$

**▸▸▸ Application**: ***intersecting homofocal conics***

In Chapter 7, we frequently need the solution of the equation (for example, in Application p. 335, Application p. 336)

$$\frac{r_1}{1+\varepsilon_1 \cos x} = \frac{r_2}{1+\varepsilon_2 \cos(x+\delta)}.$$

It describes the intersection of two conic sections with one common focal point. The constant value $\delta$ is the "twist angle" of the two conic sections. Deduce the equation by means of the addition theorems in order to obtain $P \sin x + Q \cos x + R = 0$, the solution of which is well-known.

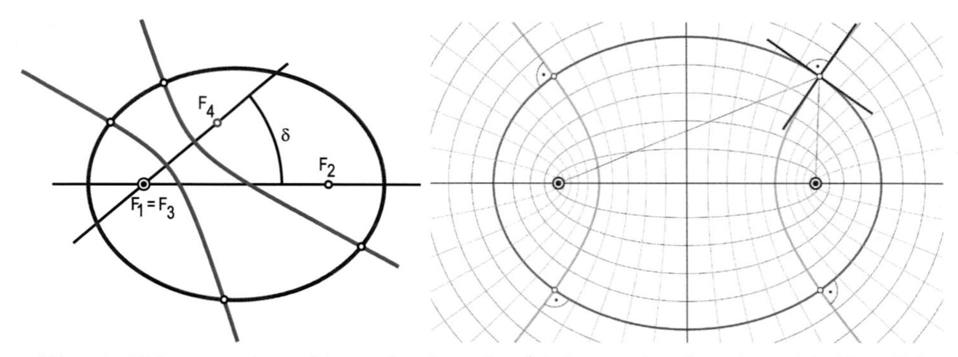


**Fig. 4.43** intersection of homofocal conics (right: confocal conics with $F_2 = F_4$)

*Solution*:

Multiplying by the common denominator, we first get

$$r_2(1+\varepsilon_1 \cos x) = r_1(1+\varepsilon_2 \cos(x+\delta)).$$

We set $\cos(x+\delta) = \cos x \cos\delta - \sin x \sin\delta = c \cos x - s \sin x$. Then, we have

$$r_2 + r_2\,\varepsilon_1 \cos x = r_1 + r_1\,\varepsilon_2\,(c \cos x - s \sin x)$$

which yields

$$\underbrace{(r_2\,\varepsilon_1 - r_1\,\varepsilon_2\,c)}_{P} \cos x + \underbrace{r_1\,\varepsilon_2\,s}_{Q} \sin x + \underbrace{r_2 - r_1}_{R} = 0.$$

⊕ *Remark*: In this context, one speaks of an orthogonal grid (Fig. 4.43, right) built by the two pencils of confocal conics (see Application p. 534). ⊕ ◂◂◂

## 4.4 The scalene triangle

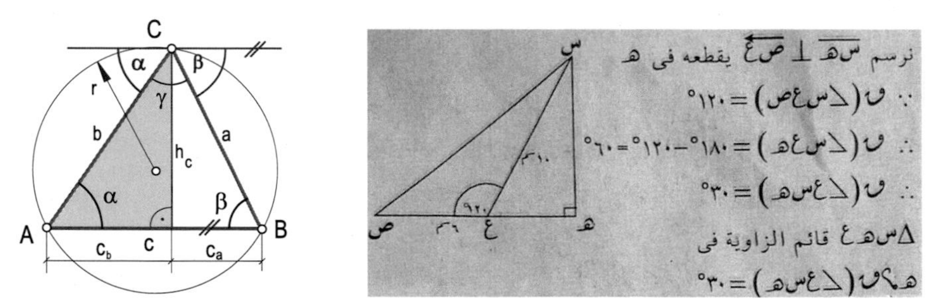

**Fig. 4.44** sum of angles

**Fig. 4.45** from an Arabic textbook ...

For an unambiguous determination of a scalene triangle, we need exactly *three specified elements.*
Exception: The specification of the three angles $\alpha$, $\beta$, and $\gamma$ is not enough, because, even among the three angles, the following relation holds: *The sum of the angles in a triangle is* $180°$, *that is,*

$$\boxed{\alpha + \beta + \gamma = 180°.} \tag{4.23}$$

***Proof***: We draw the parallel to $AB$ at $C$ (Fig. 4.44). There, $\alpha$ and $\beta$ appear once again as parallel angles. ⊙
Incidentally, this relation yields $\sin\gamma = \sin(180° - (\alpha + \beta))$, and thus,

$$\sin\gamma = \sin(\alpha + \beta). \tag{4.24}$$

⊕ *Remark*: The so-called elliptic geometry deals with spaces where one can draw no parallel to a straight line. There, the sum of angles in a triangle is always larger than 180° and not the same for all triangles. This is not just a theoretical gimmick, but plays an important role in modern physics ("curved space"). ⊕

Now, let the sides and angles of a triangle be known. We calculate the auxiliary parameters $h_c$, $c_a$, and $c_b$ (Fig. 4.44):

$$h_c = a\sin\beta,\ c_a = a\cos\beta,\ c_b = c - c_a = c - a\cos\beta. \tag{4.25}$$

Now, we apply the Pythagorean theorem:

$$c_b^2 + h_c^2 = b^2 \Rightarrow (c - a\cos\beta)^2 + (a\sin\beta)^2 = b^2$$

$$\Rightarrow\ c^2 + a^2\underbrace{(\cos^2\beta + \sin^2\beta)}_{1} - 2ac\cos\beta = b^2,$$

which leads to the

**Law of Cosines**, a generalization of the Pythagorean theorem. The following three formulas differ only by a cyclical shift of the variables:

$$\begin{aligned} b^2 &= c^2 + a^2 - 2ca\cos\beta, \\ c^2 &= a^2 + b^2 - 2ab\cos\gamma, \\ a^2 &= b^2 + c^2 - 2bc\cos\alpha. \end{aligned} \tag{4.26}$$

In order to break away from notations, we should keep the following sentence in mind:

> The square on one side is equal to the sum of the squares of the other two sides, reduced by twice the product of the other side lengths times the cosine of the enclosed angle.

From the side lengths of a triangle, the angles can be calculated. For example:

$$\cos\alpha = \frac{b^2 + c^2 - a^2}{2bc}.$$

▸▸▸ **Application**: ***forces in equilibrium*** (Fig. 4.46)
Three coplanar forces (i.e. forces acting in a plane)

$$F_1 = 60\text{N}, \; F_2 = 45\text{N}, \; F_3 = 30\text{N}$$

attached to a point are balanced. Calculate the angles which they mutually subtend.

*Solution*:

$$a = 60, \; b = 45, \; c = 30 \Rightarrow \text{Law of Cosines} \Rightarrow$$

$$\alpha = 104.5^\circ, \; \beta = 46.6^\circ \Rightarrow$$

$$\varphi = \alpha + \beta = 151.1^\circ, \; \psi = 180^\circ - \alpha = 75.5^\circ, \; \sigma = 180^\circ - \beta = 133.4^\circ.$$

◂◂◂

If two sides and the enclosed angle are given, one also uses the Law of Cosines.

**Fig. 4.46** forces in equilibrium

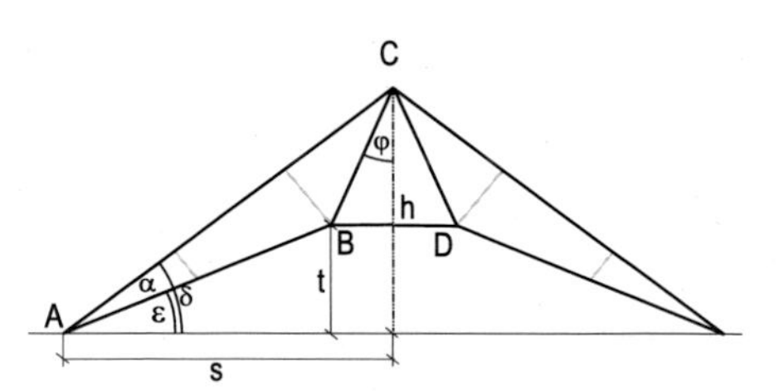


**Fig. 4.47** Belgian roof trusses

▸▸▸ **Application**: ***Belgian roof trusses*** (Fig. 4.47)
The angle $\varepsilon$ of inclination of the bottom chord and the angle $\delta$ of the roof's inclination with a total chord length $\overline{AB} = 2s$ are given. We are looking for the lengths of the other bars.

*Solution*:
Once we have found the rafter's length $\overline{AC} = \dfrac{s}{\cos\delta}$, we know the two side lengths $\overline{AB}$ and $\overline{CA}$ of the triangle $ABC$ and the enclosed angle $\alpha = \delta - \varepsilon$. The side $\overline{BC}$ is then found via the Law of Cosines:

$$\overline{BC} = \sqrt{\overline{AB}^2 + \overline{AC}^2 - 2\overline{AB}\,\overline{AC}\cos\alpha},$$

$$\overline{BD} = 2(s - \overline{AB}\cos\varepsilon).$$

◂◂◂

▸▸▸ **Application**: ***intersection of two spheres*** (Fig. 4.48)
The intersection of two spheres $\Sigma_1$, $\Sigma_2$ (centers $M_1$, $M_2$, radii $r_1$, $r_2$, central distance $d = \overline{M_1M_2}$) is a circle $k$ with radius $r$ and center $M$ in a plane normal to the axis $M_1M_2$. Compute $r$ and $e = \overline{MM_1}$. For which $d$ is $r = 0$ and $r^2 < 0$ (or $r$ not real)?

*Solution*:
In Fig. 4.48, we can see a scalene triangle $M_1M_2S$ with known side lengths $r_1$, $r_2$, and $d$. We apply the Law of Cosines and get

$$r_2^2 = r_1^2 + d^2 - 2r_1d\cos\alpha \Rightarrow \cos\alpha = \frac{r_1^2 + d^2 - r_2^2}{2r_1d}.$$

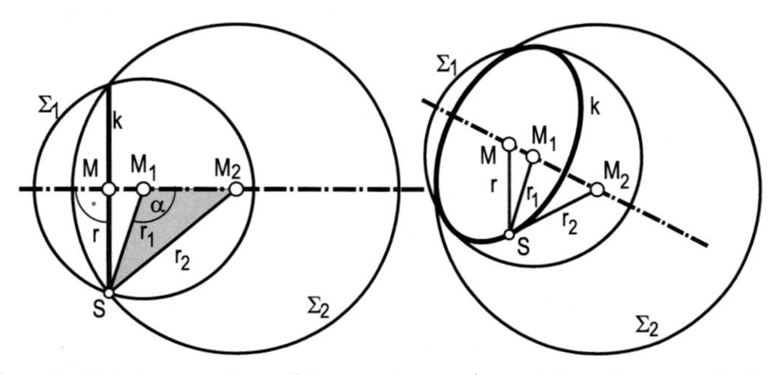


**Fig. 4.48** intersection of two spheres (special and general view)

The following holds:

$$\cos\alpha \le 1 \Rightarrow \frac{r_1^2 + d^2 - r_2^2}{2r_1d} \le 1$$

$$\Rightarrow r_1^2 + d^2 - r_2^2 \le 2r_1d \Rightarrow r_1^2 - 2r_1d + d^2 \le r_2^2 \Rightarrow (r_1 - d)^2 \le r_2^2.$$

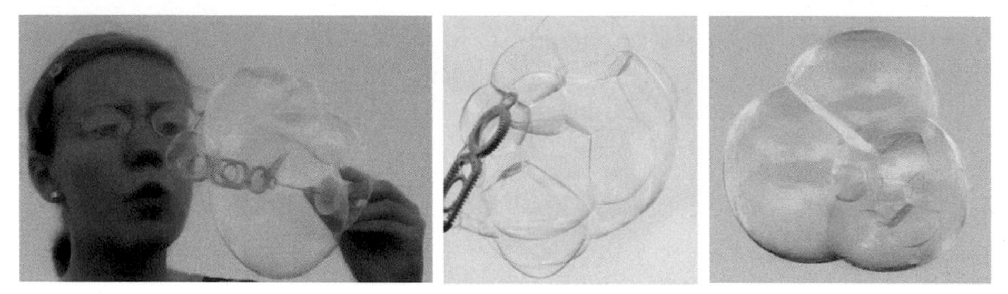

**Fig. 4.49** the coming into being of soap bubbles consisting of spheres

For $(r_1 - d)^2 > r_2^2$, the circle of intersection will not be real. For

$$r_1 - d = \pm r_2 \Rightarrow r_1 - r_2 = d \text{ or } r_1 + r_2 = d,$$

the two spheres touch at a point ("zero circle"). In the case of a real intersection, the radius $r$ and the distance $e$ from the center point $M$ to $M_1$ are related by

$$r = r_1 \sin\alpha \quad \text{and} \quad e = r_1 \cos\alpha.$$

⊕ *Remark*: The plane that carries the circle of intersection is called the *radical plane.* The intersection of two spheres is always a circle, regardless of whether it is real or not. By virtue of Application p. 534, we note that all spheres share the *absolute circle* of Euclidean geometry, which is not only *at infinity*, but also carries no real point. ⊕ ◄◄◄

## Area of a triangle:

First of all, the area $A$ of a triangle (with the notation from Fig. 4.44) equals

$$A = \frac{c\,h_c}{2} = \frac{c\,b\,\sin\alpha}{2}.$$

Since the labels can be shifted cyclically, we also get

$$\boxed{A = \frac{a\,c\,\sin\beta}{2} = \frac{b\,a\,\sin\gamma}{2} = \frac{c\,b\,\sin\alpha}{2}.} \tag{4.27}$$

From (4.27), we immediately deduce the

## Law of Sines:

$$\boxed{\frac{a}{\sin\alpha} = \frac{b}{\sin\beta} = \frac{c}{\sin\gamma}.} \tag{4.28}$$

In words:

The ratio of side length and the sine of the opposite angle is constant.
The ratio of any two sides is equal to the ratio of the sines of opposite angles.

Thus, the following tasks can be solved:
**Task 1**: Let two angles (and thus, all three) and a side (for example, $a$) of a triangle be given. Then, the side $b$ equals

$$b = \sin\beta \frac{a}{\sin\alpha}.$$

▸▸▸ **Application**: ***height of a tower*** (Fig. 4.50)
At $A$, one measures the angle $\varepsilon$ from the roadside to the base $F$ of the tower and the elevation angle $\delta$ to the top $H$. Then, one moves along the roadside to position $B$ at distance $s = \overline{AB}$ and again measures the angle $\varphi$ to the base $F$. Thus, the height of the tower is determined:
1) The angles $\varepsilon$, $\varphi$, $\psi$ sum up to 180°. Thus, $\psi = 180° - \varepsilon - \varphi$ and $\sin\psi = \sin(\varepsilon + \varphi)$.
2) We apply the Law of Sines to the triangle $ABF$:

$$\frac{x}{\sin\varphi} = \frac{s}{\sin\psi} \Rightarrow x = s\,\frac{\sin\varphi}{\sin(\varepsilon + \varphi)}.$$

3) The altitude $h$ in the right triangle $AFH$ satisfies

$$\frac{h}{x} = \tan\delta \Rightarrow h = x\,\tan\delta.$$

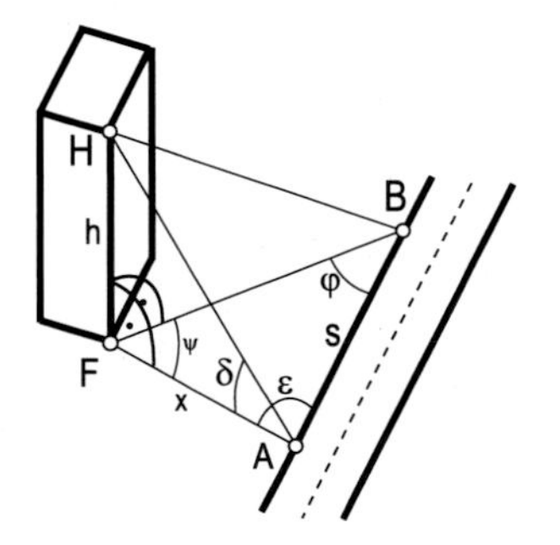


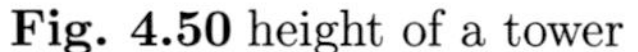
**Fig. 4.50** height of a tower

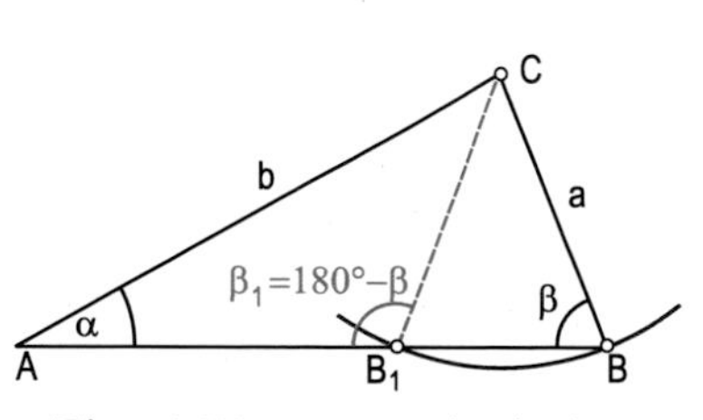


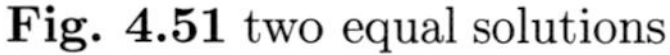
**Fig. 4.51** two equal solutions

◂◂◂

**Task 2**: An angle (for example, $\alpha$) and two sides that do not enclose this angle (for example, $a$ and $b$) are given. Then, we can calculate the angle $\beta$ by

$$\sin\beta = b\,\frac{\sin\alpha}{a}.$$

It has to be taken into account that this formula can have two solutions (or only one or none at all), namely $\beta$ and $180° - \beta$ (Fig. 4.51). In practice, one solution can often be neglected due to an additional condition.

▸▸▸ **Application**: ***forces in equilibrium***
Two forces $F_1 = 60$N and $F_2$ acting on the same point enclose the angle $\varphi = 151.1°$ and result in a force $F_3 = 30$N. Calculate $F_2$!

*Solution*:
With the usual notation, we have

$$a = 60,\ c = 30,$$

$$\frac{c}{\sin\gamma} = \frac{a}{\sin\alpha} \Rightarrow \sin\alpha = a\frac{\sin\gamma}{c} = 0.967 \Rightarrow$$

$$\alpha_1 \approx 75.2°,\ \alpha_2 \approx 180° - 75.2° = 104.8° \Rightarrow$$

$$\beta_1 = 180° - \alpha_1 - \gamma = 75.9°,\ \beta_2 = 180° - \alpha_2 - \gamma = 46.3°,$$

$$b_1 = \sin\beta_1\frac{a}{\sin\alpha_1} \approx 60.2,\ b_2 = \sin\beta_2\frac{a}{\sin\alpha_2} \approx 44.9.$$

Thus, we see that there are two equally correct solutions! An additional condition could help us eliminate one of the two solutions! ◂◂◂

An important result related to the scalene triangle is the *angle of circumference theorem* (Fig. 4.52). It is a generalization of *Thales*'s theorem:

> Angle of circumference theorem: A chord of a circle forms the angle $\gamma$ from each point on one part of the circle and the angle $180° - \gamma$ from each point on the supplementary part of the circle. The central angle for the supplementary arc is twice the central angle on the primary side.

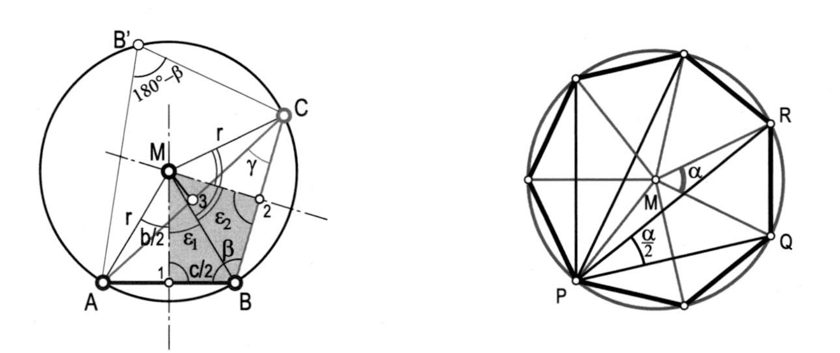


**Fig. 4.52** angle of circumference theorem **Fig. 4.53** diagonals in an $n$-gon

***Proof***: Consider a circle with the fixed chord $AC$ and another point $B$ on the circle. With the notations of Fig. 4.52, we get: The sum of interior angles in the quadrilateral $1B2M$ equals 360°, as in any quadrilateral, for it can be divided along a diagonal into two triangles. It has right angles at 1 and 2. Thus, the remaining angles sum up to 180°, $\varepsilon_1 + \varepsilon_2 = 180° - \beta$. Furthermore, $2(\varepsilon_1 + \varepsilon_2) = \angle AMC$, and therefore, $\beta = 180° - \frac{1}{2}\angle AMC$ for each point $B$ on the circle on the same side. On the other side, the angle $\angle AMC$ is obtained by taking $360° - \angle AMC$, and we have $\angle AB'C = 180° - \beta$. Since $\angle AMC = 360° - 2\beta$, we further have $\angle AMC = 2\angle AB'C$, that is, the central angle is twice the angle of circumference. ⊙

The angle of circumference theorem leads to the following statement:

In a cyclic quadrangle, i.e. a quadrangle inscribed into a circle, the sum of opposite angles equals $180°$.

Almost simultaneously, the following statement has thus been proven:

The diagonals in a regular $n$-gon make equal angles with the neighboring diagonals.

***Proof***: Let $P$ be an arbitrary point, and let $Q$ and $R$ be two neighboring points of the $n$-gon (Fig. 4.53). In the circumcircle, we can apply the theorem of the angle of circumference with the chord $QR$. Thus, $\angle QPR$ does not depend on the actual position of $P$. Further, $\angle QMR = \alpha = 360°/n$ is constant. According to the theorem of the angle of circumference, the central angle is twice the angle of circumference, and thus, $\angle QPR = \alpha/2 = 180°/n$ is constant. ⊙

▸▸▸ **Application**: ***positioning by angle measurement*** (Fig. 4.54)
From the map, we know the relative position of three distinct points $A$, $B$, and $C$ on the ground. In order to determine our own position $P$, it is enough to measure two angles (for example, $\varphi = \angle APB$ and $\psi = \angle BPC$). $P$ is located on two circles according to the theorem of the angle of circumference. ◂◂◂

It is very easy to prove the following theorem with the help of the Law of Sines (Fig. 4.55):

Each interior angle bisector of a triangle divides the opposite side in the ratio of the adjacent side lengths.

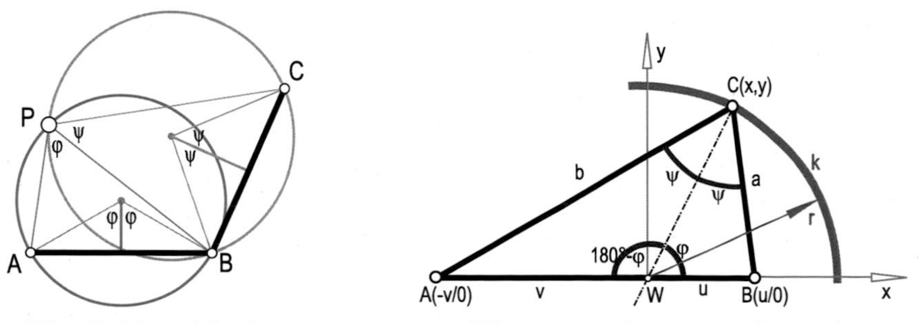


**Fig. 4.54** positioning

**Fig. 4.55** reflection in the circle

***Proof***: According to Fig. 4.55, we should have $u : v = a : b$. In the triangles $BWC$ and $AWC$, we get the following after applying the Law of Sines

$$\frac{u}{\sin\psi} = \frac{a}{\sin\varphi} \quad \text{and} \quad \frac{v}{\sin\psi} = \frac{b}{\sin(180° - \varphi)}.$$

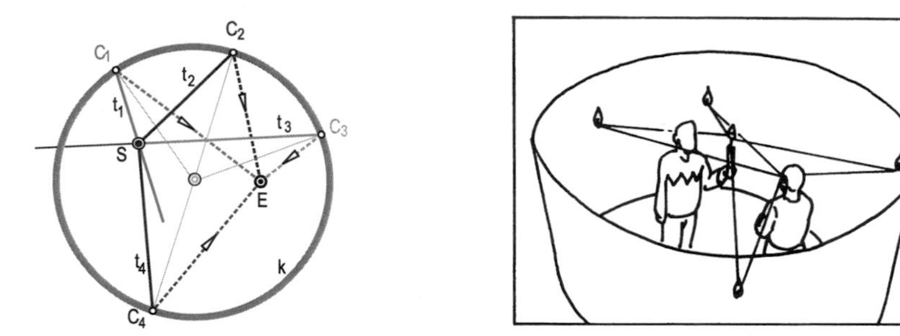


**Fig. 4.56** four reflections in a circle ... **Fig. 4.57** ... or a cylinder of revolution

Since $\sin(180° - \varphi) = \sin\varphi$, the relation $u : v = a : b$ already holds. ⊙

### ▸▸▸ Application: *reflection in a circle*

We find an application of the above theorem if we consider a "reflecting" circle centered at $W$ on the angle bisector through $C$. Then, $C$ is the reflection of $A$ in $k$ with respect to $B$. Or in other words: A laser ray emanating from $A$ is reflected at $C$ to $B$. We fix a Cartesian coordinate system at $W$. Then, the points $A$, $B$, $C$ have the coordinates $A(-v/0)$, $B(u/0)$, and $C(x/y)$ (where $x^2 + y^2 = r^2$, and thus $y^2 = r^2 - x^2$). Furthermore, according to the law of reflection (angle of incoming ray = angle of outgoing ray) and the above theorem on the bisecting line, we find

$$\frac{\sqrt{(x+v)^2+y^2}}{\sqrt{(x-u)^2+y^2}} = \frac{v}{u} \Rightarrow \frac{(x+v)^2+y^2}{(x-u)^2+y^2} = \frac{v^2}{u^2} \Rightarrow \frac{(x+v)^2+r^2-x^2}{(x-u)^2+r^2-x^2} = \frac{v^2}{u^2}$$

$$\Rightarrow \frac{2vx+v^2+r^2}{-2ux+u^2+r^2} = \frac{v^2}{u^2} \Rightarrow u^2(2vx+v^2+r^2) = v^2(-2ux+u^2+r^2)$$

$$\Rightarrow 2uv(u+v)x = r^2(v^2-u^2) \Rightarrow x = \frac{v-u}{2uv}r^2.$$

The $y$-value is calculated via the Pythagorean theorem as $y = \sqrt{r^2 - x^2}$.

**Fig. 4.58** Without knowing the possible maximum number of reflections (in this case two, because the viewer is outside the cylinder), it is not easy to understand the individual pictures of this series. The time difference in between the four photographs was 1/60 s each. This also helps us to understand the situation.

⊕ *Remark*: The determination of the reflection requires only the solution of a quadratic equation, although in the general case (Fig. 4.56), reflecting in the circle leads to an irreducible equation of degree 4. This is because, in this case, two reflections of $A$ in $k$ are already known. Trivially, these are the intersections of $k$ with the straight line $AB$. ⊕ ◂◂◂

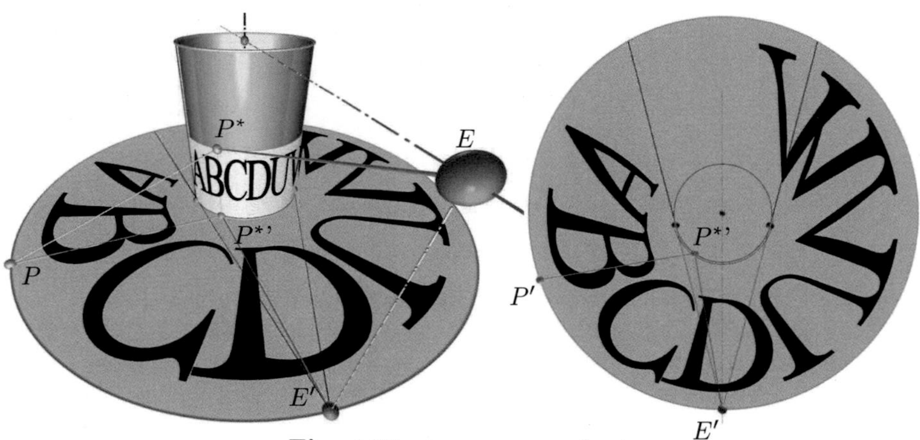


**Fig. 4.59** mirror anamorphosis

▸▸▸ **Application**: ***mirror anamorphosis*** (Fig. 4.59)
When looking at a point $P$ in a cylindrical mirror (eye point $E$), the reflection point $P^*$ is unequivocal. In principle, the task is still two-dimensional: In a top view (right), we look for the reflection point $P^{*\prime}$ of $P'$ in a circle with respect to $E'$. How can one draw distorted images in the base plane such that the reflection "makes sense"?

*Solution*:
We put the cart before the horse and start with the result $P^*$. Then, we just have to reflect the ray $P^*E$ in the cylinder (we reflect at the circle in the top view, and the inclination of the ray stays the same). Further, we intersect the reflected ray with the base plane and get the desired point $P$, the reflection of which is $P^*$. Note that points "behind" the cylinder will never have a visible reflection point. ◂◂◂

## 4.5 Further applications

In this section, you will find some interesting exercises relating to the previous sections, but they can also be skipped without affecting the understanding of the overall concept.

▸▸▸ **Application**: ***rolling cone*** (Fig. 4.60)
When a cone of revolution is rolling on the base plane, it will stay inside a circle around the apex. We now choose two parallel circles on the cone. Let the ratio of their radii be $1 : k$, and $d$ be their distance measured on the cone's generating line. What is the radius $s$ of the inner blue circle in Fig. 4.60? Compare also Application p. 154.

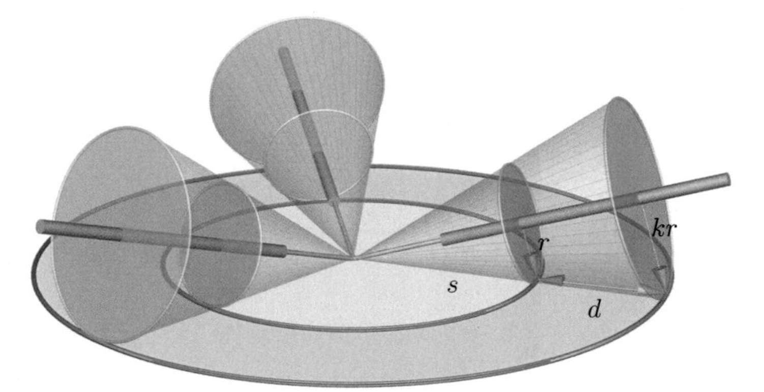


**Fig. 4.60** The rolling of the cone can be substituted by the rolling of two rigidly connected parallel circles of the cone.

*Solution*:
Due to the intercept theorem, we have $s : r = (d + s) : (kr)$, and thus, $s = d/(k - 1)$. ◂◂◂

▸▸▸ **Application**: ***Why does a train not derail?***
This application does not need calculations, but has something to do with arc lengths. The answer "The train does not derail, because it runs on rails" is too simple.

**Fig. 4.61** conical wheels

*Solution*:
The train would continually derail if pairs of co-axial wheels were not conical. Even the steel ring on the inside of the wheel cannot prevent derailment. It is mainly the slightly conical shape of the wheels (Fig. 4.61) that can do so: If the train drifts to the right, then the contact circle on the right wheel apparently increases its diameter due to the conical shape of the wheel. At the same time, the diameter of the contact circle on the left wheel decreases (causing a tilt of the axis). This is exactly what forces the axis to move back into the right and stable position.

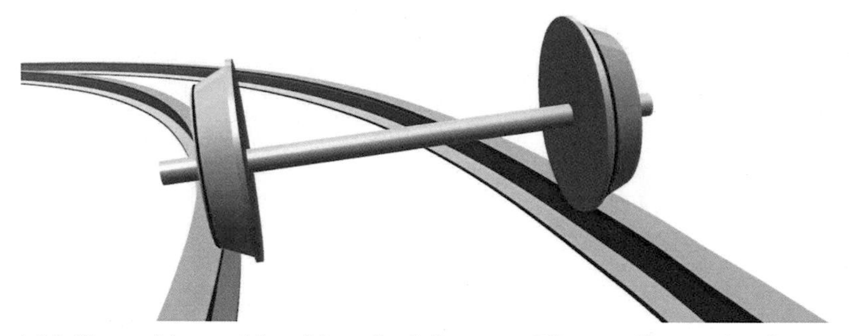

**Fig. 4.62** No problem with taking the left curve: The smaller circle of contact on the left and the larger to the right make the train stay on the tracks.

⊕ *Remark*: When the train has to take a curve, the conic wheels are also of great advantage. Imagine the left curve in Fig. 4.62: The tracks have to be inclined, otherwise the centrifugal forces would tip the train over. These centrifugal forces will still shift the train to the right. This causes the circle of contact with the right track to increase, and the corresponding circle on the left to decrease. This is good, because now the train automatically tilts to the left (Fig. 4.60). Inclining not enough to the right will push the train further right, whilst inclining too far to the right will push it back to the left. The perfect equilibrium lies somewhere in between. ⊕

◀◀◀

**Fig. 4.63** The Sun and the Moon both appear under a viewing angle of $\approx 0.5°$.

▸▸▸ **Application**: ***diameter of the Sun and the Moon*** (Fig. 4.63, Fig. 4.64)
The Sun and the Moon can be seen in the sky at almost exactly the same angle, namely a little more than half a degree (this was not always so, because the Moon was closer to the Earth in ancient times). Moreover, since the ecliptic plane of the Moon is tilted only slightly towards that of the Sun, solar eclipses frequently occur. Through precise angle measurements from various points on the Earth, the average distance between the two celestial bodies can be calculated: $d_S \approx 150,000,000\,\text{km}$, $d_M \approx 384,000\,\text{km}$. What are their diameters?

**Fig. 4.64** Partial solar eclipse in May 2003: blended pictures over the course of approx. 20 minutes (5:09–5:31). The Sun and new Moon start "leaving each other".

*Solution*:
Either visual angle can be expressed in terms of radians as $\varphi \approx \dfrac{\pi}{360}$. Now, we compare with the diameter $D$ of the Earth ($\approx 12,740\,\text{km}$) and find

$$D_S = d_S\,\varphi \approx 1,300,000\,\text{km} \approx 100\,D_E,$$

$$D_M = d_M\,\varphi \approx 3,350\,\text{km} \approx 0.26\,D_E.$$

The exact values are

$$D_S = d_S\,\varphi \approx 1,392,000\,\text{km}, \quad D_M = d_M\,\varphi \approx 3,476\,\text{km},$$

because the angle is slightly larger than 0.5° and the distances – especially that of the Moon – also vary.

⊕ *Remark*: A total solar eclipse occurs relatively infrequently. A total lunar eclipse can be observed more often, because, in this case, the Earth shadows the much smaller Moon (1/4 of the Earth's diameter). Therefore, lunar eclipses take even longer, because the Moon takes hours to "dive through" the shadow zone. From the Moon, this phenomenon must look extremely spectacular: The huge "new Earth", then, begins to glow red at the edges (more in Application p. 315). ⊕ ◂◂◂

▸▸▸ **Application**: ***landing*** (Fig. 4.65)
An airliner lands at sea level within 20 minutes of coming from an altitude of $h = 35{,}000$ (1 foot = 0.314 m $\Rightarrow h \approx 11\,\text{km}$). What is the average descent angle when the average body speed (*ground speed*) is 540 km/h?

**Fig. 4.65** smooth landing on the water surface

*Solution*:
The total distance is $s = 540\,\text{km/h} \cdot \frac{1}{3}\,\text{h} = 180\,\text{km}$. For the descent angle $\alpha$, we have $\tan\alpha = h/s$. For small angles, the tangent is equal to the angle in radians (Application p. 383). By multiplying, we get the degree value: $\alpha^\circ = \frac{11\,\text{km}}{180\,\text{km}} \cdot \frac{180^\circ}{\pi} \approx 3.5^\circ$.

⊕ *Remark*: A spectacular "jump" with special "bat suits" between the continents of Africa and Europe in the summer of 2005: The jump started at an altitude of 35,000 feet at an outdoor temperature of −50° above the Moroccan coast. Landing took place in Algecira Bay in Cadiz, with a ground distance of 20.5 km. The gliding duration was 6 minutes. The descent angle $\alpha$ was approximately 30°, since $\tan\alpha \approx 11 : 20.5$. The average speed was 240 km/h (204 km/h ground speed). The parachute was opened at the last minute (see Application p. 21). ⊕ ◂◂◂

▸▸▸ **Application**: ***"curvature" of the sea*** (Fig. 4.66)
The surface of the sea, as well as that of larger freshwater lakes, is visibly curved. By connecting two distant surface points $A$ and $B$ with a straight line, the "vault" can be defined. What is the vault $w$ of a lake with a length of $L$ km? What height $h$ must a tower at $B$ have in order to see the middle of the lake?

*Solution*:
Half the central angle $\varphi$ (Fig. 4.66) is calculated in radians from the Earth's radius $R = 6{,}370\,\text{km}$ by

$$\varphi = \frac{L/2}{R} = \frac{L}{2R}.$$

**Fig. 4.66** curvature of the water surface, curved horizon

Furthermore, $\overline{MN} = R\cos\varphi$ and $\overline{MC} = \dfrac{R}{\cos\varphi}$ and so

$$w = R\left(1 - \cos\frac{L}{2R}\right),\ h = R\left(\frac{1}{\cos\frac{L}{2R}} - 1\right) \quad (\varphi \text{ in radians!}).$$

*Numerical example*:
$L = 10\,\text{km} \Rightarrow w = 1.96\,\text{m},\ h \approx w,$
$L = 30\,\text{km} \Rightarrow w = 17.66\,\text{m},\ h \approx w,$
$L = 100\,\text{km} \Rightarrow w = 196.23\,\text{m},\ h \approx w,$
$L = 200\,\text{km} \Rightarrow w = 784.91\,\text{m},\ h = 785.01\,\text{m},$
$L = 750\,\text{km} \Rightarrow w = 11{,}035\,\text{m},\ h = 11{,}054\,\text{m}.$
Thus, we see that, for a small $L$, the curvature and the tower height are almost identical.

⊕ *Remark*: A final numerical example: An airplane at an altitude of 11 km can be seen under ideal conditions from 750 km/2 = 375 km. Conversely, it is only from this distance onwards that the aircraft can be detected by radar. Looking at the sea with a wide angle lens, one sees a slightly curved horizon of approximately 500 km in length. The curvature itself accounts for about 11 km. This view is captured by the photograph of Fig. 4.66. ⊕ ◂◂◂

### ▸▸▸ Application: *"morning star" and "evening star"*

(Fig. 4.67) The planets move on elliptical orbits around the Sun. Except for the rather eccentric elliptical orbit of Pluto, the orbital ellipses are almost in one plane, the *ecliptic.* The orbital ellipses of Venus and the Earth are almost circular (medium distance to the Sun 108 million km and 150 million km). What is the maximum viewing angle, under which one can see the Earth, the Sun, and Venus? Why is Venus the "morning star" or "evening star"?

*Solution*:

From Fig. 4.67, one recognizes a right triangle $ETS$, the sides of which are formed by the tangent to the orbital circle of Venus, the axis Earth-Sun, and the radius at the contact point $T$. For the angle $\alpha$ in this triangle, we get

$$\sin\alpha = \frac{\overline{TS}}{\overline{ES}} = \frac{108}{150} = 0.72 \Rightarrow \alpha \approx 46^\circ.$$

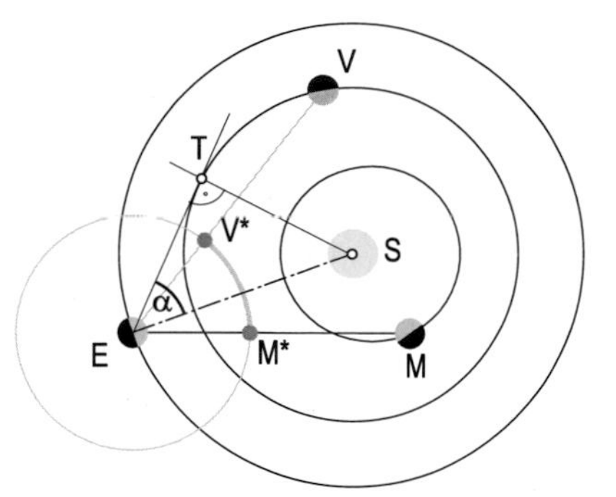


**Fig. 4.67** Sun, Venus, Mercury

Venus is, thus, always seen at an angle of at most 46° around the Sun. Therefore, you see it only just before sunrise or just after sunset.

⊕ *Remark*: Still closer to the Sun is Mercury. With an average distance of 58 million km, the calculation yields the maximum visual angle $\alpha \approx 23°$ (because of the more eccentric path, this angle may, however, be up to 27°). Yet, since Mercury is much smaller than Venus, it generally appears less bright in the firmament. The Sun, Mercury, and Venus are almost on a "straight line" (or more precisely on a great circle in the sky): the Earth, the Sun, Mercury, and Venus are, indeed, almost in one plane (the ecliptic), and this plane almost appears to "project" from the Earth. In Fig. 4.67, the positions of Venus and Mercury in the firmament are labeled by $V^*$ and $M^*$. ⊕ ◂◂◂

▸▸▸ **Application**: ***lapping a planet*** (Fig. 4.67)

The Earth has an orbital period of 365 days around the Sun. Venus has one of about 225 days, and Mars takes 1.88 years. All orbits lie almost exactly in a plane and the orbital rotations are equally oriented. How frequently do the planets have the same relative position (for example, minimum distance)?

*Solution*:

Assuming that the orbits are circular, the Earth orbits through an angle $\approx 1°$ per day around the Sun. For Venus, it is $\approx 1.6°$. That is, Venus orbits $\approx 0.6$ faster and needs about $\frac{360}{0.6} = 600$ days to return to the same relative position. Mars loses $\approx 0.47°$ per day with respect to the Earth. Therefore, a similar position is reached after 766 days.

⊕ *Remark*: In early June of 2003, two Mars probes were launched so that, when Mars reached its closest position to the Earth at the end of August, the probes had travelled about "half" of their trajectory. In that year, Mars was particularly close to the Earth, because, during its slightly elliptical orbit, the lapping occurred when Mars reached its position closest to the Sun ("perihelion"). Since Mars – in comparison with Venus – is farther away from the Sun than the Earth, we can see its fully lit side during a lapping and Mars thus becomes the third brightest celestial body after the Sun and the Moon. ⊕ ◂◂◂

**▸▸▸ Application**: ***How long does a solar eclipse last?***
When the Moon is exactly between the Sun and Earth, there may be a total solar eclipse. However, the observer on the Earth has to be in the umbra of the Moon (Fig. 4.68). This approximately circular area has a diameter of at most 269 km. Calculate the velocity with which the shadow moves over the surface of the Earth.

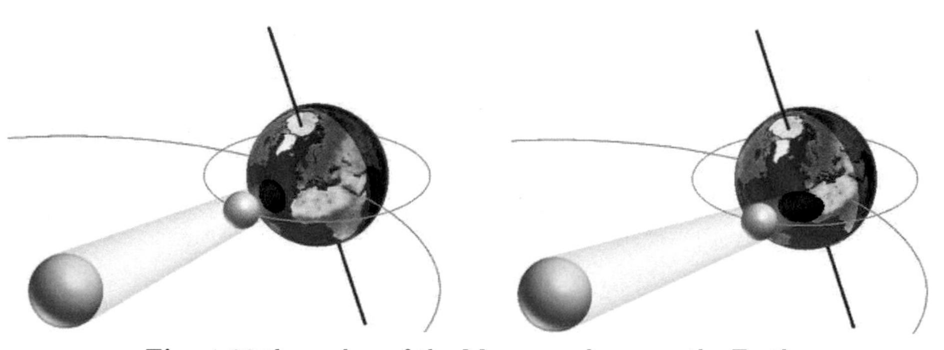

**Fig. 4.68** the umbra of the Moon wanders over the Earth

*Solution*:
Fig. 4.68 illustrates that it is essentially the speed of the Moon on its orbit around the Earth that matters. Within 27.52 days, the Moon travels $2\pi \cdot 384{,}000\,\mathrm{km}$, i.e., about 61 km per minute. Thus, the area of the umbra would rush within 4.5 minutes over the observer. However, the observer on Earth rotates relatively quickly about the Earth's axis (which is in general slightly inclined to the connection Earth-Sun). The maximum speed is the path velocity at the equator ($1{,}667\,\mathrm{km/h}$, which is around 28 km per minute, see Application p. 171). Thus, the speed of the umbra decreases to a value between $61\,\mathrm{km/min}$ and $33\,\mathrm{km/min}$. The total solar eclipse can, therefore, take $269/33 \approx 8$ minutes in extreme cases. ◂◂◂

**▸▸▸ Application**: ***determination of an inaccessible radius*** (Fig. 4.69)
In order to determine the radius $R$ of a cylinder of revolution, one can use three round bars of equal thickness with a diameter $d$. Then, we measure the height difference $h$ between the outer and inner bars. What is the formula for $R$ if $d$ and $h$ are given?
*Solution*:
We refer to Application p. 123, where we have derived (with the notation of Fig. 4.69) Formula (4.3):

$$r = \frac{s^2 + 4h^2}{8h}.$$

Since $d = \overline{M_1 M_2}$, we find the distance $s = \overline{M_1 M_3} = 2\sqrt{d^2 - h^2}$. The radius that we are looking for is thus

$$R = r + \frac{d}{2} = \frac{4(d^2 - h^2) + 4h^2}{8h} + \frac{d}{2} = \frac{d^2}{2h} + \frac{d}{2} = \frac{d}{2h}(d + h).$$

⊕ *Remark*: Instead of measuring the height difference $h$, one could also measure the total width $b$ of the three bars. Of course, it is $s = b - d$, and thus, $h = \sqrt{d^2 - \left(\frac{s}{2}\right)^2} = \sqrt{d^2 - \frac{(b-d)^2}{4}}$. A small measurement error in $b$ may already have quite a strong effect on $h$, so that the result is less reliable.
*Numerical example*:
Variant 1: $d = 20\,\text{mm}$, $h = (4.4 \pm 0.1)\,\text{mm} \Rightarrow R = (55 \pm 1)\,\text{mm}$,
Variant 2: $d = 20\,\text{mm}$, $b = (59.0 \pm 0.1)\,\text{mm} \Rightarrow R = (55 \pm 2)\,\text{mm}$. ⊕ ◂◂◂

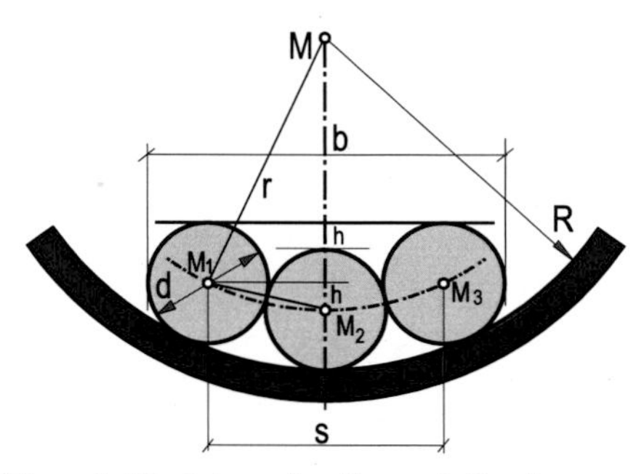


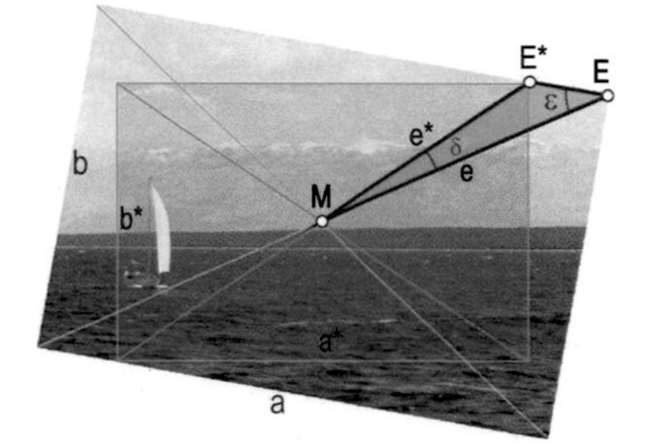


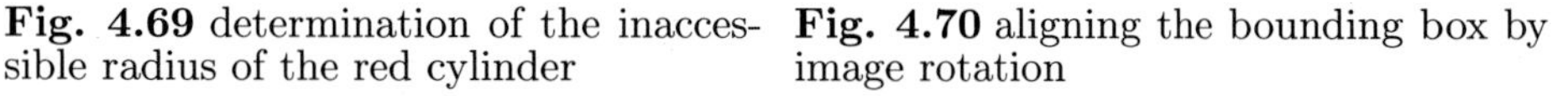

**Fig. 4.69** determination of the inaccessible radius of the red cylinder

**Fig. 4.70** aligning the bounding box by image rotation

▸▸▸ **Application**: ***tilting the horizon ...*** (Fig. 4.70)
For the aesthetics of a photograph in which the horizon can be seen, it is important that the horizon is parallel to the margin. It is best to pay attention to this at the time the photograph is taken. If necessary one could instead rotate the image through an angle of $\delta$ by using image processing software afterwards, risking a loss of quality. Subsequently, the largest possible rectangular part with the same proportions is cut out. How much space (pixels) does one lose?

*Solution*:
Let the proportions of the original photo be $a \times b$ cm$^2$, where $a : b$ is usually $3 : 2$ in SLR cameras and $4 : 3$ in compact cameras. Then, we get the following with the notation of Fig. 4.70: $e = 1/2\sqrt{a^2 + b^2}$, $\varepsilon = \arctan b/a$.
The oblique auxiliary triangle $MEE^*$ is drawn in green and uniquely defined by $e$, $\delta$, and $\varepsilon$. The scaling factor of the cropped image is found via the Law of Sines:

$$k = e^* : e = \sin\varepsilon : \sin(180^\circ - \varepsilon - \delta).$$

The dimensions of the new rectangle are, then, $a^* = k\,a$, $b^* = k\,b$ and the loss of area (loss of pixels) is $(a^* \cdot b^*) : (a \cdot b) = k^2$.

⊕ *Remark*: For a rotation angle $\delta = 10°$, the loss in the 3:2-format is 35%. In the 4:3-format, it is 32%. For turning angles of about 20°, we lose more than half of the pixels (up to 70%). So far, we have not taken into account that a rotation through an arbitrary angle will result in a loss of quality (only rotations through multiples of 90° are free of quality loss). ⊕ ◂◂◂

▸▸▸ **Application**: ***jumping high***

Compare the two photos in Fig. 4.71.

**Fig. 4.71** comparable situations

*Solution*:

The height of a jump depends almost exclusively on the take-off velocity. The following principle is well-known from physics: In the absence of air resistance, if something falls down from $h$ meters, it reaches a speed of $v = \sqrt{2gh}$. Thus, if something shoots up with a vertical speed of $v$, it reaches the height $h = v^2/(2g)$. For instance: If a motorcycle takes off at a speed of 14 m/s and at an angle of 30°, the "vertical velocity" is $14 \cdot \sin 30° \text{m/s} = 7\text{m/s}$. The motorcycle will, then, reach a height of h $= 49/20$ m ($= 2.45$ m).

**Fig. 4.72** These images show the proportion of the humpback whale and its enormous fluke, which exerts tremendous forces to speed up the giant.

⊕ *Remark*: 2.45 m is the current world record for high jumping. However, due to the jumping technique that is employed, the jumper has to lift his barycenter by about 1.50 m, which leads to a vertical speed of approx. 5.5 m/s. Therefore, he has

to initiate the jump at a velocity that allows him to reach the necessary vertical speed.
Humpback whales are famous for jumping out of the water. They usually get out by a little more than half of their length (about 8 m). Their barycenter, however, does not need to be lifted quite as much. If the whale could speed up to 10 m/s and hit the surface in a perpendicular position, it would theoretically lift its barycenter by 5 m. However, this is not enough to lift its entire body out of the water. ⊕ ◂◂◂

▸▸▸ **Application**: ***throw into deeper ground*** (Fig. 4.73)
In Application p. 142 (Fig. 4.40), we have calculated that an initial angle of 45° yields an optimal throw. This is only true on a horizontal ground. When throwing in a downhill terrain, the optimal angle may be smaller.

**Fig. 4.73** tilted throw into terrain

If one uses the time parameter $t$, a point $P(x/y)$ on the ballistic parabola has the coordinates $x = v_0\, t \cos\alpha,\ y = v_0\, t \sin\alpha - g/2t^2$.
The first equation can be solved for $t$, which can then be inserted into the second equation:

$$t = \frac{x}{v_0 \cos\alpha} \Rightarrow y = x\frac{\sin\alpha}{\cos\alpha} - \frac{g}{2}\frac{x^2}{v_0^2\cos^2\alpha}.$$

Thus, the trajectory has the form

$$y = a\,x^2 + b\,x \quad \text{with} \quad a = -\frac{g}{2v_0^2\cos^2\alpha},\ b = \tan\alpha.$$

Now, we intersect two trajectories with different launching angles:

$$y_1 = a_1x^2 + b_1x,\ y_2 = a_2x^2 + b_2x,$$
$$y_1 = y_2 \Rightarrow a_1x^2 + b_1x = a_2x^2 + b_2x \Rightarrow (a_1 - a_2)x^2 + (b_1 - b_2)x = 0.$$

The trivial solution $x = 0$ is the launch point. The non-trivial solution is obtained by dividing through by $x(a_1 - a_2) \neq 0$, which yields

$$x_s = (b_2 - b_1) : (a_1 - a_2).$$

If the terrain's slope permits such a throw at all, the flatter ballistic curve laps the steeper one. ◂◂◂

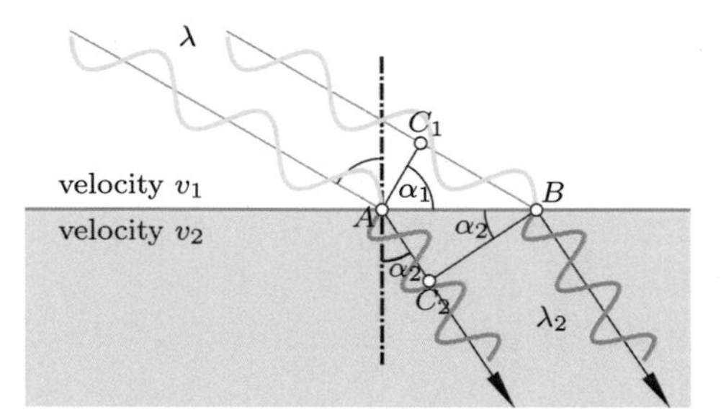


**Fig. 4.74** The wave front plausibly explains the law of refraction of light rays (the wave front tilts upon hitting the layer of separation).

▸▸▸ **Application**: **Snell's *law*** (Fig. 4.74, Fig. 4.75)
From Fig. 4.74, we obtain

$$\frac{\overline{BC_1}}{\overline{AC_2}} = \frac{v_1}{v_2} = n \Rightarrow \frac{\overline{BC_1}}{\overline{AB}} = n \cdot \frac{\overline{AC_2}}{\overline{AB}} \Rightarrow \sin \alpha_1 : \sin \alpha_2 = n.$$

According to the law of refraction (Snell's law), the angle $\alpha$ of the incoming ray and the angle $\beta$ of the outgoing ray (measured to the normal of the surface) are related by $\sin \alpha : \sin \beta = 4 : 3$ for the transition from air to water. There is, due to the atmosphere, no total reflection. That is, all light rays are at least partially deflected into water.

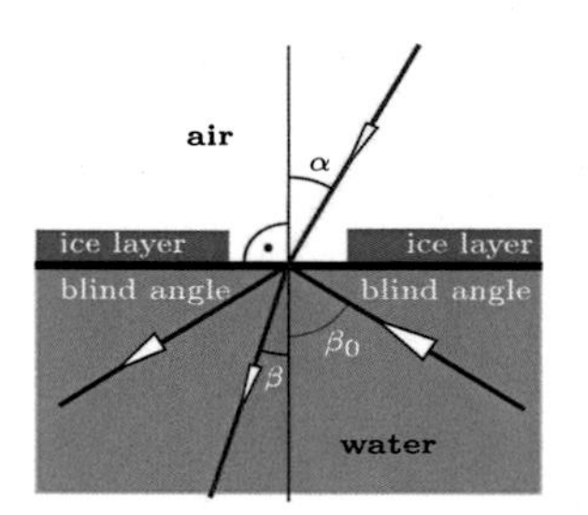


**Fig. 4.75** total reflection

**Fig. 4.76** situation in the water ...

Conversely, there is a critical angle $\beta_0$ at which no light passes from the water through the surface. The critical angle $\beta_0$ is obtained from the maximum value of 1 from $\sin \alpha$:

$$\frac{1}{\sin \beta} = \frac{4}{3} \Rightarrow \sin \beta_0 = \frac{3}{4} \Rightarrow \beta_0 \approx 48.6°.$$

⊕ *Remark*: A seal that floats on the blind spot (Fig. 4.75) cannot see the Inuit who is waiting for it at the air hole in the ice: The only light rays that lead to the hunter are shielded by the ice. Even the Inuit sees the seal only when its nostrils stretch out of the water (Fig. 4.76)! ⊕ ◂◂◂

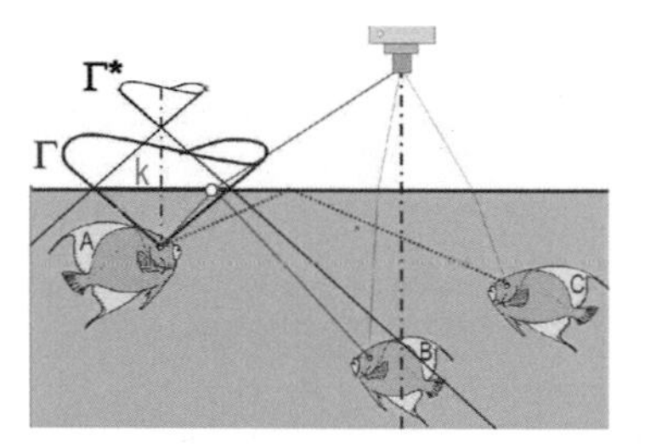


**Fig. 4.77** Who sees whom now? This question is by no means trivial!

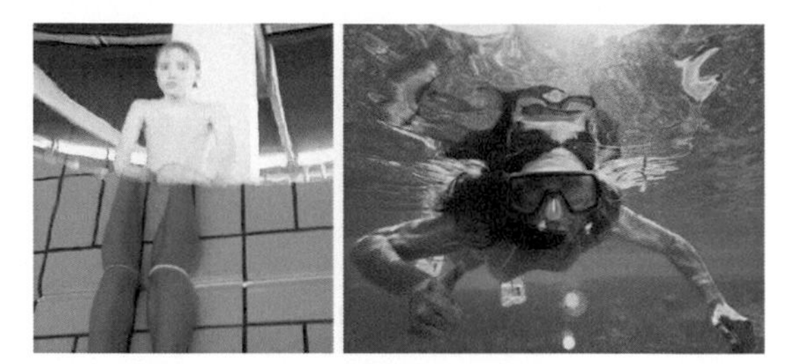

**Fig. 4.78** total reflection outside of $\Gamma^*$ (on the left the track circuit of $\Gamma^*$ is seen)

### ▸▸▸ Application: *Who sees whom?*

In the calm pool of water (Fig. 4.77, left), the fish $A$ (the diver) sees:

- "everything" outside the pool, albeit it is distorted greatly. The refracted image is inside a circle $k$ on the surface. This circle is the intersection of a cone $\Gamma$ of revolution with an aperture angle of $2 \times 48.6°$;
- the total reflections of those parts of the basin which are outside the reflected cone $\Gamma^*$, e.g. the fish $C$ (particularly clear in Fig. 4.78, left);
- reflections of the rest of the basin (Fish $B$) in the interior of $k$ – as a result of partial reflection (the calmer the surface of the water, the clearer the image);
- the fishes $B$ and $C$, even directly!

On a photograph from outside the basin (for instance, from a diving board) you can see all the fishes – to some extent severely distorted. ◂◂◂

### ▸▸▸ Application: *refraction at a spherical surface*

Use the approximation $\sin x \approx \tan x \approx x$ for small $x$ (Application p. 383) in order to deduce the Formula (4.29) for the refraction at a spherical surface.

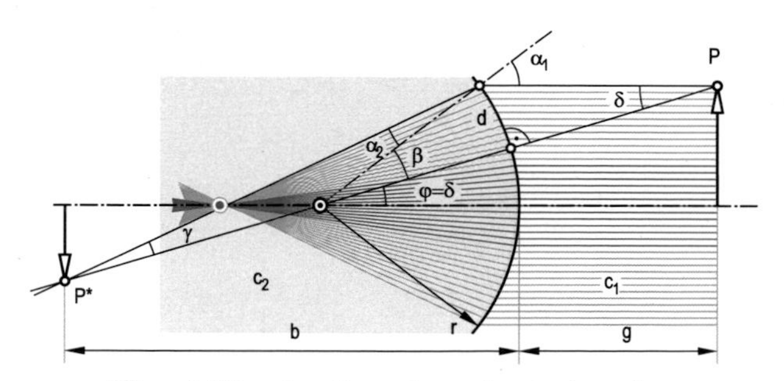


**Fig. 4.79** refraction at a spherical surface

*Solution*:
With the notation of Fig. 4.79, the relations of exterior angles $\beta$ and $\alpha_1$ are

$$(1)\quad \beta = \alpha_2 + \gamma, \qquad (2)\quad \alpha_1 = \delta + \beta.$$

If $c_1$ and $c_2$ are the velocities of the propagation of light in the two mediums, then by the law of refraction (Application p. 378), we have

$$n = c_1 : c_2 = \sin\alpha_1 : \sin\alpha_2 \approx \alpha_1 : \alpha_2 \quad \Rightarrow \quad (3)\quad \alpha_2 \approx \frac{1}{n}\alpha_1.$$

This leads to

$$\beta = \frac{1}{n}(\delta + \beta) + \gamma \;\Rightarrow\; \delta + \beta + n\gamma = n\beta \;\Rightarrow\; \delta + n\gamma = (n-1)\beta.$$

Further, the following holds:

$$\tan\delta \approx \delta \approx \frac{d}{g}, \quad \sin\beta \approx \beta \approx \frac{d}{r}, \quad \tan\gamma \approx \gamma \approx \frac{d}{b},$$

and thus, we find the approximation

$$\frac{1}{g} + \frac{n}{b} = \frac{n-1}{r}. \tag{4.29}$$

Here, $r$ is the radius of the sphere. $g$ and $b$ are the object distance and the image distance. The lens laws in this simple form are valid for only small angles $\varphi = \delta$ in the vicinity of the optical axis (Gaussian space), which are marked by green rays in Fig. 4.79. The orange rays at a greater distance from the axis envelope a "focal curve", and no longer pass through the focus of the lens (Fig. 2.10, left). ◂◂◂

▸▸▸ **Application**: ***How does a rainbow form?*** (Fig. 4.81)
A good moment to see a rainbow is when the Sun is low after a downpour (Fig. 4.80). Why? How does a rainbow form?

*Solution*:
The wavelengths of the spectrum of visible light range from 380 to 780 nanometers, i.e. slightly less than one thousandths of a millimeter. The individual spectral colors are less broken during the transition to an optically denser medium if their frequencies are smaller. Therefore, violet is refracted more strongly than red. Beyond red, we have the warming infrared rays. Beyond violet, we have the harmful ultraviolet rays.
Now, after a downpour, the air is full of (spherical) water droplets (Fig. 4.81). These, we usually see in a diffused form as mist or cloud. Let $r$ be the direction of the incoming light rays. At any point of the illuminated hemisphere, a light ray is partially reflected at the hemisphere, while the other part is partially refracted inside the hemisphere ($\rightarrow$ ray $r_1$). The ray is "fanned" into the

**Fig. 4.80** Two concentric rainbows (primary and secondary). Left: two "real" rainbows were captured from a terrace with a fish-eye lens during a low sun position. These shimmering water droplets are many kilometres away! Right: this photo was taken with a 30 mm wide angle lens. No rainbow was present in the sky – rather, a jet of water was shooting closely past the camera in the shape of a parabolic trajectory, producing two clearly visible rainbows.

different spectral colors due to the different refractive indices. The ray $r_1$ meets the sphere from the inside and is partially refracted, partially reflected ($\rightarrow$ ray $r_2$). The ray $r_2$ again meets the spherical wall and is partially refracted ($\rightarrow$ ray $r_3$), partly reflected where the fanning is partly undone, but is still maintained. The outgoing ray $r_3$ is a further fanned bundle of rays. The reflected residual light "wanders further around", whereby the light ray loses intensity by splitting into reflected and refracted light.
It now seems that $r_3$ is mainly responsible for the rainbow effect. To be more precise, it is responsible for the *primary rainbow*. A further reflection and subsequent emergence produces a *secondary rainbow* (Fig. 4.80). This was known by René *Descartes* (1596 – 1650) almost 400 years ago!

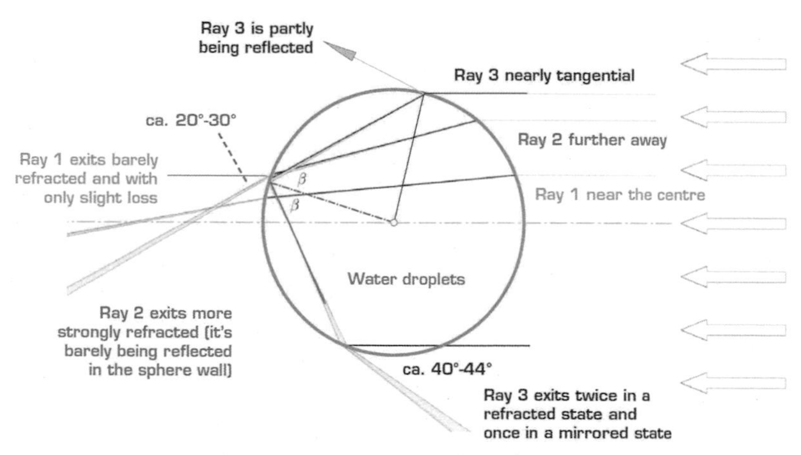


**Fig. 4.81** emergence of the rainbow

If we send not just one but an infinite number of parallel rays of light into the sphere (Fig. 4.81), most of the rays – being refracted in every direction – will leave the water droplets at the back. Those rays that arrive at the backside of the droplet approximately at the critical angle of total reflection

will return relatively highly fanned towards $r_3$. Fig. 4.81 illustrates that, at a certain exit angle ($\alpha \approx 43°$), there is a distinct maximum of such rays $r_3$.[1] For each partial spectrum in the entire spectral range – for each color – this angle is slightly different (Violet ... 42°, Red ... 44°). Viewing the droplet at this angle, an appropriate amount of color prevails.
Thus, all rain drops that "dispatch" rays $r_3$ at the angle $\alpha$ are seen in the corresponding spectral color. All these droplets are distributed on a cone of revolution whose apex is the eye and its half aperture angle equals $\alpha$. Thus, this cone of revolution appears in a "projecting" manner, and therefore, it appears as an arc (Fig. 4.81). The human eye is able to distinguish seven essentially different colors, with red on the *margin*. In the secondary rainbow – it occurs for $\alpha \approx 51°$ – the order turns back because of the additional reflection!

⊕ *Remark*: Depending on the position of the Sun, one can see at most a semicircle. Looking out of a plane (above the clouds), you can see a whole circle if the Sun is high. Since the cone is dependent on the position of the observer, one always sees a new rainbow when walking. So, it makes no sense "to look for the treasure that is buried at the foot of the rainbow". ⊕ ◂◂◂

**Fig. 4.82** red sunset with delay of five minutes ...

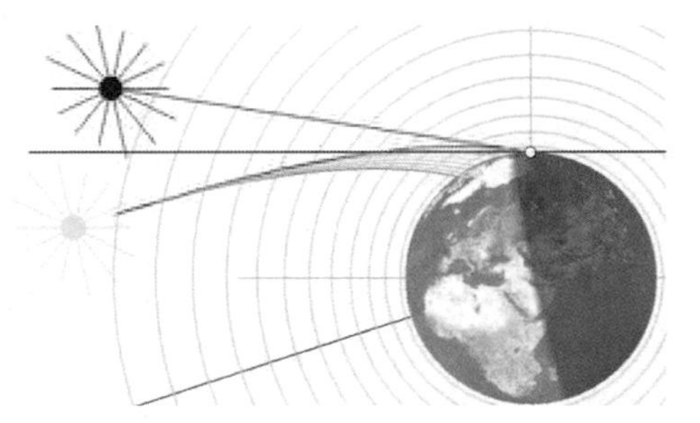

**Fig. 4.83** The underlying theory shows: We can see around the bend!

## ▸▸▸ Application: *Why does the setting Sun appear to be Red?*

*Solution*:
If the Sun is low on the horizon, the light rays, which contain the entire visible spectrum (see Application p. 166), transit the atmosphere at equal angles. The atmosphere becomes denser as it gets closer to the Earth's surface. The rays are, thereby, increasingly fanned into the different colors, with violet being the most bent towards the front and red the least. Theoretically, the

[1] Start the demo program `rainbow.exe` to see the animated situation. For more theoretical details on this issue, see the following interesting website:
`https://plus.maths.org/content/rainbows`
See also Steven *Janke*: *Modules in Undergraduate Mathematics and its Applications*, 1992.

portion of light that is blue reaches us for a longer period of time than the red component (Fig. 4.82). However, due to its shorter wavelength, this blue portion is scattered across the atmosphere (which is why the sky is blue). So, when the Sun goes down under the horizon, you can see the color red for about five minutes (Fig. 4.83)!

⊕ *Remark*: Thus, for about five minutes, we have the impression that an object is located in the extension of the light rays reaching the eye. Eventually, the red portion of light is no longer able to make it "around the bend" and the Sun goes down, seemingly deformed to an oval. ⊕ ◂◂◂

▸▸▸ **Application**: ***Fata Morgana?*** (Fig. 4.84)
This particular sunset deserves a closer explanation. Both images to the left look relatively familiar. From the third image onwards, however, something like a reflection appears to emerge. This reflection is then increasingly distorted in the subsequent images, even when the Sun has already disappeared completely.

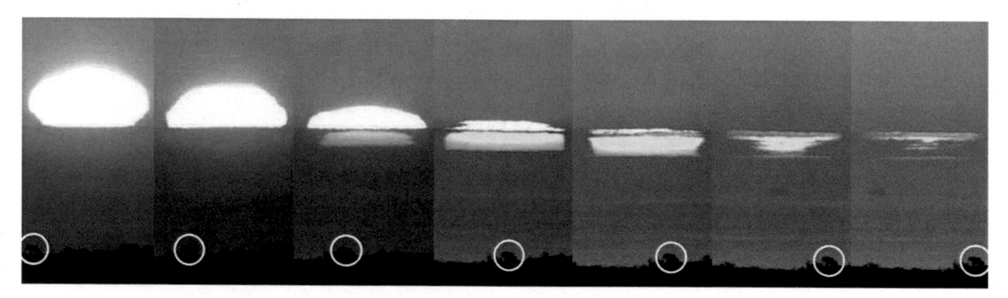

**Fig. 4.84** 3.5 exciting minutes

*Solution*:
A Fata Morgana is an optical illusion caused by the refraction of light in layers of air of differing temperature and humidity. As previously mentioned (Application p. 168), the Sun should already have disappeared from sight several minutes before the picture was taken. Due to the low angle of incidence, the blue components of light with their short wavelength were already fully absorbed by the atmosphere, while the red components still managed to go around the bend. Yet, what causes the strange reflection? The specular points are not infinitely far away, but are instead located on the surface of the water. In fact, due to the curvature of the Earth, they can be much closer than one might expect. Depending on one's height above water level, it is possible to see anywhere from several to 100 kilometres ahead (Application p. 157), though the latter distance is only possible during clear atmospheric conditions. The images on the right were taken a few seconds after the Sun had disappeared completely. The fisherman at sea, however, would still have been able to see it. The sunrays which hit the surface of the water at the fisherman's position could have been detected by us, although they might

have been slightly refracted, owing to the different atmospheric conditions of temperature and humidity. The series of photos was taken on the Cape of Good Hope at the end of February. On the northern hemisphere, at the same latitude, the Sun sets clockwise in the direction west-north-west (Application p. 305). On the same day on the southern hemisphere, the Sun sets in anti-clockwise direction towards the west-south-west. Note the encircled tree which appears to move one Sun diameter to the right within a span of 3.5 minutes. ◂◂◂

▸▸▸ **Application**: ***optical prism*** (Fig. 4.85)
The refraction of light and its dispersion into spectral colors can be beautifully demonstrated by using a three-sided glass prism. In the photo on the right, one can see three rainbows. What is the explanation?

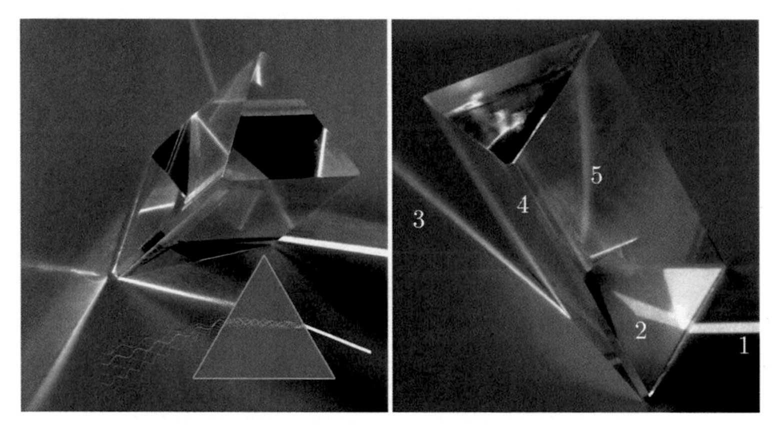


**Fig. 4.85** three different rainbows on the right

*Solution*:
At first, the white sunrays that incides on the right (1) are being dispersed as on the left. Due to refraction, this does not appear at the location where we might expect it (2). The further dispersed light rays eventually exit the object (3). This spectrum is reflected in the left surface (4), but it also reaches the eye through the right surface (5) after multiple refractions. Due to the varying angles of incidence, the spectrum appears as a curved rainbow, and unlike a common rainbow in the sky, the red colour appears inside.

⊕ *Remark*: The two almost identical reflections of the grasshopper sitting on a mirror and of the gecko on the glass window can be explained as follows: In both cases, we have two parallel planes. The light is partly reflected in a "classical" manner on the plane the animals are sitting on. Other parts of the light enter the glass zone that lies behind (refracted). If the incidence angle is small enough, a total reflection on the parallel plane occurs. The light is finally "inversely refracted" and leaves the first plane parallel to the directly reflected one. When the thickness of the reflecting material is very small, this can even create interferences. ⊕ ◂◂◂

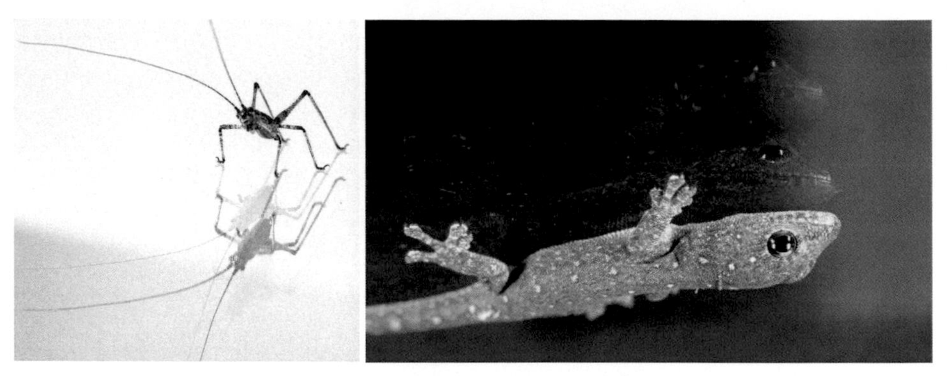

**Fig. 4.86** two almost identical reflections

**▸▸▸ Application**: ***an hour-long sunset***

On what circle of latitude is the speed of an airliner ($\approx 800\,\text{km/h}$) sufficient so that, during a flight towards the west, the apparent position of the Sun does not change?

*Solution*:

The Earth rotates over a period of 24 hours. The orbital speed $v_0$ of a point on the equator is, therefore, calculated as $\dfrac{40,000\,\text{km}}{v_0} = 24\,\text{h}$ or $v_0 = 1,667\,\dfrac{\text{km}}{\text{h}}$. (According to Application p. 127, this corresponds exactly to 900 nautical miles per hour.) An aircraft flies at this speed over the equator from east to west. The Sun's position does not change. However, $v_0$ exceeds the speed of sound, and almost all passenger aircraft fly well below this limit. The circumference, and thus also the radius $r$, of the latitude corresponding to this circle must, therefore, be smaller than that of the equator: the scaling factor $\frac{800}{1,667} = 0.48$ yields $r = 0.48\,R$. On the other hand, the radius of the circle of latitude can be calculated according to the formula $r = R\cos\varphi$, where $\varphi$ is the (northern or southern) latitude. Thus, $\cos\varphi = 0.48 \Rightarrow \varphi = 61.3°$.

⊕ *Remark*: The southern tip of Greenland is somewhat north of this geographical latitude and flights from Europe to North America – depending on the weather conditions and jet streams – operate in this passage because of the shorter distance. So, it is not unusual that during such a transatlantic flight, the Sun "stops" for hours on the horizon. The Sun can even "walk backwards" for a short time when the latitude is exceeded and/or the travel speed is increased. ⊕ ◂◂◂

**▸▸▸ Application**: ***true and mean time, time zones***

In true time, 12 noon is defined as the time when the Sun is at its peak. Accordingly, all points with the same geographical longitude should have equal local times. However, because Earth does not move uniformly due to Kepler's laws through space, the true midday thus varies by ±15 minutes. Therefore, one uses the term *mean noon* or the *average local time*. For practical reasons, time zones were introduced (average times for larger areas). So, Central European Time is the average local time for locations on 15° E longitude. Since the Earth rotates about its axis in one hour at $360°/24 = 15°$, it corresponds to the summer time in Eastern European Time (30° E longitude).

⊕ *Remark*: The People's Republic of China extends from 74° E to 135° E, i.e. more than 60 degrees of longitude. This would "normally" be enough to cover four time zones. However, for whatever reasons, China has decided to introduce "Beijing time".[2] A sunrise at a specific time (Beijing Time) "somewhere in China" can, therefore, take place within an interval of four hours of true local time. Now, if the working time of approximately 8:00–17:00 were given, it would mean that one would have to start work in the dark for most of the year in western China. ⊕ ◂◂◂

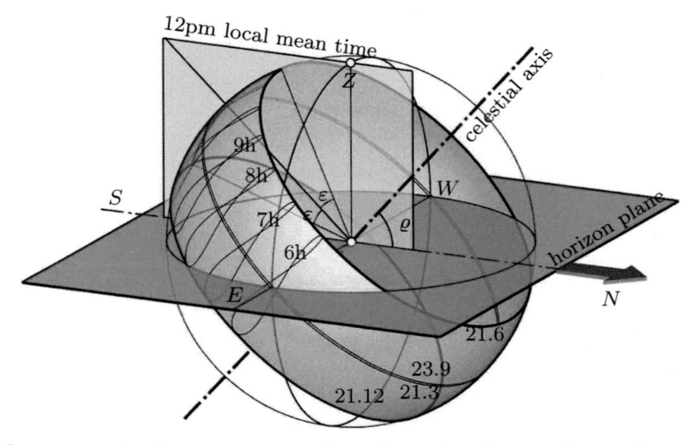


**Fig. 4.87** If you mark the points on the celestial sphere where the sun is at, at a certain hour, they form an asymmetrical 8-loop. More on this can be found on `www.uni-ak.ac.at/geom`.

The following two final applications of this chapter stem from triangle geometry and are more theoretical than applications in this book usually are. They are, however, good examples of applications of theoretical results in proofs.

### ▸▸▸ Application: *the Law of Sines in a challenging proof*

Triangle geometry is an extremely "mature" discipline of science. Making a sensational discovery in this field in our day and age is close to a miracle. In the year 1904, the American mathematician Frank *Morley* discovered the following: A theorem whose aesthetics recalls the elegant proofs in Euclid's Elements – despite the fact that the theorem is not fundamentally geometrical, as we shall soon see.

> *Morley's theorem*: If the interior angles of an arbitrary triangle are trisected, the intersections of the trisecting straight lines that are adjacent to the triangle sides form an equilateral triangle.

[2] `http://en.wikipedia.org/wiki/Time_zones_of_China`

***Proof***: The proof that follows is, in essence, due to D.J. *Newman* (1996).[3] We put the cart before the horse, so to speak, and begin with the result – an equilateral triangle $XYZ$ with side length 1. Then, we draw angles $u$, $v$, and $w$ (as shown in the lower figure), each twice. We require that $u + v + w = \frac{4\pi}{3}$. The resulting auxiliary triangles should not overlap. If we can show that $s = s^*$ and $t = t^*$, then we are almost done, because by "relabeling", we obtain equality for the remaining angles. First, we show that $s + t = s^* + t^*$.

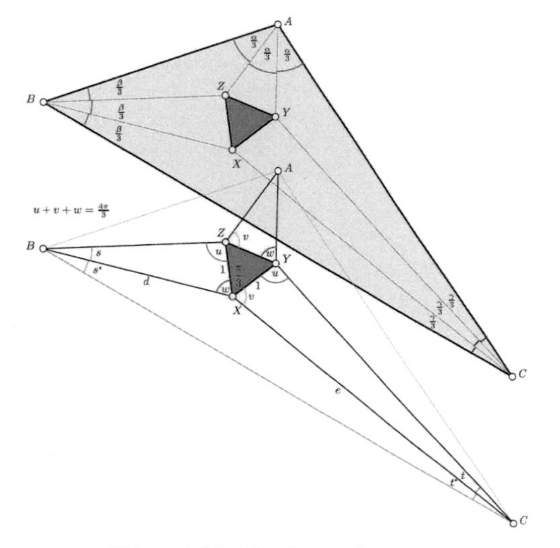

**Fig. 4.88** *Morley*'s theorem

This results in $s = \pi - u - w,\ t = \pi - u - v \ \Rightarrow\ s + t = \frac{2\pi}{3}$
and $s^* + t^* = \pi - \left(2\pi - \frac{\pi}{3} - (v + w)\right) = \frac{2\pi}{3} - u.$
Moreover, by applying the Law of Sines in the triangles $XZB$ and $XYC$, we obtain

$$\frac{1}{\sin s} = \frac{d}{\sin u},\ \frac{1}{\sin t} = \frac{e}{\sin u} \Rightarrow \frac{\sin s}{\sin t} = \frac{e}{d},$$

and by applying the Law of Sines in $BCX$, we get

$$\frac{d}{\sin t^*} = \frac{e}{\sin s^*} \Rightarrow \frac{\sin s^*}{\sin t^*} = \frac{e}{d}.$$

We may now conclude (because the sine function is monotonic, see p. 240) that

$$s + t = s^* + t^* \ \wedge\ \frac{\sin s^*}{\sin t^*} = \frac{\sin s}{\sin t} \Rightarrow s = s^* \ \wedge\ t = t^*.$$

⊙ ◂◂◂

[3]Another proof by G. *Schallenkamp* – without the use of the law of sines – can be found on `http://mathoid.de/data/documents/Morley-Theorem.pdf`.

▸▸▸ **Application**: ***The angle of circumference theorem helps to find a famous point in triangle geometry*** (Fig. 4.89)
Where is the point inside a triangle for which the sum of the distances to the vertices is minimal?

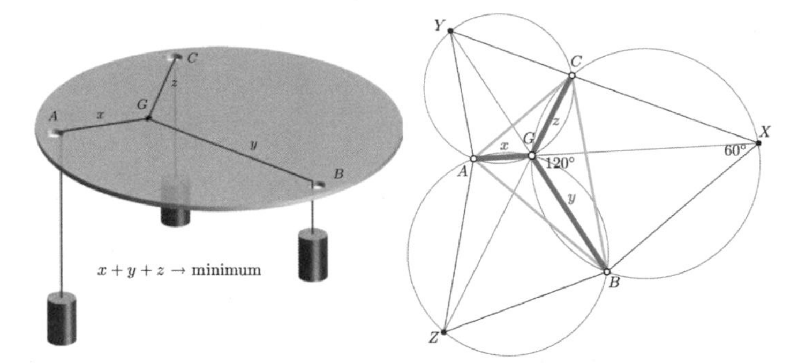


**Fig. 4.89** The Fermat point, ascribed to both Fermat and Steiner, is sometimes also referred to as Fermat–Torricelli point.

*Solution*:
Let us now imagine that we have drilled three tiny holes into a disk, through which we have pulled three strings of equal length that are tied together at the end $G$. (The triangle that is formed by the holes must not have an angle greater than 120°.) To the other ends of the strings, we attach three equal weights. The weights will immediately stretch the strings to form angles of 120°: The three equal forces are then in equilibrium. In sum, the system tries to place the weights at the lowest possible location. With a given string length $d$ and the distances $x$, $y$, and $z$ of the point $G$ from the triangle points $A$, $B$, and $C$, the weights hang at depths $d - x$, $d - y$, and $d - z$. The sum of these depths is, therefore, $3d - (x + y + z)$. If the weights have to hang as low as possible, then the expression $x + y + z$ must be a minimum. It is also possible to define $G$ as the point where the sum of distances to the triangle vertices is minimal.
The point $G$ can be constructed by intersecting the two circles or by intersecting at least two circles of circumference, or much more simply, by extending every triangle side outwards into equilateral triangles and connect the new points $X$, $Y$, and $Z$ with the opposite triangle points $A$, $B$, and $C$: The circumcircle of the triangle $ABZ$ is the circle of circumference of the segment $AB$, from whose points the segment on the side of $Z$ is visible at 60°, and the segment on the side $G$ at the supplementary angle of $180° - 60° = 120°$. In this special case, it should be obvious that the three lines actually intersect at a point: If a point $G$ is found as the intersection of two lines, then two triangle sides can be seen under 120° from $G$ – and the third triangle side must automatically be 120°, as the sum of these angles must add up to 360°.
◂◂◂

# 5 Vector analysis

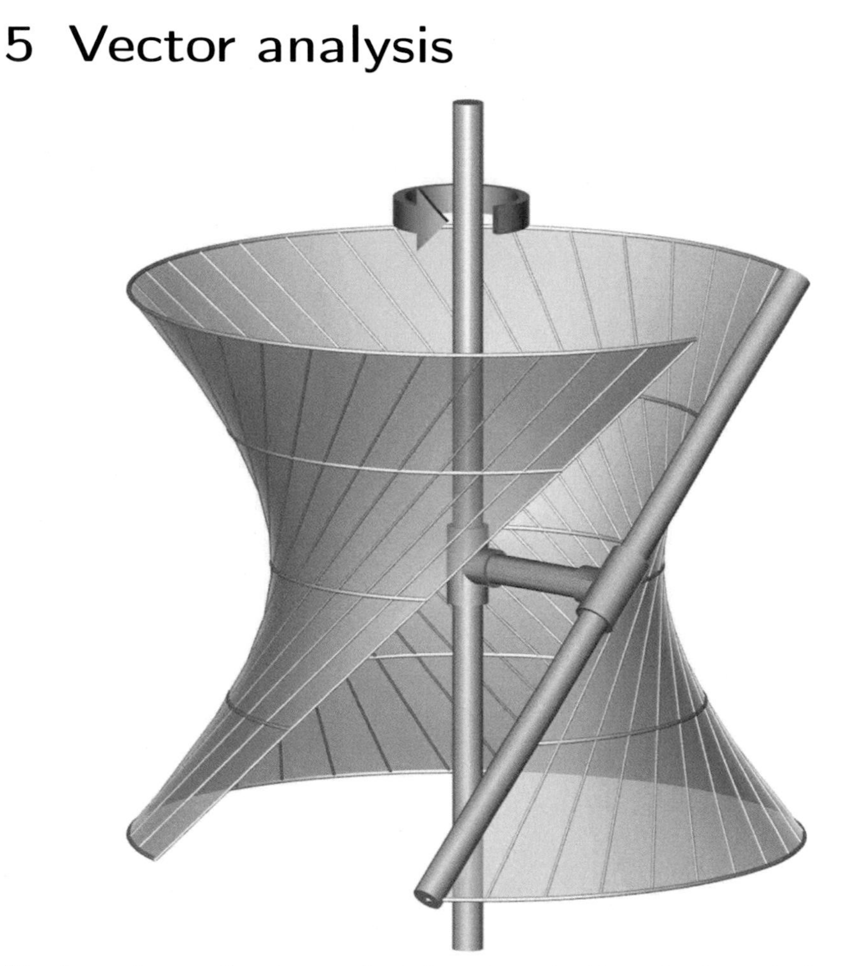

The importance of vector analysis has risen significantly through the advent of computers. Through their use, complex calculations in spatial geometry (intersection, measurement, rotation, reflection, etc.) and physics (calculations involving the parallelogram of force, center of mass, momentum of inertia, etc.) are quickly and easily accomplished.
In the following chapter, we will mostly work in three-dimensional Euclidean space $\mathbb{R}^3$. The formulae we will discuss can, if properly applied, also be used for calculations in the plane $\mathbb{R}^2$: In such cases, the third coordinate of points or the third component of vectors is simply set equal to 0. By analogy, vector analysis can be applied quite elegantly for calculations in higher dimensions. Straight lines in the drawing plane correspond to the $z$-parallel planes ("projecting" in the top view) – and not to straight spatial lines!
Interesting applications can be found even for techniques as simple as vector addition and scaling. With two further types of vector multiplication, one can also gain command over the calculation of angles, surface areas, and volumes. The wide gamut of applications ranges from problems of analytic geometry to physics, including the calculation of the position of the sun.


G. Glaeser, *Math Tools*, https://doi.org/10.1007/978-3-319-66960-1_5

## 5.1 Elementary vector operations

### Position vectors

Let us imagine a three-dimensional *Cartesian coordinate system* (Fig. 5.1) consisting of three mutually orthogonal axes $x$, $y$, and $z$. Each point $P$ in space possesses unique coordinates $P(p_x, p_y, p_z)$. The vector $\vec{p} = \overrightarrow{OP}$ emanating from the origin $O(0,0,0)$ pointing to the point $P$ is called the *position vector* of $P$, and we write

$$\vec{p} = \begin{pmatrix} p_x \\ p_y \\ p_z \end{pmatrix} \quad \text{or sometimes} \quad \vec{p}(p_x, p_y, p_z) \tag{5.1}$$

– the second version saves space. Instead of point *coordinates*, one speaks of vector *components*.

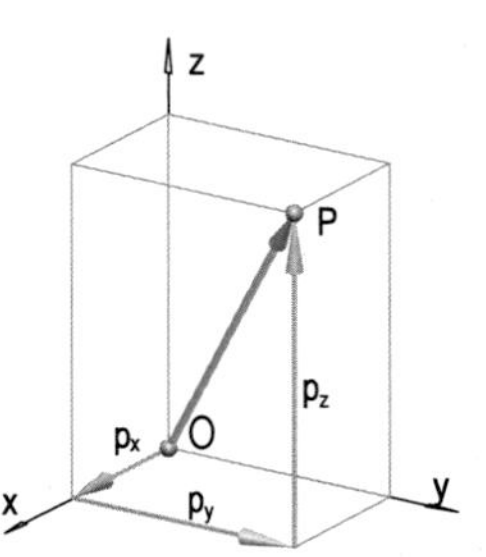


**Fig. 5.1** coordinate system and position vector

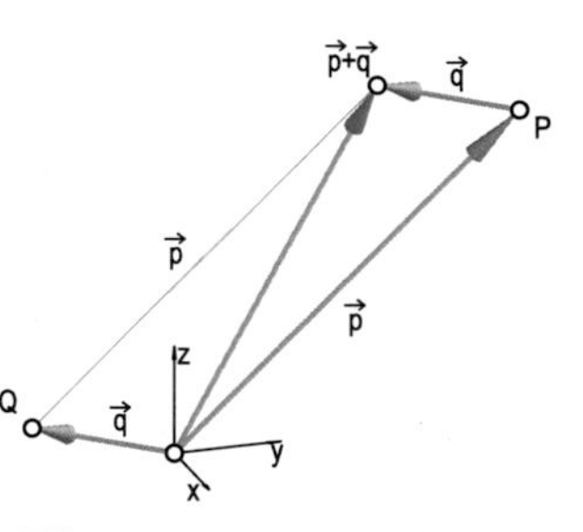


**Fig. 5.2** vector addition

### Length of a vector

According to the Pythagorean theorem, the following holds for the distance of the point $P$ to the origin of the coordinate system, and thus, for the length of its position vector $\vec{p}$:

$$|\vec{p}| = \left| \begin{pmatrix} p_x \\ p_y \\ p_z \end{pmatrix} \right| = \sqrt{p_x^2 + p_y^2 + p_z^2}. \tag{5.2}$$

### Vector addition

Let $\vec{p}(p_x, p_y, p_z)$ and $\vec{q}(q_x, q_y, q_z)$ be two vectors (Fig. 5.2). Their sum

$$\vec{p} + \vec{q} = \begin{pmatrix} p_x + q_x \\ p_y + q_y \\ p_z + q_z \end{pmatrix} = (p_x + q_x, p_y + q_y, p_z + q_z) \tag{5.3}$$

is a vector pointing towards the vertex opposite to $O$ of the parallelogram defined by $\vec{p}$ and $\vec{q}$ (Fig. 5.2).

## Translation vectors

If $\vec{p}$ is interpreted as a position vector of the point $P$ and $\vec{q}$ as a translation vector, then $\vec{p}+\vec{q}$ is the position vector of the translated point $P^t$. In physics, vector addition is applied in the "addition of forces" (Application p. 177).
*"Dualism" of a vector*:
If a vector is interpreted as a position vector, it describes a point, and if it is interpreted as a translation vector, it describes a translation. This property – to which one, admittedly, needs to get accustomed – is the secret behind the versatility of vector mathematics.

⊕ *Remark*: In physics, the word *dualism* is used to describe the fact, that radiation may be simultaneously described as a wave or as a stream of particles. The dualism of light waves↔light quanta is necessary to explain certain properties of light. ⊕

### ▸▸▸ Application: *parallelogram of forces* (Fig. 5.3)

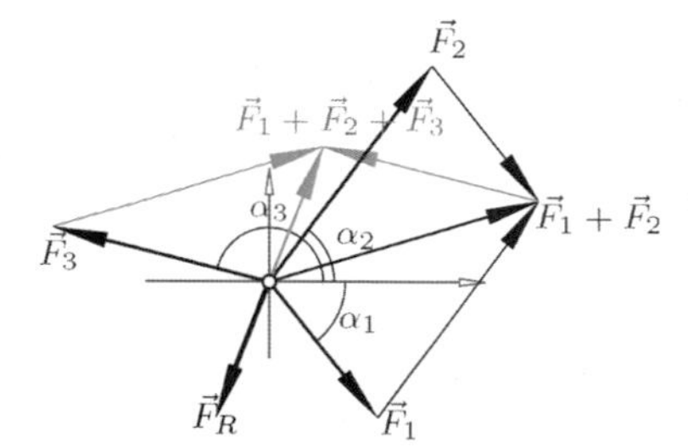

**Fig. 5.3** "Trigo" system

Two or more forces are in equilibrium if their sum is the *zero vector*. The forces themselves do not have to lie in the plane as in the figure – their sum must merely be zero.
Three forces $F_1$, $F_2$, and $F_3$ are given in the plane by their lengths $|F_i|$ and polar angles $\alpha_i$. The task is to calculate the resulting force $F_R$, which keeps these three forces in an equilibrium ("Trigo")!

*Solution*:
The resulting force is given by $\vec{F}_R = -(\vec{F}_1 + \vec{F}_2 + \vec{F}_3)$. In the plane, we may also write

$$\vec{F}_i = \begin{pmatrix} |F_i| \cos\alpha_i \\ |F_i| \sin\alpha_i \end{pmatrix} \Rightarrow \vec{F}_R = \begin{pmatrix} -\sum |F_i| \cos\alpha_i \\ -\sum |F_i| \sin\alpha_i \end{pmatrix}.$$

⊕ *Remark*: Three equally long vectors in the plane which mutually enclose an angle of 120° sum up to the zero vector. From this, Formula (4.17) from Application p. 134 immediately follows. ⊕ ◂◂◂

### ▸▸▸ Application: *Leonardo's centrifugal force experiment*

Leonardo da Vinci made spheres rotate on chains as depicted in Fig. 5.4 (the central sketch is his). Given a constant angular velocity, the spheres will move on fixed circular paths. The task is to determine the angle of inclination of the chains.

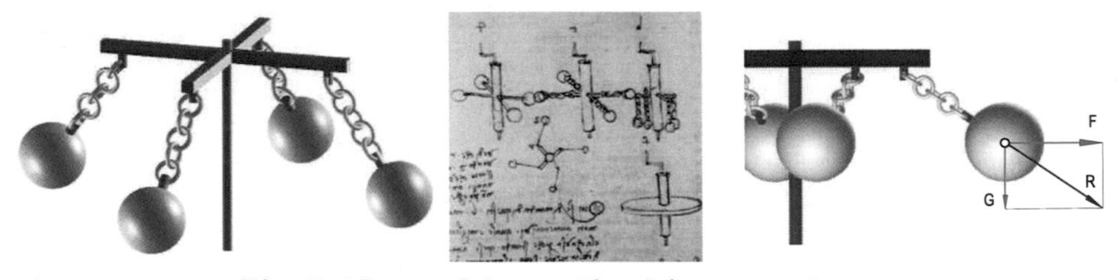


**Fig. 5.4** Leonardo's centrifugal force experiment

*Solution*:
Each sphere experiences its mass $G$ and the centrifugal force $F$ (proportional to the distance from the axis of rotation and to the square of the angular velocity). The resulting force $R$ influences the inclination of the chain (the weight of the chain is not considered).
*Numerical example*: The distance of the chain connection point is 10 cm. The length of the chain (towards the sphere's center) is 14 cm. Which angular velocity produces a chain angle of 45°?

$$F = G \Rightarrow m\,r\,\omega^2 = m\,g \Rightarrow \omega = \sqrt{\frac{g}{r}} = \sqrt{\frac{10\,\mathrm{m/s^2}}{(10 + 14\cos 45°)\,\mathrm{cm}}} \approx \sqrt{\frac{10\,\mathrm{m}}{0.2\,\mathrm{m\,s^2}}} \approx \frac{7}{\mathrm{s}}.$$

The spheres must, therefore, rotate at a rate slightly faster than one rotation about the central axis per second (with $\omega = \frac{2\pi}{\mathrm{s}}$, it would be exactly once per second). The result is obviously independent of the spheres' masses (see Application p. 138).

⊕ *Remark*: If one increases the angular velocity, the matter is more complicated. Due to the sphere's inertia, they will "lag behind", and thus, they will experience greater acceleration than the rest of the system. This causes a short-term excessive centrifugal force, which "raises" the spheres a bit too far. If the additional acceleration ends, the system returns to equilibrium. Such unstable configurations are seen in the computer simulations in Fig. 5.4. With constant angular velocity, the imaginary extensions of the chains meet the axis of rotation in a fixed point! ⊕

◂◂◂

▸▸▸ **Application**: ***Why does an airplane fly?*** (Fig. 5.5)

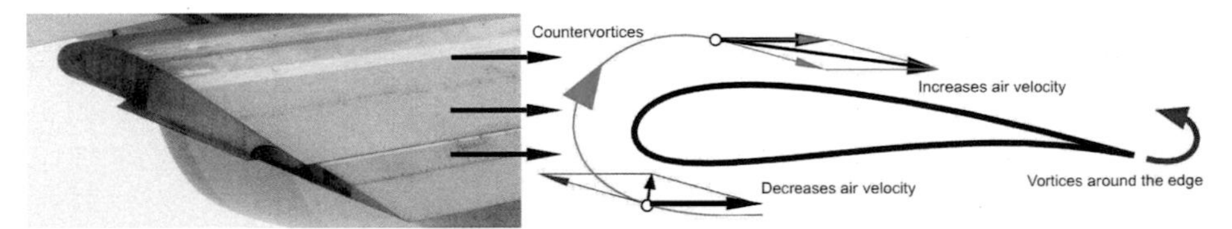


**Fig. 5.5** A non-trivial explanation of how lifting forces make an aircraft fly.

*Solution*:
The airfoil (usually almost symmetrical in the case of faster aircraft and curved upward in the case of slower ones) has a sharp edge in its rear. When "pitching" upwind, there arises a counter-clockwise velocity that is strongly

dependent on the speed $v$. (Aircraft with symmetrical profiles have extendible flaps which enhance the formation of vortices used for slow flight phases.)

**Fig. 5.6** wing motions of a butterfly (average frequency)

According to the *law of conservation of angular momentum*, a vortex arises, rotating clockwise around the profile (speed $|\Delta v|$). If the angle of approach is too large, the flow breaks off and the aircraft descends uncontrollably. At the opposite vortex, there are velocity vectors $\overrightarrow{\Delta v}$ (by definition directed and oriented) assigned to each point. The sum vector $\vec{v} + \overrightarrow{\Delta v}$ is, therefore, greater at the top than at the bottom.
Due to the *aerodynamic paradox*, negative pressure acts on the side where the air flows at greater speed, and this will generate a buoyant force that causes the aircraft to stay in the air.

**Fig. 5.7** fin motions at relatively low frequencies

⊕ *Remark*: A simplified explanation that is occasionally found reads as follows: The path of the air at the top is longer than the path at the bottom, so that the air at the top must flow at a greater velocity. This explanation is too simple.
Flying animals – like butterflies – produce the necessary vortex by synchronously twisting their wings (Fig. 5.6). At a very high wing-beat frequency, air behaves like a denser medium, and flying insects can – similar to how water animals work with water resistance (Fig. 5.7) – repel compressed air to create cushions. Aircrafts glide at high speed on air cushions. ⊕ ◂◂◂

▸▸▸ **Application**: ***increasing twisting from inside to outside*** (Fig. 5.8)
Photographs show that towards the tip, the leading edge of a bird's wing is

increasingly pulled downwards (pronated). A similar twisting can be found in an aircraft propeller. What is the impact of this wing twisting?

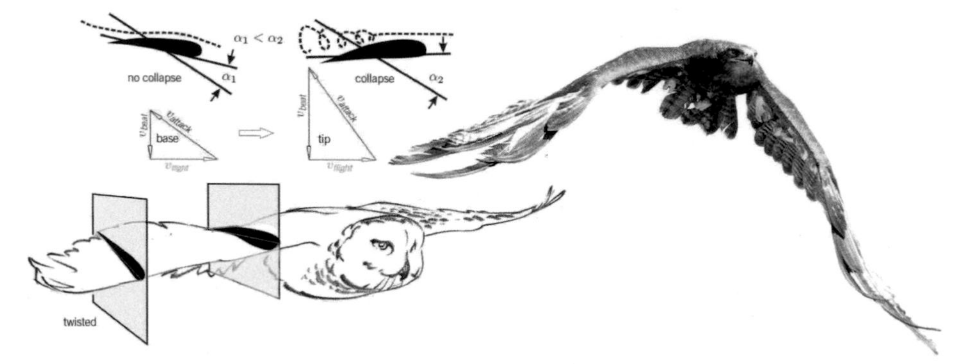


**Fig. 5.8** Wings with increasing twisting from inside to outside: $\vec{v}_{\text{flight}}$ is constant, $\vec{v}_{\text{flap}}$ increases from inside to outside. We have $\vec{v}_{\text{attack}} + \vec{v}_{\text{flight}} + \vec{v}_{\text{flap}} = \vec{0}$, and $\vec{v}_{\text{attack}}$ changes direction (and length) along the wing.

*Solution*:
A wing section can only produce beneficial lift and thrust if the resulting aerodynamic force is inclined forwards. This requires a favorable angle of attack $\alpha_1$ that is not too large, as may be the case with the proximal portion of the wing. If the wing were not twisted in this way, the distal portion of the wing would be hit by the flow at too large an angle of $\alpha_2$. Then, the airflow would collapse and this section of the wing would fail to produce any aerodynamic force. This is because, due to the constant angular velocity of the flapping wings, the absolute flapping velocity $\vec{v}_{\text{flap}}$ rises as the distance from the wing basis increases, while the flight velocity $\vec{v}_{\text{flight}}$ remains steady (see the two velocity triangles). This negative effect can completely be compensated by means of pronating wing twisting. The favourably small angle $\alpha_1$ is, thus, retained across the whole wing span. ◀◀◀

## Vector subtraction, direction vectors

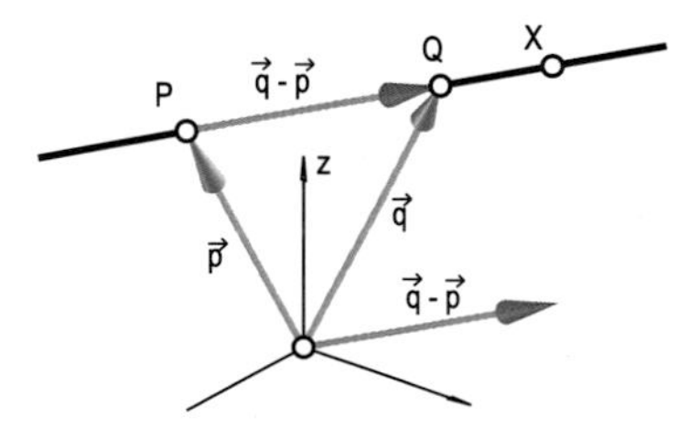

**Fig. 5.9** vector subtraction

Again, let $\vec{p}$ and $\vec{q}$ be the position vectors of two points $P$ and $Q$ (Fig. 4.8). The direction vector $\overrightarrow{PQ}$ is given by the difference vector

$$\overrightarrow{PQ} = \vec{q} - \vec{p} = \begin{pmatrix} q_x - p_x \\ q_y - p_y \\ q_z - p_z \end{pmatrix} \qquad (5.4)$$

("tip minus shaft").

▸▸▸ **Application**: ***distance between two points***
The distance $\overline{PQ}$ of two points $P$ and $Q$ equals the length of the difference vector $\overline{PQ} = \left|\overrightarrow{PQ}\right| = |\vec{q} - \vec{p}|$. ◂◂◂

## Scaling a vector

The product of the vector by a real number $\lambda$ (a "scalar") is defined as the scaling of all components:

$$\lambda\vec{p} = \vec{p}\lambda = \begin{pmatrix} \lambda\, p_x \\ \lambda\, p_y \\ \lambda\, p_z \end{pmatrix}. \tag{5.5}$$

In particular, the opposite vector $-\vec{p}$ of $\vec{p}$ is obtained by setting $\lambda = -1$.

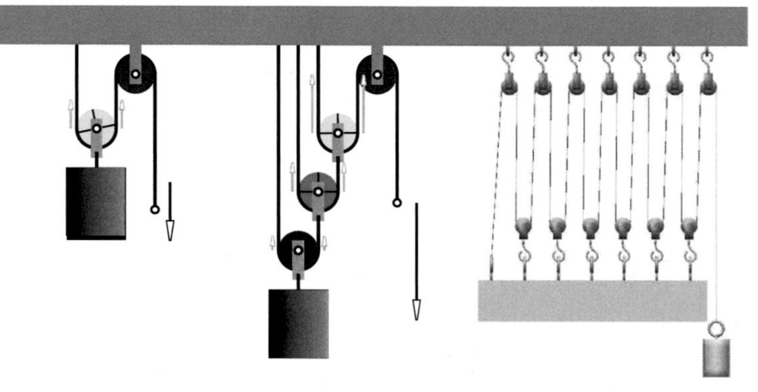

**Fig. 5.10** simple and multiple pulleys (according to *Leonardo*)

▸▸▸ **Application**: ***pulleys*** (Fig. 5.10)
For a simple pulley (incidentally invented by *Archimedes* in the 3rd century BC) where an object of weight $\vec{G}$ hangs and moves by means of a movable roller on a rope, the pulley is fixed by being mounted to a beam on one side and the rope is deflected on the other side via a guide pulley. It is sufficient to pull with a force $\vec{F} > \frac{1}{2}\vec{G}$. Find the reason for this and also explain the multiple hoists in Fig. 5.10, center and right.

*Solution*:
During lifting, kinetic energy is converted into potential energy. We have *Work* = *Force* × *Distance* = *Weight* × *Travel Distance*. Pulling the deflected rope by a length $s$ with the force $\vec{F}$, one performs an amount $s \cdot \vec{F}$ of work. Since the roll above the object is movable and will occupy the lowest position at every moment, the change in the rope length on both sides of the roll will be distributed, and the travel distance is only $\frac{1}{2}s$. Thus, the equilibrium condition reads

$$s\,\vec{F} = \frac{1}{2}\,s\,\vec{G} \Rightarrow \vec{F} = \frac{1}{2}\,\vec{G}.$$

If $\overrightarrow{F} > \frac{1}{2}\overrightarrow{G}$, then the object is pulled.
With each additional pulley, this effect occurs again: Only half the power is required allowing the load to be lifted halfway. Consequently, with $n$ pulleys, $\overrightarrow{F} > \frac{1}{2^n}\overrightarrow{G}$ provides enough force, and the load is lifted by $\frac{1}{2^n}$s. Thus, with three guide pulleys shown in the middle of Fig. 5.10, only one eighth of the weight is needed in order to keep the load in equilibrium.
The pulley on the right (after a drawing by Leonardo da Vinci) makes it possible to lift the 13-fold load: By pulling the right rope by a distance $s$, this segment is split up equally among the 13 sections. ◂◂◂

If we interpret the vector $\lambda\vec{p}$ as a position vector, it is directed towards the point that is obtained from $P$ by dilating it from the origin of the coordinate system by a factor $\lambda$. *Interpreted geometrically, multiplication of a vector by a real number describes a central dilation.*
In physics, this notation allows us to compute *centers of gravity* elegantly:

▸▸▸ **Application**: ***the center of gravity*** $S$ ***of a segment*** $AB$
Let us assume that a segment has "dead weight" if the mass is distributed homogeneously. Then, the center of gravity and the midpoint of the segment $AB$ coincide:

$$\boxed{\vec{s} = \frac{1}{2}(\vec{a} + \vec{b}).} \tag{5.6}$$

Therefore, the center of gravity divides the segment in the ratio $1:1$. ◂◂◂

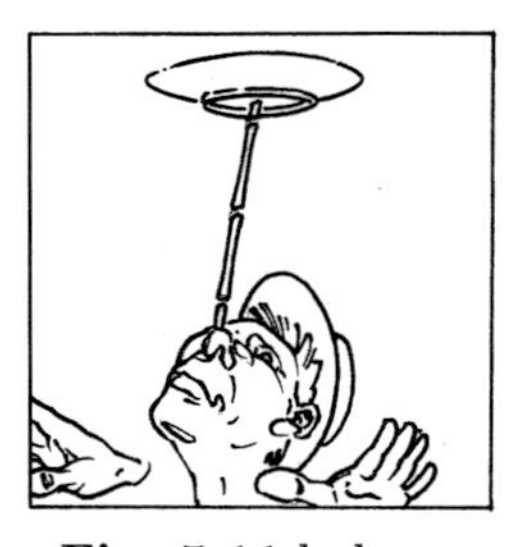

**Fig. 5.11** balance

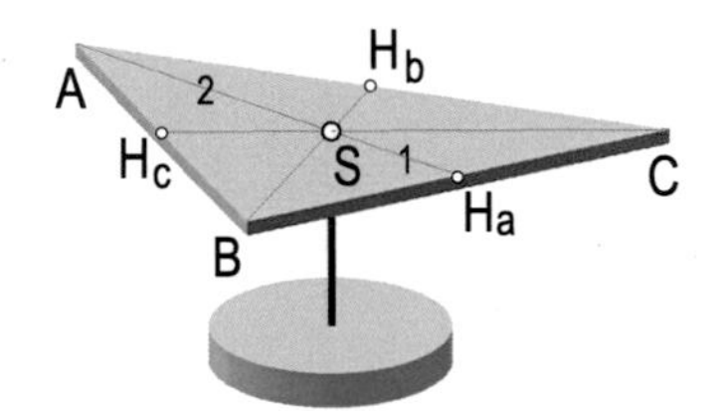


**Fig. 5.12** center of gravity of a triangle

▸▸▸ **Application**: ***center of gravity*** $S$ ***of a triangle*** $ABC$ (Fig. 5.12)
The geometric construction of the center of gravity of a triangle (centroid) is just the intersection of the medians of the triangle. The following important theorem holds:

The center of gravity of a triangle is obtained by the formula

$$\vec{s} = \frac{1}{3}(\vec{a} + \vec{b} + \vec{c}). \tag{5.7}$$

The centroid of a triangle divides the medians in the ratio of $1:2$.

***Proof***: Let $H_a = \frac{1}{2}(\vec{b}+\vec{c})$ and $H_c = \frac{1}{2}(\vec{a}+\vec{b})$ be the midpoints of $BC$ and $AB$. Then, we use the ansatz

$$\vec{s} = \vec{c} + u \cdot \overrightarrow{CH_c} = \vec{a} + v \cdot \overrightarrow{AH_a}$$

$$\Rightarrow \; \vec{c} + u \cdot (\frac{1}{2}(\vec{a}+\vec{b}) - \vec{c}) = \vec{a} + v \cdot (\frac{1}{2}(\vec{b}+\vec{c}) - \vec{a})$$

$$\Rightarrow \; \frac{u}{2}\vec{a} + \frac{u}{2}\vec{b} + (1-u)\vec{c} = (1-v)\vec{a} + \frac{v}{2}\vec{b} + \frac{v}{2}\vec{c}.$$

A comparison of the scalars of $\vec{b}$ shows $\frac{u}{2} = \frac{v}{2} \Rightarrow u = v$.
The scalars of $\vec{a}$ result in $\frac{u}{2} = 1 - v = 1 - u \Rightarrow u = v = \frac{2}{3}$. These solutions for $u$ and $v$ also induce the equality of the scalars of $\vec{c}$. Thereby, it is already proven that the medians are divided by the centroid in the ratio of $1:2$. If we now set $u = v = \frac{2}{3}$ in the first equation, we obtain the desired formula (5.7). ⊙ ◂◂◂

The center of gravity of a general planar $n$-gon is not obtained by simply adding the position vectors and multiplying the sum vector by $\frac{1}{n}$! One must instead disassemble the polygon into triangles ("triangulate"), calculate the individual centers of gravity and their areas, and combine them. How this is done is shown in Formula (5.16). In Application p. 218, we will calculate the center of gravity of a general quadrangle.
In the special case of a *regular polygon with an even number of vertices* (Fig. 5.13), it is obvious what the center of gravity is. The same applies to all polygons, resulting from the above through stretching or compression (that is, by an "affine mapping"), in particular this holds for all parallelograms.

⊕ *Remark*: If you draw a triangle in which a straight line cuts the triangle into two equal parts (in general, a triangle and a quadrangle), these generalized focal lines do not go through the center, but envelope a curve. See Application p. 346). ⊕

▸▸▸ **Application**: ***the centroid $S$ of a tetrahedron*** $ABCD$ (Fig. 5.14):
By no means must a tetrahedron be regular. The geometric center of gravity is the intersection of the "medial planes". In terms of vectors, the centroid equals

$$\vec{s} = \frac{1}{4}(\vec{a} + \vec{b} + \vec{c} + \vec{d}). \tag{5.8}$$

The center of gravity divides the "medial planes" in the ratio of $1:3$.

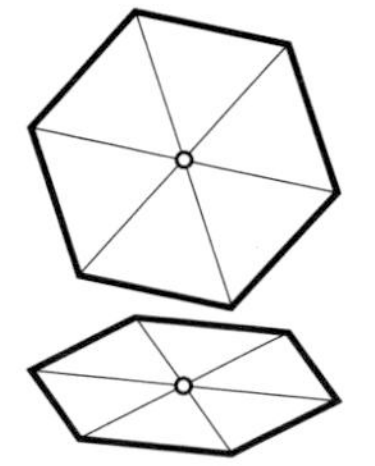

**Fig. 5.13** center of gravity of a hexagon ...

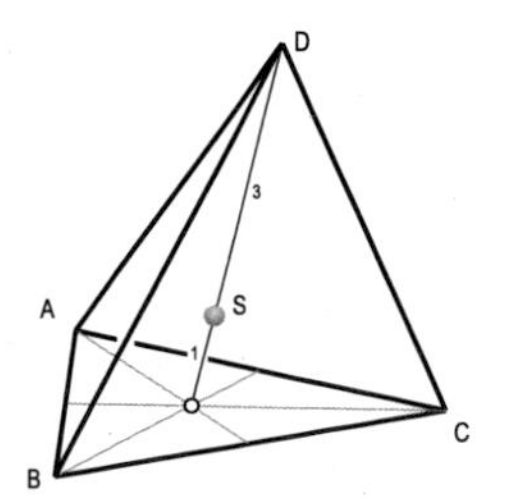


**Fig. 5.14** ... and a tetrahedron

The proof is derived analogously to the one above by choosing the centers of gravity of two triangles instead of midpoints $H_a$ and $H_c$. This yields $u = v = \frac{3}{4}$, which shows the specified division of the medial lines.

⊕ *Remark*: The calculated centroid is the center of gravity of the four vertices and at the same time the center of gravity of the tetrahedron considered as a full solid. It is different from the center of gravity of the six edges and also different from the center of gravity of the four triangular faces. ⊕ ◂◂◂

Again, the formula applies *only to a tetrahedron*, but not to general polyhedra with $n > 4$ vertices. The point obtained by computing the "average" of the position vectors $\vec{a}_i$ of the vertices $A_i$

$$\vec{s} = \frac{1}{n}(\vec{a}_1 + \vec{a}_2 + \cdots + \vec{a}_n) = \frac{1}{n}\sum_{i=1}^{n}\vec{a}_i \tag{5.9}$$

is the center of gravity of the $n$ vertices $A_i$. It is also the center of $n$ spheres of equal weight centered at the points $A_i$.
In addition, there is the *center of gravity of the edges*, which occurs when the polyhedron is considered only as a wireframe. In the following, however, we are most interested in the *center of gravity of full solids*.

For *centrally symmetric polyhedra*, it is intuitively clear that the different kinds of centers of gravity coincide. In particular, this applies to regular prisms and cubes and to affine copies of them (where the edges are not perpendicular to the base surface).

## Parametric representation of a straight line

An arbitrary point $X$ in space (Fig. 5.9) with the position vector $\vec{x}$ on the line $PQ$ can now be described by the vector equation

$$\boxed{\vec{x} = \vec{p} + \lambda\,\overrightarrow{PQ} \quad (\lambda \text{ real}).} \tag{5.10}$$

We give two examples:

▸▸▸ **Application**: ***dividing a segment***
The points $P$ and $Q$ correspond to the values $\lambda = 0$ and $\lambda = 1$. The points between $P$ and $Q$ are described by $0 < \lambda < 1$. In particular, the *center* $M$ corresponds to the value

$$\lambda = \frac{1}{2} \Rightarrow \vec{m} = \vec{p} + \frac{1}{2}(\vec{q} - \vec{p}).$$

This results in

$$\boxed{\vec{m} = \frac{1}{2}(\vec{p} + \vec{q}).} \tag{5.11}$$

◂◂◂

▸▸▸ **Application**: ***reflection of a point $Q$ in a point $P$***
For the reflected point $Q^*$, we must set $\lambda = -1$:

$$\vec{q^*} = \vec{p} - (\vec{q} - \vec{p}) = 2\vec{p} - \vec{q}. \tag{5.12}$$

◂◂◂

An important physical application of the parametric representation of a straight line is again the calculation of the center of gravity:

▸▸▸ **Application**: ***center of gravity of a complex solid with homogeneous mass distribution*** (Fig. 5.15)
We assume that a solid is composed of two simple building blocks with masses $M_1$ and $M_2$ and respective centers of gravity $S_1$ and $S_2$. Then, $S$ is the overall center of gravity that lies on the line $S_1S_2$

$$\vec{s} = \vec{s_1} + \lambda\overrightarrow{S_1S_2}. \tag{5.13}$$

According to the *lever principle*, for the parameter $\lambda$, we have

$$\lambda\, M_1 = (1 - \lambda)\, M_2 \quad \Rightarrow \quad \lambda = \frac{M_2}{M_1 + M_2}. \tag{5.14}$$

Together with equation (5.13), this gives

$$\boxed{\vec{s} = \frac{1}{M_1 + M_2}\,(M_1\vec{s_1} + M_2\vec{s_2})\,.} \tag{5.15}$$

When dealing with homogeneous solids, we can compute with volumes instead of masses. This process can be repeated if there are more than two partial solids, where we apply the same technique:

$$\vec{s} = \frac{1}{\sum M_i}\sum M_i\vec{s_i}. \tag{5.16}$$

◂◂◂

▸▸▸ **Application**: ***the common center of gravity of the Earth and the Moon*** (Fig. 5.16)
According to Formula (5.15) and with $M_1 = 81M_2$, we have $\vec{s} = \frac{1}{82}(81\vec{s_1} + \vec{s_2})$. Because $d \approx 384,000\,\text{km}$, the common center of gravity, thus, lies at a distance $\frac{d}{82} \approx 4,700\,\text{km}$ from the center of the Earth, that is still within the Earth (Earth's radius, $r \approx 6,370\,\text{km}$). The Earth and the Moon rotate around this common center of gravity, that is, the Earth "wobbles".

⊕ *Remark*: Analogously, the Sun "wobbles" around the common center of our planetary system. We now know that there are other suns (="stars") which also "wobble". With the fact that the double planet Earth-Moon rotates around the common center of gravity $S$, one can also explain why there are high tides and low tides *twice* a day

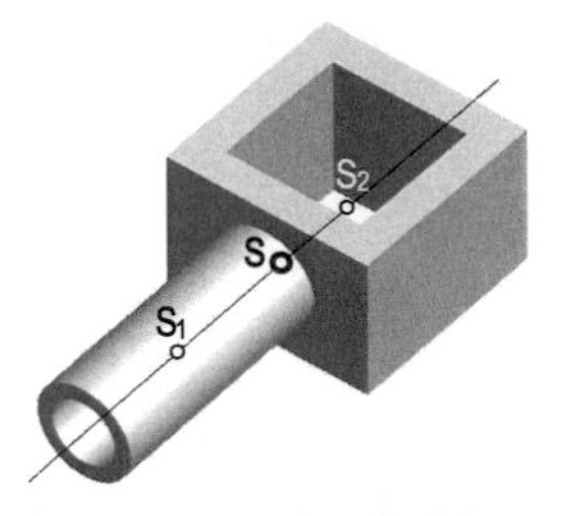


**Fig. 5.15** multiple blocks

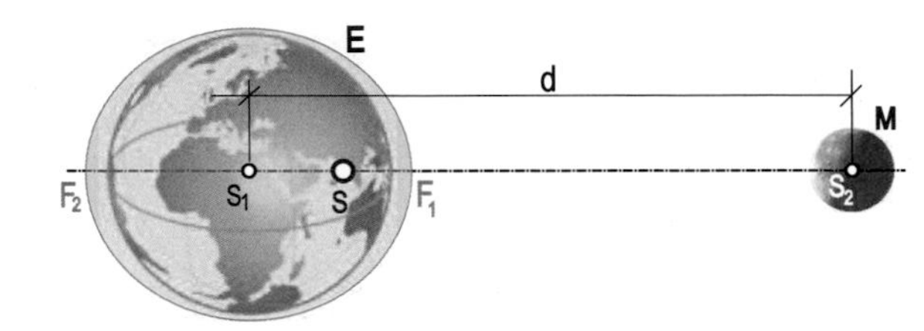


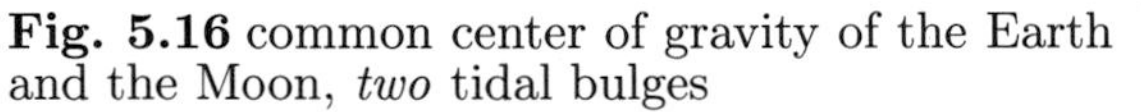
**Fig. 5.16** common center of gravity of the Earth and the Moon, *two* tidal bulges

(Fig. 5.16): The first tidal bulge is at the point $F_1$ that is closest to the Moon. This point on Earth is affected the most by lunar attraction, while it is only minimally affected by the centrifugal force during the rotation around $S$. The second tidal bulge occurs at the exact opposite point $F_2$ – we get the maximum centrifugal force due to the rotation around $S$ because it is so far away from this and more than 6 times as far as the Moon's next point. The two tide mountains move due to the rotation of the Earth over the course of almost 25 hours (the Earth has to "overturn" in order to "catch" the next migrated moon, Fig. 8.56). ⊕ ◂◂◂

**▸▸▸ Application**: ***center of gravity of a regular pyramid or a cone of revolution***

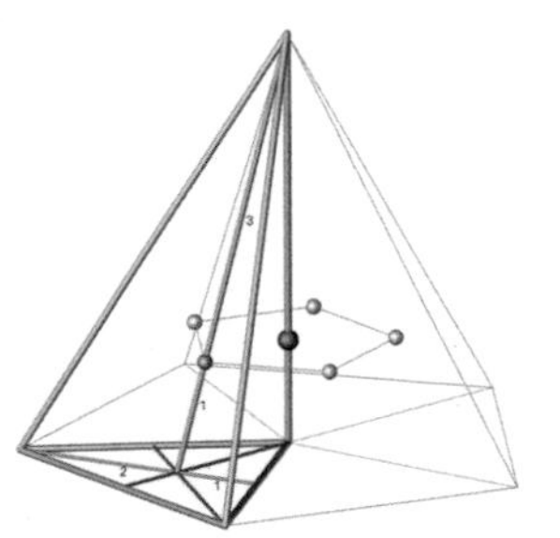

**Fig. 5.17** center of gravity of a pyramid

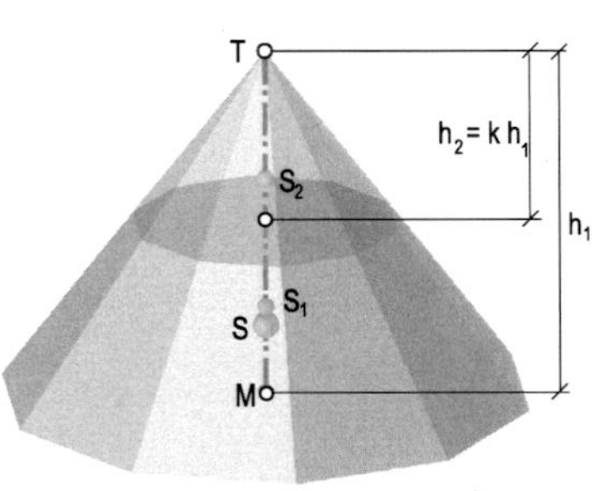


**Fig. 5.18** truncated pyramid

Let $P_1, P_2, \ldots, P_n$ be the vertices of the base and let $T$ be the tip of a regular $n$-sided pyramid with height $h$ (Fig. 5.17). (For very large $n$, the pyramid is close to a cone of revolution.) The pyramid can now be subdivided into $n$ tetrahedra $MP_iP_{i+1}T$. Their volume centers of gravity $S_i$ are the arithmetic means of the four vertices according to formula (5.8). The centers $S_i$ form a regular $n$-gon at height $\frac{h}{4}$ above the base plane. Since all sub-tetrahedra have the same mass, the overall (volume) center of gravity coincides with the center of the regular $n$-gon, and thus, we have (indeed, also for the tetrahedron)

$$\vec{s} = \vec{m} + \frac{1}{4}(\vec{t} - \vec{m}) = \frac{1}{4}(3\vec{m} + \vec{t}).$$ ◂◂◂

Formula (5.15) also works when we "subtract" one solid from a second solid. In this case, the extracted mass is negative.

►►► **Application**: ***volume center of gravity of a regular truncated pyramid***
A truncated pyramid (frustrum) is obtained when we cut off a (scaled but smaller version) $\Pi_2$ of the initial pyramid $\Pi_1$ (Fig. 5.18).
The pyramid $\Pi_1$ has height $h_1$, mass $M_1$, and volume center of gravity $S_1$ with $\overline{S_1T} = \frac{3}{4}h_1$. The pyramid $\Pi_2$ has height $h_2 = kh_1$. Then, its mass is $M_2 = k^3 M_1$. For its center of gravity $S_2$, we get $\overline{S_2T} = \frac{3}{4}h_2 = \frac{3k}{4}h_1$, and thus,

$$\Rightarrow \overline{ST} = \frac{1}{M_1 - M_2}(M_1 \overline{S_1T} - M_2 \overline{S_2T}) =$$

$$= \frac{1}{M_1 - k^3 M_1}(M_1 \frac{3}{4}h_1 - k^3 M_1 \frac{3k}{4}h_1)\overline{ST} = \frac{3(1-k^4)}{4(1-k^3)}h_1.$$

Special cases occur for $k = \frac{1}{2} \Rightarrow \overline{ST} \approx 0.80\, h_1$ and $k = \frac{3}{4} \Rightarrow \overline{ST} \approx 0.89\, h_1$. ◄◄◄

►►► **Application**: ***Stretch those legs backwards!*** (Fig. 5.19)
Flamingos fly with extended necks and legs. Their center of mass is located in the middle of the trunk, around the sternum. The bird can thus rotate about its lateral axis. If the depicted flamingos did not stretch out their legs, they would rotate in a counter-clockwise direction (G. Glaeser, H.F. Paulus, W. Nachtigall: *The Evolution of Flight.* Springer Nature, Heidelberg 2017).

**Fig. 5.19** Torque compensation avoids tilting

In order to eliminate these tilting tendencies, flamingos must move their wings in a complex manner, which would require additional metabolic energy. However, torque compensation allows them to focus their flight performance on moving forward and generating lift. Their flight muscles are relatively weak in comparison to those of other birds. Yet it is possibly due to their balanced flight that flamingos are still capable of flying distances of up to 500 km. ◄◄◄

## 5.2 Dot product and cross product

Vectors can be multiplied in two different ways. The dot product results in a *real number*. The cross product results in a *vector*. Both products play a central role in vector analysis. In particular, they are used for the determination of lengths, angles, areas, and volumes. Further, the intersection of lines/planes, distances between points/lines/planes, and reflections can be computed.

### Dot product

Let $\vec{a}(a_x, a_y, a_z)$ and $\vec{n}(n_x, n_y, n_z)$ be two vectors. Then their dot product is defined by

$$\vec{n}\cdot\vec{a} = \vec{a}\cdot\vec{n} = \begin{pmatrix} n_x \\ n_y \\ n_z \end{pmatrix} \cdot \begin{pmatrix} a_x \\ a_y \\ a_z \end{pmatrix} = n_x\, a_x + n_y\, a_y + n_z\, a_z. \tag{5.17}$$

By definition, two vectors $\vec{a}$ and $\vec{n}$ (neither being zero) are perpendicular to each other if their dot product vanishes:

$$\boxed{\vec{n} \perp \vec{a} \quad \Leftrightarrow \quad \vec{a}\cdot\vec{n} = 0.} \tag{5.18}$$

The latter equation is actually a special case of Formula (5.36) and is verified there.

For a vector $\vec{a}$ in the plane (two-dimensional), all possible normal vectors can be immediately found by

$$\vec{a} = \begin{pmatrix} a_x \\ a_y \end{pmatrix} \quad \Rightarrow \quad \vec{n} = \lambda \begin{pmatrix} -a_y \\ a_x \end{pmatrix}. \tag{5.19}$$

Using vector analysis, we can elegantly prove the following fundamental theorem of Descriptive Geometry dealing with the images of right angles:

> *Theorem of the right angle:* In a normal projection, a right angle appears as a right angle if at least one of the two legs is perpendicular to the projection direction (parallel to the plane of projection), and neither is projecting.

***Proof***: We choose a coordinate system such that the image plane coincides with the $xy$-plane. The two legs of the right angle are described by the vectors $\overrightarrow{a}$ and $\overrightarrow{n}$, where Formula (5.17) is valid because $\overrightarrow{a} \perp \overrightarrow{n}$. In the normal projection onto the $xy$-plane, the $z$-component simply disappears. If the projections of $\overrightarrow{a}$ and $\overrightarrow{n}$ again form a right angle, then we have $a_x\, n_x + a_y\, n_y = 0$. If we subtract Formula (5.17) from the latter equation, we obtain $a_z\, n_z = 0$. This is only satisfied if either $a_z = 0$

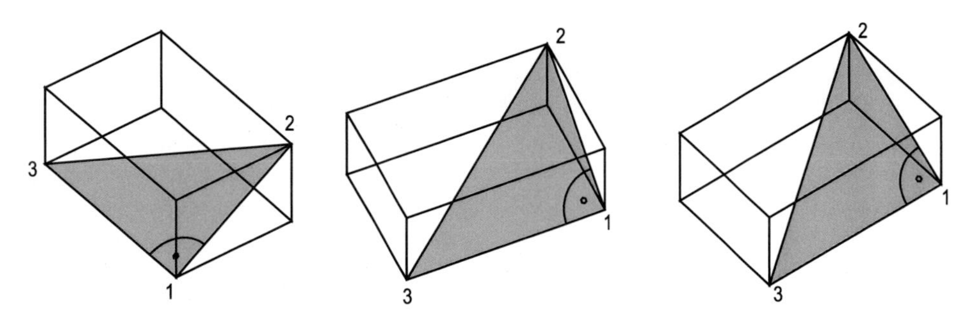


**Fig. 5.20** Application of the theorem of right angles: A cuboid is displayed in different ways in a normal projection so that the drawn right angle can be seen "undistorted", i.e. as a right angle. The direction of the projection has to be perpendicular to the edge 12, which can be done in an infinite number of ways.

and/or $n_z = 0$, that is, if at least one of the two vectors is horizontal (i.e. parallel to the image plane). ⊙

**▸▸▸ Application**: ***When exactly do spring and the other seasons begin?*** The beginnings of spring and fall are characterized by the fact that day and night have the same duration (*equinox*). Summer and winter begin when the nights are longest or shortest respectively. How can we find the exact moment?

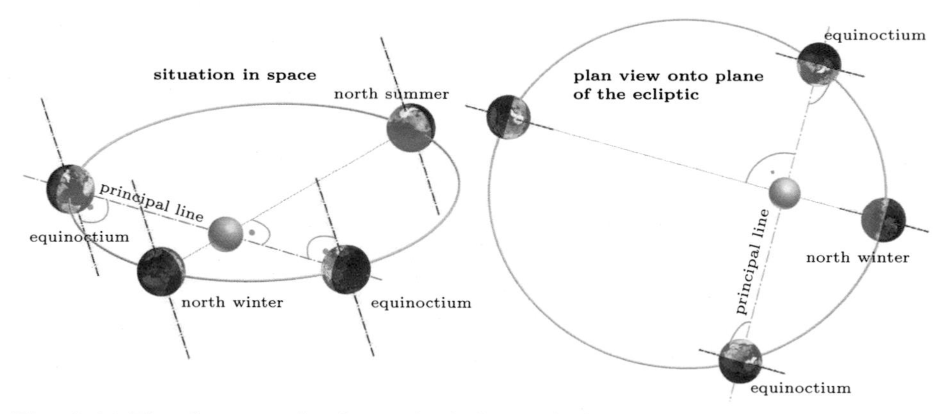


**Fig. 5.21** The theorem of right angles helps to find the beginnings of the seasons. Left: generic normal projection, right: plan view. (The shape of the elliptic orbit is exaggerated to show the construction more clearly.)

*Solution*:
Due to *Kepler*'s law, the orbit of the Earth (*ecliptic*) is an ellipse with the Sun in one focus (Fig. 5.21). The axis direction of the Earth stays constant during the year. If we consider a plan view onto the plane of the ecliptic, the Sun rays that reach the Earth are always principal lines.
The equinox happens when the terminator of the Earth goes through the poles. At this moment, the Earth axis and the sun rays are perpendicular. Thus, the desired moment can be determined accurately by means of the

theorem of right angles: The plan views of the corresponding Earth positions lie on the normal to the plan view of the axis direction.
Perpendicular to this direction, we immediately get the positions where the angle $\sigma$ between the sun rays and the axis direction is extreme – this leads to the longest days and nights.

***Proof***: To prove this, we again need the theorem of right angles (Fig. 5.22): Let $E$ be the position of the Earth where the plan view of the axis direction runs through the sun $S$. $\sigma$ is determined by a right triangle $EAC$, where $A$ is a fixed point on the axis ($\overline{EA}$ is constant) and $C$ is its foot on $s$.

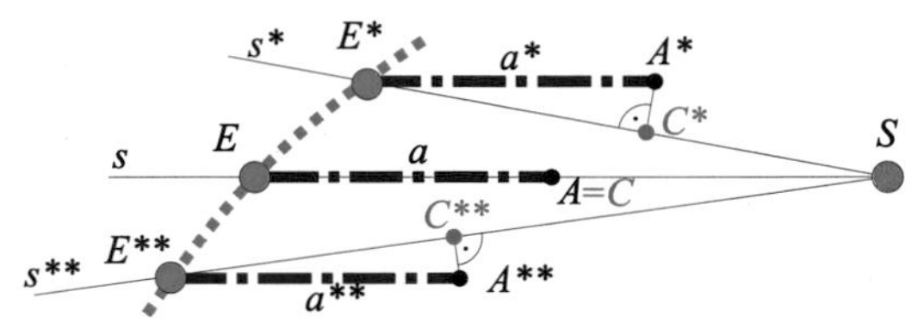


**Fig. 5.22** the extreme angle between the Sun rays and the Earth axis

In the given position, the triangle appears as an edge. In any other neighboring position $E^*$ or $E^{**}$, the foot $C$ can be constructed as the planar foot of $A$, since $s$ is a principal line. This means $\overline{AC}$ grows longer, and thus, $\sigma$ becomes smaller in both directions, so that $\sigma$ has its maximum in $E$. ⊙ ◂◂◂

## Cross product

If $\vec{n}$ is normal to two vectors $\vec{a}$ and $\vec{b}$, its components have to satisfy additionally the condition $n_x\,b_x + n_y\,b_y + n_z\,b_z = 0$. The so-called *cross product*

$$\vec{n} = \vec{a} \times \vec{b} = \begin{pmatrix} a_x \\ a_y \\ a_z \end{pmatrix} \times \begin{pmatrix} b_x \\ b_y \\ b_z \end{pmatrix} = \begin{pmatrix} a_y\,b_z - a_z\,b_y \\ a_z\,b_x - a_x\,b_z \\ a_x\,b_y - a_y\,b_x \end{pmatrix} \tag{5.20}$$

gives a vector that fulfils both conditions and is, therefore, normal to both $\vec{a}$ and $\vec{b}$:

$$\boxed{\vec{n} = \vec{a} \times \vec{b} \quad \Rightarrow \quad \vec{a} \cdot \vec{n} = \vec{b} \cdot \vec{n} = 0.} \tag{5.21}$$

***Proof***: We check the condition for the vector $\vec{a}$ (analogously for $\vec{b}$) by calculation. We have

$$\vec{n} \cdot \vec{a} = \begin{pmatrix} a_y\,b_z - a_z\,b_y \\ a_z\,b_x - a_x\,b_z \\ a_x\,b_y - a_y\,b_x \end{pmatrix} \cdot \begin{pmatrix} a_x \\ a_y \\ a_z \end{pmatrix} = (a_y\,b_z - a_z\,b_y)a_x + (a_z\,b_x - a_x\,b_z)a_y + (a_x\,b_y - a_y\,b_x)a_z = 0.$$

⊙

Note that the order of the vectors is important:

$$\vec{a} \times \vec{b} = -\vec{b} \times \vec{a}. \tag{5.22}$$

If one likes to compute (2,2)-determinants, Formula (5.20) can also be re-written as follows:

$$\vec{a} \times \vec{b} = \begin{pmatrix} \begin{vmatrix} a_y & b_y \\ a_z & b_z \end{vmatrix} \\ - \begin{vmatrix} a_x & b_x \\ a_z & b_z \end{vmatrix} \\ \begin{vmatrix} a_x & b_x \\ a_y & b_y \end{vmatrix} \end{pmatrix}. \tag{5.23}$$

▸▸▸ **Application**: ***buildings with a pitched roof*** (Fig. 5.23)
Compute the normal vector of the roof surface $PQR$ and check the result.

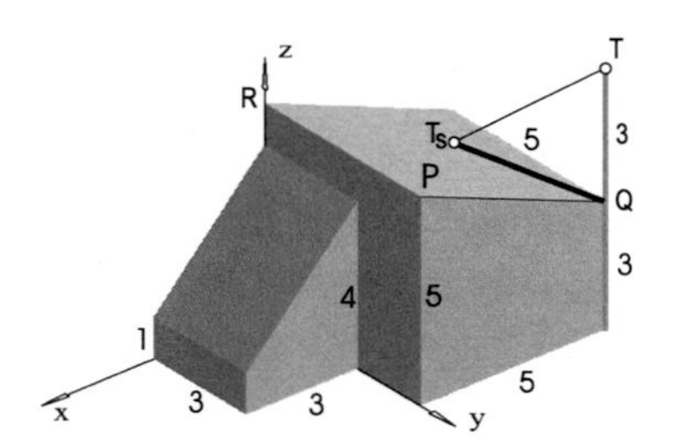


**Fig. 5.23** pitched roof

*Solution*:
From the drawing in Fig. 5.23, we read the coordinates of the points $P$, $Q$, $R$: $P(0/5/5)$, $Q(-5/5/3)$, $R(0/0/5)$. We get $\vec{n} = \overrightarrow{PQ} \times \overrightarrow{PR} =$

$$= \begin{pmatrix} -5 \\ 0 \\ -2 \end{pmatrix} \times \begin{pmatrix} 0 \\ -5 \\ 0 \end{pmatrix} = \begin{pmatrix} 0 \cdot 0 - (-2) \cdot (-5) \\ (-2) \cdot 0 - (-5) \cdot 0 \\ (-5) \cdot (-5) - 0 \cdot 0 \end{pmatrix} = \begin{pmatrix} -10 \\ 0 \\ 25 \end{pmatrix}.$$

Check: $\vec{n} \cdot \overrightarrow{PQ} = (-10) \cdot (-5) + 0 \cdot 0 + 25 \cdot (-2) = 0$,
$\vec{n} \cdot \overrightarrow{PR} = (-10) \cdot 0 + 0 \cdot (-5) + 25 \cdot 0 = 0$.

⊕ *Remark*: For many applications, it is important that the normal vector is *oriented outwards*. If necessary, the vector can be "reversed". ⊕ ◂◂◂

## Implicit equation of a plane

In general, three points $P$, $Q$, $R$ determine a plane (Fig. 5.23) with a uniquely determined normal vector

$$\vec{n} = \overrightarrow{PQ} \times \overrightarrow{PR}. \tag{5.24}$$

For any other point $X$ (position vector $\vec{x}$) in the plane, the following holds:

$$\overrightarrow{PX} \perp \vec{n} \Leftrightarrow \overrightarrow{PX} \cdot \vec{n} = 0.$$

With $\overrightarrow{PX} = \vec{x} - \vec{p}$, we have

$$\vec{n} \cdot (\vec{x} - \vec{p}) = 0 \Rightarrow \vec{n} \cdot \vec{x} = \vec{n} \cdot \vec{p}$$

where $\vec{n} \cdot \vec{p} = c$ is constant. The equation of the plane can, therefore, be written as follows:

$$\boxed{\vec{n} \cdot \vec{x} = c.} \tag{5.25}$$

►►► **Application**: ***equation of a plane***

What is the equation of the plane $\varepsilon = PQR$ from Fig. 5.23?

*Solution*:

With the results from Application p. 191, we get

$$\varepsilon : \begin{pmatrix} -10 \\ 0 \\ 25 \end{pmatrix} \bullet \vec{x} = \begin{pmatrix} -10 \\ 0 \\ 25 \end{pmatrix} \bullet \begin{pmatrix} 0 \\ 0 \\ 5 \end{pmatrix} = 125. \tag{5.26}$$

We have hereby inserted the point $R$. We could also have used $P$ or $Q$ to obtain the same result:

$$\begin{pmatrix} -10 \\ 0 \\ 25 \end{pmatrix} \bullet \begin{pmatrix} 0 \\ 5 \\ 5 \end{pmatrix} = \begin{pmatrix} -10 \\ 0 \\ 25 \end{pmatrix} \bullet \begin{pmatrix} -5 \\ 5 \\ 3 \end{pmatrix} = 125.$$

⊕ *Remark*: The constant $c = \vec{n} \bullet \vec{p} = \vec{n} \bullet \vec{q} = \vec{n} \bullet \vec{r}$ still depends on the length of the normal vector and this, in turn, on the points $P$, $Q$, and $R$. Instead of $\begin{pmatrix} -10 \\ 0 \\ 25 \end{pmatrix} \bullet \vec{x} = 125$, we can write $\begin{pmatrix} -2 \\ 0 \\ 5 \end{pmatrix} \bullet \vec{x} = 25$. Notated differently, the equation is, then, $-2\,x + 5\,z = 25$. ⊕ ◄◄◄

►►► **Application**: ***plane of symmetry between two points***

*Solution*:

The plane of symmetry (bisector) $\sigma$ can be interpreted as the locus of points in space that have the same distance to $P$ and $Q$.

The normal vector $\vec{n_\sigma}$ of $\sigma$ between $P$ and $Q$ is the vector $\vec{n_\sigma} = \overrightarrow{PQ} = \vec{q} - \vec{p}$. As a point of $\sigma$, we choose the center $\vec{m} = \frac{1}{2}(\vec{q} + \vec{p})$ of $\overline{PQ}$. Thus, $\sigma$ has the equation $\vec{n_\sigma} \bullet \vec{x} = \vec{n_\sigma} \bullet \vec{m} = (\vec{q} - \vec{p}) \bullet \frac{1}{2}(\vec{q} + \vec{p})$ or

$$\sigma : \; (\vec{q} - \vec{p}) \bullet \vec{x} = \frac{1}{2}(\vec{q}^{\,2} - \vec{p}^{\,2}). \tag{5.27}$$

⊕ *Remark*: In $\mathbb{R}^2$, equation (5.25) can be interpreted as a parameter-free equation of a line. Formula (5.27) is the equation of the *bisector* of the segment $\overline{PQ}$. ⊕ ◄◄◄

## 5.3 Intersecting lines and planes

In space, the following three intersection tasks often occur:

### Intersection of a line and a plane

Let a line $g = PQ$ be given by its parametric representation $\vec{x} = \vec{p} + \lambda\,\overrightarrow{PQ}$, and let a plane $\varepsilon$ have the (implicit) equation $\vec{n} \cdot \vec{x} = c$. Then, the point of intersection $S = g \cap \varepsilon$ has to satisfy the condition $\vec{n} \cdot (\vec{p} + \lambda\,\overrightarrow{PQ}) = c$. Therefore, the parameter value of the point $S$ equals

$$\lambda = \frac{c - \vec{n} \cdot \vec{p}}{\vec{n} \cdot \overrightarrow{PQ}}. \tag{5.28}$$

For $\vec{n} \cdot \overrightarrow{PQ} = 0$ ($g \parallel \varepsilon$ or $g \subset \varepsilon$), Formula (5.28) fails, because there is either no intersection or all points of $g$ are contained in $\varepsilon$.

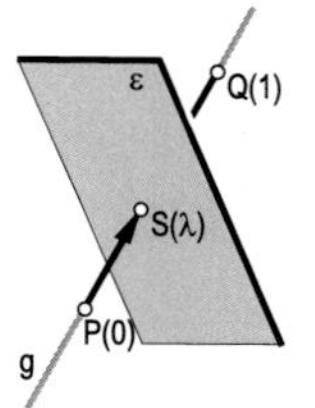


**Fig. 5.24** intersection of plane and line

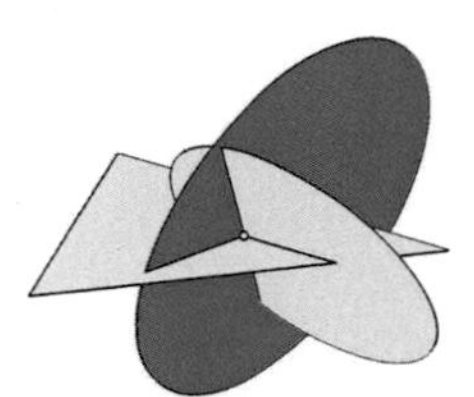

**Fig. 5.25** intersection of three planes

▸▸▸ **Application**: ***shadow of a point*** (Fig. 5.23)
Determine the shadow $T_s$ of the point $T(-5/5/6)$ sketched in Fig. 5.23 on the roof plane $PQR$ with the light direction $(1, -2, -3)$.

*Solution*:
We have to intersect the straight line $s$ through $T$ with the direction vector $\vec{s}$ with the roof plane in order to get $T_s$.

⊕ *Remark*: The light ray $\vec{s}$ reads $\vec{x} = \begin{pmatrix} -5 \\ 5 \\ 6 \end{pmatrix} + \lambda \begin{pmatrix} 1 \\ -2 \\ -3 \end{pmatrix}$. Inserted into Formula (5.26), we get $\begin{pmatrix} -10 \\ 0 \\ 25 \end{pmatrix} \cdot \vec{x} = 125$ and

$$\begin{pmatrix} -10 \\ 0 \\ 25 \end{pmatrix} \cdot \left[ \begin{pmatrix} -5 \\ 5 \\ 6 \end{pmatrix} + \lambda \begin{pmatrix} 1 \\ -2 \\ -3 \end{pmatrix} \right] = 125$$

which leads to

$$200 - 85\lambda = 125 \Rightarrow \lambda = \frac{75}{85} \approx 0.882,$$

and therefore,

$$\vec{t}_s = \begin{pmatrix} -5 \\ 5 \\ 6 \end{pmatrix} + 0.882 \begin{pmatrix} 1 \\ -2 \\ -3 \end{pmatrix} = \begin{pmatrix} -4.12 \\ 3.24 \\ 3.35 \end{pmatrix}.$$

⊕ ◂◂◂

▸▸▸ **Application**: ***anamorphoses***

Consider Fig. 5.26, right: A (fictional) hemisphere is projected from an eye point $E$ (top view $E'$) onto the ground plane $\gamma$. For this, the lines of sight $EP$ are to be intersected with $\gamma$ in order to get $P^c = EP \cap \gamma$. From the position $E$, we can – with one eye – see no difference between the point in space and the image point (Fig. 5.26, left). Thus, one can easily produce optical illusions which are called anamorphoses.

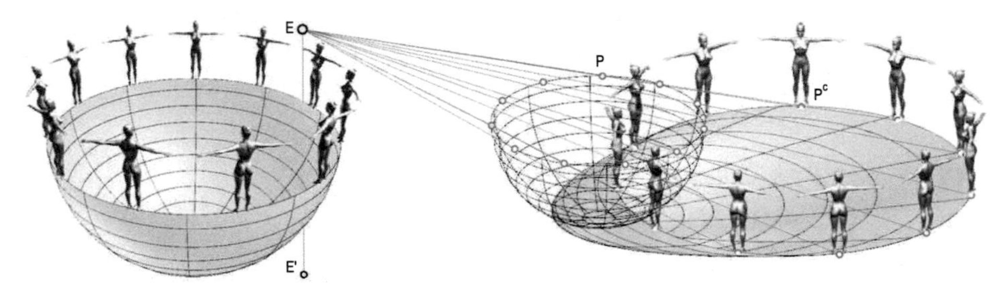


**Fig. 5.26** From the correct position, the characters merge with the projection.

*Numerical example*: The eye point $E(0/0/a)$ shall lie on the $z$-axis at the height $a$. $P(p_x/p_y/r)$ is a point at the height of the sphere's radius $r$. The line of sight $EP$ has the equation $\vec{x} = (0,0,a) + \lambda(p_x, p_y, r - a)$. For the intersection $P^c$ with $\gamma$ ($z = 0$), this results in the constant $a + \lambda(r - a) = 0 \Rightarrow \lambda = a/(a - r)$. Therefore, $P^c(\lambda\, p_x/\lambda\, p_y/0)$. Since the affine ratio $(EPP^c)$ is constant, we see figures of constant size, as shown in Fig. 5.26, established right over the points $P^c$, as well as figures of constant size on the spherical boundary. ◂◂◂

## Intersection of three non-parallel planes

Let $\vec{n_1} \cdot \vec{x} = c_1$, $\vec{n_2} \cdot \vec{x} = c_2$, $\vec{n_3} \cdot \vec{x} = c_3$ be the equations of three planes (Fig. 5.25). Then, the coordinates of the intersection point $S(s_x, s_y, s_z)$ is the solution of a system of three linear equations with three unknowns (Formula (5.29)). This can be solved by *Cramer*'s rule.

$$\begin{aligned} n_{1x}s_x + n_{1y}s_y + n_{1z}s_z &= c_1, \\ n_{2x}s_x + n_{2y}s_y + n_{2z}s_z &= c_2, \\ n_{3x}s_x + n_{3y}s_y + n_{3z}s_z &= c_3. \end{aligned} \tag{5.29}$$

If we are not able to solve the above system, then two planes are either parallel or identical. The main application of this is to solve the following tasks:

## Intersection of two non-parallel planes

The line $g$ of intersection of the two planes

$$\vec{n_1} \bullet \vec{x} = c_1, \ \vec{n_2} \bullet \vec{x} = c_2$$

has an equation of the form $\vec{x} = \vec{s} + \lambda \vec{r}$. The direction vector $\vec{r}$ is perpendicular to both of the planes' normals

$$\vec{r} = \vec{n_1} \times \vec{n_2}. \tag{5.30}$$

Now, we need a point $S(s_x, s_y, s_z)$ on $g$. In general, $\vec{r}$ is not parallel to the $xy$-plane, and so, $r_z \neq 0$. Then, $S$ can be found in the plane $z = 0$: We intersect the three planes

$$\vec{n_1} \bullet \vec{x} = c_1, \quad \vec{n_2} \bullet \vec{x} = c_2, \quad z = 0$$

and obtain the $(3,3)$-system

$$\begin{aligned} n_{1x}s_x + n_{1y}s_y + n_{1z}s_z &= c_1, \\ n_{2x}s_x + n_{2y}s_y + n_{2z}s_z &= c_2, \\ s_z &= 0, \end{aligned}$$

which is really just a $(2,2)$-system

$$\begin{aligned} n_{1x}s_x + n_{1y}s_y &= c_1, \\ n_{2x}s_x + n_{2y}s_y &= c_2, \end{aligned} \tag{5.31}$$

and can be solved using *Cramer*'s rule. In the case $r_z = 0$, one can use the plane $x = 0$ for $r_x \neq 0$. Otherwise, the plane $y = 0$ is used.

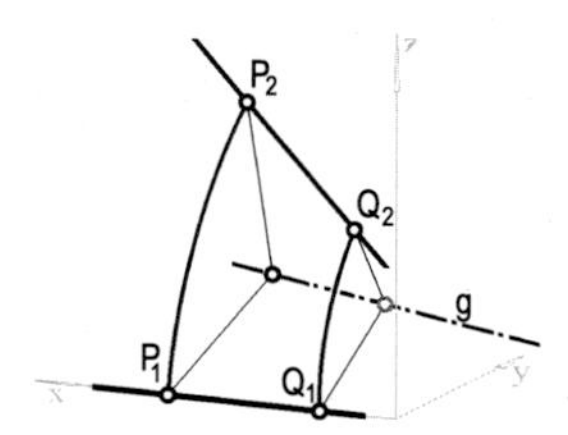


**Fig. 5.27** generic rotation

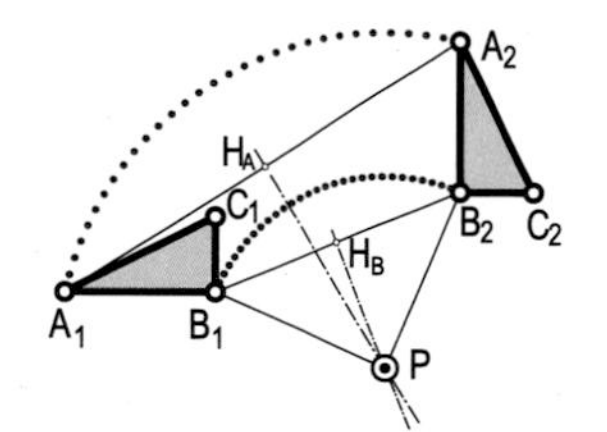


**Fig. 5.28** center of rotation

▸▸▸ **Application**: ***generic rotation in space*** (Fig. 5.27)
A line segment $PQ$ is to be moved from one position $P_1Q_1$ into another position $P_2Q_2$ by rotation. The axis of rotation is found as the intersection of the planes of symmetry of $P_1P_2$ and $Q_1Q_2$ using Formula (5.27). A detailed calculation with specific values can be found in Application p. 228.

⊕ *Remark*: In general, two congruent triangles cannot be transformed into each other by rotation. This requires a *helical motion* (the main theorem of spatial kinematics, see p. 309). ⊕ ◂◂◂

In $\mathbb{R}^2$, there is only one task related to intersections, namely the *intersection of two non-parallel lines.* Let $\vec{n_1} \cdot \vec{x} = c_1$ and $\vec{n_2} \cdot \vec{x} = c_2$ (with two-dimensional vectors!) be the equations of the two lines. Then, the coordinates of the intersection point $S(s_x, s_y)$ satisfy the system of linear equations (5.31).

▸▸▸ **Application**: ***center of a rotation*** (Fig. 5.28)
Find the center ("pole") of the rotation that transforms the triangle $A_1B_1C_1$ (system $\Sigma_1$) into the triangle $A_2B_2C_2$ (system $\Sigma_2$).

*Solution*:
$P$ is the intersection of the bisectors of $A_1A_2$ and $B_1B_2$ (again found with Formula (5.27)).

*Numerical example*:
Let $A_1(-3/0)$, $B_1(0/0)$, $A_2(1/0)$, $B_2(1/-3)$ be given. Then, we have the normal vectors $\overrightarrow{A_1A_2} = \begin{pmatrix} 4 \\ 0 \end{pmatrix}$, $\overrightarrow{B_1B_2} = \begin{pmatrix} 1 \\ -3 \end{pmatrix}$. The bisectors (Formula (5.27)) are

$$s_1: \quad \begin{pmatrix} 4 \\ 0 \end{pmatrix} \cdot \vec{x} = \frac{1}{2}\left[\begin{pmatrix} 1 \\ 0 \end{pmatrix}^2 - \begin{pmatrix} -3 \\ 0 \end{pmatrix}^2\right] = -4,$$

$$s_2: \quad \begin{pmatrix} 1 \\ -3 \end{pmatrix} \cdot \vec{x} = \frac{1}{2}\left[\begin{pmatrix} 1 \\ -3 \end{pmatrix}^2 - \begin{pmatrix} 0 \\ 0 \end{pmatrix}^2\right] = 5.$$

The (2, 2)-system has the form

$$\begin{aligned} 4 \cdot s_x + 0 \cdot s_y &= -4, \\ 1 \cdot s_x - 3 \cdot s_y &= 5, \end{aligned} \tag{5.32}$$

and, therefore, the coordinates of the center of rotation are $(-1, -2)$.
The computation of the rotation angle $\varphi$ can be done with Formula (5.36).
The fact that two directly congruent triangles in a plane can be transformed into each other by a rotation is of fundamental importance for planar kinematics (see p. 299 and the following). ◂◂◂

▸▸▸ **Application**: ***the circumcircle of a triangle***
Determine a point which is equidistant to all the three vertices of a triangle.

*Solution*:
Let $ABC$ be a triangle and let $U$ be the circumcenter. Because $U$ is equidistant to all vertices $A$, $B$, $C$, it has to lie on the bisectors of $AB$ and $BC$ (or $AC$). The radius $r$ of the circumcircle is equal to the length $\overline{UA}$.

*Numerical example*:
For $A(-3/0)$, $B(2/-4)$, $C(1/5)$, we obtain $U(0.8/-0.8)$ and $r = 3.88330$. ◂◂◂

## 5.4 Distances, angles, areas, volumes

### Normalizing a vector, unit vector

According to Formula (5.2), a vector $\vec{v}$ has length $|\vec{v}| = \sqrt{v_x^2 + v_y^2 + v_z^2}$. Scaling $\vec{v}$ by the factor $\lambda = \frac{1}{|\vec{v}|}$ returns the associate unit vector, i.e. the vector with length 1

$$\vec{v_0} = \frac{1}{|\vec{v}|}\,\vec{v} \tag{5.33}$$

parallel to $\vec{v}$ with length 1.
This process is called "normalization" of the vector. Unit vectors are usually labeled with the index 0. For each unit vector $\vec{v_0}$, we have

$$\boxed{\vec{v_0} \bullet \vec{v_0} = \vec{v_0}^{\,2} = 1.} \tag{5.34}$$

*Numerical example*: The vector $\vec{v} = \begin{pmatrix} 1 \\ -2 \\ 2 \end{pmatrix}$ has length $\sqrt{1^2 + (-2)^2 + 2^2} = \sqrt{9} = 3$. The associated unit vector is, therefore, $\vec{v_0} = \frac{1}{3}\begin{pmatrix} 1 \\ -2 \\ 2 \end{pmatrix}$.

### Drawing a segment on a line

Using normalized vectors (unit vectors), one can draw a segment of prescribed length on a given line. If we find, for example, a point $R$ on the line $PQ$ at a distance $d$ from $P$, then this point is determined by the vector equation

$$\vec{r} = \vec{p} + d\,\overrightarrow{PQ}_0. \tag{5.35}$$

### Measuring angles

Unit vectors are also the key to measuring angles in space. Let $\vec{a_0}$ and $\vec{b_0}$ be two unit vectors enclosing the angle $\varphi$. Then, we have the following important formula:

$$\boxed{\cos\ \varphi = \vec{a_0} \bullet \vec{b_0} = \frac{\vec{a} \bullet \vec{b}}{|\vec{a}| \bullet |\vec{b}|}.} \tag{5.36}$$

***Proof***: With the notation in Fig. 5.29, we have $\overrightarrow{ON} + \overrightarrow{NB} = \overrightarrow{OB}$ where $\overrightarrow{ON} = \cos\varphi\,\vec{a_0}$ and $\overrightarrow{NB} = \sin\varphi\,\vec{n_0}$ ($\vec{n} \perp \vec{a}$). This leads to the ansatz

$$\cos\varphi\,\vec{a_0} + \sin\varphi\,\vec{n_0} = \vec{b_0}. \tag{5.37}$$

a) Squaring both sides in (5.37) yields

$$\cos^2\varphi \underbrace{\vec{a_0}^2}_{1} + \sin\varphi^2 \underbrace{\vec{n_0}^2}_{1} + 2\sin\varphi\cos\varphi\,\vec{a_0}\bullet\vec{n_0} = \underbrace{\vec{b_0}^2}_{1}.$$

Together with $\cos^2\varphi + \sin^2\varphi = 1$, we deduce the condition of "orthogonality"

$$\vec{a_0}\bullet\vec{n_0} = 0.$$

b) Multiplying (5.37) by $\vec{a_0}$ gives

$$\cos\varphi \underbrace{\vec{a_0}^2}_{1} + \sin\varphi \underbrace{\vec{n_0}\bullet\vec{a_0}}_{0} = \vec{a_0}\bullet\vec{b_0} \Rightarrow \cos\varphi = \vec{a_0}\bullet\vec{b_0}.$$

⊙

With Formula (5.36), one can calculate the angle between two lines or two planes, or the angle between a line and a plane.

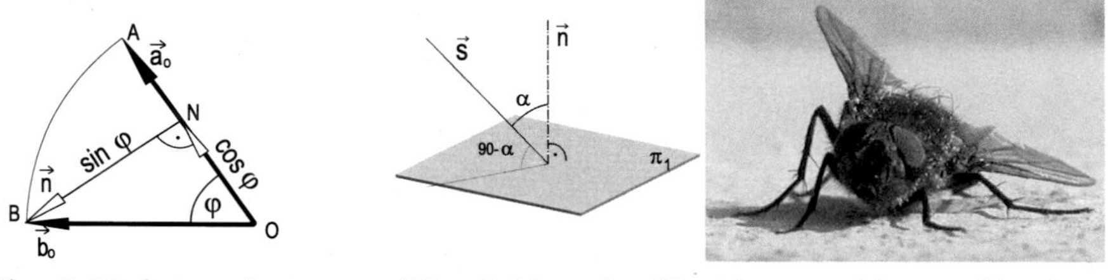


**Fig. 5.29** dot product

**Fig. 5.30** angle of incidence and best utilization

▸▸▸ **Application**: ***the angle of incidence*** (Fig. 5.30)

What is the angle of incidence $\alpha$ of the light ray $\vec{s}(1, 2, -1)$ measured to the normal of the horizontal base plane $\pi_1$? The next question would be, how bright is the base plane, or rather, what is the associated energy input (see Application p. 226)?

*Solution*:

The light ray points downwards. So, we measure the angle to the *negative* $z$-axis:

$$\vec{n} = \vec{n_0} = \begin{pmatrix} 0 \\ 0 \\ -1 \end{pmatrix}, \quad \vec{s_0} = \frac{1}{\sqrt{6}}\begin{pmatrix} 1 \\ 2 \\ -1 \end{pmatrix},$$

$$\cos\alpha = \vec{n_0}\bullet\vec{s_0} = 0.408 \Rightarrow \quad \alpha \approx 66°.$$

Following *Lambert*'s law, the brightness of the illuminated surface $h$ is proportional to the cosine of the angle of incidence ($0 \le h \le 1$). How bright is the base plane?

From the above calculation, we get

$$h = 0.408 \approx 41\%.$$

⊕ *Remark*: Without knowing *Lambert*'s law, one can guess that the fly seen on the right in Fig. 5.30 quickly warms up in the morning Sun when the angle of incidence between the wings to the light is as steep as possible. ⊕ ◂◂◂

▸▸▸ **Application**: ***a designer-lamp*** (Fig. 5.31)
An aluminium-bracket serves as a reflector of a lamp. Compute
1. the angle $BCD$ ($\cos\gamma = \overrightarrow{CB}_0 \cdot \overrightarrow{CD}_0$),
2. the angle $ADB$ ($\cos\delta = \overrightarrow{DA}_0 \cdot \overrightarrow{DB}_0$), and
3. the bending angle (angle $\beta$ enclosed by the planes $ACD$ and $BCD$). Possible additional questions: What are the lengths of the edges? What is the coat surface of the bracket?

*Solution*:
The coordinates of the relevant points can be read off from Fig. 5.31 and are

$$A(4/0/0),\quad B(0/3/0),\quad C(4/3/0),\quad D(0/0/-8).$$

Then, with $\cos\gamma = \overrightarrow{BC}_0 \cdot \overrightarrow{CD}_0$, we get the angle $\gamma \approx 65°$ and with $\cos\delta = \overrightarrow{DA}_0 \cdot \overrightarrow{DB}_0$, we find the angle $\delta \approx 33°$.
Now, we calculate the bending angle: The angle subtended by the planes $ACD$ and $BCD$ is measured as the angle of the normalized normal vectors $\vec{n}_1$ and $\vec{n}_2$. From this, the supplementary angle (i.e. the difference from 180°) is given, because the bracket is to be bent to about $\beta \approx 81°$. ◂◂◂

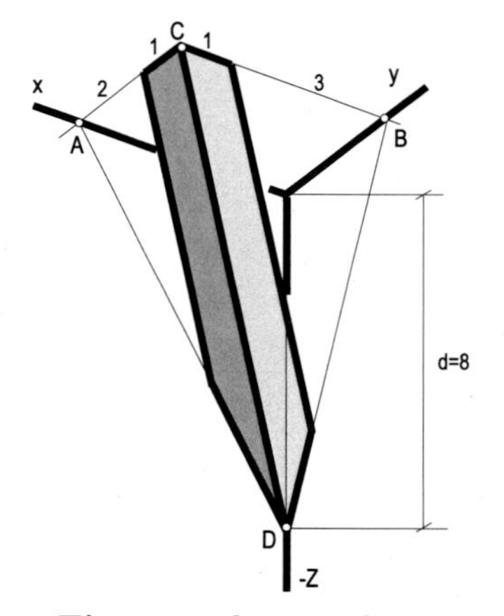

**Fig. 5.31** designer-lamp

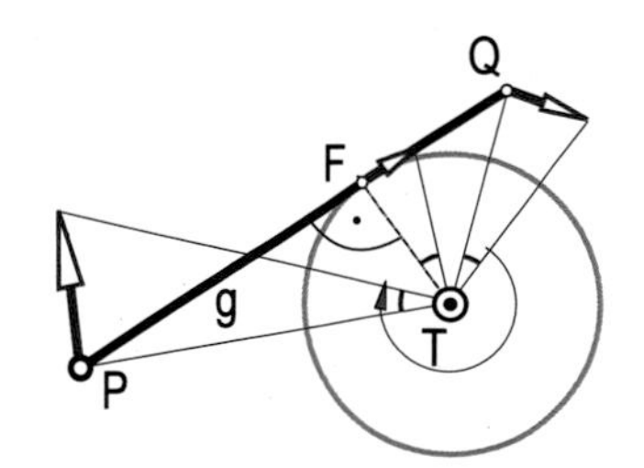

**Fig. 5.32** point with minimum speed

## Distance of a point to a plane

Let $T$ be a point in space and let $\vec{n}_0 \cdot \vec{x} = c$ be the equation of a plane $\varepsilon$. Here, $\vec{n}_0$ is already normalized. Then, $T$ has the (signed) distance from $\varepsilon$

$$d = \overline{T\varepsilon} = \vec{n}_0 \cdot \vec{t} - c. \tag{5.38}$$

## Pedal point in a plane

In particular, the pedal point $F$ of $T$ on the plane $\varepsilon$ can be easily determined:

$$\vec{f} = \vec{t} - d\,\vec{n}_0 = \vec{t} - (\vec{n}_0 \cdot \vec{t} - c)\,\vec{n}_0. \tag{5.39}$$

▸▸▸ **Application**: ***the brightest point in a plane***
Let $T$ be a light source. We will look for the point $F$ of the planar roof which is illuminated the most (Fig. 5.23).

*Solution*:
The brightest point is the pedal point of $T$ on the roof plane $\varepsilon$, since the angle of incidence of the light rays measured to the perpendicular is minimal. ◂◂◂

▸▸▸ **Application**: ***rotating a straight line*** (Fig. 5.32)
A straight line $g$ rotates about a point $T$ and envelopes a circle. Find the point $F$ on $g$ with the lowest orbital speed (i.e. the point of contact of the circle and the line $g$).

*Solution*:
Since we are in the plane, the third component of the vectors will, therefore, be zero. We use Formula (5.39) with $\vec{n}_0 \perp \overrightarrow{PQ}$ (cf. Formula (5.19)). ◂◂◂

## Distance of a point to a straight line

This problem has already been dealt with when discussing the intersection tasks: In order to find the point $F$ on a straight line $g = PQ$ which has the smallest distance from a given point $T$ (the pedal point of the normal), one intersects the normal plane $\nu$ of $g$ through $T$ with $g$ (Fig. 5.33):

$$g = PQ: \ \vec{x} = \vec{p} + \lambda \vec{r}_0 \quad (\vec{r}_0 \ \ldots \ \text{normalized vector } \overrightarrow{PQ}),$$

$$\nu: \ \vec{r}_0 \cdot \vec{x} = \vec{r}_0 \cdot \vec{t} = c,$$

$$\nu \cap g: \ \vec{r}_0 \cdot (\vec{p} + \lambda \vec{r}_0) = c \Rightarrow \vec{r}_0 \cdot \vec{p} + \lambda \underbrace{\vec{r}_0^{\,2}}_{1} = \vec{r}_0 \cdot \vec{t}.$$

This yields the parameter $\lambda_F$ of $F$, and consequently, the distance $d = \overline{Tg}$ equals

$$\boxed{\lambda_F = \vec{r}_0 \cdot (\vec{t} - \vec{p}), \quad d = \overline{TF}.} \tag{5.40}$$

▸▸▸ **Application**: ***shortest distance***
A plane flies along a straight line $PQ$ and passes an observer $T$. When is the observer closest to the plane and what is this distance if the points are $P(3/0/1)$, $Q(0/4/1)$, $T(0/0/0)$.

*Solution*:

$$\overrightarrow{PQ} = \vec{r} = \begin{pmatrix} -3 \\ 4 \\ 0 \end{pmatrix} \Rightarrow \vec{r}_0 = \frac{1}{5} \begin{pmatrix} -3 \\ 4 \\ 0 \end{pmatrix}, \quad \vec{t} - \vec{p} = \begin{pmatrix} -3 \\ 0 \\ -1 \end{pmatrix},$$

$$\lambda_F = \vec{r}_0 \cdot (\vec{t} - \vec{p}) = \frac{9}{5} \Rightarrow \vec{F} = \begin{pmatrix} 3 \\ 0 \\ 1 \end{pmatrix} + \frac{9}{5} \cdot \frac{1}{5} \begin{pmatrix} -3 \\ 4 \\ 0 \end{pmatrix} = \begin{pmatrix} 1.92 \\ 1.44 \\ 1 \end{pmatrix},$$

$$d = |\vec{f} - \vec{t}| = 2.6.$$

◂◂◂

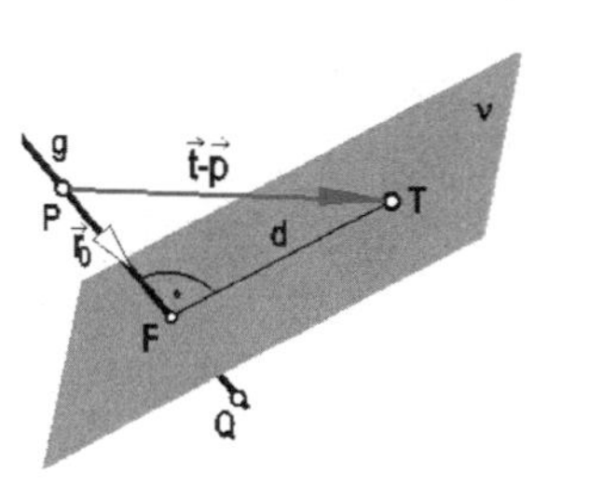


**Fig. 5.33** the pedal point of a straight line

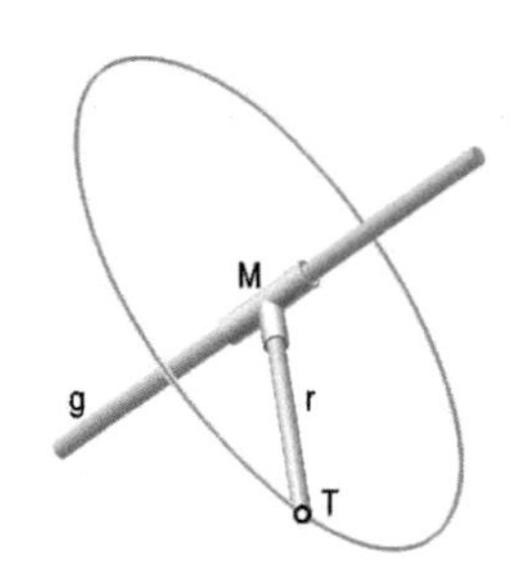


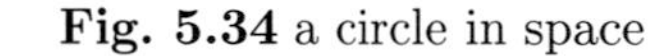

**Fig. 5.34** a circle in space

▸▸▸ **Application**: ***circle in space*** (Fig. 5.34)
A point $T$ rotates about a line $g = PQ$. Where is the center $M$ of the circle and what is the radius $R$?

*Solution*:
The center $M$ of the circle is the pedal point of $P$ on $g$ and $R = \overline{MT}$. ◂◂◂

## Distance between two skew lines

Two lines $g = PQ$ and $h = RS$ which do not lie in a plane, in general, have no intersection (that is, they are "skew"). It makes sense to define the length of the common normal of $g$ and $h$ as the distance $d = \overline{gh}$. The common normal has the direction $\vec{n} = \overrightarrow{PQ} \times \overrightarrow{RS}$ (normalized $\vec{n_0}$). The following vector equation holds (Fig. 5.35): $\vec{p} + \overrightarrow{PF_g} + \overrightarrow{F_gF_h} = \vec{r} + \overrightarrow{RF_h}$, thus, $\vec{p} + \lambda\overrightarrow{PQ} + d\vec{n_0} = \vec{r} + \mu\overrightarrow{RS}$ , and

$$\lambda\overrightarrow{PQ} + d\vec{n_0} = \vec{P}R + \mu\overrightarrow{RS}. \tag{5.41}$$

**Fig. 5.35** common normal **Fig. 5.36** approximations of a 1-sheeted hyperboloid

We take the dot product of the equation with $\vec{n_0}$. Because $\vec{n_0} \perp \overrightarrow{PQ} \Rightarrow \vec{n_0} \cdot \overrightarrow{PQ} = 0$, $\vec{n_0} \perp \overrightarrow{RS} \Rightarrow \vec{n_0}\overrightarrow{RS} = 0$, and $\vec{n_0}^2 = 1$, we find

$$d = \vec{n_0} \cdot \overrightarrow{PR}. \tag{5.42}$$

If we define $d$ as the absolute value of this dot product, then $d$ is always positive.
The parameters $\lambda$ and $\mu$ of the pedal points $F_g$ and $F_h$ are obtained by taking the dot product of the equation (5.41) with the vectors $\vec{n}_h = \overrightarrow{RS} \times \vec{n}_0$ and $\vec{n}_g = \overrightarrow{PQ} \times \vec{n}_0$:

$$\lambda = \frac{\vec{n}_h \cdot \overrightarrow{PR}}{\vec{n}_h \cdot \overrightarrow{PQ}}, \quad \mu = -\frac{\vec{n}_g \cdot \overrightarrow{PR}}{\vec{n}_g \cdot \overrightarrow{RS}}. \tag{5.43}$$

▸▸▸ **Application**: ***one-sheeted hyperboloid*** (Fig. 5.37)
A straight line $g$ rotates about an axis $h$ and sweeps a one-sheeted hyperboloid. Calculate the center $M$ of the hyperboloid and the radius $r$ of the smallest circle (waistline).

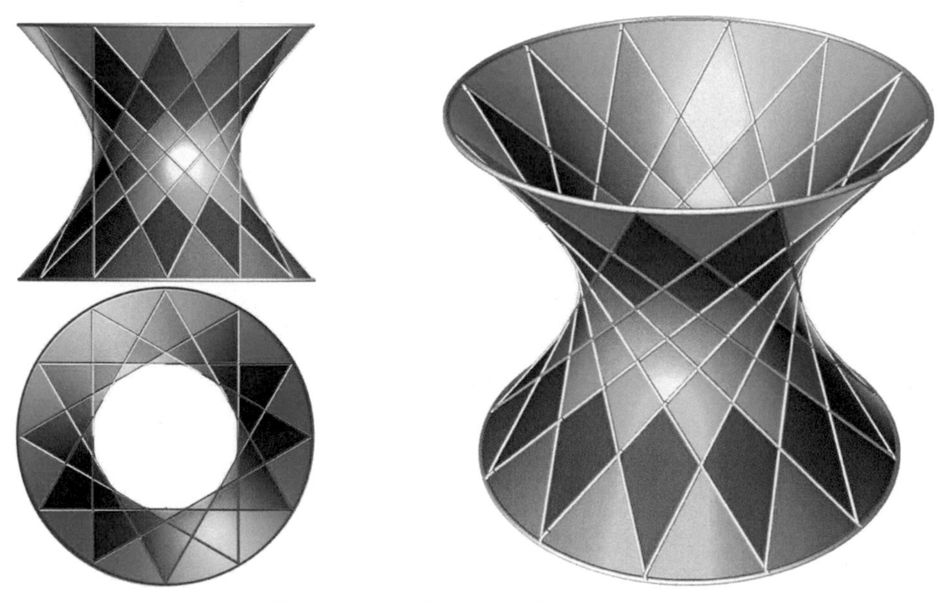

**Fig. 5.37** two families of straight lines

*Solution*:
$M \in h$ is the pedal point of the common normal $n$ of $g$, and $h$, $r$ is the shortest distance between the two lines.
The reflection of $g$ in the plane determined by $h$ and $n$ results in a straight line $\overline{g}$ which sweeps the same hyperboloid if rotated about $h$. Thus, there are *two* families of straight lines on the surface. This allows a sturdy construction of a giant walled structure in the form of a hyperboloid.

⊕ *Remark*: The cooling towers of nuclear power plants (Fig. 5.38) are examples of this. The nozzle form of the towers causes an acceleration of the water vapour in the upper third section of the tower. ⊕ ◂◂◂

**Fig. 5.38** practical application

▸▸▸ **Application**: ***hyperbolic paraboloid (HP-surface)*** (Fig. 5.39)
If one draws the lines meeting one axis $a$ at a right angle from all points on a line $g$, then these lines sweep an HP-surface $\Phi$. The line $a$ and the common normal $n$ of $a$ and $g$ are the generators at the vertex $S$ of $\Phi$.

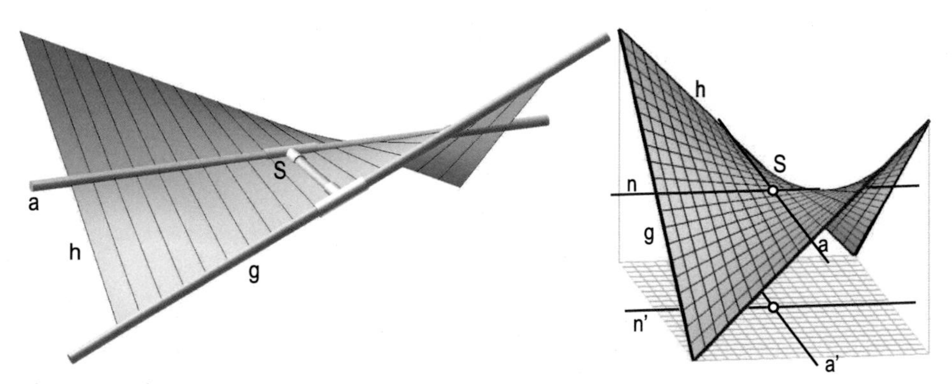


**Fig. 5.39** An HP-surface as the set of lines meeting two lines (one is met at a normal). Right: There are two families of straight lines on the HP-surface!

Conversely, one can take any arbitrary line meeting these two – in Fig. 5.39, for example, the boundary $h$ – and then collect the horizontal lines meeting $n$ in order to define the surface. Thus, the HP-surface also carries *two* different families of straight lines. Except for the one-sheeted hyperboloid (Application p. 202) and the hyperbolic paraboloid, there is no other surface with this property! According to the *Theorem of Right Angles* (p. 188), the two families of lines form a rectangular grid when projected orthogonally onto the plane spanned by $a$ and $n$. Therefore, HP-surfaces are sometimes used for covering rectangular floor plans. ◂◂◂

## Area of a triangle (parallelogram)

For the calculation of areas (area of a triangle), the cross product can be used: Let $P$, $Q$, and $R$ be the vertices of a triangle. Then, the area $A$ of the triangle is computed via

$$\boxed{A = \tfrac{1}{2}\,|\overrightarrow{PQ} \times \overrightarrow{PR}|.} \tag{5.44}$$

The area of the *parallelogram* spanned by $\overrightarrow{PQ}$ and $\overrightarrow{PR}$ is twice the area of the triangle $PQR$.

***Proof***: We give just a brief sketch: For the area of a triangle ("height times base divided by 2"), we have

$$A = \tfrac{1}{2}\,|\vec{a}|\,|\vec{b}|\,\sin\varphi.$$

Now, we have to show that

$$A = \tfrac{1}{2}\,|\overrightarrow{PQ} \times \overrightarrow{PR}|,$$

holds, and so,

$$|\vec{a}|\,|\vec{b}|\,\sin\varphi = |\overrightarrow{PQ} \times \overrightarrow{PR}|.$$

Squaring both sides of the equation, (Formula (5.36)) $\sin^2\varphi = 1 - \cos^2\varphi$ and

$$\cos\varphi = \vec{a}_0 \bullet \vec{b}_0 = \frac{1}{|\vec{a}||\vec{b}|}\vec{a} \bullet \vec{b}$$

allows us to obtain a true statement with some laborious calculations. ⊙

▸▸▸ **Application**: ***the "Spatial Pythagoras"***

In the plane, the Pythagorean theorem holds. If you intersect a right angle (vertex $C$) with a straight line, then the chord $\overline{AB} = c$ associated to the legs $\overline{AC} = a$ and $\overline{BC} = b$ can be calculated via $c^2 = a^2 + b^2$.

This theorem admits a generalization (in fact many) to a spatial version: If you intersect a rectangular tripod (vertex $D$ of a cuboid) with a plane, then the area $d$ of the triangular cross-section $ABC$ can be expressed in terms of the areas $a$, $b$, and $c$ of the triangles $DAB$, $DBC$, and $DAC$ as $d^2 = a^2 + b^2 + c^2$.

***Proof***: Let $D$ be the origin of a Cartesian coordinate system with the points $A(u/0/0)$, $B(0/v/0)$, and $C(0/0/w)$. Then, the triangles $ASB$, $BSC$, and $CSA$ have areas

$$a = \frac{uv}{2}, \quad b = \frac{vw}{2} \quad c = \frac{wu}{2}.$$

The area $d$ of the triangle $ABC$ is calculated using the cross product

$$d = \frac{1}{2} \cdot \left|\overrightarrow{AB} \times \overrightarrow{AC}\right| = \frac{1}{2} \cdot \left|\begin{pmatrix} -u \\ v \\ 0 \end{pmatrix} \times \begin{pmatrix} -u \\ 0 \\ w \end{pmatrix}\right| = \frac{1}{2} \cdot \left|\begin{pmatrix} vw \\ wu \\ uv \end{pmatrix}\right| = \frac{1}{2} \cdot \sqrt{(vw)^2 + (wu)^2 + (uv)^2}.$$

Therefore, we have $d^2 = \dfrac{1}{4}\left[\,(vw)^2 + (wu)^2 + (uv)^2\,\right] = b^2 + c^2 + a^2$. ⊙ ◂◂◂

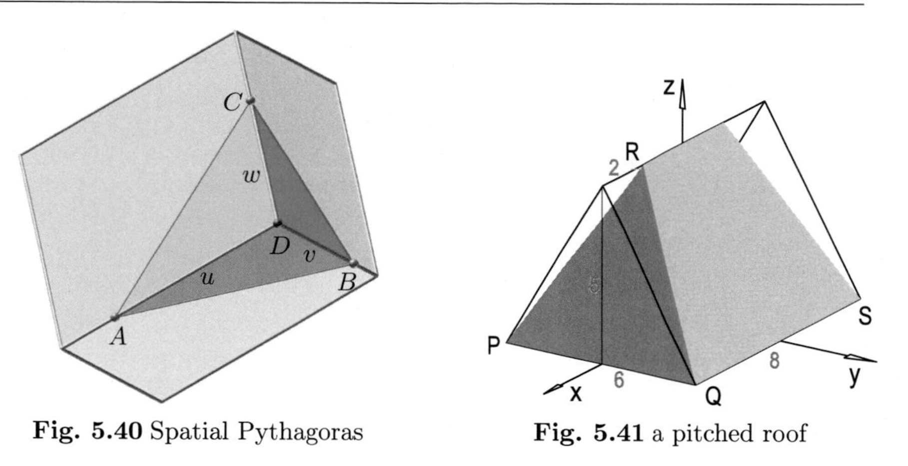


**Fig. 5.40** Spatial Pythagoras

**Fig. 5.41** a pitched roof

▸▸▸ **Application**: ***Determine the area of the triangular face of the given pitched roof.*** (Fig. 5.41)

*Solution*:

$$\overrightarrow{PQ} = \begin{pmatrix} 0 \\ 6 \\ 0 \end{pmatrix}, \ \overrightarrow{PR} = \begin{pmatrix} -2 \\ 3 \\ 5 \end{pmatrix},$$

$$A = \frac{1}{2}\left|\begin{pmatrix} 0 \\ 6 \\ 0 \end{pmatrix} \times \begin{pmatrix} -2 \\ 3 \\ 5 \end{pmatrix}\right| = \frac{1}{2}\left|\begin{pmatrix} 30 \\ 0 \\ 12 \end{pmatrix}\right| = \frac{\sqrt{1,044}}{2} \approx 16.2.$$ ◂◂◂

## Volume of a tetrahedron

If $S$ is a further point in space which together with $P$, $Q$, and $R$ spans a tetrahedron, then we obtain the *volume* of the tetrahedron with

$$V = \frac{1}{6}\overrightarrow{PS} \cdot (\overrightarrow{PQ} \times \overrightarrow{PR}). \tag{5.45}$$

The volume of the associated *oblique parallelepiped* ("spar") is six times as large.

▸▸▸ **Application**: ***volume of convex bodies*** (Fig. 5.42)
Determine the volume of an arbitrary convex body.

*Solution*:
If the body is not a polyhedron, one must first triangulate it (and thus, approximate it with arbitrary precision by means of a polyhedron). Then, look for a point $S$ inside the body (for example, the center of gravity as the arithmetic mean of all vertices) and "connect" each triangle $PQR$ of the body with $S$ in order to form tetrahedra. The total volume is the sum of all the tetrahedral volumes.
In special cases – such as the pentagon dodecahedron in Fig. 5.43 – we use regular pyramids instead of arbitrary tetrahedra.

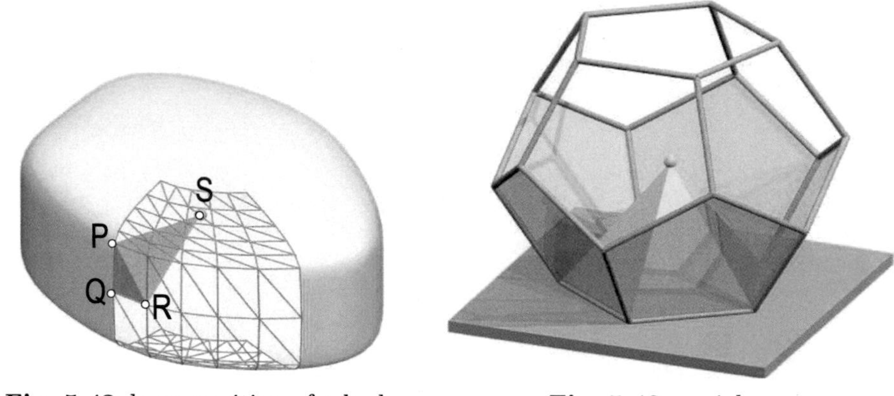


**Fig. 5.42** decomposition of a body

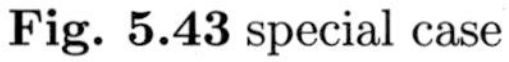

**Fig. 5.43** special case

⊕ *Remark*: The body can also be "harmlessly non-convex". The important thing is to a find a point $S$ so that the tetrahedral facets do not overlap. ⊕ ◂◂◂

## Calculating torques by means of the cross product

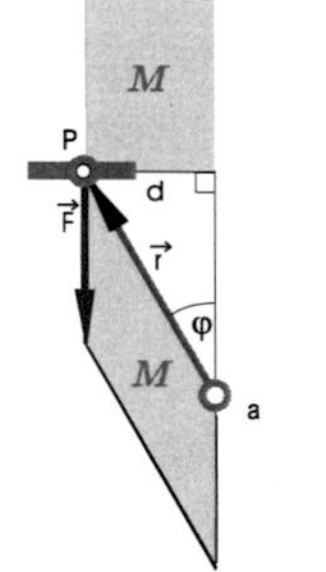


**Fig. 5.44** torque relative to an axis ...

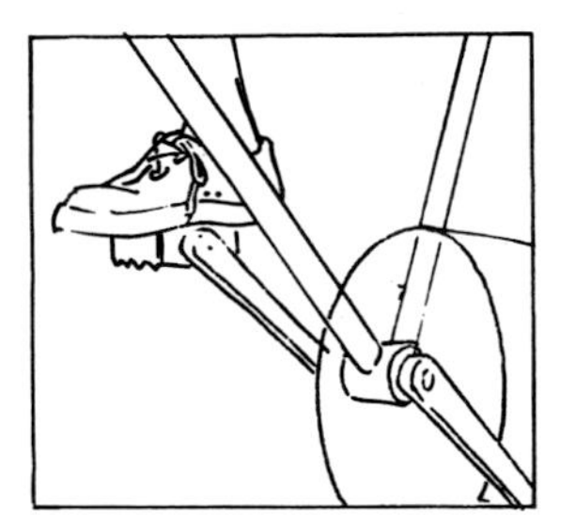

**Fig. 5.45** ... in practice

▸▸▸ **Application**: ***torque relative to an axis*** (Fig. 5.44)
By means of the cross product, the torque can be calculated. The formula

$$\textit{Torque} = \textit{Force} \times \textit{Lever}$$

is valid according to the ordinary *law of the lever*. If the force is not perpendicular to the lever arm, then the torque is calculated with the notation of Fig. 5.44 so that the lever equals

$$\mathbf{M} = \underbrace{F \cdot d}_{\text{rectangle}} = |\vec{F}|\,|\vec{r}|\sin\varphi = \underbrace{\left|\vec{F} \times \vec{r}\,\right|}_{\substack{\text{parallelogram}\\ \text{of equal area}}}.$$

Note that the order of the vectors in the cross product matters. Reversing the order changes the sign. ◂◂◂

# 5.5 Reflection

## Reflection in a plane

Let $P(p_x, p_y, p_z)$ be a point in space and $\vec{n_0} \cdot \vec{x} = c$ be the equation of the specular plane $\sigma$ ($\vec{n_0}$ normalized). Then, according to equation (5.38), $d = \vec{n_0} \cdot \vec{p} - c$ is the oriented distance $\overline{P\sigma}$. For the reflection $P^*$ of $P$ in $\sigma$, one obtains the vector equation

$$\overrightarrow{p^*} = \vec{p} - 2d\,\vec{n_0} = \vec{p} - 2(\vec{n_0} \cdot \vec{p} - c)\vec{n_0}. \tag{5.46}$$

In $\mathbb{R}^2$, one can use Formula (5.46) for the reflection in the line $\vec{n_0} \cdot \vec{x} = c$.

▸▸▸ **Application**: ***highlight point in a plane*** (Fig. 5.46)
Which point $R$ (the "highlight") of a specular tetrahedron $\sigma$ shows the position of the light source from the position of the observer at $P$?

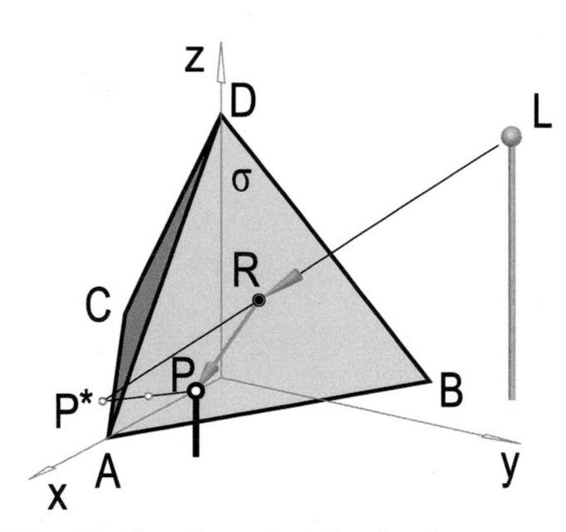


**Fig. 5.46** reflected point in theory ...

**Fig. 5.47** ... and in practice

*Solution*:
One reflects $P$ in $\sigma$, obtains $P^*$, and intersects the line $g = LP^*$ with $\sigma$, which results in $R$.

⊕ *Remark*: In glass panes, the reflection of the Sun is usually quite clear to see (Fig. 5.47, left). With reflections in the water, the water surface must be perfectly smooth. Otherwise, the reflection is elongated (Fig. 5.47). ⊕ ◂◂◂

▸▸▸ **Application**: ***strained rope*** (Fig. 5.48)
A rope of given length $s = 15$ is strained between two contact points $A$ and $B$ with a weight. Calculate the position of the buckle point $C$.

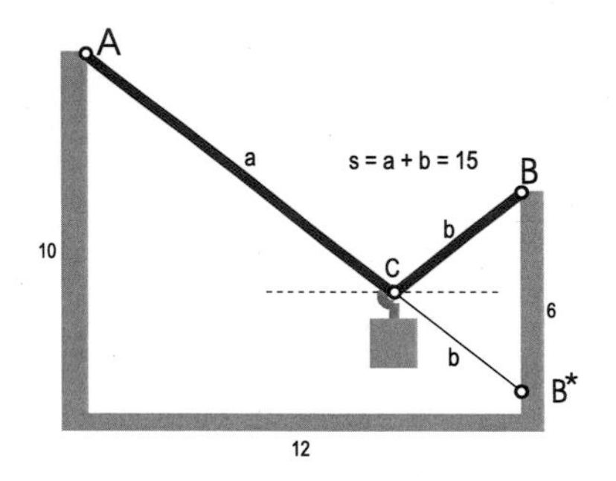


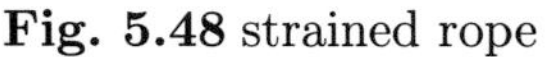
**Fig. 5.48** strained rope

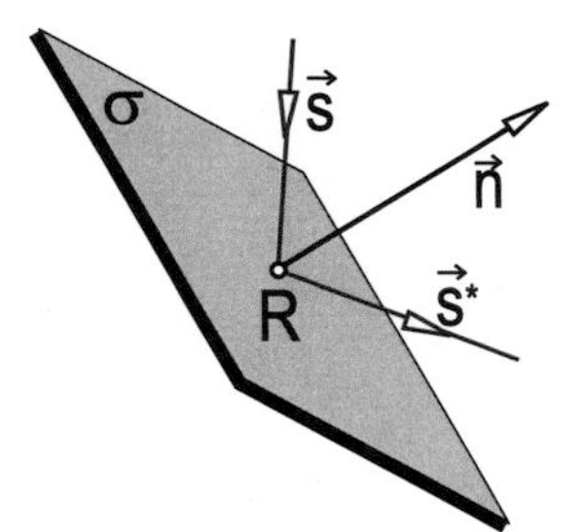


**Fig. 5.49** reflection of the ray of light

*Solution*:
$C$ lies on $AB^*$ where $B^*$ is obtained as the reflection of $B$, and thus, $\vec{c} = \vec{a} + \lambda(\vec{b}^* - \vec{a})$. With $A(0,10)$, $B(12,6)$, and $\overline{AB^*} = s$, the height difference equals $AB^* = \sqrt{s^2 - 12^2} = 9$ and, consequently, the height of $C$ is $\frac{6+1}{2} = 3.5$. Then,

$$\begin{pmatrix} 12\lambda \\ 10 - 9\lambda \end{pmatrix} = \begin{pmatrix} x \\ 3.5 \end{pmatrix} \Rightarrow \lambda = \frac{6.5}{9} \Rightarrow C(8.67/3.5).$$

◂◂◂

If we reflect "only" one vector $v$ in the plane $\sigma$, we obtain the same result by reflecting the plane $\vec{n_0} \cdot \vec{x} = 0$ that is obtained from $\sigma$ by translating it parallel into the origin of the coordinate system. It is, therefore, necessary to set $c = 0$ in equation (5.46):

$$\overrightarrow{v^*} = \vec{v} - 2(\vec{n_0} \cdot \vec{v}) \cdot \vec{n_0}. \tag{5.47}$$

▸▸▸ **Application**: ***reflection of a ray of light*** (Fig. 5.49)
At a point $R(3/0/0)$, a mirror is mounted with a prescribed normal direction $\vec{n}(0/1/1)$. In which direction will the incident solar rays $\vec{s}(1/1/-2)$ be reflected?

*Solution*:

$$\vec{n_0} = \frac{1}{\sqrt{2}}\begin{pmatrix} 0 \\ 1 \\ 1 \end{pmatrix}, \quad d = \vec{n_0} \cdot \vec{s} = \frac{1}{\sqrt{2}}\begin{pmatrix} 0 \\ 1 \\ 1 \end{pmatrix} \cdot \begin{pmatrix} 1 \\ 1 \\ -2 \end{pmatrix} = -\frac{1}{\sqrt{2}},$$

$$\overrightarrow{s^*} = \vec{s} - 2d\vec{n_0} = \begin{pmatrix} 1 \\ 1 \\ -2 \end{pmatrix} + \frac{2}{\sqrt{2}}\frac{1}{\sqrt{2}}\begin{pmatrix} 0 \\ 1 \\ 1 \end{pmatrix} = \begin{pmatrix} 1 \\ 2 \\ -1 \end{pmatrix}.$$

◂◂◂

▸▸▸ **Application**: ***stereo image with one mirror*** (Fig. 5.50)
If one looks suitably at two photos of a scene (one mirrored) in such a way that one focuses with both eyes on one photo, say, the right one (with a

vertical mirror placed between the eyes), then the scene appears stereoscopic. Explain this "space image" effect.

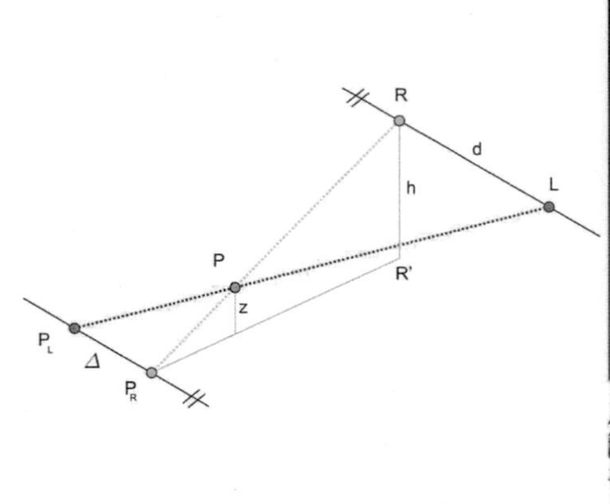

**Fig. 5.50** a stereo image without and with a mirror

*Solution*:
If we view a planar or spatial object $\Phi$ in a mirror, we see a virtual object $\Phi^*$ which is symmetric with respect to the mirror through the "mirror window". So, if we look at the left image in *the mirror* with the left eye, we no longer see a mirror-reversed image. The right eye sees the right – in any case reflected – image. Under certain conditions, the associated lines of sight from the right eye to the right image and from the left eye to the left image intersect *in space*: The two photographs of the object must have been created so that the straight line connecting the two camera positions (specifically with respect to the lens centers) were parallel to the photo-plane. This allows us to place the photos in parallel so that all of its points $P$ and $P^*$ lie on lines parallel to the bottom margin of the image.

⊕ *Remark*: If you look at stereo images without a mirror, and these, for example, should appear separated on the screen, then the spatial impression can be achieved by squinting or using optical glasses, which forces the lines of sight to "cross". ⊕

◂◂◂

## Multiple reflections

▸▸▸ **Application**: ***specular corners*** (Fig. 5.51, Fig. 5.53)
If you look into a mirror, you see yourself "reversed". In a two-way mirror with an *acute* angle $\alpha$ (Fig. 5.51: $\alpha = 60°$), you can see how others see you. What happens if $\alpha = 90°$ or $\alpha > 90°$.

*Solution*:
If $\alpha = 90°$, then the doubly-reflected ray is "exactly parallel to the incident ray" (Fig. 5.52). This has the consequence that one eye sees the other, and,

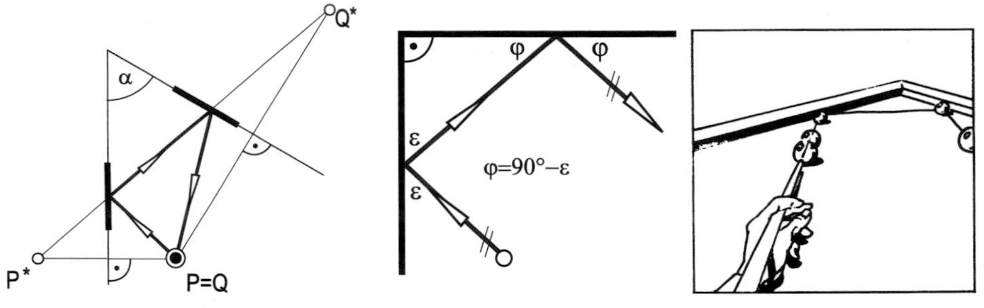


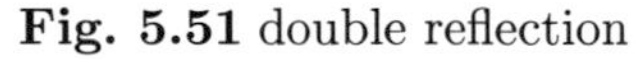

**Fig. 5.51** double reflection

**Fig. 5.52** $\alpha = 90°$, billiards

vice versa. The same applies to the two halves of the face. The distance from the mirror does not matter. When looking at the edge of a rectangular double mirror, one sees – separated by the edge – both sides of the face, and they do not appear mirror-inverted.

At obtuse angles $\alpha > 90°$ and with increasing distance, you will not see yourself when looking into the mirror's corner.

⊕ *Remark*: The above example with $\alpha = 90°$ and $P \neq Q$ can be interpreted in terms of billiards (rail shot). ⊕ ◂◂◂

**Fig. 5.53** Confusion results from using three mirror planes, each constituting approximately 60°. Since the angles are not accurate, some artefacts come into being.

▸▸▸ **Application**: ***reflecting "cuboid corner"*** (Fig. 5.55)

The special case $\alpha = 90°$ has an important practical application: Consider a reflecting "cuboid corner", which consists of three concurrent faces of a cube (Fig. 5.55). Prove that if one shoots a laser ray into this corner, the ray is redirected parallel to the incoming ray.

***Proof***: We look at the situation from above to get a top view: There, the reflections in the vertical planes appear as planar reflections in the respective normals. However, the reflection in the horizontal plane is not seen in the top view because the normal

Fig. 5.54 cuboid corner in shipping

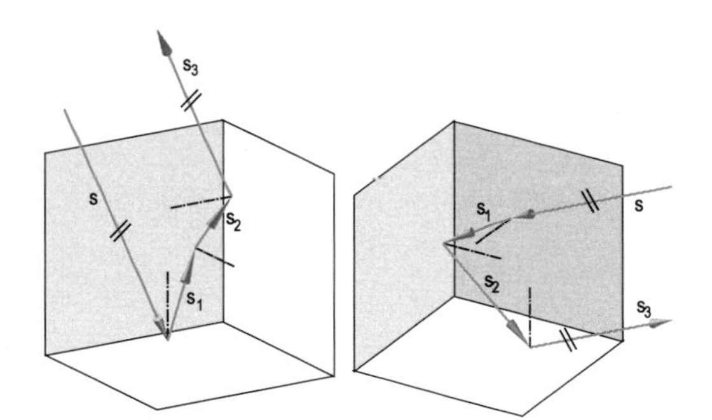


**Fig. 5.55** reflecting cuboid corner

"projects" to the base plane. Thus, following the above (Fig. 5.52) considerations in two-dimensions, the incoming ray appears in the top view parallel to the outgoing ray. The same holds for the views from the front and right (front view, right-side view). So, the outgoing ray is parallel to the incoming ray in space.

Another proof is derived from the fact that, with each reflection in a coordinate plane, the sign of the corresponding component of the direction vector switches. Since all coordinate planes are affected, the vector is just reoriented. ⊙

⊕ *Remark*: This property of the specular cuboid corner allows us to measure distances. Reflecting prisms are, for example, mounted on the Moon to carry out certain experiments (verification of light speed and distance measuring). Such devices are also mounted on bridges to allow ships to determine their distance from the bridge's pillars. ⊕

**Fig. 5.56** Reflectors for bicycles consist of many small corners!

⊕ *Remark*: Fig. 5.56 shows how reflecting cuboid corners are applied in daily life: The reflectors which are mounted on bicycles – viewed under a microscope – consist of nothing but small cuboid corners. When they are illuminated by a car, they accurately reflect back to the car driver. Along with the rotation, the reflectors produce an extremely effective and bright warning. ⊕ ◂◂◂

▸▸▸ **Application**: ***multiple reflections in two orthogonal mirrors***

With multiple reflections in two perpendicular planes, a remarkable effect occurs, whereby left and right – and not, as in the ordinary mirror, the front and the rear – are interchanged. What exactly is happening here?

*Solution*:
For the sake of simplicity (Fig. 5.57), the two planes $\xi$ and $\eta$ are assumed to be vertical coordinate planes with the equations $x = 0$ and $y = 0$. Each of the planes initially generates a virtual – usually reversed – counterpart $\Omega_x$ (change of signs in the $x$-values) or $\Omega_y$ (change of sign in the $y$-values). The two objects are indirectly congruent and – from the standpoint of perception – are of equal rank as the original $\Omega$.

**Fig. 5.57** Several mirrors in two mutually perpendicular planes of symmetry

Consequently, $\Omega_x$ has a virtual counterpart $\Omega_{xy}$ through reflection in $\eta$, and $\Omega_y$ has a virtual counterpart $\Omega_{yx}$ through reflection in $\xi$. However, because of the special position of the mirror planes ($\xi \perp \eta$), the two new objects are identical: $\Omega^* = \Omega_{xy} = \Omega_{yx}$. Furthermore, they can only be partly seen in the respective "mirror window". The part labeled with $\Omega_{yx}$ is produced through reflection in the right mirror window of the visible virtual object $\Omega_y$ in the left mirror window.

The two mirror windows are touching each other along the edge $s = \xi \cup \eta$ (according to our choice, the $z$-axis), so they merge into a single window, which consists of two mutually perpendicular rectangles. This allows us to view our object $\Omega^*$. It was created by double-reflection and is, therefore, directly congruent to $\Omega$. The two directly congruent objects $\Omega$ and $\Omega^*$ can be transformed into each other by a half-turn about the intersection $s$ of the mirror planes or by an axial reflection in $s$. Thereby, left and right are interchanged firstly, and the characters are still readable as usual.
In the photo in Fig. 5.57, a wide-angle lens was used in order to make the "auxiliary objects" $\Omega_x$ and $\Omega_y$ entirely visible.
In practice (also in human vision), the visual angle is smaller, and then, one only sees $\Omega^*$ in the image – comparable to an ordinary mirror image, but *left and right are interchanged.* ◂◂◂

▸▸▸ **Application**: ***What exactly happens between two parallel mirrors?***
Consider the confusion with two parallel mirrors in Fig. 5.58: The photographer was obviously able to focus through different distance settings on any of the available originals or reflections. Visually, one is unable to distinguish between original and virtual objects, so that it is not even guaranteed that

the person depicted is actually seen as such – this depends on the focal length that is actually used. That is to say, it depends essentially on the aperture of the viewing cone.

**Fig. 5.58** Three similar scenes, but in each another reflection is in focus. Can you see where "the original" is?

*Solution*:
Let $A$ be the position of the observer (Fig. 5.59) standing between two mirrors $\sigma_1$ and $\sigma_2$ (distance $d$ apart). The camera, which is directed towards the axis $a$, can only capture those things that are within the cone of vision $\Delta$.

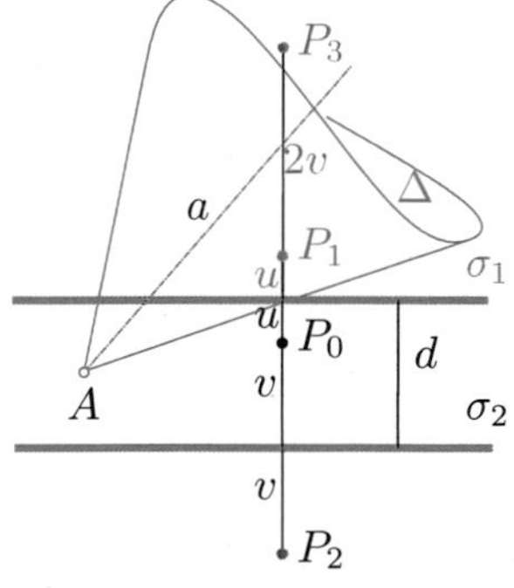


**Fig. 5.59** For each virtual point a new one arises at the "speed of light".

Now, assume that another person is standing (symbolized by a point $P_0$) between the mirrors, at a distance $u$ from $\sigma_1$ and at a distance $v$ from $\sigma_2$: $u + v = d$. Thus, two virtual mirror images $P_1$ and $P_2$ arise at the distances $u$ and $v$ behind $\sigma_1$ and behind $\sigma_2$. The two points are a distance $2u + 2v = 2d$ apart. $P_2$ is visually indistinguishable from a real point and is reflected in $\sigma_1$ to a point $P_3$. This is at distance $2d$ from $P_0$. If one continues, one obtains two congruent series of points $P_0$, $P_3$, ... and $P_1$, $P_5$, ... whose points are separated by a constant distance of $2d$.

In Fig. 5.58, these series are clearly visible (in one of them, the person can be seen from the front, in another from behind). In both Fig. 5.58 and Fig. 5.59, the observer and the multiple reflections are not contained in the cone of vision. It is noteworthy that the autofocus of the camera has no problem focusing on any reflection, which again demonstrates that a purely optical mirror image is indistinguishable from reality. The distance to which the autofocus adjusts equals exactly that of the "reflected" person in the room – and not the distance to the mirror. The repeated reflection causes a loss of brightness, so that further apart reflections appear weaker (darker). ◄◄◄

▸▸▸ **Application**: ***mirror eyes*** (Fig. 5.60)
Nearly all animals with lens eyes bundle light on their retina by means of refracting lenses. Some crayfish and shrimp, however, manage this trick by using compound eyes and *reflection.*

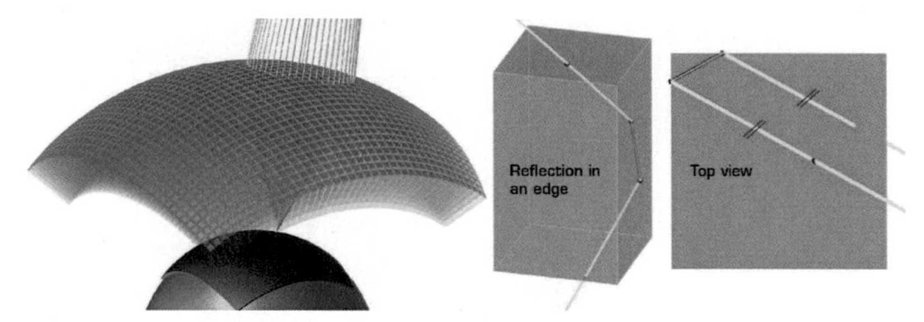


**Fig. 5.60** mirror eyes

**Fig. 5.61** "normal" and "rectangular" pseudopupils

*Solution*:
The underlying principle is purely geometrical: What happens to a light ray if it enters a hollow quadratic prism with reflecting faces? The light ray either passes through without a reflection, or bounces around between faces until it exits the prism. An interesting case occurs when the light bounces around between two subsequent – and thus perpendicular – faces: As one can see in the top view (Fig. 5.60, right), the incoming ray is parallel to the outgoing one in this projection. That is, the first two components of the directional vector are swapped. However, the third component (the $z$-value) is not affected. Now, we arrange thousands of tiny prisms on a sphere (Fig. 5.60, left), and investigate what happens to light rays that arrive from a certain radial direction. We consider all prisms whose axes are at a constant angle in relation to the incoming light. Their axes are distributed on a cone of revolution through the sphere's center. Then, we consider the intersection circle of this cone with the sphere. If the prism's diameter is very small, one might estimate that the ray exits the prism if it has been reflected in a single plane

through the axis of the prism perpendicular to the projection of the incoming ray in the direction of the axis. All light rays that hit the sphere within this circle and that are also reflected in an edge of the tiny corresponding prism enclose the same angle with the prism's axis (a generating line of the cone). Thus, the reflected rays form another cone with its vertex on the radial ray (at approximately half the radius of the sphere). This vertex is a point on the retina, which is itself a concentric sphere. The locus of all quadratic facets (compounds) that allow at least a certain percentage of light rays to focus on the retina is a "spherical rectangle". Therefore, the "pseudopupil" appears to be rectangular (Fig. 5.61, right). ◂◂◂

### ▸▸▸ Application: *the viewfinder of an SLR*

In each SLR, one can find an ingeniously simple invention: Through the viewfinder, one sees exactly what appears on the sensor when the mirror is folded up in the moment of admission, namely upright and not inverted. Explain the so-called "pentaprism" (Fig. 5.62).

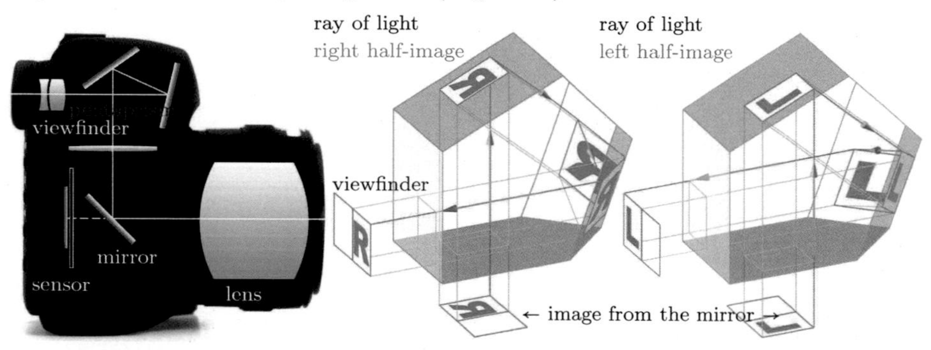


**Fig. 5.62** This shows the operation of the pentaprism. The image that is incident to the mirror is "processed" in two different ways. The result is a composite of two fields of an upright image, as if you looked right through the lens.

*Solution*:
Through the lens, the light rays reach the sensor. The image is rotated by 180°, so it seems to be upside-down behind the lens. Standing upright, the observer would like to see the image through the viewfinder. Between the lens and the sensor, there is a tilted mirror inclined at an angle less than 45°. It is folded up only at the actual moment of taking a picture. A parallel lying mirror would draw the image back to the viewfinder, but in its rotated position. This rotation is instead corrected by means of the pentaprism inside the camera. The folding mirror deflects the incident light rays upwards in this prism. There, a double-reflection takes place. After a total of three reflections, the image appears mirror-inverted in the viewfinder. In order to turn it upright in the "last minute", the pentaprism is inserted (two orthogonal planes which deliver the additional two reflections). ◂◂◂

## 5.6 Further Applications

▸▸▸ **Application**: ***downpour*** (Fig. 5.63)
During a downpour, every one of us has asked him- or herself: How fast do I have to run to the nearest shelter to remain as dry as possible? Must it be "as fast as possible"?

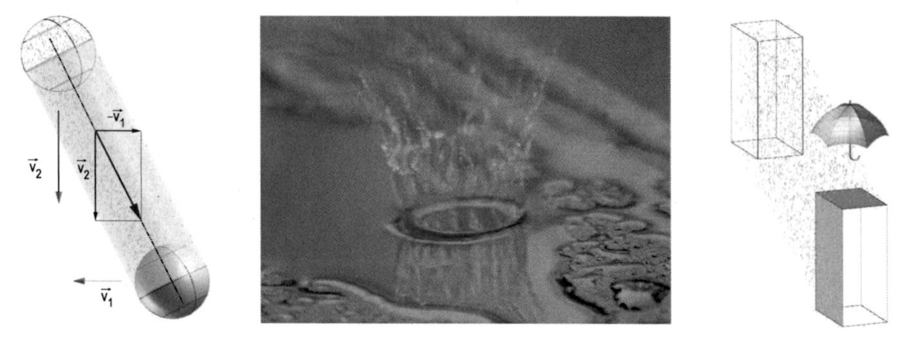


**Fig. 5.63** Run away or endure the downpour?

*Solution*:
First, we move a sphere (radius $r$) horizontally by a distance $s$ (velocity vector $\vec{v_1}$, Fig. 5.63, left). The raindrops fall vertically (velocity vector $\vec{v_2}$). If we assume that both motions have constant speed, the sphere can only be met by droplets that are located within the volume that is generated by the sphere during the relative motion in the direction of $\vec{v_1} - \vec{v_2}$. This is a cylinder of revolution with radius $r$ and the axis direction $\vec{v_1} - \vec{v_2}$. The projection of the axis length in the horizontal plane always equals $s$, whereby the volume of the cylinder (that is proportional to the number of "strikes") is the smaller, the greater the speed of the sphere. So, the motto is: "As soon as possible!"
Does the shape of the moving object play a role? We consider the simple case of a cuboid, and we move it again horizontally by a distance $s$ (Fig. 5.63, right). This time, the volume where the various raindrop strikes occur is a six-sided oblique prism whose volume is not so easily specified. Let us now imagine the following special case: The top of the cuboid is protected against the rain. (The green drops shown in Fig. 5.63 (right) have no effect.) We move the cube so that only the front side (of area $A$) becomes wet. Then, the space of drops striking the prism is an oblique prism with *constant* volume. (If the drops fall vertically, then the volume equals $A \times s$.) The velocity of the object does not matter! Therefore, the shape of the object is probably very important. In a downpour, it is recommended to hold a newspaper over one's head when seeking the nearest shelter. When the rain – lashed by wind – is incident at an oblique angle, hold it *against* the wind.

⊕ *Remark*: One thing seems clear: If we do not move, we get wet the most. Yet, not even this is self-evident: Mosquitoes and other small insects are rarely, if ever, hit by raindrops when flying through the air: Every drop has a "bow wave" in the air, and this pushes the insect aside in less than one millisecond! The middle of Fig. 5.63 illustrates the impact of a raindrop. See also Application p. 484. ⊕ ◂◂◂

▸▸▸ **Application**: ***Why does a shark not sink?*** (Fig. 5.64)
The following question seems to be related to that of Application p. 178: Why does an airplane fly? Surprisingly, the answer is very similar.

**Fig. 5.64** An oceanic shark bears a striking resemblance to an aircraft. The pectoral fins work like the wings of an aircraft.

*Solution*:
Big sharks like the oceanic whitetip shark (*Carcharhinus longimanus*) have a density greater than water. So, if the shark stops swimming, it will sink. As it swings, the stiff pectoral fins, which are slanted upwards, generate lift like an aircraft wing. Since the pectoral fins are positioned before the animal's center of mass, the generation of lift results in a torque that tilts the shark's head upward.
To repeat it again (this time adapted to the fluid water): The sharp edges of the pectoral fins create vortices that, according to the *law of conservation of angular momentum*, generate counter-rotating vortices around the profile (speed $|\Delta v|$). The vector sum $\vec{v} + \overrightarrow{\Delta v}$, therefore, has greater magnitude at the top than at the bottom.
Because of the *hydrodynamic paradox*, negative pressure is generated on the side where the water flows at greater speed, which will, in turn, produce a buoyant force that causes the shark to ascend.
The shark's asymmetrical dorsal fin, however, simultaneously produces a torque that tilts the head downwards. Both torques compensate each other, and thus, it is possible for the shark to swim straight ahead without sinking. ◂◂◂

▸▸▸ **Application**: ***centroid of a quadrangle*** (Fig. 5.65)
Consider the quadrangle $ABCD$. Compute its centers of gravity with homogeneous mass distribution (areal centroid), as well as with equal masses centered at the four vertices.

*Solution*:
We divide the quadrangle into two triangles $ABC$ and $ACD$. For these, we calculate the centroids $S_1$ and $S_2$ and the "masses" (= areas) $m_1$ and $m_2$. The (areal) center of gravity $S$ is computed with Formula (5.15):

$$\vec{s} = \frac{1}{m_1 + m_2}(m_1\vec{s_1} + m_2\vec{s_2}).$$

The (ordinary) centroid is simply the "average" of the vertices

$$\vec{s_E} = \frac{1}{4}(\vec{a} + \vec{b} + \vec{c} + \vec{d}).$$

Of course, this point does generally not coincide with the areal centroid. ◂◂◂

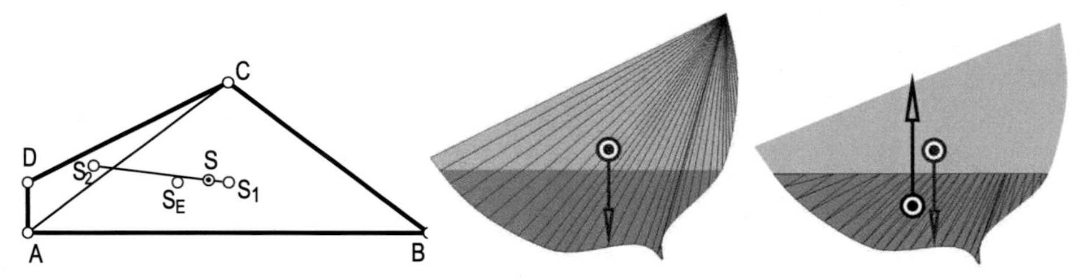


**Fig. 5.65** center of gravity of a quadrangle

**Fig. 5.66** positive or negative buoyancy

▸▸▸ **Application**: ***positive or negative buoyancy*** (Fig. 5.66)
A hull sinks into water, and depending on the weight of the ship, it will go down to a certain level. How far does it sink? Which forces occur, and when are they balanced? What will happen if the body topples over?

*Solution*:
We simplify the reasoning by only looking at a typical cross-section of a ship. The implementation of the following calculations, of course, is left to a computer.

- One can now consider the weight force acting at the center of gravity (Fig. 5.66, left). For the computation, we triangulate the cross-section. Then we use Formula (5.16) and the weight is proportional to the total area.

- According to *Archimedes*, the buoyancy is equal to the weight of the displaced fluid. The force of gravity acts on the center of gravity of the displaced water (Fig. 5.66, right). We intersect the partial triangles of the cross-section with the surface of the water. If a triangle is completely outside the water, it does not count. If it is completely under the water, the area and centroid remain unchanged. For a triangle that intersects the surface of the water such that the part under the water is again a triangle, one can immediately calculate the area and the centroid. Otherwise, the part underwater is a quadrangle, which can be decomposed into two triangles.

- When the buoyant force is greater than the weight force, the ship is lifted. If the centers of gravity are not lying precisely above each other, they induce a torque. The cross-sections of vessels are designed so that the ship is automatically lifted "upright". Observe this in the demo program on the website that accompanies this book. See also Application p. 346 and Application p. 347. ◂◂◂

▸▸▸ **Application**: ***Cardan (universal) joint*** (Fig. 5.67)
We want to study the transmission of a rotation about an axis $a$ to a rotation about an axis $b$ intersecting $a$. A fork with shaft $a$ includes a universal joint, which, in turn, drives a fork with axis $b$. The angle between the two axes is $\gamma$ $(0° < \gamma < 90°)$.
"Input angle" $\alpha$ and "output angle" $\beta$ are obviously different.

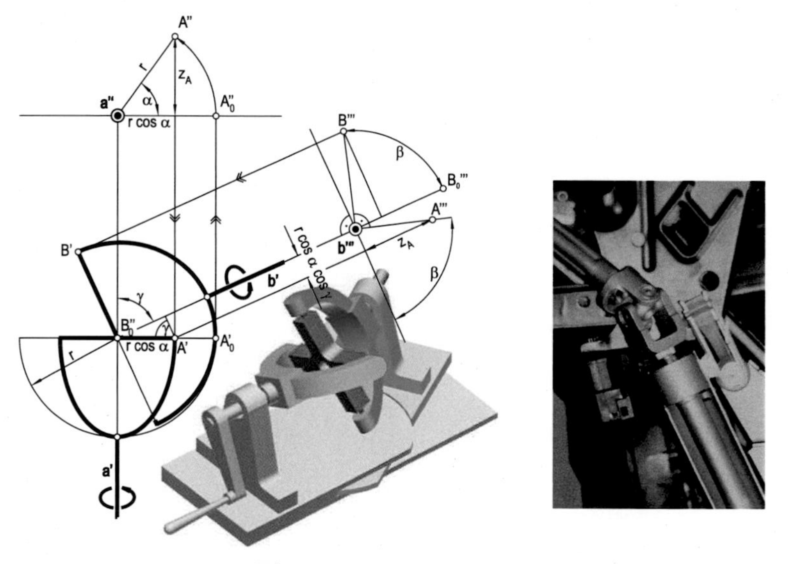


**Fig. 5.67** universal joint

Let us first consider the construction of a general configuration: We assume that the axes $a$ and $b$ lie in a horizontal plane. From the front view and the side view, the following relation can be derived:

$$\boxed{\tan\beta = \frac{1}{\cos\gamma}\tan\alpha.} \tag{5.48}$$

The construction is not trivial and requires solid knowledge in the field of Descriptive Geometry. Now, we solve the problem by means of vector analysis:
We associate a Cartesian coordinate system with the joint ($xy$-plane = $ab$-plane, $a$ is the $x$-axis). The fork has a radius of 1. In the initial position, the end points $A$ and $B$ of the Cardan joint are located such that $A$ has the coordinates $A_0(0/1/0)$ and $B$ has the coordinates $B_0(0/0/1)$ (Fig. 5.67). Now, $A$ is rotated about $a$ ($x$-axis) by the given angle $\alpha$ which gives

$$A(0/\cos\alpha/\sin\alpha).$$

In order to determine the new location of $B$, we apply two superimposed rotations:

1. The rotation about $a$ through the unknown angle $\beta$ moves $B_0$ to the point $B(0/\sin\beta/\cos\beta)$.

2. The rotation about the $z$-axis through the angle $-\gamma$ assigns to $B$ the final coordinates
$$B(\sin\gamma\sin\beta/-\cos\gamma\sin\beta/\cos\beta).$$

The position vectors of $A$ and $B$ enclose a right angle. So, the following holds

$$\begin{pmatrix} 0 \\ \cos\alpha \\ \sin\alpha \end{pmatrix} \cdot \begin{pmatrix} \sin\gamma\sin\beta \\ -\cos\gamma\sin\beta \\ \cos\beta \end{pmatrix} = 0.$$

This leads to the condition

$$-\cos\alpha\cos\gamma\sin\beta + \sin\alpha\cos\beta = 0, \tag{5.49}$$

which is equivalent to (5.48). ◂◂◂

▸▸▸ **Application**: ***Where does the Sun rise and set in the solstice?***
Stone Age monumental buildings such as Stonehenge (Fig. 5.68, left) and Chichén Itzá (Fig. 5.68, center and right) are aligned over many hundreds of meters up to tenths of a degree precisely in the direction of the sunrise or sunset on June, 21st.
At that time, the corresponding directions were determined empirically as the minimum deviation of the rising or setting sun from the northern direction. The directions were measured so accurately that one can draw the conclusion that, 4,500 years ago (Stonehenge), the inclination angle $\delta$ of the Earth's axis deviated by more than half a degree from the current angle of $\delta = 23.44°$.
Investigate the corresponding directions for a given latitude by means of vector analysis.
We use two simple "fit-for-everyday-use" and obvious lemmas.

**Fig. 5.68** Stonehenge (left) and Chichén Itzá (middle and right)

Proposition 1: You can find the North Star, and thus the direction of the Earth's axis, by looking to the north at an angle of elevation $\varphi$ equal to the current latitude. At the Chephren Pyramid, one had to look towards the pyramid's vertex from a point $O$ which is south of the pyramid's base at a distance equal to the pyramid's height $h$ in order to see the ancient North Star Thuban (Fig. 5.69).

Proposition 2: In the course of a day, the Earth rotates once about its axis. That is, relatively to the Earth, the Sun revolves around the Earth's axis. Especially at the solstices, the angle enclosed by the Sun's rays and the Earth's axis remains nearly constant. So, the light rays sweep out a cone of revolution. At the summer solstice, this angle is $\sigma = 90° - \delta$, i.e. currently $\approx 66.6°$ ($4,500$ years ago it was $\approx 66.0°$).

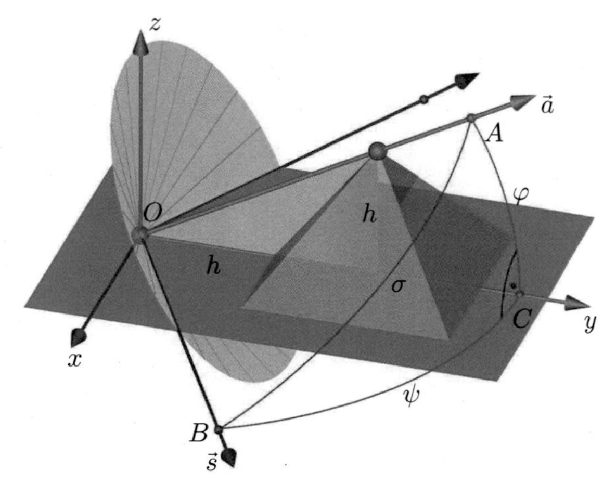


**Fig. 5.69** Calculating the direction of the sunset (actually carried out at the Pyramid of Chephren at the summer solstice 4,500 years ago).

Now, we define a Cartesian coordinate system with origin $O$ and the horizontal $xy$-plane with the $y$-axis pointing in the northern direction (Fig. 5.69), so that the Earth's axis lies in the $yz$-plane inclined under the angle $\varphi$ and has the aforementioned normalized direction vector $\vec{a} = (0, \cos\varphi, \sin\varphi)$ with

$\varphi$ as its latitude.
Let $\psi$ be the angle between the direction of the setting Sun and the North. The corresponding normalized direction vector then equals $\vec{s} = (\sin\psi, \cos\psi, 0)$. The angle $\sigma$ between the two directions satisfies $\cos\sigma = \vec{a}\cdot\vec{s}$, resulting in the simple condition

$$\cos\sigma = \cos\varphi\cos\psi. \tag{5.50}$$

Specifically, one obtains the geographic coordinates of Stonehenge (approximately 4,500 years old) $\varphi = 51.2°, \sigma = 66.0°$ or Chichén Itzá (ca. 1,000 years old) $\varphi = 20.7°$, $\sigma = 65.4°$. The respective values for $\psi$ are $\approx 49.5°$ or $\approx 63.6°$. The result can be verified in the images provided by Google Earth (see Fig. 5.68, center). ◂◂◂

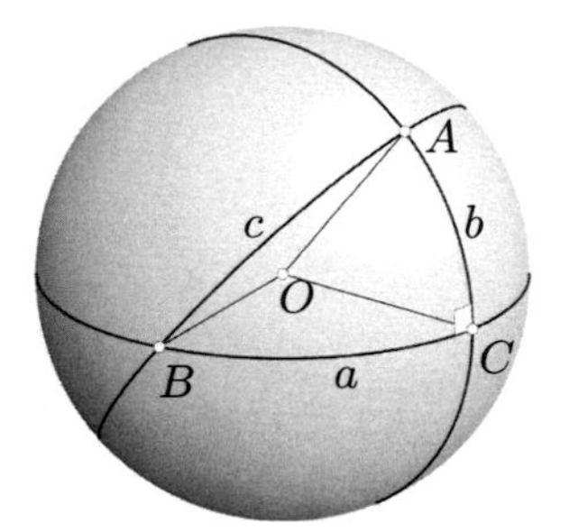


**Fig. 5.70** A right triangle on the sphere shows a simple relation.

▸▸▸ **Application**: ***the spherical Pythagoras***
Consider the unit sphere centered at the origin $O$ in Fig. 5.69. Then the angles $\psi$, $\varphi$, and $\sigma$ correspond to lengths $a$, $b$, and $c$ of the arcs on the great circles (Fig. 5.70) forming a spherical right triangle $ABC$. Moreover, equation (5.50) corresponds to the formula for the Pythagorean Theorem on the sphere:

$$\boxed{\cos a\cos b = \cos c.}$$

◂◂◂

## Switching from spherical to Cartesian coordinates

On a sphere (e.g. the Earth), points are often given in spherical coordinates $\lambda$ (the longitude) and $\varphi$ (the latitude). This is quite practical for beings living on a sphere, since we only need two numbers to know our location. For calculations with vectors, however, it is sometimes necessary to switch to Cartesian coordinates.
Let a point be given by latitude and longitude: $P(\lambda, \varphi)$. The radius of the sphere shall be $r$. According to Fig. 5.71, we immediately have

$$x = r\cos\lambda\cos\varphi,\ \ y = r\sin\varphi\cos\varphi,\ \ z = r\sin\varphi. \tag{5.51}$$

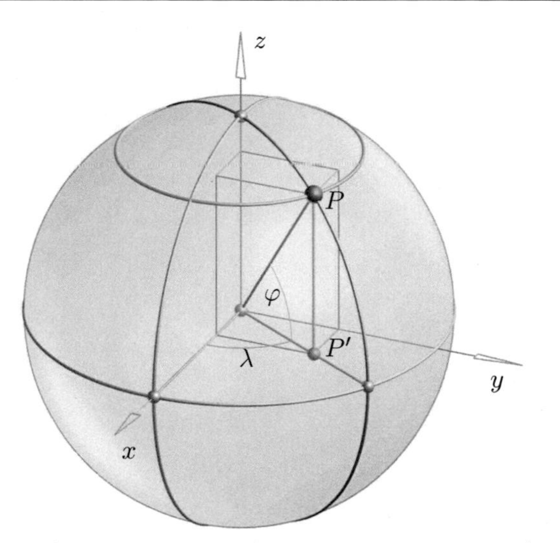


**Fig. 5.71** switching from spherical to Cartesian coordinates: $P(\lambda, \varphi) \to P(x, y, z)$

▸▸▸ **Application**: ***shortest distance between two points on the Earth***

*Solution*:

Let $r = 1$ for the time being and the two points be given by their spherical coordinates $A(\lambda_A, \varphi_A)$ and $B(\lambda_B, \varphi_B)$. Then, the two points have – according to (5.51) – the Cartesian coordinates $A(x_A, y_A, z_A)$ and $B(x_B, y_B, z_B)$. Their coordinates can be interpreted as the components of unit vectors (length 1) $\vec{a}$ and $\vec{b}$. They enclose the angle $\psi = \arccos \vec{a} \cdot \vec{b}$.

The shortest path $\widehat{AB}$ from $A$ to $B$ is part of a great circle (Fig. 5.72, left). All great circles have the same circumference as the equator (40,000 km), and we have $\widehat{AB} = 40,000 \cdot \psi^\circ / 360^\circ$.

Numerical example:

$A \ldots$ Anchorage (N61.2°, W149.9°) $\Rightarrow \vec{a} = (-0.417, -0.242, 0.876)$,

$B \ldots$ Baltimore (N39.3°, W76.6°) $\Rightarrow \vec{b} = (0.179, -0.7528, 0.633)$,

$\psi = 48.535^\circ \Rightarrow \widehat{AB} = 40,000 \cdot 48.535^\circ / 360^\circ \approx 5,400$ km. ◂◂◂

▸▸▸ **Application**: ***initial course angle, when flying from $A$ to $B$***

*Solution*:

This application is a classical application of vector calculus. It is quite useful and must, of course, be implemented in any navigation system, especially on ships and aircrafts.

Let again $r = 1$ and $A(\lambda_A, \varphi_A)$ and $B(\lambda_B, \varphi_B)$ be two points on the unit sphere with corresponding unit vectors $\vec{a} = (x_A, y_A, z_A)$ and $\vec{b} = (x_B, y_B, z_B)$. Again, the shortest path $\widehat{AB}$ from $A$ to $B$ is part of a great circle (Fig. 5.72, left). Its carrier plane $\gamma$ is defined by the three points $A$, $B$, and the center of the sphere (the origin of the coordinate system). Its normal vector is perpendicular to both $\vec{a}$ and $\vec{b}$, and it is, thus, given by $\vec{g} = \vec{a} \times \vec{b}$. We have

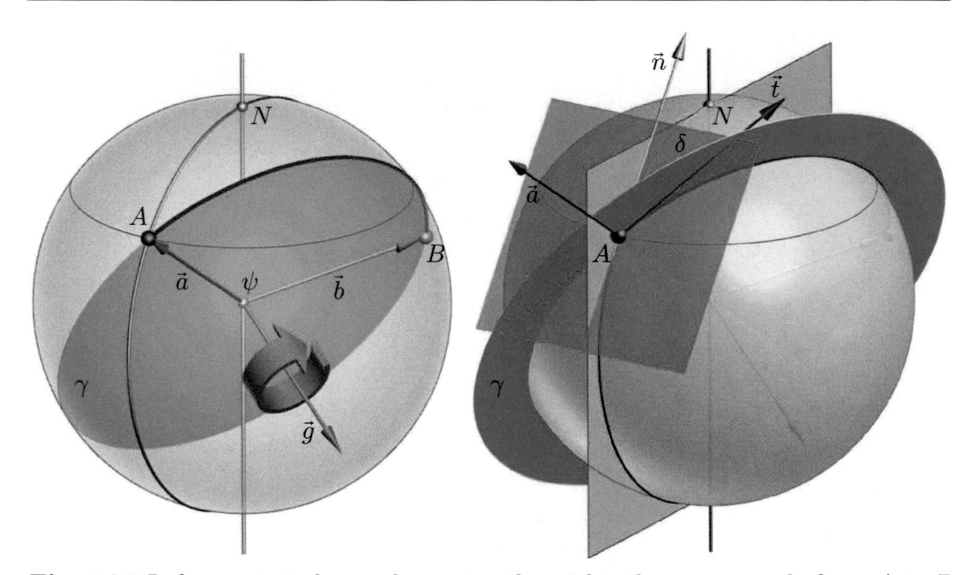


**Fig. 5.72** Left: great circles as shortest paths, right: the course angle from $A$ to $B$.

to intersect $\gamma$ with the tangent plane in $A$ with normal vector $\vec{a}$ (Fig. 5.72, right). The direction vector $\vec{t}$ of the intersection line is given by $\vec{t} = \vec{g} \times \vec{a}$. The north direction in $A$ is given by $\vec{n} = \vec{a} \times (-y_A, x_A, 0)$. The course angle $\delta$ can finally be determined after normalizing $\vec{n}$ and $\vec{t}$ via $\cos\delta = \vec{n}_0 \cdot \vec{t}_0$.
Numerical example:
For $A$ and $B$, we choose again Anchorage (N61.2°, W149.9°) and Baltimore (N39.3°, W76.6°) $\Rightarrow$ $\vec{a} = (-0.417, -0.242, 0.876)$, $\vec{b} = (0.179, -0.753, 0.633)$. Then, we have $\vec{g} = (0.507, 0.421, 0.357)$, $\vec{n}_0 = (0.758, 0.440, 0.482)$, and $\vec{t}_0 = (0.608, -0.791, 0.071)$. This yields $\cos\delta = 0.147$ and $\delta = 81.5°$, which corresponds more or less to the direction EbN, but eventually leads southwards. That is to say, the airplane first goes slightly north, though the final destination is 22° further south! ◂◂◂

▸▸▸ **Application**: ***calculating the Sun's position in the generic case***
Making the following simplified assumptions, we calculate the incidence of sunlight in a given latitude $\varphi$, time $z$ and date:

1. Time is taken to be "true solar time". The Sun reaches its culmination at exactly $12^h$ noon.
2. The Earth's orbit is approximated by a circle (the maximum deviation of the orbital ellipse is below 1.5% of the circle's diameter).
3. The Sun's rays are assumed to be parallel because of the Sun's distance from the Earth.
4. Each month is said to have 30 days (year = 360 days). Then, the Earth rotates by 1° around the Sun per day.

Calculate and discuss the direction vector to the Sun in:
a) Vienna ($\varphi = 48°$) on December 21st, $16^h20$;

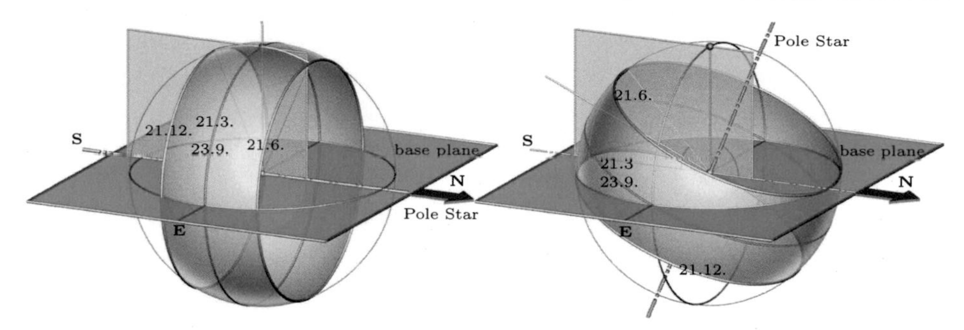


**Fig. 5.73** where the Sun can reside (Equator or the Arctic Circle)

b) Nairobi ($\varphi = 0°$) on May 20th, $12^h$ or $18^h$;
c) Rio de Janeiro ($\varphi = 23°$ south) on June 21st, $12^h$,
using the following formula:

$$\vec{s} = c\begin{pmatrix} 0 \\ \cos\varphi \\ \sin\varphi \end{pmatrix} + \cos\omega\begin{pmatrix} 0 \\ -\sin\varphi \\ \cos\varphi \end{pmatrix} + \begin{pmatrix} \sin\omega \\ 0 \\ 0 \end{pmatrix}. \tag{5.52}$$

**Fig. 5.74** sunsets that are one month apart

The angle of rotation from the south direction is (with $z$ as the true solar time) $\omega° = 15°(12 - z)$. The day-specific constant $c$ is calculated from the rotation angle $\alpha$ in the "orbit circle" of the Earth: $c = \tan(23.44° \sin\alpha)$. In the accompanying coordinate system, the east direction is the $x$-direction, the north direction coincides with the $y$-direction, and the $z$-axis always points to the zenith. In this system, the Pole Star lies in the direction $(0, \cos\varphi, \sin\varphi)$.[1]

*Solution*:

a) We have $\alpha = -90°$, $z = 16.33$, and $\varphi = 48°$. For this typical Central European latitude and at winter solstice, one would expect that the Sun has already set. However, the calculation shows that the Sun sets relatively

[1] More on this can be found in *Geometry and Its Applications.*

close to the $WSW$ direction $\vec{s} = (-0.91, -0.57, 0)$. This is because the Sun has already started to descend *before* 12am Central European winter time due to its irregular orbit (but we compute with the *true* time).

b) At the Equator, only 12-hour days are to be expected in the course of a year (Fig. 5.73, left). For $\alpha \approx 60°$, the Sun at noon is in the direction $(0, 0.37, 1)$, i.e. at an elevation angle of almost 70° in the North (and not, as one might suppose naively, at the zenith). At 6pm, it sets towards the $WNW$ direction $(-1, 0.37, 0)$.

c) In the southern hemisphere, we do not see the North Star. However, at the elevation angle of the actual southern latitude, we see the Southern Cross. On June, 21st ($\alpha = 90°$), the Sun reaches an elevation angle of slightly less than 45°: $(0, 0.79, 0.75)$ and is located in the north.

**Fig. 5.75** sunrise in the temple of Abu Simbel

⊕ *Remark*: Fig. 5.73 shows that the Sun actually sets in the west at the solstice. The photomontage in Fig. 5.74 shows two sunsets one month apart. ⊕

⊕ *Remark*: The ancient Egyptians had a profound knowledge of the positions of the Sun. Fig. 5.75 shows a sunrise at the temple of Abu Simbel. Only for two days of the year, the leftmost nearly hidden statue of the God of Darkness is illuminated in the inner sanctum. ⊕ ◂◂◂

### ▸▸▸ Application: *energy input by solar radiation*

Formula (5.52) allows us to calculate the normalized direction vector $\vec{s}$ pointing towards the Sun. According to Lambert's law, the brightness, and thus also the energy input, is proportional to the cosine of the angle of incidence (Application p. 198). Using a computer program in order to sum the energy input values at certain latitudes and at short time intervals, say, every five minutes, every day, we can determine the corresponding theoretical energy input with relative precision. Interpret the diagram in Fig. 5.76.

*Solution*:

We assume that the program presumes clear blue skies at each day and at each latitude.

Let us first consider the energy input at the Equator. As expected, it varies only slightly within a calendar year.

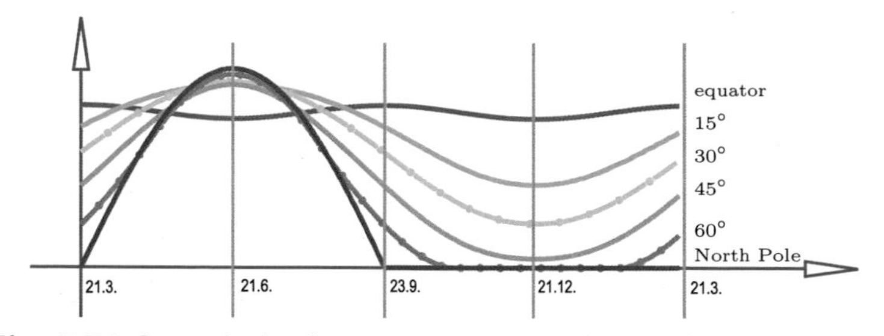


**Fig. 5.76** theoretical solar energy input on the northern hemisphere

At the North Pole, there is no energy input in the winter time. Yet, there is quite a lot during the summer months. In fact, during the period from late May to late July, the input exceeds that at the Equator! The explanation for this: The Sun is at the North Pole on June 21st at 23.44° above the horizon, and this will last for 24 hours a day. At this time, the Sun's elevation angle at the Equator fluctuates between 0° (in the morning and evening) and 66.56° = 90° − 23.44° (during lunch), and it only does so for 12 hours a day. In the summer, the temperate latitudes also have more energy input than the region at the Equator (but less than the North Pole!). In the winter time, the supply is significantly lower than in the tropical belt. ◂◂◂

▸▸▸ **Application**: ***illumination of a telephone box*** (Fig. 5.59)

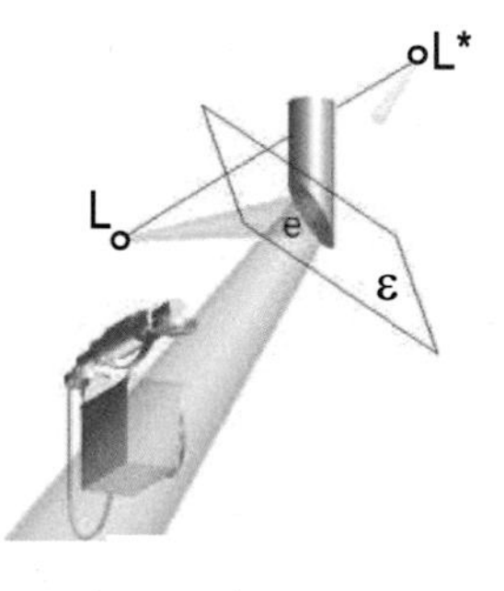


A point-shaped light source $L$ emits a light cone with vertex $L$ (spotlight). This cone intersects an inclined plane $\varepsilon$ along an ellipse $e$. The surface of the elliptical area is specular (comparable to an oblique section of a cylinder of revolution). The reflected rays now form a cone of revolution (spotlight) through $e$, with vertex $L^*$ which is the reflection of $L$ in $\varepsilon$. Here, the center of the keyboard (dial) should be on the axis of the reflected spotlight.

◂◂◂

▸▸▸ **Application**: ***solar power plant*** (Fig. 5.77)

With the help of mirrors, light rays are bundled at a center (focal point). Let $\vec{s} = (4,0,3)$ be the direction of the Sun; $\vec{s}$ is to be redirected from the point $P(6,8,0)$ to the focal point $F(0,0,24)$. What is the normal vector of the tiltable mirror plane?

*Solution*:

The new direction of the light ray is $\vec{s}^* = \overrightarrow{PF} = (-6,-8,24)$. We form both the two unit vectors $\vec{s}_0 = \frac{1}{5}(4,0,3)$ and $\vec{s}_0^* = \frac{1}{13}(-3,-4,12)$. Their sum $\vec{s}_0 + \vec{s}_0^*$ is the normal vector of the desired plane. ◂◂◂

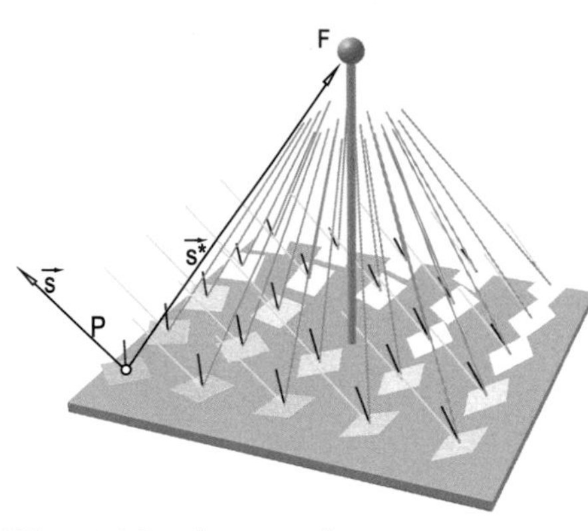


**Fig. 5.77** solar panel

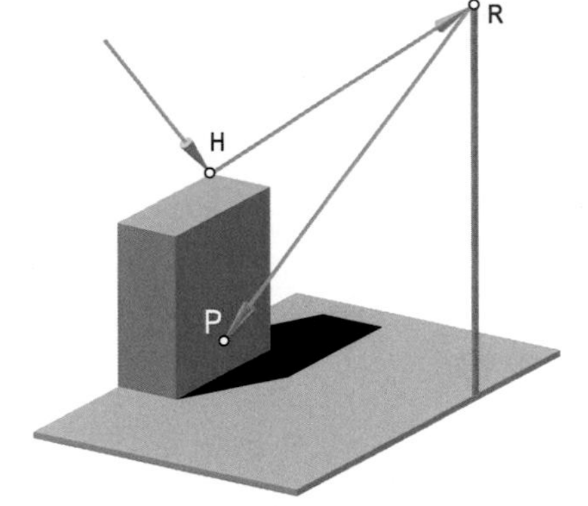


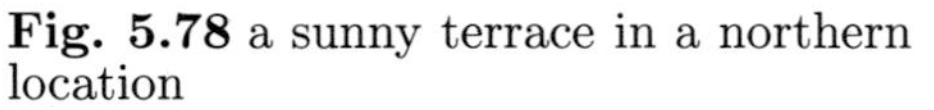

**Fig. 5.78** a sunny terrace in a northern location

▸▸▸ **Application**: ***morning sun despite the north exposure*** (Fig. 5.78)
At point $H$, there is a heliostat (a mirror rotated by clockwork toward direction of the North Star). The light ray $\vec{s}$ is, thereby, reflected through a fixed point $R$ on a tower. The ray $HR$ points to the north in the top view and is inclined at $\varphi°$ (latitude) to the base plane. At the point $R$, the light ray can be reflected to an arbitrary point $P$. The (fixed) normal direction $\vec{n}$ of the mirror in $R$ is the angle bisector of $\overrightarrow{RP}$ and $\overrightarrow{RH}$.

⊕ *Remark*: The village Viganella in Piedmont used to be known as the "darkest village of Italy", because there was no Sun for almost three months due to the steep mountains around. However, in 2006, a computer-controlled mirror with a surface of 40 m$^2$ size was mounted 400 meters above the village to redirect the Sun's light exactly to the piazza during the winter months. ⊕ ◂◂◂

▸▸▸ **Application**: ***rotation about a generic line*** (Fig. 5.79)
Two aircraft wheels are to be moved by rotation from the operating position into the aircraft body. Calculate the rotation axis and the rotation angle.

**Fig. 5.79** rotation of aircraft wheels in theory and practice

*Solution*:
Let $M_1$ be the center of the wheel in the starting position and $A_1$ a point on the wheel axle. In the final position, the wheel center is at $M_2$ and the axis point has reached $A_2$. Then, the axis of rotation $d$ is the intersection of the bisectors (planes of symmetry) $\sigma_M$ and $\sigma_A$ of $M_1M_2$ and $A_1A_2$.
The circular orbit of the wheel's center is located in the normal plane $\nu$ (center $M$) to $d$. The rotation angle $\alpha$ is the angle $M_1MM_2$.

In order to practice, we will compute a numerical example in detail: For example, suppose the centers are $M_1(3/0/0)$, $M_2(1/2/2)$ and the points on the axes are $A_1(4/0/0)$, $A_2(1/2/1)$.

1. To determine the plane of symmetry, we use Formula (5.27), according to which

$$\sigma_A: \ (\vec{a}_2 - \vec{a}_1)\bullet\vec{x} = \frac{1}{2}(\vec{a}_2{}^2 - \vec{a}_1{}^2) \Rightarrow$$

$$\Rightarrow \begin{pmatrix} -3 \\ 2 \\ 1 \end{pmatrix} \bullet\vec{x} = \frac{1}{2}(6-16) = -5 \Rightarrow \sigma_A: \ -3x + 2y + z = -5,$$

$$\sigma_M: \ (\vec{m}_2 - \vec{m}_1)\bullet\vec{x} = \frac{1}{2}(\vec{m}_2{}^2 - \vec{m}_1{}^2) \Rightarrow$$

$$\Rightarrow \begin{pmatrix} -2 \\ 2 \\ 2 \end{pmatrix} \bullet\vec{x} = \frac{1}{2}(9-9) = 0 \Rightarrow \sigma_M: \ -2x + 2y + 2z = 0.$$

2. The axis of rotation is the intersection of $\sigma_M$ and $\sigma_A$, i.e. $d = \sigma_M \cap \sigma_A$. The position of the point $P \in d$ with $z = 0$:

$$-2x + 2y = 0 \wedge -3x + 2y = -5 \Rightarrow x = 5, \ y = 5 \Rightarrow P(5/5/0).$$

The direction vector $\vec{r}$ of $d$ is normal to $\vec{m}_2 - \vec{m}_1$ and $\vec{a}_2 - \vec{a}_1$. Thus, we find

$$\vec{r} = \begin{pmatrix} -2 \\ 2 \\ 2 \end{pmatrix} \times \begin{pmatrix} -3 \\ 2 \\ 1 \end{pmatrix} = \begin{pmatrix} -2 \\ -4 \\ 2 \end{pmatrix}, \text{ and } d: \ \vec{x} = \begin{pmatrix} 5 \\ 5 \\ 0 \end{pmatrix} + t\begin{pmatrix} -2 \\ -4 \\ 2 \end{pmatrix}.$$

3. The normal plane $\nu \perp d$, $\nu \ni M_1$ (and automatically $\nu \ni M_2$) has the equation

$$\vec{r}\bullet\vec{x} = \vec{r}\bullet\overrightarrow{M_1} = \begin{pmatrix} -2 \\ -4 \\ 2 \end{pmatrix} \bullet \begin{pmatrix} 3 \\ 0 \\ 0 \end{pmatrix} = -6 \Rightarrow \ \nu: \ \begin{pmatrix} -2 \\ -4 \\ 2 \end{pmatrix} \bullet\vec{x} = -6.$$

4. Intersecting the normal plane $\nu$ with the axis of rotation yields $M = \nu \cap d$:

$$\begin{pmatrix} -2 \\ -4 \\ 2 \end{pmatrix} \bullet \left[\begin{pmatrix} 5 \\ 5 \\ 0 \end{pmatrix} + t\begin{pmatrix} -2 \\ -4 \\ 2 \end{pmatrix}\right] = -6,$$

$$-30 + 24t = -6 \Rightarrow t = 1 \Rightarrow \ \overrightarrow{M} = \begin{pmatrix} 5 \\ 5 \\ 0 \end{pmatrix} + 1\begin{pmatrix} -2 \\ -4 \\ 2 \end{pmatrix} = \begin{pmatrix} 3 \\ 1 \\ 2 \end{pmatrix}.$$

5. The angle of rotation $\alpha$ is measured between $\overrightarrow{MM_1}$ and $\overrightarrow{MM_2}$:

$$\overrightarrow{MM_1} = \begin{pmatrix} 0 \\ -1 \\ -2 \end{pmatrix}, \ \overrightarrow{MM_2} = \begin{pmatrix} -2 \\ 1 \\ 0 \end{pmatrix} \Rightarrow \cos\alpha = \frac{\overrightarrow{MM_1} \bullet \overrightarrow{MM_2}}{\sqrt{5}\cdot\sqrt{5}} = -\frac{1}{5} \Rightarrow \alpha \approx 101.54°.$$

◂◂◂

▸▸▸ **Application**: ***angles enclosed by the heights of a regular tetrahedron***
In nature, one hardly finds exact mathematical forms. More often, we find approximations. Among the few exact examples, there are various crystal lattices, such as that of diamond or germanium (perfect tetrahedral), as well as table salt (cubical). Designers also apply these highly symmetric forms in original ways. In order to produce the lamp shown in Fig. 5.80 (left), determine the angles enclosed by the altitudes of a regular tetrahedron.

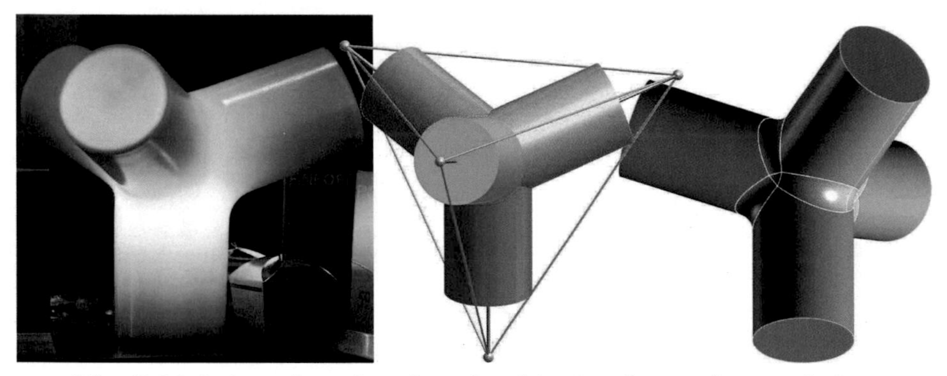

**Fig. 5.80** designer lamp based on the altitudes of a regular tetrahedron

*Solution*:
We start with the angles formed by the diagonals of a cube (Fig. 5.81, middle):

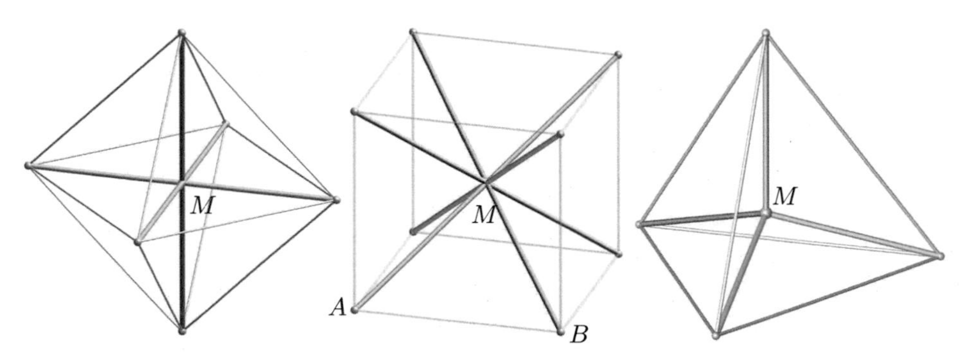


**Fig. 5.81** The three diagonals of an octahedron trivially form an orthogonal tripod (left). The four diagonals of a cube do not form right angles.

Assume the cube is centered at $M(0/0/0)$, and its vertices are given by the coordinates $A(1/-1/-1)$ (position vector $\vec{a}$), $B(1/1/-1)$ (position vector $\vec{b}$), etc. Then, we build, for example, the direction vectors $\overrightarrow{MA} = \vec{a}$ and $\overrightarrow{MB} = \vec{b}$ of the diagonals, and the angle $\varphi$ is determined by

$$\cos\varphi = \frac{\vec{a}\cdot\vec{b}}{|\vec{a}|\,|\vec{b}|} = \frac{1}{3}.$$

All other angles are either equal or, due to symmetry, they complement each other to 180° ($\cos\varphi = -1/3$).
For the tetrahedron (Fig. 5.81, right), computation is not as easy at first, because the coordinates of the vertices "in the standard position" are complicated to write down and the direction vectors of the altitudes cannot be read off so easily.

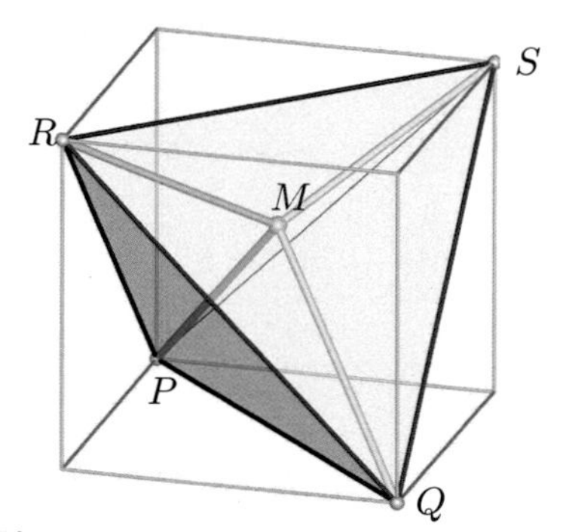

**Fig. 5.82** tetrahedron in the cube

A much simpler coordinate representation of the vertices $P$, $Q$, $R$, $S$ of the tetrahedron is obtained by "carving" the tetrahedron out of a cube as shown in Fig. 5.82. We have already assigned coordinates to the vertices of the cube. The calculation of the angle between the heights needs no additional work, and we get $\cos\varphi = -1/3$, thus, $\varphi \approx 109.5°$.

At this point, we shall try to determine the angle subtended by the two faces of a tetrahedron. For this purpose, we calculate the angle between the faces' normals. The interior angles are, therefore, supplementary angles to $\varphi$ (and hence – precisely up to two decimal places – $180° - 109.47° = 50.53°$). ◂◂◂

▸▸▸ **Application**: ***dihedral angle on an icosahedron*** (Fig. 5.83)
Among the five Platonic solids (tetrahedron, hexahedron = cube, octahedron, dodecahedron, icosahedron), the icosahedron has the largest number of faces (Greek: *ikosi* = 20). What is the angle between two faces?
*Solution*:
We can fairly easily sketch an icosahedron as follows: We draw a harmonic rectangle (sidelengths 1 and $\Phi = (1+\sqrt{5})/2 \approx 1.618$, Application p. 37) in the $xy$-plane and then redraw it two more times as shown on the left-hand side of Fig. 5.83. The twelve points of the polyhedron then have the coordinates $(0, \pm1, \pm\Phi)$, $(\pm1, \pm\Phi, 0)$, and $(\pm\Phi, 0, \pm1)$. The "convex hull" of these points is the desired icosahedron. Now, consider the midpoint $A(0, 0, \Phi)$ of the uppermost edge, and $B(0, \Phi, 1)$ as a neighboring vertex. The angle $\alpha$ of the vector $\overrightarrow{AB}$ to the negative $z$-axis is half the dihedral angle of the

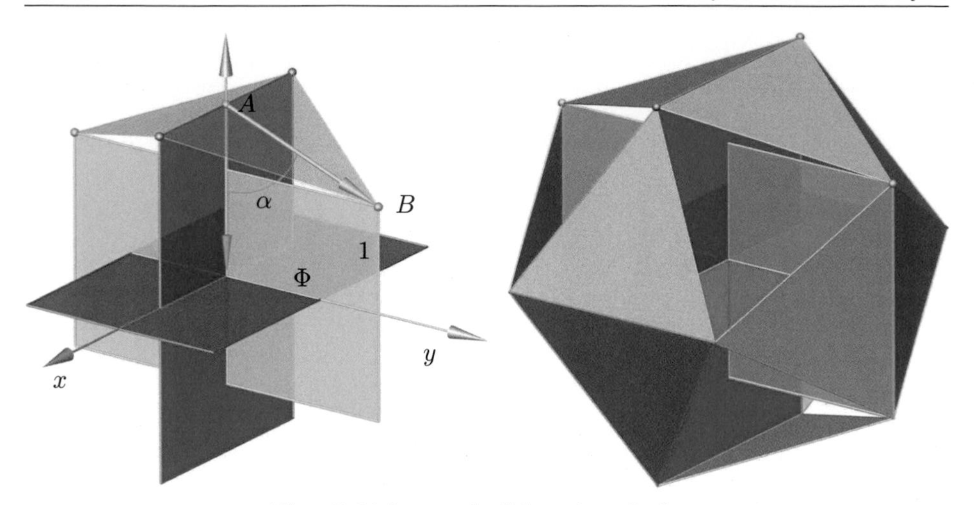


**Fig. 5.83** how to build an icosahedron

icosahedron. We have

$$\cos\alpha = \frac{1}{\sqrt{\Phi^2 + (1-\Phi)^2}} \begin{pmatrix} 0 \\ \Phi \\ 1-\Phi \end{pmatrix} \cdot \begin{pmatrix} 0 \\ 0 \\ -1 \end{pmatrix} = \frac{\Phi - 1}{\sqrt{\Phi^2 + (1-\Phi)^2}} \Rightarrow \alpha \approx 69.1^\circ$$ ◂◂◂

▸▸▸ **Application**: ***Artists and astronomers*** (Fig. 5.84)
The wonderful Stone Age paintings of Lascaux are some 17,000 years old. The *Hall of the Bulls* is very famous. Besides the artwork, there are dots on the walls, and these dots were most likely painted there on purpose. In particular, the pattern of the Pleiades appears several times. This is the case in many other ancient findings: The Navajo Indians also painted the same patterns, and on the sky disc of Nebra, the Pleiades are quite dominant. For the Mayas as well, the Pleiades were of extreme importance. Could Fig. 5.84 also be a star-map?

*Solution*:
*Taurus* spans the 30–60th degree of the zodiac. The Pleiades are presently in the same position with respect to the other dominant star of the zodiac, Aldebaran. Furthermore, to the left of Aldebaran, there is the prominent Orion belt – three stars that more or less point to the Pleiades. So far, so good. Yet, unfortunately, the artist painted four dots in the Orion belt, instead of three. Does this mean that the speculation was wrong?
The first question is: Did the sky 17,000 years ago resemble the sky today? Most stars stay rather stable in their relative position over many thousands of years (Application p. 233), because they are so far away that any – even fast – motion is almost negligible, especially when they have similar distances. Some closer stars like Sirius, however, move comparatively fast. So, one has to check whether the mysterious fourth star might have been Sirius. Indeed, it turns out that Sirius was closer to the Orion belt at that time, but –

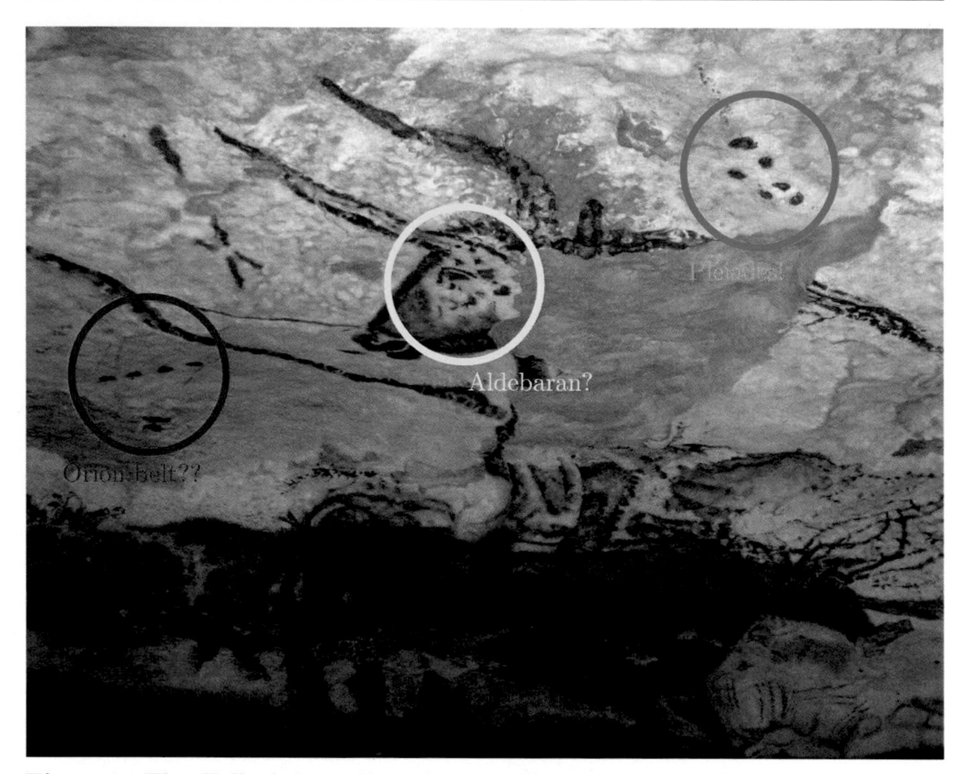


**Fig. 5.84** The *Hall of the Bulls* in Lascaux: Is it also a map of the nocturnal sky? The Pleiades (encircled in green) are depicted on many ancient paintings all over the world. Photo: Michael A. Rappenglück, 2012.

unfortunately – not as close as one might have hoped. In order to conclude this discussion: The supposed star may still have been a comet (we know too little about comets at that time). Neither should we underestimate the skills of the artists. They knew exactly what they were drawing. Picasso, who was one of the first to see the discoverd paintings, declared, "We have learned nothing since then". ◂◂◂

▸▸▸ **Application**: ***fixed stars or not?*** (Fig. 5.85)
Stars like our Sun and all visible stars on the firmament move quite fast with respect to the center of the galaxy, for instance, or also with respect to our Sun. However, since they are so far away from us, this will make no difference to the celestial map during a human life span. Within tens of thousands of years, there are some noticeable changes. Star clusters that are comparably far away from us will barely change their mutual positions on the sky. Some stars, like Sirius – "only" $d_S = 8.6$ light-years away – move differently, since they are *much* closer in comparison. It has a relatively high "proper motion". Where was Sirius at the time when the artist in Lascaux was watching the sky (Application p. 232)?

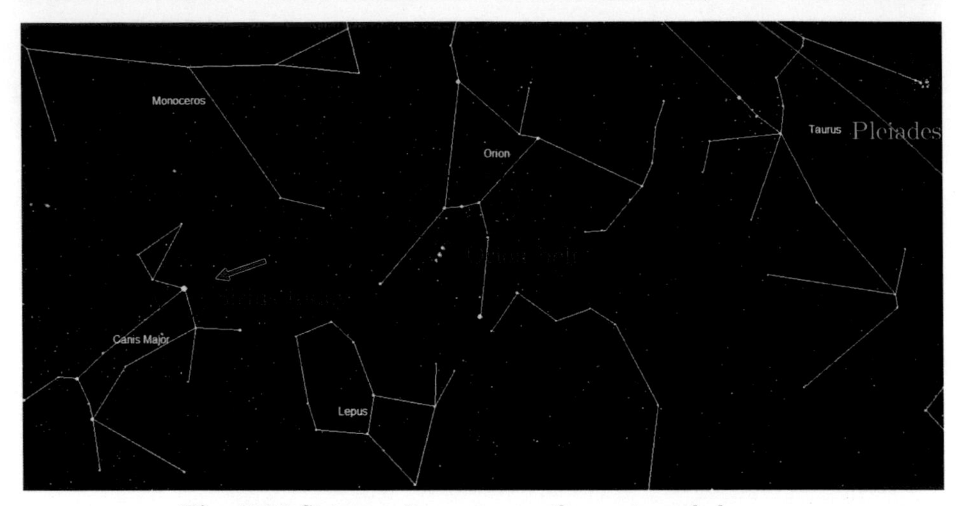


**Fig. 5.85** Sirius on its way over the nocturnal sky ...

*Solution*:
From tables found on the Internet, we get the following data for Sirius:

- radial velocity $R_v = -5.5$ km/s.
  Sirius decreases its radial distance to our Sun by 5.5 km per second.
- Right ascension $\alpha = -546$ mas/yr,
  i.e. longitude difference on the celestial sphere:
- Declination $\delta = -1,223$ mas/yr,
  i.e. latitude difference on the celestial sphere.

The abbreviation mas stands for milliarcsecond.
Thus, 1 mas $= 1/1,000 \cdot 1/3,600°$. To convert into radians, we have to multiply by $\pi/180$: 1 mas $\approx 4.85 \cdot 10^{-9}$. Thus, we have (measured in radians) $\overline{\alpha} \approx 0.00000265$ and $\overline{\delta} \approx 0.00000593$.
Let us take 1 light-year (1 Ly $\approx 30 \cdot 10^6$ s $\cdot$ 300,000 km/s $\approx 9.5 \cdot 10^{12}$ km, Application p. 13) as our length unit. Then, the distance of Sirius is $d_S = 8.6$ and the radial speed $R_v \approx 0.000018$ Ly/s. Sirius's current position shall be defined by the direction $\vec{s}_0 = (1, 0, 0)$ (Cartesian coordinate system with our Sun as the center). Provided that Sirius moves on a straight line, its position in $t$ years is given by the vector

$$\vec{s}(t) = \begin{pmatrix} d_S \\ 0 \\ 0 \end{pmatrix} + t \begin{pmatrix} R_v \\ d_S \cdot \overline{\alpha} \\ d_S \cdot \overline{\delta} \end{pmatrix}.$$

In order to measure angles, we have to normalize $s(t)$ ($\rightarrow s_0(t)$). The angle $\varphi(t)$ between those two vectors is given by $\cos\varphi(t) = \vec{s}_0 \bullet \vec{s}_0(t)$.
For $t = -17,000$ years (Lascaux), Sirius was $\Delta \approx 6°$ off the current position (in the direction towards Orion's belt). This is not close enough to justify the claim that the artist might have seen Sirius as part of the belt (the distance is three times as large). Of course, one must now also consider the

relative motion of the belt stars. They are, however, so far away that the corresponding angles are *much* smaller. If the painter would have lived some 40,000 years earlier, the hypothesis that Sirius is the fourth star in the belt would have been a good theory. ◂◂◂

▸▸▸ **Application**: ***How many points, lines, and planes are in three-space?*** This question may seem strange at first. The answer is, of course, "infinitely many". Let us now investigate whether there are different categories of "infinite".[2] By $\infty^1$, we denote the number of points on the real line. Then, we can also say: There are $\infty^1$ real numbers. The position vector of a point in space is uniquely defined by three real components. For each component, we have $\infty^1$ possibilities. Thus, there are, in our terminology, $\infty^3$ vectors and also

- $\infty^3$ points in space.

Now we want to determine the number of *unit vectors* in three-space. So, we impose the condition $n_x^2 + n_y^2 + n_z^2 = 1$ on a vector $\vec{n}(n_x, n_y, n_z)$, and it becomes a unit vector. Therefore, we can prescribe only two components $n_x$ and $n_y$. The third component is generated (aside from the sign) as $n_z = \sqrt{1 - n_x^2 - n_y^2}$. Thus, there are "only" $\infty^2$ unit vectors in three-space. Interpreted as position vectors, these $\infty^2$ vectors describe the points on the "unit sphere".

Now for the planes: They are uniquely determined by a unit normal vector $\vec{n}$ ($\infty^2$ possibilities) and a real constant $c$ ($\infty^1$ possibilities), because $\vec{n} \cdot \vec{x} = c$ is the equation of a plane. We can, therefore, define a plane by specifying the three real numbers $n_x$, $n_y$, and $c$. Thus, there are

- as many planes as there are points in three-space, namely $\infty^3$!

Still, the lines are to be counted. Are there as many, more, or fewer lines than points in three-space?

A straight line $g$ is uniquely determined by a point $P$ ($\infty^3$ possibilities) and a normalized direction vector $\vec{n_0}$ ($\infty^2$ possibilities). That would be an inconceivable number of $\infty^5$ lines. On the other hand, the point $P$ can be chosen freely on the line $g$ ($\infty^1$ possibilities) – $g$ remains unchanged. Thus, there are only

- $\infty^4$ straight lines in three-space!

⊕ *Remark*: This sounds less dangerous than it is. One must first rephrase: There are infinitely more straight lines in space than there are points or planes. This is the reason why it is so hard to imagine a rotation in space about a "generic" line. Recall the example of the aircraft wheels (Application p. 228) or the fact that very few people know that the position of the Sun rotates about the imaginary axis through the Polar Star during a day (Fig. 5.73). Furthermore, in Descriptive Geometry, two

[2] On such counting questions, one could give a whole lecture, and one easily falls into fallacies. In the end *dimensions of manifolds* (for example, all lines in space) are to be determined. It is possible to find a one-to-one and invertible mapping that assigns to each point in the plane precisely one line in the plane. In order to be precise, instead of posing the question "how many points", one should ask instead "how many parameters" determine the points – this number equals the dimension of the manifold.

side views are necessary in order "to cause a line to project", i.e. to make it appear as a point. In contrast, with only a single side view, you can make each plane project in such a way that it appears as a straight line!
The advantages of vector analysis become evident in the treatment of the straight lines in space: Both the auxiliary point on the line and the direction of the line are, indeed, described by vectors, namely by a position vector and a direction vector. ⊕
◂◂◂

### ▸▸▸ Application: *How many points and lines are in the plane?*

With the considerations of Application p. 235, we obtain:
Points in the plane are described by position vectors with two components. So, there are
- $\infty^2$ points in the plane.

There are only $\infty^1$ normalized unit vectors, namely the position vectors of the points of the "unit circle".
Lines in the plane are – like the planes in three-space – defined by an equation of the form $\vec{n}_0 \cdot \vec{x} = c$. So, they are specified by $n_x$ and $c$. Thus, there are
- as many lines as points in the plane!

The lines in the plane somehow correspond to the planes in three-space.

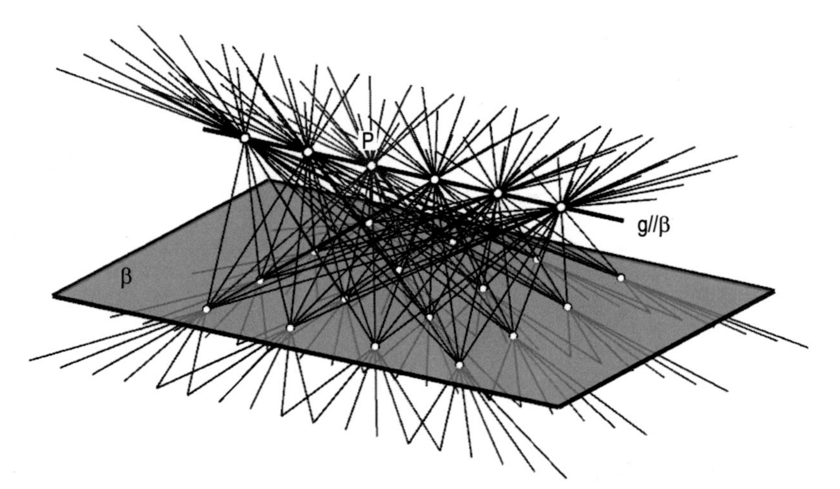


**Fig. 5.86** Consider all straight lines determined by a fixed point $P$ and an arbitrary point in a horizontal plane $\beta$ ($P$ not in $\beta$). There exist as many lines as there are points in this plane, because any point in $\beta$ can be joined with $P$.

⊕ *Remark*: Fig. 5.86 illustrates that through each point in space we have $\infty^2$ straight lines and that we already have $\infty^3$ straight lines that meet a single straight line. ⊕
◂◂◂

# 6 Functions and their derivatives

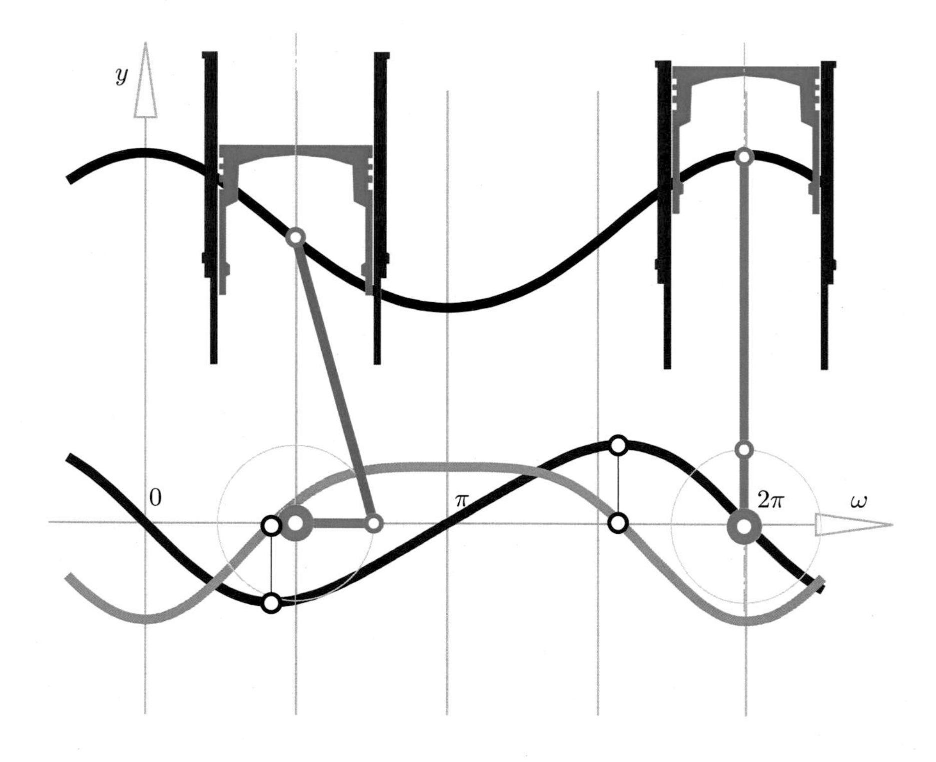


In very generalized terms, a real function $y = f(x)$ can be understood as a rule for assigning a real number $y$ to a real number $x$. Some functions play a central role in mathematics and their numerous applications, as well as in nature. Among others, these are the power functions of the form $y = x^n$; the trigonometric functions $y = \sin x$, $y = \tan x$, etc.; the exponential functions $y = a^x$; and also their inverses.
Besides elementary functions, which are given with mathematical exactness, the so-called empirical functions also play an essential role in applied mathematics. They are mostly given by a number of measurement points (sampling points), and intermediate points are interpolated.
When solving problems of a physical, technical, or geometric nature, it is often necessary to employ the "derivative function" $y' = f'(x)$ of the original function. The derivative is computed by means of simple rules of differential calculus, and it is indispensable in the solution of many problems.
In the age of computing, digital techniques are not only used to display functions, but also to differentiate them. This often leads to numerical problems, which will be discussed in this chapter. In general, the use of computers makes great sense, and it even allows us to solve so-called transcendental (non-algebraic) equations with sufficient precision.


G. Glaeser, *Math Tools*, https://doi.org/10.1007/978-3-319-66960-1_6

## 6.1 Real functions and their inverses

In order to get acquainted with the problem, let us first discuss a diagram showing a series of measurements.

▸▸▸ **Application**: ***amount of lactate in blood related to physical work***
The lactate value describes the relation between the concentration of lactic acid (lactate concentration) in blood and the intensity of physical work of a human body (Fig. 6.1).

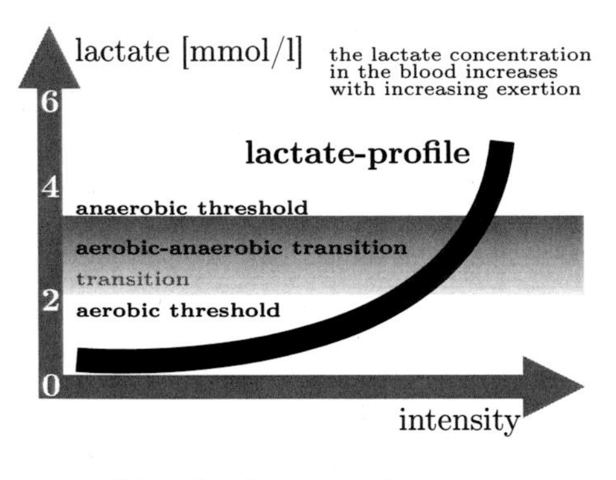

**Fig. 6.1** lactate values ...

Lactate values in blood during physical work: During each motion, muscles produce both energy and lactate. Persistent intense physical work causes over acidification of the muscles due to a lack of sufficient oxygen supply. At a lactate value of up to 2 mmol per liter, the organism is working "aerobically", receiving a large portion of the required energy through the oxygen-intensive burning of body fat – without depleting the valuable glycogen storage of the muscles and the liver too quickly (which can only be replenished over longer periods of time).

In the "anaerobic range", the body increasingly needs to make use of these energy storages. Once the glycogen is fully depleted, the body becomes completely slack. Many long-distance runners experience this phenomenon if they do not hold back their tempo during the initial phase of their run, which causes them to reach the anaerobic phase too soon. On the axis of abscissas (usually the $x$ axis), we notate the intensity of the physical work. On the axis of ordinates (usually the $y$ axis), we notate a sufficient quantity of measured lactate values and connect the data points. ◂◂◂

After this "warm-up example", let us take a moment to clarify several terms so that we may approach the next considerations more exactly.
The real numbers $\mathbb{R}$ are the set of all "decimal numbers". Calculators and computers, for instance, usually work with finite decimal numbers – in other words, with a mere subset of the rational numbers and with functions that act on them.
Real numbers may be visualized on the *real line*, as the points on a straight line correspond exactly and unambiguously to real numbers. A segment corresponds to an interval of real numbers. The set $\{x|\ a \leq x \leq b\}$ is called a

*closed interval* $[a, b]$, and the set $\{x|\ a < x < b\}$ is called an *open interval* $]a, b[$.

## Definition and graph of a function

A real function $y = f(x)$ defined on $[a, b]$ describes a rule that assigns a single real number $y$ to each real number $x \in [a, b]$.
The variable $x$ is called the independent variable or the argument, and $y$ is called the dependent value or the function value. The interval in which $x$ is defined is called the domain **D** ($\{x\} = [a, b]$), and the interval onto which $x$ is mapped is called the codomain **B**. ($\{y\} = [c, d]$).

The function value can be the result of a mathematical calculation or simply a measurement value.
Let us return once more to our "warm-up example", and this time, let us be a little more precise:

▸▸▸ **Application**: ***lactate concentration as a function graph*** (Fig. 6.1)
In the diagram Fig. 6.1, $x$ represents the intensity of physical work done by the body, which may be rather subjective, but which can, nevertheless, be approximated by measuring the current heart frequency. Naturally, this intensity is subject to constraints, depending on the age and physical fitness of the athlete in question. Each heart frequency can be associated (in a relatively constant correlation) with a lactate value in the bloodstream. This means that an athlete who knows his or her individual "lactate function" may, with considerable accuracy, infer the current lactate concentration from the current heart frequency.

⊕ *Remark*: Sports medicine makes use of this insight when constructing training schedules. We may immediately see: A rise in heart frequency correlates with a rise of lactate concentration in the blood. However, this relation is not linear. Instead, lactate concentration increases in an exponential manner. We will soon discover what this means.
From repeated measurement, it follows that it is sufficient to focus on the ranges $\mathbf{D} = [40, 230]$ (boundary values for pulse) and $\mathbf{B} = [0.3, 10]$ (boundary values for lactate). Any statements that go beyond these ranges amount to speculations, about which one should be skeptical. ⊕ ◂◂◂

In order to draw the graph of the function (Fig. 6.2), we inscribe the corresponding value $y = f(x)$ for each value $x \in \mathbf{D}$ in the Cartesian coordinate system. In general, the resulting graph is a line, which does not, however, look as in the picture to the right, because, by definition, a function is only given if the relation $x \to y$ is unambiguous: Each *y-parallel* can intersect the function graph *no more than once*!

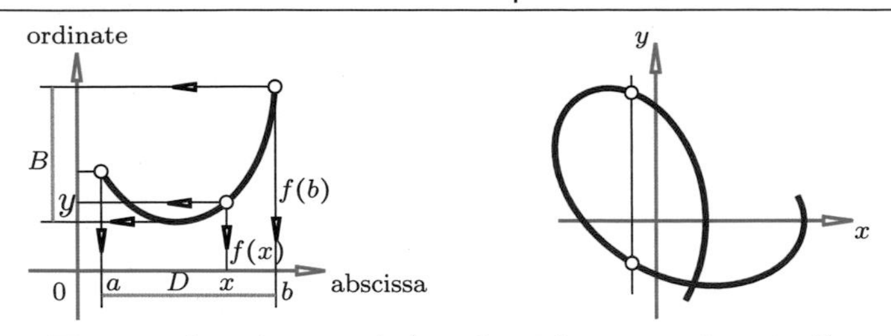


**Fig. 6.2** function graph (on the right: not a function!)

## Inverse function

Let $y = f(x)$ be defined in $\mathbf{D} = [a, b]$. Then, for every $x \in \mathbf{D}$, exactly one $y \in \mathbf{B}$ (Fig. 6.3, left) is assigned. Often, we need the concept of the *inverse function*:

> If $y \in \mathbf{B}$ matches exactly one $x \in \mathbf{D}$, then $y = f(x)$ is invertible.

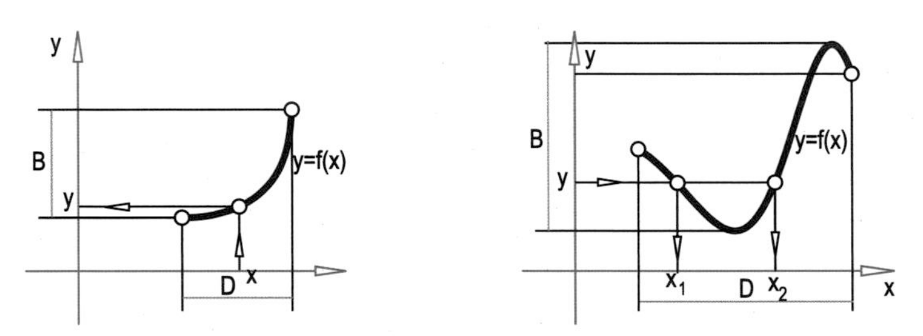


**Fig. 6.3** invertible and non-invertible function

The immediate and obvious theorem holds:

> A function is invertible if it constantly increases or decreases, that is, if it is "monotonic".

**►►► Application**: ***conclusions from the lactate values***

Our lactate function is invertible: It increases monotonously, which means that, at a higher heart rate, the body produces more lactate. Since lactate does not readily disappear from the blood, one can infer the impact of the lactate levels after an extreme athletic performance. ◄◄◄

Now, we want to show by means of simple examples how the inverse function can be calculated when a mathematical formula exists for the function or how it can be found graphically if it is only given empirically.

⊕ *Remark*: In mathematics, the inverse function of a function $f$ is denoted by $f^{-1}$ or also by $f^*$ where the first term should not be confused with $\frac{1}{f}$. The notation for the inverse function is not uniform for typical calculators. For example, the names of the inverse of the tangent function are found on most calculators as

INV TAN, TAN-1, ATAN, ARCTAN. ⊕

**▸▸▸ Application**: ***linear functions and their inverses***
Determine the inverse function of the linear function $f(x) = kx + d$

*Solution*:
We "solve" for $x$:

$$x = \frac{y-d}{k} = \frac{1}{k}y - \frac{d}{k} = f^{-1}(y).$$

Now, $y$ is the independent variable and $x$ depends on $y$. In mathematics, it is common that the independent variable is denoted by $x$ and the dependent variable is denoted by $y$. Thus, the variables in $x = f^{-1}(y)$ are replaced by $(x \leftrightarrow y)$, and we get $y = f^*(x)$ as the inverse function of $y = f(x)$

$$y = \frac{1}{k}x - \frac{d}{k}.$$ ◂◂◂

The inverse of a linear function is, thus, again a linear function. This becomes immediately apparent since an exchange of the coordinates of a point results in a reflection of the point on the line $y = x$, which is called the "first median" (Fig. 6.4). If we apply this reflection to all points $P^{-1}$ of $x = f^{-1}(y)$, we obtain the inverse function $x = f^{-1}(y)$. Since $y = f(x)$ has the same graph as $x = f^{-1}(y)$, we get:

| The graphs of $y = f(x)$ and $y = f^{-1}(x)$ are symmetric with respect to the first median. |
|---|

For empirically defined functions, one has to be satisfied with this reflection.

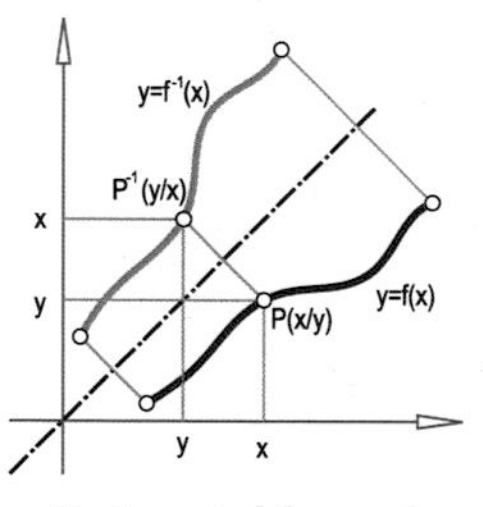

**Fig. 6.4** variable exchange

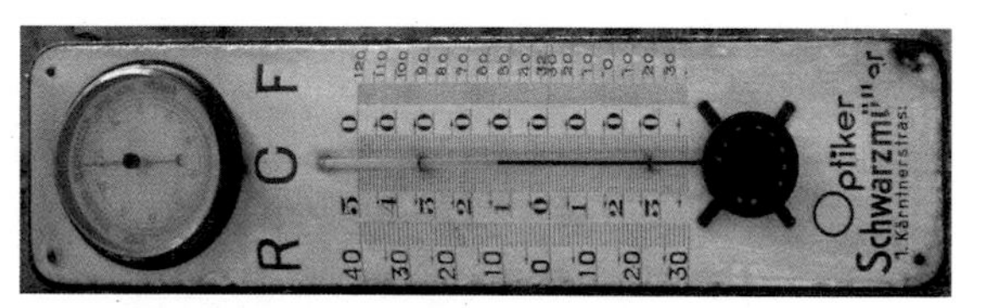

**Fig. 6.5** three temperature scales

**▸▸▸ Application**: ***conversion between* Celsius *and* Fahrenheit** (Fig. 6.5)
Compute the function $f : C \to F$ and its inverse $f^{-1} : F \to C$. Which temperature in $C$ corresponds to $451°F$?[1] What temperature in $C$ and $F$ has the same numerical value?

*Solution*:
In Application p. 24, we derived the formulas for the conversion of the two linear temperature scales. If we call $x$ the temperature in $C$, then $y$ is the temperature in $F$.

[1]For film buffs: In the film adaptation of a novel by Ray Bradbury it is the temperature which leads to the ignition of paper.

With the considerations of Application p. 241, we ensure that the following holds for the function and its inverse function:

$$f:\ y=\frac{9}{5}x+32,\quad f^{-1}:\ y=\frac{5}{9}(x-32). \tag{6.1}$$

From this, we see that $451°F$ corresponds to $\frac{5}{9}(451° - 32°)C \approx 233°C$. An easy to remember pair of corresponding values is obtained by swapping digits: $16°C$ correspond to $61°F$.
If $C = F$, then $x$ equals $y$ and thus 0.76

$$x=\frac{9}{5}x+32 \Rightarrow -\frac{4}{5}x=32 \Rightarrow x=-40 \Rightarrow -40°C=-40°F.$$

We get the same result either by equating the inverse function $\frac{5}{9}(x-32)$ with $x$ or, in a more complex manner, by equating (= intersecting) the function and the inverse function in Formula (6.1). ◂◂◂

▸▸▸ **Application**: ***parabola and inverse function*** (Fig. 6.6)
"In standard position", the "unit parabola" has the equation $y = x^2$. The graph is not monotonic in the entire domain $-\infty \le x < \infty$. Therefore, the function is not invertible. If we restrict ourselves to positive values of $x$, the graph is monotonically increasing and the function is, therefore, invertible.
*Solution*:
The function $y = x^2$ with $0 \le x < \infty$ has the inverse $x = +\sqrt{y}$ with $0 \le y < \infty$. We write the inverse function (interchange $x$ and $y$):

$$y=+\sqrt{x}\quad \text{with}\quad 0\le x<\infty.$$

⊕ *Remark*: All parabolas are similar to each other (this is not true for the other conic sections, i.e. for ellipses and hyperbolas). The graph of the parabola

$$y=a\,x^2\ (0\le x<\infty)$$

looks the same as the unit parabola when the unit length is multiplied by the factor $\sqrt{a}$. The inverse function is, then,

$$y=\sqrt{\frac{x}{a}}\quad \text{with}\quad 0\le x<\infty.$$

⊕ ◂◂◂

▸▸▸ **Application**: ***equilateral hyperbola and inverse function*** (Fig. 6.7)
The asymptotes of an equilateral hyperbola are perpendicular. One can, therefore, rotate them in such a way that the axes of the coordinate system coincide with their asymptotes. Then, the equation of the function is

$$y=\frac{a}{x}\quad (-\infty<x<\infty,\ x\ne 0).$$

What is the inverse function?

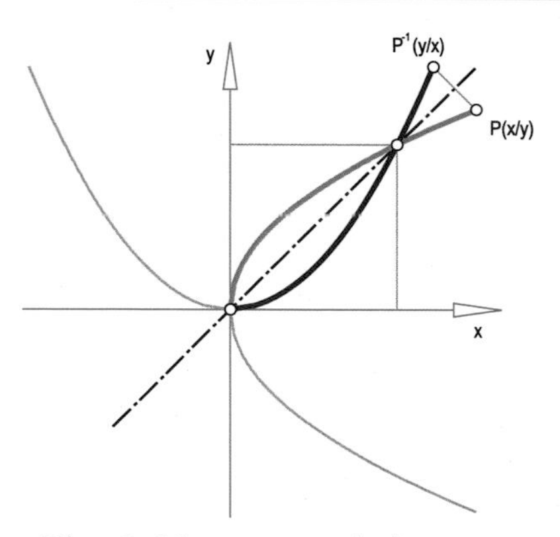


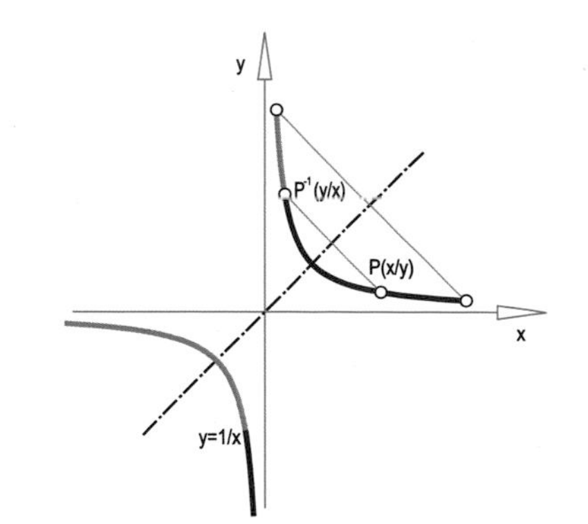


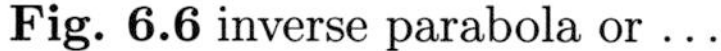
**Fig. 6.6** inverse parabola or ...

**Fig. 6.7** ... equilateral hyperbola

*Solution*:

Given the function $y = \frac{a}{x}$, we change the variables: $x = \frac{a}{y}$ and the inverse function reads $y = \frac{a}{x}$ $(x \neq 0)$. Therefore, the function is self-inverse.

⊕ *Remark*: The vertices lie on the axis of symmetry: $x = \frac{a}{x} \Rightarrow x^2 = a \Rightarrow x = y = \pm\sqrt{a}$. Among all hyperbolas, the equilateral hyperbolas play a special role as circles do among ellipses. The equilateral hyperbolas are – like parabolas and circles – all similar to each other.

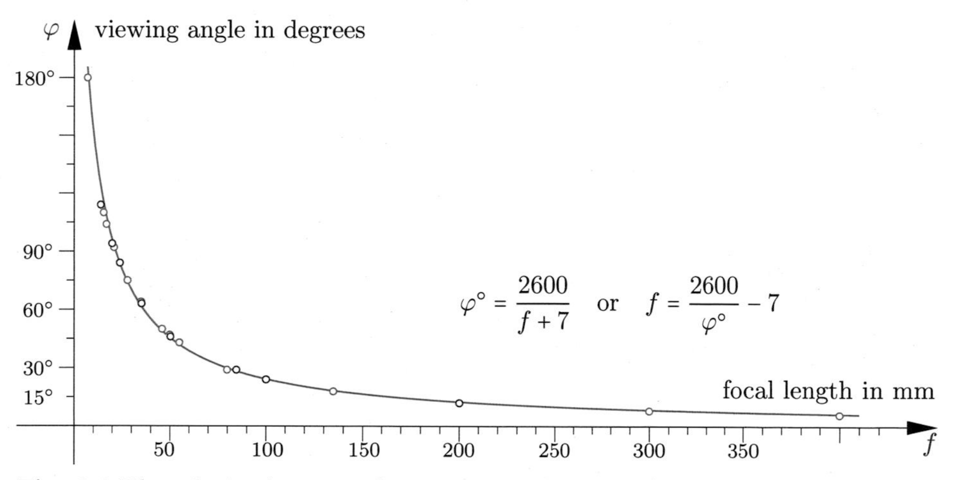


**Fig. 6.8** The relation between the viewing angle $\varphi$ and focal length $b$ is indirectly proportional: The larger the focal length, the smaller the viewing angle. Based on information provided by the manufacturers, the given approximation formulas can be calculated. On the graph of the function, the manufacturers' data are marked with small circles.

Fig. 6.8 shows a practical application where the function graph is an equilateral hyperbola: The relation between the optical angle $\varphi$ (measured diagonally in degrees) and focal length $f$ of a lens is indirectly proportional. The inverse function is of the same type. ⊕ ◂◂◂

## 6.2 Power, exponential, and logarithmic functions

A function of the type $y = x^n$ with a constant exponent and a variable base is called a *power function*.

We have seen special power functions on many occasions. For $n = 1$, the function is linear (the graph is a straight line). For $n = 2$, the function is quadratic (the graph is a parabola). For $n = 3$, the function is "cubic" (the graph is a *cubic parabola*). The power functions can also be equipped with negative exponents, and the rule $x^{-n} = \frac{1}{x^n}$ comes into play. As a simple example, we have already considered the function $y = \frac{1}{x}$ with an equilateral hyperbola for its graph.
Consequently, $n$ is allowed to be any rational or real number. The following rule applies: $x^{\frac{p}{q}} = \sqrt[q]{x^p}$.
It is obvious that the *power functions* $y = x^n$ *and* $y = \sqrt[n]{x}$ *are each other's inverses. Their graphs are, therefore, symmetric with respect to the first median.*

▸▸▸ **Application**: ***calculation of points in the decathlon*** (Fig. 6.9)
The calculation of points in the decathlon uses the formula

$$y = a|b - x|^c$$

to derive the number of points $y$ ($a$, $b$, and $c$ are constants where $b$ is always the limit above and below which there are no more points). The value $x$ is the output that is actually achieved. Discuss the function for $a = 25.44$, $b = 18$, and $c = 1.81$ (100 m sprint), as well as for $a = 51.39$, $b = 1.5$, and $c = 1.05$ (shot put). Which time in the 100 m sprint achieves 500 points?

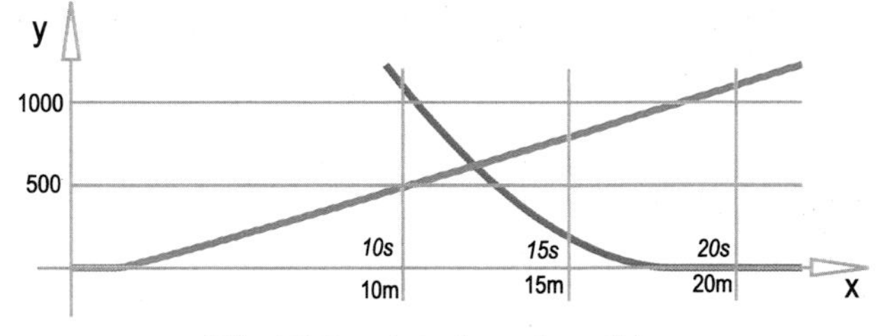


**Fig. 6.9** points in a decathlon

*Solution*:
In the case of the 100 m sprint, the formula reads $y = 25.44(18 - x)^{1.81}$. In the case of shot put, it reads $y = 51.39(x - 1.5)^{1.05}$. The function graphs can be

seen in Fig. 6.9. In shot put, the number of points increases almost linearly with the achieved distance. Extremely good performances in the 100 m sprint (the magical 10 s) can thus be said to be "rewarded disproportionately".
In order to calculate $x$ from a prescribed $y$, we use the inverse function:

$$y = 25.44(18-x)^{1.81} \Rightarrow \left(\frac{y}{25.44}\right)^{\frac{1}{1.81}} = 18 - x \Rightarrow x = 18 - \left(\frac{y}{25.44}\right)^{0.5525}.$$

It needs $x \approx 12.8$ seconds in order to reach $y = 500$ points. ◂◂◂

▸▸▸ **Application**: ***uniformly inflating a balloon*** (Fig. 6.10)
The volume of a balloon will uniformly increase by inflating, wherein the balloon substantially retains its shape. How do the length scale or the surface area change?

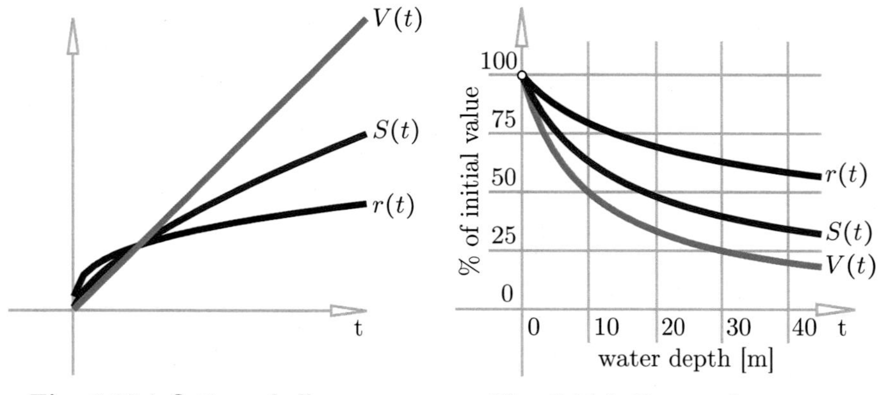


**Fig. 6.10** inflating a balloon

**Fig. 6.11** balloon under water

*Solution*:
We know from the section on similarities that volumes change with the cube of the length scale $r$ and surface areas change with the square of $r$. We will now apply the time $t$ on the $x$-axis ("abscissa"). By assumption, the volume increases linearly with respect to time. Then, the length of the scale varies with the cubic root of the time: $r(t) = t^{1/3}$. The surface $S$ changes with the square of the length scale $r$: $S(t) = \left(t^{1/3}\right)^2 = t^{2/3}$. Fig. 6.10 shows the corresponding diagrams.

⊕ *Remark*: The equation does not contain any of the gas laws (see next Application). The air pressure inside the balloon will rise slightly until it bursts, thus reducing the actual volume and the actual surface. When inflating a bicycle tube, the outer shell prevents the tube from expanding. This does not increase the volume, but, as desired, only the internal pressure. ⊕ ◂◂◂

▸▸▸ **Application**: ***submersion of a balloon*** (Fig. 6.11)
A balloon is pushed under water. How do the volume, the scale of length, and the surface change?

*Solution*:
Let $V_0$, $r_0$, and $S_0$ be the volume, the radius, and the surface area of the balloon, and let $p_0 = 1\,\text{bar}$ be the initial pressure at the water's surface.
This time, the abscissa is the depth of the water $t$. At constant temperature, *Boyle*'s Law is valid: $p\,V = p_0\,V_0 = \text{const.}$ The external pressure $p$ increases by 1 bar with every 10 m of depth:

$$p = 1 + \frac{t}{10} = x\,\text{bar} \quad \text{with} \quad x = 1 + \frac{t}{10}.$$

This applies to the volume, radius, and surface area of the balloon:

$$V = \frac{V_0}{p} = V_0\,x^{-1},\ r = r_0\,x^{-1/3},\ S = S_0\,x^{-2/3}.$$

Fig. 6.11 illustrates quite clearly the dependencies; along the ordinate axis, the dependent variables are marked by the percentages of the various initial values. ◂◂◂

In Application p. 245, the dependence of the volume $V$ on the initial volume $V_0$ was indirectly proportional to the "depth factor" $t_1$. Now consider the following questions:

▸▸▸ **Application**: ***decrease in brightness*** (Fig. 6.12)
In a lake, the intensity of light decreases by $p = 7\%$ with every 1 m of water depth. What is the intensity at a depth of $x$ meters? At what depth is 50% of the original intensity left?

**Fig. 6.12** Left: Deep Blue. Right: Red is the camouflage under water!

*Solution*:
Let $L_0$ be the light intensity at the surface. At a depth of 1 m, we have only $L(1) = 0.93\,L_0$. At a depth of 2 m, we have $L(2) = 0.93\,(0.93\,L_0) = 0.93^2\,L_0$. At a depth of 3 m, we have $L(3) = 0.93\,(0.93\,L(1)) = 0.93^3\,L_0$. So, at a depth of $x$ m, we obviously have

$$L(x) = 0.93^x\,L_0.$$

For $x = 9$, we thus obtain a light intensity of $L(9) = 0.520\,L_0$. For $x = 10$, we obtain $L(10) = 0.484\,L_0$. So, it is somewhere between a depth of 9 m and 10 m that the light intensity is only half of the original intensity.

⊕ *Remark*: The various light components lose their intensity at different rates. For example, blues and greens prevail at greater depths. At a depth of 30 meters, blood appears greenish to the diver! The red parts disappear first. This is why red provides a good camouflage under water. It can only be seen clearly when it is photographed with a flash as in Fig. 6.12 on the right! ⊕ ◂◂◂

Although the tasks in Application p. 245 and Application p. 246 appear to be very similar at the beginning, they lead to substantially different functions. Note that in the formula $L(x) = 0.93^x\,L_0$, the base is 0.93, and thus, it is constant. The exponent is the variable $x$.

A function of the type $y = a^x$ with a constant base and an exponent as a variable is called an *exponential function* of base $a$.

A typical example of an exponential function is the relation between the interest and capital:

▸▸▸ **Application**: ***monthly interest***

The formula for the interest on an initial capital of $K_0$ over $x$ years with an annual capitalization (with an interest rate of $p$ percent) is

$$K = K_0\,a^x \quad \text{with} \quad a = 1 + \frac{p}{100}.$$

What is the refined formula for monthly interest if the capital at the end of the year should be the same?

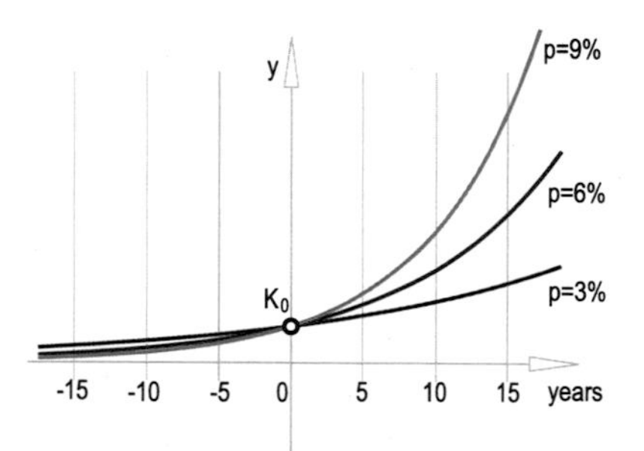


**Fig. 6.13** different interest rates

*Solution*:

Let $q$ be the monthly interest in percent. Then, after $x$ months we have $K = K_0\,(1 + \frac{q}{100})^x$. In fact, after 12 months, it must be true that

$$K_0\left(1+\frac{q}{100}\right)^{12} = K_0\left(1+\frac{p}{100}\right) = K_0\,a \Rightarrow 1+\frac{q}{100} = \sqrt[12]{a} \Rightarrow q = 100(\sqrt[12]{a} - 1).$$

This yields

$$K = K_0\, b^{12x} \quad \text{with} \quad b = \sqrt[12]{a}.$$

◂◂◂

Application p. 247 shows that a monthly interest return (or any other interest) leads to an exponential function. Further refinement allows us to compute interest rates at every second with

$$K = K_0 \left( \sqrt[\sigma]{a} \right)^{\sigma x} = K_0\, a^x.$$

Here, $\sigma \approx 3 \cdot 10^7$ is the number of seconds per year. Now, we have a function that gives us the value of the capital "at every time" $x \in \mathbb{R}$. Fig. 6.13 shows the associated exponential functions for $p = 3\%$ ($y = 1.03^x$), $p = 6\%$ ($y = 1.06^x$), and $p = 9\%$ ($y = 1.09^x$). We choose the current time to be $x = 0$. This calculation of capital can also be used for negative periods of time. We can, for example, ask how much capital we could save up to a certain date in order to get paid the capital $K_0$ at that moment.

⊕ *Remark*: The study of the exponential graph in Fig. 6.13 shows that, in the long run, a doubled or even tripled interest rate "generates" disproportionately more capital. This can be good for savings but can be potentially ruinous for borrowers. ⊕

Without proof, the following quite non-trivial theorem applies (see Fig. 6.14 or Application p. 286):

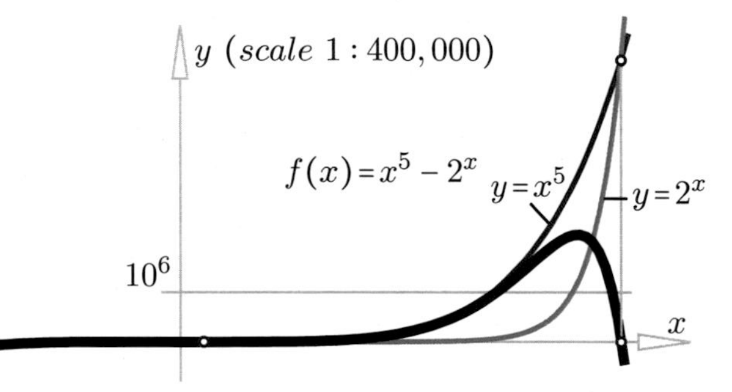


**Fig. 6.14** exponential vs. power function

Each exponential function $a^x$ with even smaller base $a > 1$ becomes "obsolete" at some point compared to any other power function $x^n$ with a large exponent $n$.

This means that it is not possible to catch up with the exponential function in the "long run".

▸▸▸ **Application**: ***return over a very long time***

Just as a theoretical juggling with numbers: Suppose you had one monetary unit ($1\,MU$) given to the bank 2,000 years ago and the bank is obliged to reimburse at a lower but annual interest rate of $p = 0.5\%$ ($1\%$, $1.5\%$) over the years. How many units of money would you have at the bank now?

*Solution*:

$$p = 0.5\% \implies 1.005^{2,000} \cdot 1\,MU \approx 2 \cdot 10^4\ MU,$$
$$p = 1.0\% \implies 1.010^{2,000} \cdot 1\,MU \approx 4 \cdot 10^8\ MU,$$
$$p = 1.5\% \implies 1.015^{2,000} \cdot 1\,MU \approx 9 \cdot 10^{12}\ MU.$$

⊕ *Remark*: Just to give an idea of these magnitudes: If we think of a monetary unit as a distance of one kilometer, then the last number corresponds to about $9{\cdot}10^{12}$ km, i.e. a light-year. ⊕ ◀◀◀

An interesting example in which this is applied can be found in the Appendix chapter on Mathematics in Music: The length of the air column in wind instruments grows exponentially with the tonal depth (Application p. 517).

## Logarithms and exponential equations

In practice, a common equation of the form $a^x = c$ is called an *exponential equation*. In order to solve it, we need the inverse of the exponential function. One has to "bring down" the exponent. We define:

| The inverse function of the exponential function $y = a^x$ is a logarithm to the base $a$: $y = \log_a x$. The logarithm of $x$ to the base $a$ is a number $y$ that is obtained by raising $a$ to the power $x$. |
|---|

Now, we have

$$\boxed{a^y = x \Rightarrow y = \log_a x.} \tag{6.2}$$

In practice, one is satisfied with the "logarithm" $\log_{10} x$ (the base 10 is usually indicated), and the "natural logarithm" $\ln x$ (sometimes called $\mathrm{Log}\,x$). In these cases, the base is 10 or Euler's number

$$\mathrm{e} \approx 2.71828183,$$

which is discussed in greater detail in the Appendix. Both logarithms are available on calculators, but, on PCs, only the natural logarithm is usually implemented. In the section on power series, we will see how computers calculate the logarithm. (However, we do not need to worry about it.)
For logarithms, the following important rules hold:

$$\boxed{\begin{array}{rcl} \log_a x_1 + \log_a x_2 &=& \log_a (x_1 \cdot x_2), \\ \log_a x_1 - \log_a x_2 &=& \log_a \dfrac{x_1}{x_2}, \\ n \cdot \log_a x &=& \log_a x^n, \\ \frac{1}{n} \cdot \log_a x &=& \log_a \sqrt[n]{x}. \end{array}} \tag{6.3}$$

***Proof***: The proof is purely formalistic and uses the rules of calculations with powers. We find it sufficient to prove only the first statement. The remaining statements can be proven in an analogous way: Since the logarithm and the exponential function "compensate" each other, we always have $y = \log_a a^y$. So, we get

$$\log_a x_1 + \log_a x_2 = \log_a \left[a^{(\log_a x_1 + \log_a x_2)}\right] =$$

$$= \log_a \left[a^{\log_a x_1} \cdot a^{\log_a x_2}\right] = \log_a (x_1 \cdot x_2). \qquad \odot$$

With logarithms, we can reduce the multiplication and division of numbers to the addition and subtraction of the respective logarithms. The power and root function can be reduced to multiplying or dividing numbers. This is still one of the most important applications of logarithms. A computer is constantly expected to do this internally. The invention of logarithmic tables was a major breakthrough in applied mathematics. (*John Napier* (1550–1617) invented calculation with logarithms and published the first logarithmic tables in 1614.)
The relation between different logarithms is given by constant factors. For example, if the $a$-logarithm of a number $x$ is known, then we also know every other logarithm, such as the $b$-logarithm. We have the formula

$$\boxed{\log_b x = k \cdot \log_a x \quad \text{with} \quad k = \frac{1}{\log_a b} = \text{constant.}} \tag{6.4}$$

***Proof***: Given the definition $x = b^{\log_b x}$, we can also deduce

$$\log_a \left(x\right) = \log_a \left(b^{\log_b x}\right).$$

Following the rules of calculation for logarithms, this equation can be transformed into

$$\log_a x = \log_b x \cdot \log_a b,$$

which results in

$$\log_b x = \frac{1}{\log_a b} \log_a x. \qquad \odot$$

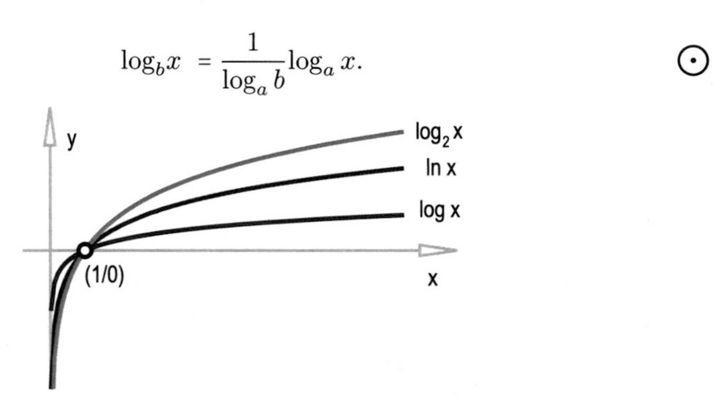


**Fig. 6.15** logarithms with different bases

⊕ *Remark*: The graphs of the various logarithmic functions differ by a scaling in the $y$-direction (Fig. 6.15). Programmers sometimes need to be familiar with Formula (6.4), because the computer only "knows" the natural logarithm $\ln x$. The adjusted formula is

$$\log_b x = \frac{\ln x}{\ln b}. \qquad \oplus$$

▸▸▸ **Application**: ***cooling a liquid***

A liquid with a temperature of 97° cools down to a room temperature of 22°. After ten minutes, it has a temperature of 76°. What is the temperature at $n$ minutes? When does it reach body temperature 37°?

*Solution*:

The initial temperature difference is $\Delta = 75°$. After ten minutes, it is only $\Delta_{10} = 54°$. When $\Delta$ is reduced by a factor of $k$ per minute, it has decreased by a factor of $k^{10}$ in ten minutes: $\Delta_{10} = k^{10}\Delta$. Thus,

$$54 = 75 \cdot k^{10} \Rightarrow k = \sqrt[10]{\frac{54}{75}} = 0.96768.$$

After $n$ minutes, the temperature $t_n$ is, thus,

$$t_n = 22° + 0.96768^n \cdot 75°.$$

If $t_n = 37°$, then we have

$$37° = 22° + 0.96768^n \cdot 75° \Rightarrow 0.96768^n = 0.20000.$$

We take the natural logarithm on both sides of the equation and obtain

$$\ln(0.96768^n) = \ln 0.20000 \Rightarrow n \cdot \ln 0.96768 = \ln 0.20000 \Rightarrow n \approx 49.$$

Therefore, the liquid has reached body temperature after about 49 minutes.

⊕ *Remark*: The time it takes for a liquid to equilibrate to the outdoor temperature depends on the insulating capacity of the container and also on the size of the area of the contact surface with the air: Deep lakes, therefore, heat up at a slower rate in early summer than shallower ones. ⊕ ◂◂◂

▸▸▸ **Application**: ***discounting***

At what cash value $B$ can a debt of 60,000 MU, which is overdue in 5 years, be superseded today when an annual rate of 7% is applied? At what value must an annual repayment rate (due at the end of the year) be set? Use and prove the following formula, which is important in financial mathematics:

$$1 + q + q^2 + \ldots + q^{n-1} = \frac{1 - q^n}{1 - q}. \tag{6.5}$$

*Solution*:

After one year, the present value of $B$ is "worth" $1.07\,B$. After 5 years, it is, therefore, $1.07^5\,B = 60{,}000 \Rightarrow B = 42{,}779$ MU.

Let $R$ be the annual rate. Then,

$$1.07^4\,R + 1.07^3\,R + 1.07^2\,R + 1.07^1\,R + 1.07^0\,R = 60{,}000.$$

We calculate the sum using Formula (6.5). The "proof" of the formula is the simple calculation (multiplication) of the equivalent formula

$$(1 + q + q^2 + \cdots + q^{n-1})(1 - q) = 1 - q^n.$$

Therefore, we have

$$\frac{1 - 1.07^5}{1 - 1.07} R = 60,000 \Rightarrow R = 10,433.$$ ◂◂◂

**▸▸▸ Application**: ***an unfair offer***

The following text is taken from an advertisement in a daily newspaper: "*penthouse, ...,* € 456,000. *Financing: single premium* € 150,000*, monthly* € 570 (Note: 30 years) ...".

What is striking here? What happens if the interest rate is 5%?

*Solution*:

After a rough calculation, any buyer will opt for the offered financing. The buyer may already have €456,000 in cash, but paying in instalments is still much cheaper. Even computing without interest and compound interest will yield a much smaller amount

$$150,000 + 30 \cdot 12 \cdot 570 = 355,200.$$

Which bank would give a *negative* interest rate?

The seller's declaration: The whole thing is financed by means of cheap yen loans. If the yen loans were no longer so low, one could, indeed, "roll over" by swapping these yen debts into debts in other currencies. No problem!?

Suppose you have to roll over and, therefore, take a "local" loan with an annual interest of $p = 5\%$, which is still a bargain. We calculate the monthly rate $M$ as follows: After the single premium is paid, we still have $456,000 - 150,000 = 306,000$ € due. According to Application p. 247, we have $b = \sqrt[12]{1 + p/100} = 1.00407$. As in Application p. 251, we relate all amounts to the same time, for instance, after about 30 years or $12 \cdot 30$ (= 360) months. We use Formula (6.5):

$$306,000\, b^{360} = M\,(b^{359} + b^{358} + \cdots + b^2 + b + 1) = M\,\frac{1 - b^{360}}{1 - b} \Rightarrow M = 1,615€.$$

In such a case, which is not unlikely to happen, the monthly payment is, therefore, nearly three times as high as what was given to the buyer at the time of buying – and this over a period of 30 years! ◂◂◂

**▸▸▸ Application**: ***exponential growth***

A certain bacterial culture multiplies by 12% per minute. After how long will it have increased tenfold or thousandfold?

*Solution*:
$10 = 1.12^x \Rightarrow \ln 10 = x \ln 1.12 \Rightarrow \dfrac{\ln 10}{\ln 1.12} \approx 20$ and
$1{,}000 = 1.12^x \Rightarrow x = \dfrac{\ln 1{,}000}{\ln 1{,}12} \approx 60.$
After 20 minutes, a tenfold increase will have occurred. After one hour, the culture will have already increased a thousandfold! The immune system of the body can, by the way, "neutralize" an infection with up to 150 million germs.

⊕ *Remark*: This particular calculation would work better with a logarithm to the base 10, because $\log 10 = 1$ and $\log 1{,}000 = 3$. However, this function is not always available on a PC. ⊕ ◂◂◂

▸▸▸ **Application**: ***the*** $^{14}C$***-method to determine the age***
After the death of an animal or plant, the radioactive carbon isotope $^{14}C$ disintegrates in its body with a half-life period of $5{,}730 \pm 40$ years. What is the age of a mummy that contains 40% of the initial amount of $^{14}C$?
*Solution*:
Let $G_0$ be the amount of $^{14}C$ of a living human, and let $G$ be the amount of $^{14}C$ in the mummy in question. Let now $x$ be the age of the mummy, measured in units of 5,730 years. Then,

$$G = 0.5^x G_0 \Rightarrow 0.4 = 0.5^x \Rightarrow x = \frac{\log 0.4}{\log 0.5} = 1.3219.$$

The age of the mummy is, thus, $1.3219 \cdot (5{,}730 \pm 40) = 7{,}575 \pm 53$ years. ◂◂◂

▸▸▸ **Application**: ***decrease of air pressure***
Atmospheric pressure $b$ decreases by half of the previous value after $h_0 = 5.5\,\text{km}$ height. What is $b$ at a height of $h$ km if $b_0 (\approx 1\,\text{bar})$ is measured at sea level? What is the air pressure at Mount Everest (almost 9 km above sea level)? What is the height at $0.8\, b_0$?
*Solution*:
We measure the height $h$ again as a multiple $x$ of the "unit" $h_0$: $x = \dfrac{h}{h_0}$. The corresponding air pressure $b$ is then

$$b = 0.5^x\, b_0.$$

At a height of approximately 9 km ($\Rightarrow x \approx 1.6$), the pressure is $b \approx \dfrac{1}{3} b_0$.
From $0.5^x\, b_0 = 0.8\, b_0$, we infer $x = \dfrac{\ln 0.8}{\ln 0.5} \approx 0.32 \Rightarrow h = 1{,}770\,\text{m}$. At this height, the relative air pressure is only 80%.

⊕ *Remark*: The altitude of 9 km is considered the absolute top barrier in which a human can survive (only for hours) without artificial oxygen. Therefore, it is always

pointed out in airplanes that, when there is a drop in cabin pressure, oxygen masks must be used.
When diving in mountain lakes at high altitudes, one has to take into account that the decompression is much more powerful than at sea level. At 40 m depth, an external pressure of 0.8 bar prevails, and there is only a total pressure of 4 bar + 0.8 bar = 4.8 bar. The ratio when surfacing is $\frac{4.8}{0.8} = 6$, which occurs only when diving from sea level to a depth of 50 m. While the short-term plunge at 40 m is still classified as harmless, you need a "decompression stop" when diving to 50 m (usually several minutes akinetic and hovering at 5 m) to prevent the "bends" (see Application p. 77). ⊕ ◂◂◂

▸▸▸ **Application**: ***air pressure vs. water pressure***

Why does air pressure decrease exponentially, while water pressure decreases linearly?

*Solution*:

Liquids (water) cannot be compressed like gas (air). A "water column" of 10 m height always has the same weight (⇒ linear growth), while the weight of a 10 m "air column" increases proportionally to the pressure (weight) of the amount of gas lying above it. Thus, air becomes more compressed and is getting heavier (⇒ exponential growth!). ◂◂◂

▸▸▸ **Application**: ***population growth*** (Fig. 6.16)

Exponential growth in nature must eventually lead to a collapse. For example, an increase of population cannot be exponential in the long run. Interpret the graph in Fig. 6.16.

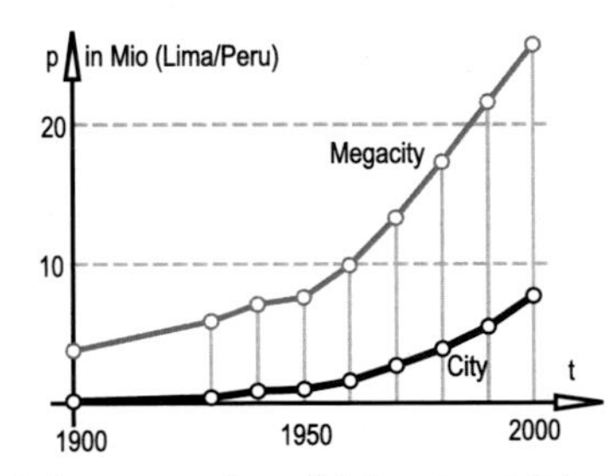


**Fig. 6.16** demography of Lima and Lima-Megacity

*Solution*:

The growth of the city almost seems to be an exponential function. However, the growth of the megacity is "only" linear in the last decades, and from 1900 onwards, it has even weakened. The "difference" between the two graphs – i.e. the number of persons who live in Greater Lima, but not in the city proper – shows that there is still a rising trend but one can recognize a "bend down". See also Application p. 259. ◂◂◂

## Logarithmic Graphs

In textbooks of physics, chemistry, or biology, one can often find graphs showing exponential changes that appear linear. This is accomplished by the so-called *logarithmic scale*. Repeated multiplication of a number by the same factor will always cause the same addition in the logarithmic chart, and the problem becomes "linearized". So, one can – at least graphically – obtain fast solutions to specific tasks. See Application p. 255, 6.24.

►►► **Application**: ***exponential growth with logarithmic scale***
In 2001, Nigeria had approximately 125 million inhabitants and an annual growth rate of 3.5%, whereas Brazil had 160 million inhabitants and a growth rate of 2.1%. When will the population of Nigeria exceed the population of Brazil? Draw a logarithmic graph.

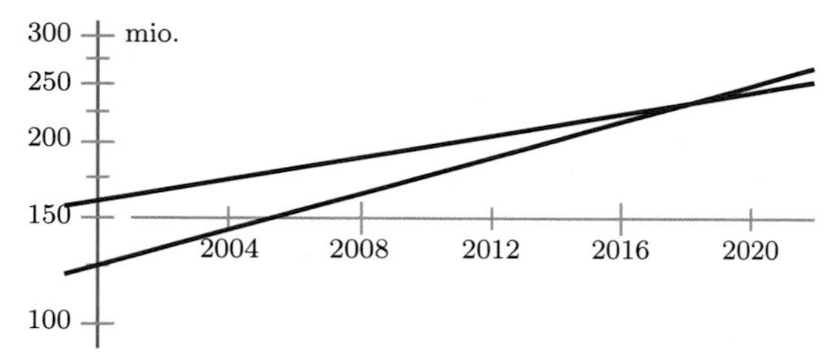


**Fig. 6.17** logarithmic population chart

*Solution*:
Mathematically, this problem is easy to solve:

$$125 \cdot 10^6 \cdot 1.035^x = 160 \cdot 10^6 \cdot 1.021^x \Rightarrow 1.014^x = 1.28$$

$$\Rightarrow x = \frac{\log 1.28}{\log 1.014} \approx 18.$$

Thus, after about 18 years, Nigeria's population will exceed that of Brazil. In the logarithmic graph (Fig. 6.17), we do not plot the number $y$ of individuals but the logarithm $\log y$ of the number of individuals along the ordinate (with arbitrary base). Since the logarithm increases linearly, the instant of equal population appears as the intersection of two straight lines.

⊕ *Remark*: Population forecasts are always subject to inaccuracies (in this case, the data are partly estimated) and assumptions (constant growth rate). The future will show whether the figures coincide reasonably with reality. The small angle formed by the two lines in the diagram indicates that this result is not very accurate. Yet, there are predictions from the past that turned out to be right: In 1950, about 2.4 billion people were living on Earth, and the annual growth was 43 million, corresponding to a growth rate of 1.78%. Thereupon, one predicted almost 6 billion people (more precisely, 5.85) for 2000, which later turned out to be fairly accurate. ⊕ ◄◄◄

▸▸▸ **Application**: ***the good old slide rule . . .*** (Fig. 6.18)
The slide rule is based on logarithmic scales and allows the mechanical optical completion of basic arithmetic operations (and more). For example, in order to perform the multiplication $2 \times 3$, one pushes the 1 of the movable "tongue" over the 2 on the scale below, and moves to the 3 on the display. Thus, one has added the logarithms of 2 and 3: $\log 2 + \log 3 = \log 6$. On the lower scale, we naturally read 6.

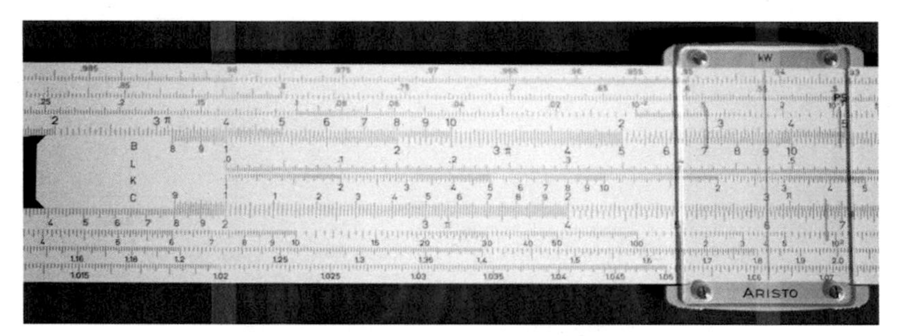


**Fig. 6.18** $2 \times 3 = 6$, carried on slide

⊕ *Remark*: Until the invention of the pocket calculator and the widespread use of PCs, slide rules were indispensable in school, science, and technology. ⊕ ◂◂◂

▸▸▸ **Application**: ***double logarithmic scales*** (Fig. 6.19, left)
Occasionally, you will also have diagrams in which the values are plotted logarithmically on both coordinate axes. Interpret the two graphs in Fig. 6.19, where the energy consumption of different animals is registered for the various locomotions ($a$ . . . swimming, $b$ . . . flying, $c$ . . . running).

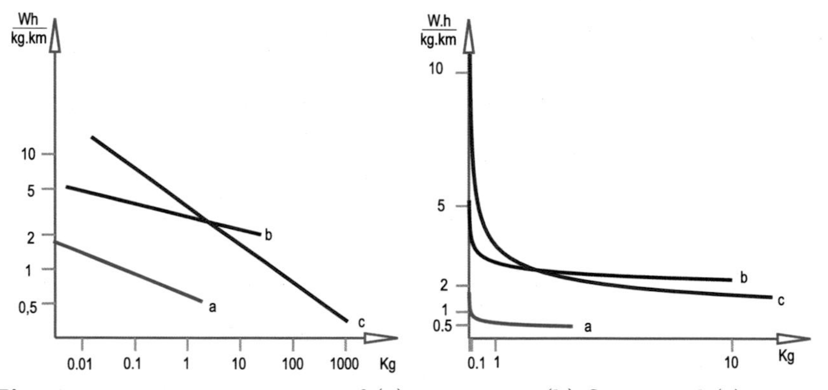


**Fig. 6.19** energy consumption of (a) swimming, (b) flying, and (c) running

*Solution*:
The non-trivial graph on the left[2] is difficult to understand for non-mathematicians:

[2] According to *McMahon* and *Bonner*: *On Size and Life*. Scientific American Books. New York, 1983.

The abscissa shows the applied body mass. In order to account for both small animals of only a few grams of weight and large land mammals, a logarithmic scale has been chosen.
On the ordinate axis, the energy consumption per kilogram of body mass and kilometer travelled is plotted. It is apparent that there is a linear relation to the mass when this energy expenditure is logarithmically distorted.
For the untrained chart reader, the image on the right is just as meaningful without distorting the data. The relevant information can clearly be seen here, though one can still imagine large land mammals with a mass of 500 kg and more quite effortlessly:

**Fig. 6.20** running or flying?

1. Swimming is (at least for fish – here, salmon were examined), by far, the most efficient locomotion. This is probably because gravity is compensated by buoyancy.

2. With more than 10 kg of body mass, running on land is more efficient than flying. (Flightless ostriches have a mass of 50 – 150 kg!)

3. In relation, small animals consume significantly more energy to perform an action. In order to keep the energy levels stable (see Application p. 85), a shrew has to eat high quality (animal) food of more than its own weight every day, while the elephant survives on inferior food (grass), with the daily amount equalling about 7% of its own weight.

⊕ *Remark*: Even when mammals are in "idle motion", a lot of energy is required in order to maintain their metabolism. Reptiles are essentially more moderate. Therefore, mammals are "ready for action" at low temperatures and have more stamina than reptiles. Science is not yet sure about whether dinosaurs, which are commonly associated with reptiles, were poikilothermal or whether they were warm-blooded like birds , which are their only surviving descendants. Anyway, after the extinction of the large dinosaur species, only mammals are left in the "upper weight class" of terrestrial animals. ⊕ ◂◂◂

## 6.3 The derivative function of a real function

If the function graph $y = f(x)$ has no kinks or jumps, then there is a line $t$ touching the graph called the *tangent* at the point $P(x/f(x))$. The tangent can be defined as a limit position of a chord $s$ that joins $P$ with a "neighboring point" $Q(x+h/f(x+h))$ (Fig. 6.21):

$$t = \lim_{Q \to P} s = \lim_{h \to 0} s.$$

The slope of the tangent tells us something about the shape of the curve in the vicinity of $P$. If, for example, the tangent is horizontal (slope 0), then the function graph has a local extremal value (minimum or maximum).
If the tangent has the inclination angle $\alpha$, then the following relation holds:

$$\tan \alpha = \lim_{h \to 0} \frac{f(x+h) - f(x)}{h}. \tag{6.6}$$

This value is referred to as the *derivative* of the function $f(x)$ at the point $x$ and is denoted by $f'(x)$. We have:

The derivative $f'(x)$ of a function $f(x)$ at $x$ yields the tangent of the angle of inclination of the tangent: $f'(x) = \tan \alpha$.

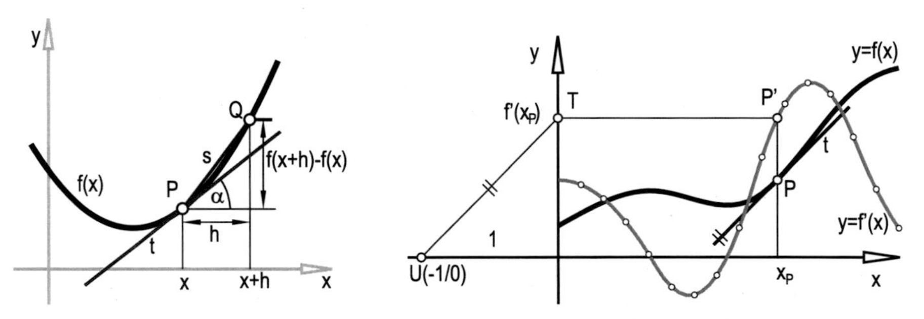


**Fig. 6.21** chord and tangent **Fig. 6.22** graphical differentiation

If $x$ traverses the entire definition interval, then the term $f'(x)$ provides a new function, namely the *derivative function*. We shall soon see that for mathematically defined functions, the *derivative function* can be calculated directly following a set of simple rules. For empirically defined functions, one has to be satisfied with "graphical accuracy", and the investigations will be limited to qualitative questions: Where does the function have a maximum or minimum?

The computation (or construction) of the derivative function is also called *differentiation.* The computation of derivatives falls within the scope of *differential calculus.*

## Graphical differentiation

In all cases, *graphical differentiation* works within graphical accuracy (Fig. 6.22). The derivative function $y' = f'(x)$ is constructed pointwise as follows: One chooses an arbitrary point $P$ of $f$ (the function to be differentiated) and draws the tangent line $t$ with graphical precision. If one shifts $t$ in parallel through the auxiliary point $U(-1/0)$ on the $x$-axis, then, in accordance with the definition, this straight line intersects the $y$-axis at a point $T$ with height $f'(x)$. The corresponding point $P'$ of the derivative function lies above (or below) $P$. Thus, the corresponding point on the derivative function equals $P'(x_P/y_T)$.

⊕ *Remark*: If you select $U(-\lambda/0)$, then the derivative curve is superelevated. ⊕

▸▸▸ **Application**: ***trends in population development*** (Fig. 6.23)
Using graphical differentiation, one can analyze population trends more efficiently. Discuss Fig. 6.23 (cf. Application p. 254). The function $f(x)$ shows the population development of the city center of Lima. The function $g(x)$ shows that of the entire city.

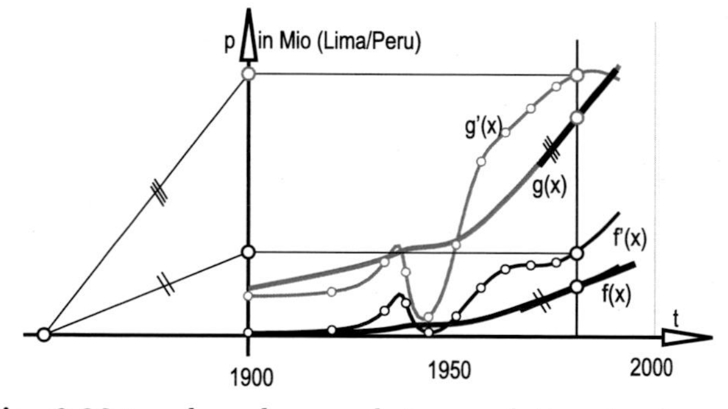


**Fig. 6.23** trends and reversals in population development

*Solution*:
Zeros of the derivative function indicate stagnation, extreme points (maxima and minima), and reversals. So, you can clearly see that, between 1940 and 1950, there was almost zero growth in both the Megacity and the city proper. Furthermore, the Megacity's peak of population growth has already been passed around 1990, whereas the growth of the town center continues unabated. ◂◂◂

▸▸▸ **Application**: ***biking in the mountains*** (Fig. 6.24)
A cyclometer counts a wheel's number of turns $n$, the corresponding time, and the altitude with an altimeter. How can it, thus, create an altitude profile

or slope profile of the route (regardless of whether one inserts intervening breaks)?

*Solution*:

Let $u$ be the constant circumference of the wheel. In smaller intervals $\Delta t$, we measure the number $n$ of revolutions of the wheel and the height difference $\Delta y$. Hence, the distance $\Delta s$ and the average speed are

$$\Delta s = n\,u, \quad v = \frac{\Delta s}{\Delta t}$$

in the interval in question. Only $\Delta s > 0$ matters, otherwise the biker is just having a break. The corresponding distance $\Delta x$ in the normal projection (in the top view) is calculated as

$$\Delta x = \sqrt{(\Delta s)^2 - (\Delta y)^2}.$$

Graphically, we find a neighboring point $(x + \Delta x / y + \Delta y)$ to any point $(x/y)$ on the upper curve in Fig. 6.24. Constant repetition results in a polygon that is an approximation of the mountain profile of the road. The smaller $\Delta t$ is chosen, the more accurate the curve is. Due to the highly inaccurate measurable difference in height, a compromise has to be reached (measured approximately every 20 seconds).

The average angle $\alpha$ of inclination is obtained from the relation $\tan\alpha = \Delta y/\Delta x$ (multiplied by one hundred, this equals the slope specified as a percentage). The lower curve in Fig. 6.24 is the slope profile whose ordinates are proportional to $\tan\alpha$ and whose points have, thus, the coordinates $(x/c \cdot \tan\alpha)$ ($c$ is the constant scaling factor). Therefore, it is essentially the derivative curve of the mountain profile.

⊕ *Remark*: Cyclists who want to evaluate "their" curves, prefer graphs which are very similar to the above described, but they do not have the normal projections of the distance drawn on the abscissa. These diagrams show the actual distance on the abscissa: Then, the $\Delta x$-intervals are replaced by $\Delta s = \Delta x \,/\cos\alpha$. Yet, even for slopes with an angle $\alpha < 8°$, we still have $\cos\alpha > 0.99$. Visually, the difference is barely recognizable. ⊕

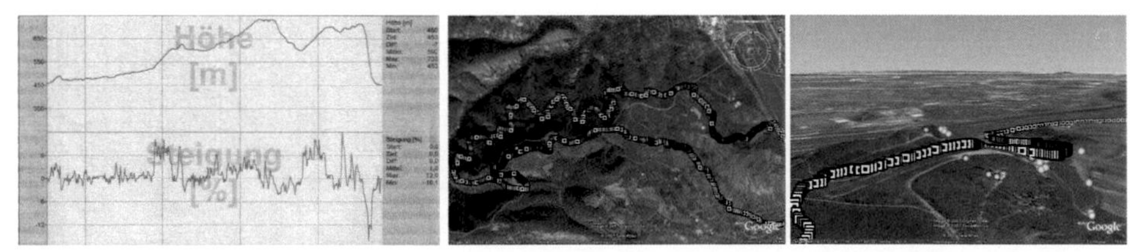


**Fig. 6.24** Profile of a bike tour in the mountains. Detailed 3D-information about your own bike or those of a virtual cyclist.

⊕ *Remark*: The *Garmin Edge Series* is a GPS-based cycling computer that offers the possibility to race against a virtual partner. For that purpose, your own track information or these of other bikers are displayed on the computer. This allows one to race against others in real time and under the same terrain conditions (slope, ground, etc.). The respective position of the virtual partner (lead, lag) is shown on the graphic display. One can, for example, upload his/her own data to a corresponding website and invite others to ride/compete in the virtual simulation. In addition, you can cycle along downloaded tracks (also with/against a virtual partner). Software download at www.garmin.com. ⊕ ◀◀◀

## Notations

Engineers and physicists like to use the so-called *differential notation*. Instead of $f'(x)$, they write

$$f'(x) = \frac{d}{dx} f(x) \quad \text{or, even shorter,} \quad f'(x) = \frac{df}{dx}.$$

The main argument in favor of this type of notation is that, in the various applications, the independent variable is not always called $x$. This is often the time $t$. In order to indicate that one does not differentiate with respect to $x$ but with respect to another variable (for example $t$), the following notation has become common:

$$\dot{f} = \frac{d}{dt} f(t) = \frac{df}{dt}.$$

## Physical interpretations of the derivative

In addition to the geometric interpretation of the derivative (slope of the tangent), there are various physical interpretations. Here are two important examples:

1. The instantaneous velocity is the derivative of the path-time function. From physics, we know the basic formula

$$s = v \cdot t \quad \text{or} \quad v = \frac{s}{t}.$$

So, the speed is the quotient of distance and required time. This formula is valid only if the speed throughout the motion is constant – or one obtains the *average speed* as a quotient. The shorter the selected time interval is, the more accurately the average speed will match the instantaneous velocity. The latter is obtained through a limit process, and the tangent is likewise obtained from the limit position of a chord of a curve. It is this limit process that is performed when differentiating.

If we interpret the distance as a function $s = s(t)$ of time, its derivative $v = v(t)$ is the "velocity function" $v(t) = \dot{s}(t) = \dfrac{ds}{dt}.$

2. The instantaneous acceleration is the derivative of the speed-time function.

Analogous considerations apply to the acceleration as the derivative of the velocity function with respect to the time.

> If we interpret the speed as a function $v = v(t)$ of time, its derivative $a = a(t)$ is the "acceleration function" $a(t) = \dot{v}(t) = \dfrac{dv}{dt}$.

▸▸▸ **Application**: ***free fall***

Thanks to physics, the formulas for free fall are known:

$$s = \frac{g}{2}t^2, \quad v = g\,t, \qquad a = g \ (= 9.81\text{ms}^{-2}).$$

Show that $v = \dot{s}$ and $a = \dot{v}$ hold true.

*Solution*:

Soon, we will see formulas that reduce the calculation to a single line. However, in order to promote understanding of the limit process, we want to perform one:

After $t$ seconds, a body will have fallen $\frac{g}{2}t^2$ meters (neglecting drag). At the end of a "short" period of time $dt$, the distance is $\frac{g}{2}(t + dt)^2$ meters. In the time $dt$, the body has, thus, fallen a distance of

$$ds = \frac{g}{2}(t{+}dt)^2 - \frac{g}{2}t^2 = \frac{g}{2}\left[t^2 + 2tdt + (dt)^2 - t^2\right] = \frac{g}{2}\left[2tdt + (dt)^2\right] = g\left[t + \frac{dt}{2}\right]dt$$

meters. For the average speed $v$ in the short time interval $dt$, we, thus, have

$$v = \frac{g\left[t + \frac{dt}{2}\right]dt}{dt} = g\left[t + \frac{dt}{2}\right].$$

Now, we let $dt$ be "infinitely small". Then, for the instantaneous velocity, we get $v = g\,t$.

To compute the instantaneous acceleration: After $t$ seconds, the body increases its speed in the short time $dt$ from $g\,t$ to $g\,(t{+}dt)$. The average acceleration (= change in velocity / time) is, therefore,

$$a = \frac{g\,(t + dt) - g\,t}{dt} = \frac{g\,dt}{dt} = g.$$

Here, we do not even need to perform a limit process: The acceleration is already independent of $dt$.

Obviously, the acceleration is obtained by differentiating the path-time function twice. Symbolically, we write $a = \ddot{s}$. ◂◂◂

## 6.4 Differentiation rules

In a symmetrical manner, we want to derive rules that makes differentiation easy.
The derivation of a simple class of functions can be calculated immediately: The derivative of a linear function $f(x) = kx + d$ equals the constant value $k$, and the derivative of the constant function $f(x) = d$ vanishes:

$$\boxed{(k\,x + d)' = k.} \tag{6.7}$$

***Proof***: a) Purely graphically: The graph of the function is, indeed, a straight line with a constant slope. So, the tangent of the angle $\alpha$ of inclination is also constant.
b) From the limit process, we obtain

$$f'(x) = \lim_{h\to 0} \frac{\overbrace{[k(x+h)+d]}^{f(x+h)} - \overbrace{[kx+d]}^{f(x)}}{h} = \lim_{h\to 0} \frac{kh}{h} = \lim_{h\to 0} k = k.$$

For $k = 0$, we have $f'(x) = 0$. ⊙

⊕ *Remark*: As an example, the function $s(t) = v_0\,t + s_0$ describes the distance at constant speed and, consequently, $\dot{s} = v_0$. ⊕

Quite often, one needs the derivative of the sine or cosine function (Fig. 6.26):

$$\boxed{(\sin x)' = \cos x, \quad (\cos x)' = -\sin x.} \tag{6.8}$$

***Proof***: We prove the first part:

$$(\sin x)' = \lim_{h\to 0} \frac{\sin(x+h) - \sin x}{h} = \cos x.$$

We use the well-known formula

$$\sin\alpha - \sin\beta = 2\cos\frac{\alpha+\beta}{2}\sin\frac{\alpha-\beta}{2}$$

and derive Formula (4.21) in Application p. 141. Then, we get:

$$(\sin x)' = \lim_{h\to 0} \frac{\sin(x+h) - \sin x}{h} = \lim_{h\to 0} \frac{2\cos(x+\frac{h}{2})\sin\frac{h}{2}}{h} =$$

$$= \lim_{h\to 0} \cos(x+\frac{h}{2}) \underbrace{\frac{\sin\frac{h}{2}}{\frac{h}{2}}}_{1} = \cos x.$$

⊙

⊕ *Remark*: An illustrative and easily comprehensible geometric proof of the relation

$$y = \sin x \Rightarrow y' = \frac{dy}{dx} = \cos x$$

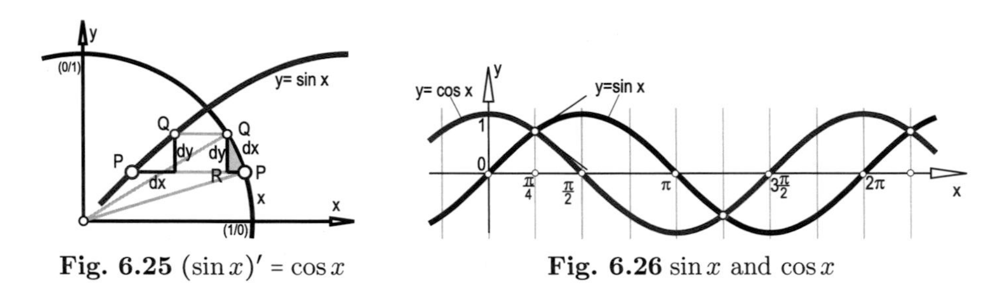


**Fig. 6.25** $(\sin x)' = \cos x$ **Fig. 6.26** $\sin x$ and $\cos x$

could look like this (Fig. 6.25):
We consider the unit circle and the two points $P$ and $Q$ on it, which correspond to the track angles $x$ and $x + dx$. The arc (and in limiting cases, thus, the chord) from $P$ to $Q$ has a length of $dx$. We consider the $x$-parallel through $P$ and the $y$-parallel through $Q$. The two lines intersect at a point $R$. At a small angle, the triangle $PQR$ is "almost" a right triangle with hypotenuse $dx$ and the angle $x + dx$ (angle between the circle tangent at $Q$ and the $y$-direction) at $Q$. The difference in height $dy$ of $P$ and $Q$ equals $\overline{RQ} = dy$. This value can, obviously and by definition, be interpreted as the growth of the function $y = \sin x$, when $x$ is increased by $dx$. Initially, we have the approximation, but in a limit process, it is exactly the relation to be proved: $\frac{dy}{dx} = \cos x$.
Such a proof is not very much appreciated by pure mathematicians. In technical books, however, it is quite common. The arithmetic of "infinitely small quantities" works surprisingly well. The only risk is the difficulty when it comes to identifying errors – one must be careful (Application p. 369). ⊕

▸▸▸ **Application**: ***At what angle do the functions $y = \sin x$ and $y = \cos x$ intersect?*** (Fig. 6.26)

*Solution*:
The two functions are "periodic", i.e. they repeat on and on (*period length* $2\pi$). In the interval $[0, 2\pi]$, the graphs intersect twice:

$$\sin x = \cos x \Rightarrow \frac{\sin x}{\cos x} = 1 \Rightarrow \tan x = 1 \Rightarrow x_1 = \frac{\pi}{4},\ x_2 = \frac{5\pi}{4}.$$

⊕ *Remark*: At $x_1$, the sine curve has a tangent with the slope $y'(\pi/4) = \cos \pi/4 = 1/\sqrt{2}$. The angle of inclination is, therefore, 35.264°. Analogously, the graph of the cosine at $x_1$ has a tangent with the slope $y'(\pi/4) = -\sin \pi/4 = -1/\sqrt{2}$. The inclination angle is −35.264°. The intersection angle is, thus, 70.53°. ⊕ ◂◂◂

## The "queen" of functions

There is – up to a multiplicative constant – just one function that does not change when it is differentiated, namely the Euler function $y = e^x$:

$$\boxed{(e^x)' = e^x.} \tag{6.9}$$

***Proof***: It must be proven that

$$(\mathrm{e}^x)' = \lim_{h\to 0} \frac{\mathrm{e}^{x+h} - \mathrm{e}^x}{h} = \mathrm{e}^x \lim_{h\to 0} \frac{\mathrm{e}^h - 1}{h} = \mathrm{e}^x.$$

Therefore, one has to show:

$$\lim_{h\to 0} \frac{\mathrm{e}^h - 1}{h} = 1. \tag{6.10}$$

Without going into the details of the proof (which would require some lemmas that we have not yet deduced): It is highly likely that *Euler* had defined "his number" $e$ so that the limit in Formula (6.10) attains the value 1, namely by

$$\lim_{n\to\infty} \left(1 + \frac{1}{n}\right)^n = 2.71828183\ldots.$$

Then,

$$\mathrm{e}^h = \lim_{n\to\infty} \left(1 + \frac{1}{n}\right)^{nh} = \lim_{n\to\infty} \left(1 + \frac{h}{nh}\right)^{nh} = \lim_{m\to\infty} \left(1 + \frac{h}{m}\right)^m.$$

From now on, briefly: Using the *binomial theorem* (see p. 468), we have

$$(a+b)^m = a^m + m\,a^{m-1}b + \frac{m(m-1)}{2} a^{m-2}b^2 + \ldots + m\,a\,b^{m-1} + b^m$$

and with $a = 1$ and $b = h/m$, we infer for $\mathrm{e}^h$:

$$\mathrm{e}^h = 1 + m\frac{h}{m} + \frac{m(m-1)}{2}\left(\frac{h}{m}\right)^2 + \ldots = 1 + h + \ldots$$

from which

$$\frac{\mathrm{e}^h - 1}{h} = 1 + h + \ldots$$

follows. Let $h \to 0$, then the limit is actually 1. ⊙

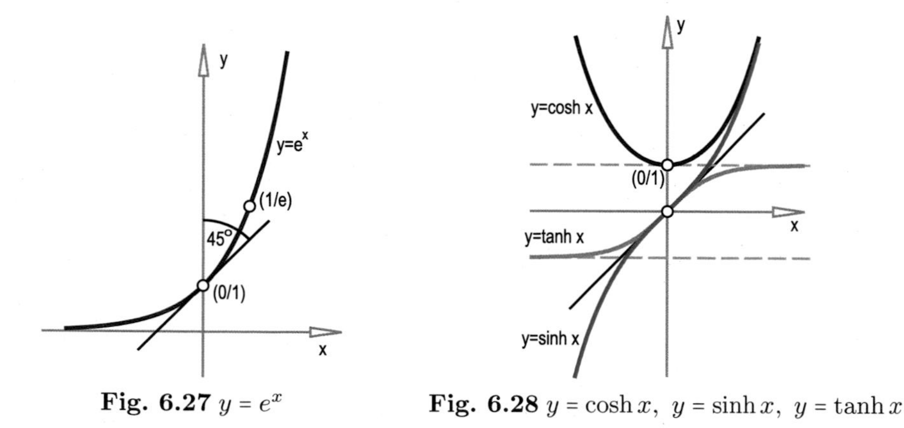


**Fig. 6.27** $y = e^x$ **Fig. 6.28** $y = \cosh x,\ y = \sinh x,\ y = \tanh x$

⊕ *Remark*: This beautiful property of the natural exponential function has enormous implications for all of mathematics. One might call $y = \mathrm{e}^x$ the "queen of the functions" for this reason (see Appendix B). Many other functions can be created from this function, like the *hyperbolic functions* (Fig. 6.28, see also Application p. 273):

$$\cosh x = \frac{e^x + e^{-x}}{2}, \quad \sinh x = \frac{e^x - e^{-x}}{2}, \quad \tanh x = \frac{\sinh x}{\cos h} = \frac{e^x - e^{-x}}{e^x + e^{-x}}$$

which we shall meet in connection with the "catenary" (Application p. 430). Via the detour of complex numbers, there is even a connection to trigonometric functions (Application p. 273). The proof can be done by means of the so-called power series expansion, which we will discuss in Chapter 8. ⊕

**▸▸▸ Application**: ***At which angle does $y = e^x$ intersect the $y$-axis?***

⊕ *Remark*: For all points on the $y$-axis, $x = 0$ holds. The function $y = e^x$ intersects the axis at the point $(0/e^0) = (0/1)$. The derivative there is also $e^0 = 1$. That is, the tangent has the slope $\tan\alpha = 1$ and, thus, $\alpha = 45°$. The angle made with the $x$-axis equals that made with the $y$-axis (Fig. 6.27). ⊕ ◂◂◂

Next, we show a few basic rules with which one can differentiate quite complex functions. In the following, $f(x)$ and $g(x)$ are two functions with the derivatives $f'(x)$ and $g'(x)$. Therefore, we have

$$f'(x) = \lim_{h\to 0} \frac{f(x+h) - f(x)}{h}, \quad g'(x) = \lim_{h\to 0} \frac{g(x+h) - g(x)}{h}.$$

It is immediately clear that *multiplicative constants* can be "factored out" while *additive constants* disappear:

$$\boxed{(k \cdot f + d)' = k \cdot f'.} \tag{6.11}$$

***Proof***: With $F(x) = k \cdot f(x) + d$, we get

$$F' = \lim_{h\to 0} \frac{[k\, f(x+h) + d] - [k\, f(x) + d]}{h} = k \lim_{h\to 0} \frac{f(x+h) - f(x)}{h} = k\, f'(x). \quad \odot$$

*Numerical example*: The derivative of the function $y = 6\cos x + 2$ reads $y' = -6\sin x$. Equally simple is the

$$\boxed{\text{summation rule: } (f+g)' = f' + g'.} \tag{6.12}$$

***Proof***:

$$(f+g)'(x) = \lim_{h\to 0} \frac{[f(x+h) + g(x+h)] - [f(x) + g(x)]}{h} =$$

$$= \lim_{h\to 0} \frac{[f(x+h) - f(x)] + [g(x+h) - g(x)]}{h} = f' + g'. \quad \odot$$

⊕ *Remark*: This formula is valid only if $f'$ and $g'$ exist, a fact which also applies in the cases of Formula (6.13) and Formula (6.17)! ⊕

▸▸▸ **Application**: ***Find the "first" positive maximum of*** $f(x) = \frac{x}{2} + \sin x$. (Fig. 6.29)

*Solution*:
A maximum occurs if the function has a horizontal tangent, that is, if the inclination angle $\alpha$ of the tangent is 0. So, we must have

$$\tan\alpha = f'(x) = 0.$$

With the summation rule, we have

$$f' = \left(\frac{x}{2}\right)' + (\sin x)' = \frac{1}{2} + \cos x = 0 \Rightarrow \cos x = -\frac{1}{2} \Rightarrow x = \frac{2\pi}{3}.$$

In the neighborhood of the point $\left(\frac{2\pi}{3}/\frac{\pi}{3} - \frac{\sqrt{3}}{2}\right)$, the sign of $f'$ changes. Left of this point, we have $y' > 0$; right of it, $y' < 0$ holds. Therefore, it is, indeed, a maximum. The curve "winds" about the line $y = x/2$ (Fig. 6.29).

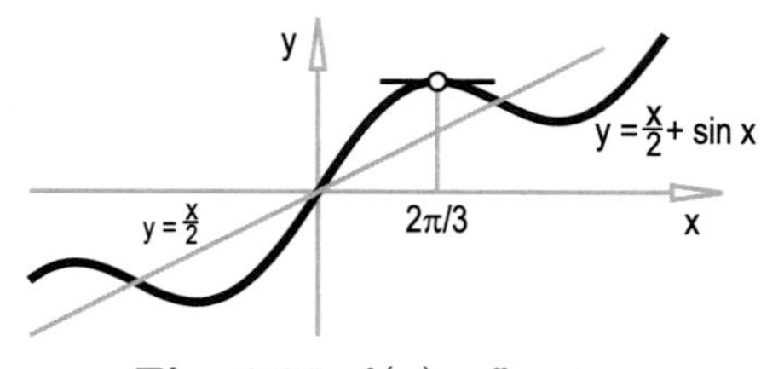


**Fig. 6.29** $f(x) = \frac{x}{2} + \sin x$

⊕ *Remark*: In this example, it is important to calculate in *radians*, rather than in degrees. So, your calculator must be set to RAD and not to DEG. The setting GRAD refers to the so-called *gradians* (a right angle equals 100 gradians), which we will not use here (they are still used in surveying). The identification of the right angle with 90° (degrees) is due to the Babylonians who did not calculate in the decimal system. The latter was "exported" by Arabian mathematicians from India to Europe some 1,000 years ago. ⊕ ◂◂◂

Furthermore, the differentiation rule for products (Leibniz's rule) reads

$$\boxed{\text{product rule:} \quad (f \cdot g)' = f' \cdot g + f \cdot g'.} \tag{6.13}$$

***Proof***: Since $F(x) = f(x) \cdot g(x)$,

$$F'(x) = \big(f(x) \cdot g(x)\big)' = \lim_{h\to 0} \frac{f(x+h) \cdot g(x+h) - f(x) \cdot g(x)}{h}.$$

Now we use a trick: We subtract an expression and add it again to "keep the balance". By cleverly combining terms, this seemingly more complicated expression yields new "constellations":

$$F' = \lim_{h\to 0} \frac{f(x+h) \cdot g(x+h) \overbrace{- f(x) \cdot g(x+h) + f(x) \cdot g(x+h)}^{\text{That's the point!}} - f(x) \cdot g(x)}{h} =$$

$$= \lim_{h \to 0} \frac{f(x+h) - f(x)}{h} g(x+h) + f(x) \lim_{h \to 0} \frac{g(x+h) - g(x)}{h} = f'g + fg'. \quad \odot$$

Using the product rule, we can prove the following basic differentiation rule for the *derivative of the power function* $y = x^n$:

$$\boxed{(x^n)' = n \cdot x^{n-1}} \tag{6.14}$$

***Proof***: We first assume that $n$ is a natural number. Subsequently, we will prove the assertion for arbitrary real numbers, see Formula (6.25).
The proof uses the method of "mathematical induction": We already know that, for $n = 1$, we have $(x^1)' = 1 \cdot x^0 = 1$ ("induction base"). Now let us assume the theorem is true for some $n$. If we can show that the theorem is also true for $n + 1$, then it is valid in any case. So, we suppose that $(x^n)' = n \cdot x^{n-1}$. We have to show that $(x^{n+1})' = (n+1) \cdot x^n$. This is done by using the product rule

$$(x^{n+1})' = (x \cdot x^n)' = 1 \cdot (x^n) + x \cdot (n \cdot x^{n-1}) = x^n + n \cdot x^n = (n+1) \cdot x^n. \quad \odot$$

**▸▸▸ Application**: ***bomb trajectory***
A parabola has the equation $y = a\,x^2 + b\,x + c$. Determine the coefficients $a$, $b$, and $c$ so that the throwing distance is 20 m and the throwing height is 8 m in a horizontal terrain. How large is the launch angle $\alpha$?

*Solution*:
The origin lies on the parabola $(0 = 0^2\,a + 0\,b + c \Rightarrow c = 0)$ for the sake of simplicity. For reasons of symmetry, the apex must have the coordinates $(10/8)$. Therefore, $8 = 10^2\,a + 10\,b + 0 \Rightarrow b = 0.8 - 10\,a$. Furthermore, the tangent must be horizontal there, i.e. $y'(10) = 0$. Applying the previous rules, we get

$$y' = 2a\,x + b.$$

Therefore, we find $20\,a + b = 0 \Rightarrow 20\,a + 0.8 - 10a = 0 \Rightarrow a = -0.08 \Rightarrow b = 1.6$. Further, $y'(0) = \tan\alpha = 1.6$. So, a launch angle of $\alpha = 58.0°$ is required. ◂◂◂

The next rule is called the *quotient rule*. The proof uses the product rule and the differentiation rule for the power functions:

$$\boxed{\text{quotient rule: } \left(\frac{f}{g}\right)' = \frac{f' \cdot g - f \cdot g'}{g^2}.} \tag{6.15}$$

**▸▸▸ Application**: ***Calculate the derivative of the function*** $y = \tan x$ ***and show that its graph intersects the*** $x$***-axis at*** $45°$.

*Solution*:
⊕ *Remark*: The function $\tan x$ can be written as a fraction:

$$\tan x = \frac{\sin x}{\cos x}.$$

Now, we can apply Formula 6.15 and find

$$(\tan x)' = \frac{\cos x \cos x - \sin x(-\sin x)}{\cos^2 x} = \frac{\cos^2 x + \sin^2 x}{\cos^2 x}.$$

Depending on whether we split the fraction into its summands or we use $\cos^2 x + \sin^2 x = 1$, we obtain

$$(\tan x)' = 1 + \tan^2 x \quad \text{or} \quad (\tan x)' = \frac{1}{\cos^2 x}. \tag{6.16}$$

The function $\tan x$ is again periodic (with period $\pi$). The zeros arise from the equation $\tan x = 0$ and are, thus, $x = 0, \ \pm\pi, \ \pm 2\pi, \ \ldots$. At $x = 0$, the derivative equals $(\tan x)'(0) = 1$, i.e. the tangent is inclined at $\alpha = 45°$ ($\tan\alpha = 1$). ⊕ ◂◂◂

▸▸▸ **Application**: ***Witch of Agnesi*** (Fig. 6.30)
Consider the function $y = 2/(1 + x^2)$. Show that the point $(0/2)$ is a maximum and the tangents at the points $(\mp 1/1)$ subtend angles of $\pm 45°$ with the $x$-axis.

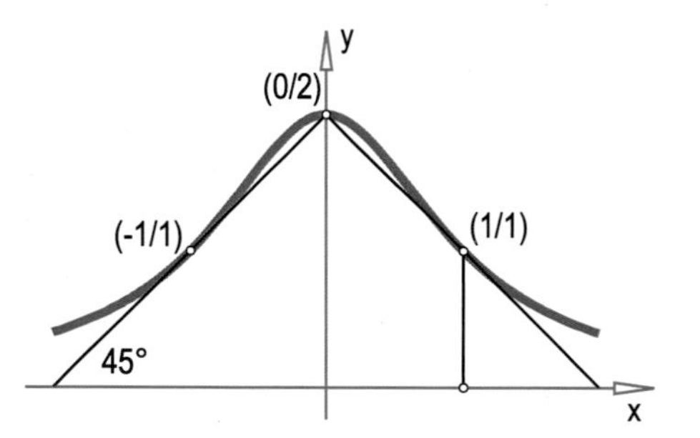

**Fig. 6.30** Witch of Agnesi (bell-shaped curve), and not a real "bell"

*Solution*:
⊕ *Remark*: Applying the quotient rule, we find

$$y' = \frac{0 \cdot (1 + x^2) - 2 \cdot 2x}{(1 + x^2)^2} = -\frac{4x}{(1 + x^2)^2} \Rightarrow y'(0) = 0.$$

For $x < 0$, we have $y' > 0$. That is, the curve is monotonically increasing until $x = 0$, where it has a maximum.
For $x = \pm 1$ ($\Rightarrow y = 1$), we have $y' = \tan\alpha = \mp 1 \Rightarrow \alpha = \mp 45°$. ⊕

⊕ *Remark*: The curve is bell-shaped, but should not be confused with the famous *Gauss*ian bell curve (Fig. 8.4). ⊕ ◂◂◂

For "nested functions", we need the so-called

$$\boxed{\text{chain rule: } \big(f[g(x)]\big)' = f'[g(x)] \cdot g'(x).} \tag{6.17}$$

The term $f'[g(x)]$ is frequently called the "exterior derivative". The expression $g'(x)$ is called the "inner derivative". Technicians or physicists prefer the differential notation

$$\boxed{\frac{df}{dx} = \frac{df}{dg} \cdot \frac{dg}{dx}.} \tag{6.18}$$

***Proof***: The differential notation is actually an abbreviation for the following limit:

$$\left(f[g(x)]\right)' = \lim_{h \to 0} \frac{f[g(x+h)] - f[g(x)]}{h} =$$

$$= \lim_{h \to 0} \frac{f[g(x+h)] - f[g(x)]}{g(x+h) - g(x)} \cdot \frac{g(x+h) - g(x)}{h} = f'[g(x)] \cdot g'(x). \qquad \odot$$

Examples are the best way to memorize the chain rule.

**▸▸▸ Application**: ***some derivatives***

Determine the derivative of the functions

$$y = \mathrm{e}^{2x}, \; y = \mathrm{e}^{\sin x}, \; y = \cos(3x^2 + 1), \; y = \tanh x = \frac{e^x - e^{-x}}{e^x + e^{-x}}.$$

*Solution*:

$$y = \mathrm{e}^{2x} \Rightarrow y' = \mathrm{e}^{2x} \cdot 2 = 2y,$$

$$y = \mathrm{e}^{\sin x} \Rightarrow y' = \mathrm{e}^{\sin x} \cdot \cos x,$$

$$y = \cos(3x^2 + 1) \Rightarrow y' = [-\sin(3x^2 + 1)] \cdot (6x) = -6x \sin(3x^2 + 1),$$

$$y = \frac{e^x - e^{-x}}{e^x + e^{-x}} \Rightarrow y' = \frac{(e^x + e^{-x})(e^x + e^{-x}) - (e^x - e^{-x})(e^x - e^{-x})}{(e^x + e^{-x})^2}$$

$$\Rightarrow (\tanh x)' = 1 - \tanh^2 x. \tag{6.19}$$

◂◂◂

In order to *differentiate the inverse function*, we use the following expression:

$$\boxed{\left(f^{-1}\right)'(x) = \frac{1}{f'(x)}.} \tag{6.20}$$

Its main application is the elegant computation of derivatives of new function types.

***Proof***: If $g$ is the inverse function of $f$, then the corresponding graphs are symmetric with respect to the first median. If the tangent to $f$ has the inclination angle $\alpha$, then the symmetric tangent to $g$ has the complementary angle of inclination, and we have

$$\tan\left(\frac{\pi}{2} - \alpha\right) = \frac{1}{\tan \alpha}. \qquad \odot$$

▸▸▸ **Application**: ***derivative of the arctangent*** (Fig. 6.31, left)
The inverse of the function $y = \tan x$ is called the arc tangent – $\arctan y$ ("the arc of the tangent"). Compute the derivative of $y = \arctan x$.

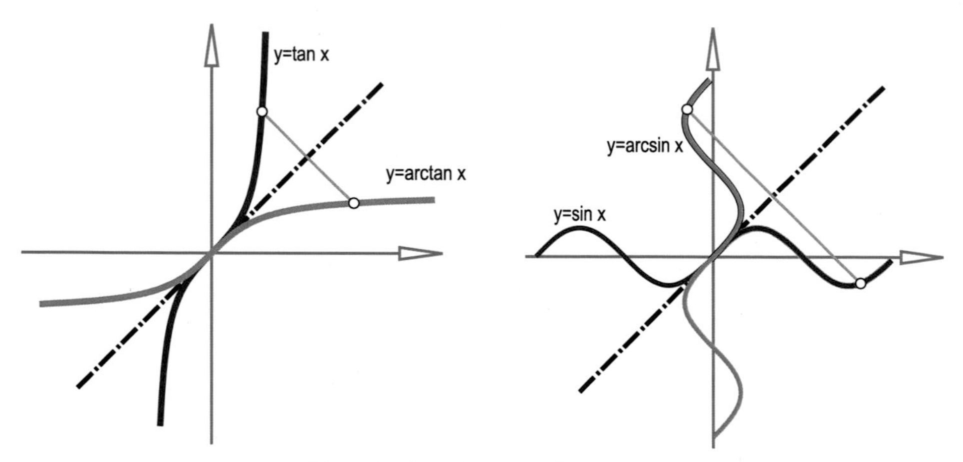


**Fig. 6.31** $\arctan x$ and $\arcsin x$

*Solution*:

We have $\tan u = x$ and $(\tan u)' = 1 + \tan^2 u$, and therefore,

$$(\arctan)'(\tan u) = (\arctan)'(x) = \frac{1}{1+x^2}. \tag{6.21}$$

⊕ *Remark*: Analogously and according to Formula (6.19), the derivative of the inverse function of tanh is

$$(\text{artanh})'(x) = \frac{1}{1-x^2}. \tag{6.22}$$

This function is called the "area hyperbolic tangent", and it is frequently used in integral calculus. ⊕ ◂◂◂

▸▸▸ **Application**: ***derivative of the arcsine*** (Fig. 6.31, right)
The inverse of the function $y = \sin x$ is called the *arcsine* ("the arc of the sine"). Calculate the derivative of the arcsin function.
*Solution*:
We find

$$(\arcsin)'(\sin u) = \frac{1}{(\sin u)'} = \frac{1}{\cos u} = \frac{1}{\sqrt{1-\sin^2 u}}.$$

Now, we set $\sin u = x$ and find

$$(\arcsin x)' = \frac{1}{\sqrt{1-x^2}.} \tag{6.23}$$

⊕ *Remark*: We may reflect the sine curve only in the interval $[0, 2\pi]$. Otherwise, $y = \arcsin x$ is not one-to-one. ⊕ ◂◂◂

▸▸▸ **Application**: ***derivative of the logarithmic function***

The inverse of the exponential function $y = \mathrm{e}^x$ is the *natural logarithm* (*logarithmus naturalis*). Compute the derivative of $y = \ln x$.

*Solution*:

We compute

$$(\ln)'(\mathrm{e}^u) = \frac{1}{(\mathrm{e}^u)'} = \frac{1}{\mathrm{e}^u}.$$

Now, we set $\mathrm{e}^u = x$ and arrive at the simple formula

$$\boxed{(\ln x)' = \frac{1}{x}.} \tag{6.24}$$

◂◂◂

Using Formula (6.20), we want to prove the

$$\boxed{\text{power rule: } (x^n)' = n \cdot x^{n-1} \quad (\text{valid for all } n \in \mathbb{R}).} \tag{6.25}$$

***Proof***: According to Formula (6.20), we can also differentiate the "root functions", since the inverse function of the power function $y = x^n$ is the root function

$$y = x^{\frac{1}{n}} = \sqrt[n]{x}.$$

By the theorem on the derivatives of the inverse functions, we find

$$\left(\sqrt[n]{u^n}\right)' = \frac{1}{n \cdot u^{n-1}}.$$

We substitute $u^n = x$ and apply the rules for power functions:

$$u^{n-1} = x^{\frac{n-1}{n}} = x^{1-\frac{1}{n}}.$$

Therefore, we have

$$\left(x^{\frac{1}{n}}\right)' = (\sqrt[n]{x})' = \frac{1}{n \cdot x^{1-\frac{1}{n}}} = \frac{1}{n} x^{\frac{1}{n}-1}.$$

⊕ *Remark*: We do not need to learn this rule by heart: If we set $\frac{1}{n} = m$, then the formula takes on the well-known form

$$(x^m)' = m \cdot x^{m-1} \quad \text{with} \quad m = \frac{1}{n}.$$ ⊕

By means of the product rule, one can easily show that this formula also holds true if one sets $m = \frac{k}{n} = k\frac{1}{n}$. So, the rule applies to all rational numbers – and, since each real number may be approximated with arbitrary precision by a fraction, even for all real exponents (for $n = 0$, because $x^0 = 1$ is constant, the derivative is 0). ⊙

⊕ *Remark*: Relatively often, one has to differentiate $y = \sqrt{x} = x^{\frac{1}{2}}$. By Formula (6.14), we find $y' = \frac{1}{2} \cdot x^{-\frac{1}{2}} = \frac{1}{2\sqrt{x}}$. We want to keep this in mind:

$$\boxed{(\sqrt{x})' = \frac{1}{2\sqrt{x}}.} \tag{6.26}$$

⊕

▸▸▸ **Application**: ***hyperbolic functions*** (Fig. 6.28)
Using the function $\mathrm{e}^x$, the so-called hyperbolic functions are defined:

$$\sinh x = \frac{\mathrm{e}^x - \mathrm{e}^{-x}}{2}, \quad \cosh x = \frac{\mathrm{e}^x + \mathrm{e}^{-x}}{2}. \tag{6.27}$$

Show that $(\sinh x)' = \cosh x$ and $(\cosh x)' = \sinh x$.
*Solution*:

$$\left(\frac{\mathrm{e}^x - \mathrm{e}^{-x}}{2}\right)' = \frac{1}{2}\left(\mathrm{e}^x - \mathrm{e}^{-x}(-1)\right) = \cosh x,$$
$$\left(\frac{\mathrm{e}^x + \mathrm{e}^{-x}}{2}\right)' = \frac{1}{2}\left(\mathrm{e}^x + \mathrm{e}^{-x}(-1)\right) = \sinh x.$$

⊕ *Remark*: By means of the imaginary unit $i$ ($i^2 = -1$) (see Section B), one is able to obtain quite analogous formulas for the derivative of the ordinary sine or cosine function:

$$\sin x = \frac{\mathrm{e}^{ix} - \mathrm{e}^{-ix}}{2i}, \quad \cos x = \frac{\mathrm{e}^{ix} + \mathrm{e}^{-ix}}{2}. \tag{6.28}$$

Again, you can easily show the already known relations $(\sin x)' = \cos x$ and $(\cos x)' = -\sin x$. ⊕ ◂◂◂

▸▸▸ **Application**: ***derivative of an exponential function***
Show the following formula for the derivative of general exponential functions:

$$(a^x)' = \ln a \cdot a^x. \tag{6.29}$$

*Solution*:
Obviously, $a = \mathrm{e}^{\ln a}$, and thus,

$$a^x = \left(\mathrm{e}^{\ln a}\right)^x = \mathrm{e}^{\ln a \cdot x} \Rightarrow (a^x)' = \mathrm{e}^{\ln a \cdot x} \cdot \ln a = a^x \cdot \ln a.$$ ◂◂◂

If the nesting of functions becomes more complex, the chain rule must be applied repeatedly. Even for that, we should do some exercises.

▸▸▸ **Application**: ***more examples using the chain rule***
Through repeated application of the chain rule, determine the derivatives of the functions

$$y = \mathrm{e}^{\sin 2x}, \quad y = \ln(1 + \mathrm{e}^{2x}), \quad y = \cos(3\sin x^2 + 1), \quad y = \arcsin(\cos 2x).$$

*Solution*:

$$y = \mathrm{e}^{\sin 2x} \Rightarrow y' = \mathrm{e}^{\sin 2x} \cdot \cos 2x \cdot 2,$$

$$y = \ln(1 + \mathrm{e}^{2x}) \Rightarrow y' = \frac{1}{1 + \mathrm{e}^{2x}} \cdot \mathrm{e}^{2x} \cdot 2,$$

$$y = \cos(3 \sin x^2 + 1) \Rightarrow y' = -\sin(3 \sin x^2 + 1) \cdot (3 \cos x^2) \cdot (2x),$$

$$y = \arcsin(\cos 2x) \Rightarrow y' = \frac{1}{\sqrt{1 - \cos^2 2x}} \cdot (-\sin 2x) \cdot 2 = -2.$$ ◂◂◂

▸▸▸ **Application**: ***harmonic oscillation*** (Fig. 6.32)
We move a point $P$ with constant angular velocity $\omega$ on a circular orbit (radius $r$) and consider its normal projection $Q$ onto the $y$-axis. The projected point $Q$ oscillates harmonically. If $\varphi$ is the *polar angle* of the point $P$ and $t$ is the elapsed time, then the $y$-coordinate of $Q$ equals

$$y = r \sin \varphi = r \sin \omega t.$$

Find the instantaneous velocity and the instantaneous acceleration of $Q$.
*Solution*:
The path of $Q$ is given as a function of the time $t$. Consequently, we need to form only the "first and the second derivative" with respect to time ($\dot{y}$ and $\ddot{y}$). Once the chain rule is applied, we find

$$y = r \sin \omega t \Rightarrow \dot{y} = r\,\omega \cos \omega t \Rightarrow \ddot{y} = r\,\omega^2 (-\sin \omega t).$$

Thus, we see that the orbital speed and acceleration of $Q$ vary harmonically. The instantaneous velocity increases linearly with the angular velocity of $P$. However, the acceleration increases with the square of $\omega$.

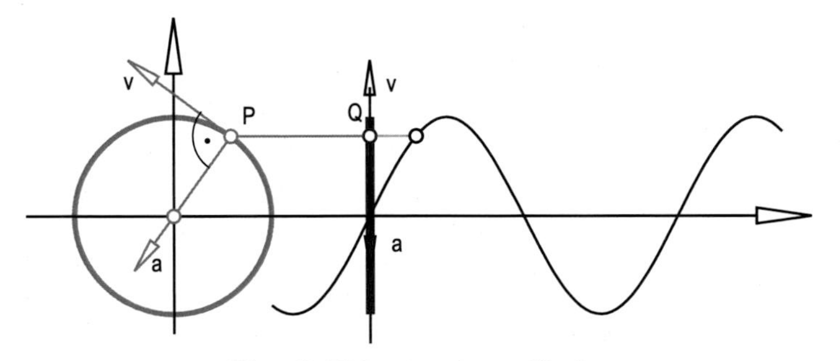


**Fig. 6.32** harmonic oscillation

⊕ *Remark*: The orbital speed of the point $P$ is $v = r\,\omega$. The circumference of $P$'s orbit is $U = 2\pi\, r$. This is the orbital period of $P$:

$$T = \frac{U}{v} = \frac{2\pi}{\omega}$$

which is the time that the point $Q$ needs to carry out a period of the harmonic oscillation.
In practice, this time is often given. From this, the angular velocity $\omega = \dfrac{2\pi}{T}$ follows and one obtains

$$y = r\,\sin\omega t = r\,\sin\frac{2\pi\,t}{T},$$

$$\dot{y} = r\,\omega\,\cos\omega t = \frac{2\pi\,r}{T}\cos\frac{2\pi\,t}{T} = v\,\cos\frac{2\pi\,t}{T},$$

$$\ddot{y} = -r\,\omega^2\,\sin\omega t = -\frac{4\pi^2 r}{T^2}\sin\frac{2\pi\,t}{T} = -\frac{4\pi^2}{T^2}y.$$

The orbital speed and the acceleration vector of $Q$ are the normal projections of the corresponding vectors of the point $P$ onto the $y$-axis. The *centripetal acceleration* of the point $P$ is obtained with the expression $\dfrac{4\pi^2 r}{T^2}$. ⊕ ◂◂◂

▸▸▸ **Application**: ***At what moment do the two shafts of a universal joint have the same angular velocity?*** (Fig. 6.33, Fig. 5.67)

*Solution*:
⊕ *Remark*: In Application p. 219, we derived a formula for the output angle $\beta$ for a given drive angle $\alpha$ of a universal joint (where $\gamma$ is the constant angle between the two axes):

$$\tan\beta = \frac{1}{\cos\gamma}\tan\alpha \Rightarrow \beta = \arctan\frac{\tan\alpha}{\cos\gamma}.$$

We assume that the drive axle rotates with constant angular velocity $\omega$. After time $t$, the drive axle has rotated through the angle $\alpha = \omega\cdot t$. Then, for the output angle, we have

$$y(t) = \arctan\frac{\tan\omega t}{c} \quad \text{with} \quad c = \cos\gamma.$$

The derivative $\dot{y} = \dfrac{dy}{dt}$ is now the instantaneous change of the output angle $dy$ in time $dt$. Therefore, it is the angular velocity of the output shaft. We want to examine only the *ratio* of the two angular velocities, and therefore, we set $\omega = 1$.
Now, we compute the derivative $\dot{y}$ according to the chain rule. Formula (6.21) is used for the derivative of the arctangent, and Formula (6.16) is used for the derivative of the tangent:

$$\dot{y} = \left(\arctan\frac{\tan t}{c}\right)'\cdot\left(\frac{\tan t}{c}\right)' = \frac{1}{1+\left(\dfrac{\tan t}{c}\right)^2}\cdot\frac{1}{c}(1+\tan^2 t) = c\,\frac{1+\tan^2 t}{c^2+\tan^2 t}.$$

At the initial point $t = 0$ ($\alpha = \beta = 0$), we have $\dot{y}(0) = \dfrac{1}{c} = \dfrac{1}{\cos\gamma}$. At time $t = \dfrac{\pi}{2}$ ($\alpha = \beta = \dfrac{\pi}{2} = 90°$), we have $\dot{y}(\dfrac{\pi}{2}) = c = \cos\gamma$, because the fraction

$$\frac{1+\tan^2 t}{c^2+\tan^2 t} = \frac{\dfrac{1}{\tan^2 t}+1}{\dfrac{c^2}{\tan^2 t}+1}$$

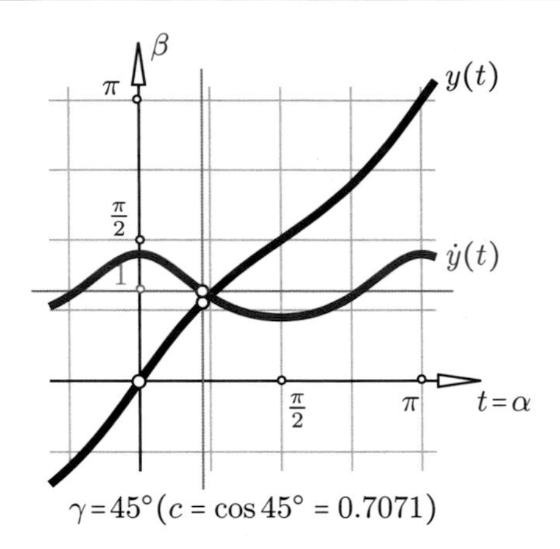


**Fig. 6.33** the drive angle and ratio of the angular velocities

converges towards 1 for $t \to \frac{\pi}{2}$ ($\Rightarrow \tan t \to \infty$).
Now, we can calculate when the instantaneous angular velocities are equal: We have to obtain $\dot{y} = 1$:

$$c\,\frac{1+\tan^2 t}{c^2+\tan^2 t} = 1 \Rightarrow c + c\,\tan^2 t = c^2 + \tan^2 t \Rightarrow \tan^2 t(c-1) = c(c-1)$$

$$\Rightarrow \tan t = \tan\alpha = \pm\sqrt{c} \Rightarrow \tan\beta = \pm\frac{\tan\alpha}{c} = \pm\frac{1}{\sqrt{c}}.$$

*Numerical example*: $\gamma = 45° \Rightarrow c = \cos 45° = 0.7071$ (Fig. 6.33).
The ratio of the angular velocities in the starting position $\alpha = 0°$ takes the value $\dot{y}(0) = 1.4142$. For $\alpha = 40.06°$, $\beta = 49.94°$, it equals 1, and for $\alpha = 90°$, it decreases to $\dot{y}(\frac{\pi}{2}) = 0.7071$. ⊕ ◂◂◂

▸▸▸ **Application**: ***piston speed for gasoline engines*** (Fig. 6.34)
Analyze the speed ratios of the given slider-crank mechanism.
In Application p. 139, we have determined the distance $y$ of the piston pin in the gasoline engine as a function of the rotation angle $\alpha$ of the crankshaft:

$$y = r\,\cos\alpha + \sqrt{s^2 - r^2\sin^2\alpha}.$$

During some rotations with constant angular velocity $\omega$, the angle $\alpha$ can be assumed to be proportional to the time $t$. Thus,

$$\alpha = \omega\, t.$$

Therefore, we have

$$y = r\,\cos(\omega\, t) + \sqrt{s^2 - r^2\sin^2(\omega\, t)}.$$

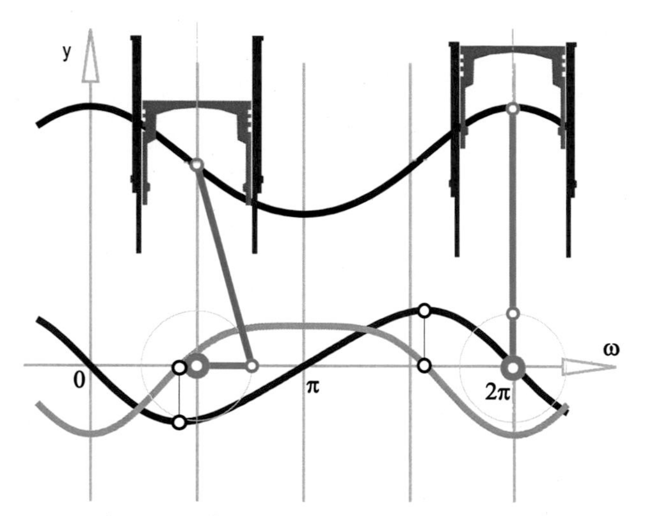


**Fig. 6.34** velocity and acceleration analysis for gasoline engines

Since we want to analyze qualitatively, we may set $\omega = 1$. The first derivative with respect to $t$ provides the instantaneous speed of the piston:

$$\dot{y} = -r\,\sin t + \frac{2\,r^2 \sin t \cos t}{2\sqrt{s^2 - r^2 \sin^2 t}} = r\,\sin t \left( \frac{r\,\cos t}{\sqrt{s^2 - r^2 \sin^2 t}} - 1 \right).$$

Therefore, for $\sin t = 0$ ($t = \alpha = 0,\, \pi,\, 2\pi, \ldots$), i.e. at the highest and lowest position respectively, the piston stands still for a moment. From this fact, one easily deduces that $\dot{y}$ can have no other zeros. Then, it would need to be

$$\frac{r\,\cos t}{\sqrt{s^2 - r^2 \sin^2 t}} - 1 = 0$$

which is never the case if $s > r$.

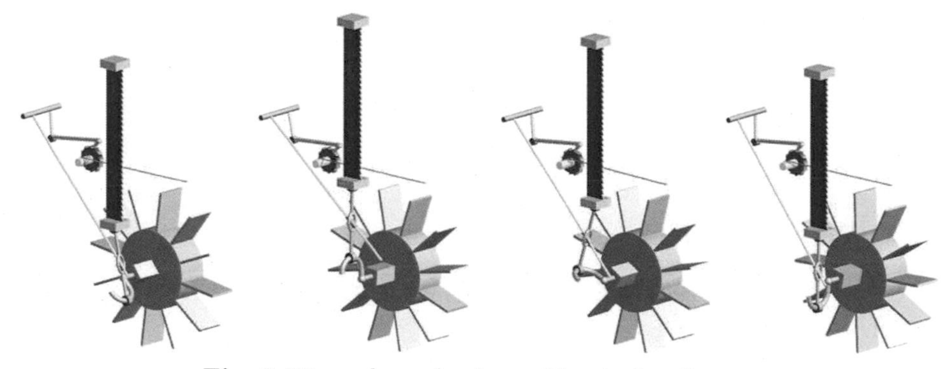

**Fig. 6.35** crank mechanism with a hydraulic saw

It becomes more interesting when the piston has maximum velocity. There, the acceleration has to change sign so that it equals zero. To this end, we

would have to differentiate again and then set $\ddot{y} = 0$. In this specific case, this is already a very arduous task.
Fig. 6.34 shows the graphs of the functions $y$, $\dot{y}$, and $\ddot{y}$, which were created with the program `otto_analysis.exe`. Only the equation $y = y(t)$ was entered. The computer can "automatically" differentiate and also finds the zeros of these functions, as we will discuss soon.

⊕ *Remark*: Slider-cranks allow an interplay between an up-down motion and a rotation. Therefore, they are used whenever a transmission between a rotation and a harmonic oscillation is to be performed. One can drive the blade of a hydraulic saw via a water wheel (Fig. 6.35). The technology of the saw developed by Leonardo da Vinci in detail is still used today. ⊕ ◂◂◂

## 6.5 Differentiating with a computer

Differentiating is – as we have seen – a relatively simple process, because only a few rules are needed and they can be applied systematically. However, when differentiating repeatedly, the results are often quite complex. Proceeding to make exact calculations with such results (like in Application p. 276) often requires a lot of mathematical experience. If these results seem to be too complex, we will sensibly use a computer to carry them out.
A computer "can" differentiate if the equation $y = f(x)$ is an explicitly known function. We distinguish between the exact differentiation that occurs with algebraic methods and approximate differentiation. Both methods are of great importance.
When approximately differentiating, one can make use of the formula

$$f'(x) = \frac{df}{dx} \approx \frac{f(x+\varepsilon) - f(x)}{\varepsilon} \tag{6.30}$$

with a "very small" $\varepsilon$, for example, $\varepsilon = 10^{-12}$. However, this involves the risk that a "loss of precision" occurs, because although a computer is very accurate, it may not be sufficiently accurate for such tiny differences.
The formula can be modified, and then, it generally delivers much more stable results:

$$f'(x) = \frac{df}{dx} \approx \frac{f(x+\varepsilon) - f(x-\varepsilon)}{2\,\varepsilon}. \tag{6.31}$$

Now, a comparatively "much larger" $\varepsilon$ is sufficient, approximately $\varepsilon = 10^{-7}$, in order to achieve an equally good approximation. Consider the following example:

**▸▸▸ Application**: ***loss of accuracy with numbers dwindling***
The exact derivative of the function $f(x) = \tan x$ is the function $f'(x) = 1 + \tan^2 x$. The value $f'(0.5)$ shall be calculated precisely or with the stated approximations given in Formula (6.30) or Formula (6.31).

*Solution*:
The "exact" value is: $f'(0.5) = 1.298\,446\,410\,409\,5$. The approximations for $\varepsilon = 10^{-7}$ are $f'(0.5) \approx 1.298\,446\,481\,154$ and $f'(0.5) \approx 1.298\,446\,410\,333$. Formula (6.30) is accurate approximately to only 7 decimal places. However, Formula (6.31) is accurate to 9 places. Yet, if one believes that greater accuracy may be achieved by decreasing $\varepsilon$, one is mistaken: For $\varepsilon = 10^{-10}$, the approximations agree only to 5 or 6 decimal places. This gets worse for even smaller $\varepsilon$: For $\varepsilon = 10^{-13}$, only 3 decimal places are exact, and for $\varepsilon = 10^{-14}$ only 2. For $\varepsilon = 10^{-15}$, both cases result in 1.30, and for $\varepsilon = 10^{-16}$, "the system collapses", which means that the results are completely meaningless.

Specifically, Formula (6.30) with $\varepsilon = 10^{-8}$ (still only 7 decimal places) and Formula (6.31) with $\varepsilon = 10^{-6}$ (10 decimal places) yield the best results. ◂◂◂

Application p. 279 and also the following example show that one should not blindly trust a computer. It does not compute with "arbitrary precision", and losing numerical precision will lead to strange results!
The second derivative can be calculated approximately from the first:

$$f''(x) = (f'(x))' \approx \frac{f'(x+\varepsilon) - f'(x-\varepsilon)}{2\varepsilon}. \tag{6.32}$$

We point out that such approaches are numerically not very reliable. In fact, the choice of $\varepsilon$ is very critical, and sometimes it depends on the nature of the function. An alternative to Formula (6.32) is the easily derivable formula

$$f''(x) \approx [f(x+\varepsilon) - 2f(x) + f(x-\varepsilon)]/\varepsilon^2.$$

Here you can see that the square of $\varepsilon$ appears in the denominator and should, therefore, not be chosen too small.

▸▸▸ **Application**: ***strong loss of accuracy in the second derivative***
Calculate the exact second derivative of the function $f(x) = \tan x$ and compute $f''(0.5)$ exactly or with the given approximation Formula (6.32).
*Solution*:

We have $f'(x) = 1 + \tan^2 x$, and by differentiating once again according to the chain rule, we obtain

$$f''(x) = 0 + \underbrace{2\tan x}_{\text{ext. der.}} \cdot \underbrace{(1 + \tan^2 x)}_{\text{inn. der.}}.$$

The "exact" value is $f''(0.5) = 1.418\,689\,014$. The approximation with $\varepsilon = 10^{-4}$ equals $f''(0.5) \approx 1.418\,689\,069$ and agrees up to 6 decimal places. If $\varepsilon = 10^{-6}$, the approximation agrees only up to 3 decimals. For $\varepsilon < 10^{-8}$, the result is a collection of irrelevant digits. ◂◂◂

# 6.6 Solving equations of the form f(x) = 0

The intersections of a function graph $y = f(x)$ with the $x$-axis $y = 0$ are called zeros or solutions of the equation $f(x) = 0$. Such solutions can only be calculated in special cases with formulas. In general, one has to be satisfied with approximations. Computer algebra systems (CAS) have modules with which one can calculate the solutions with arbitrary precision.

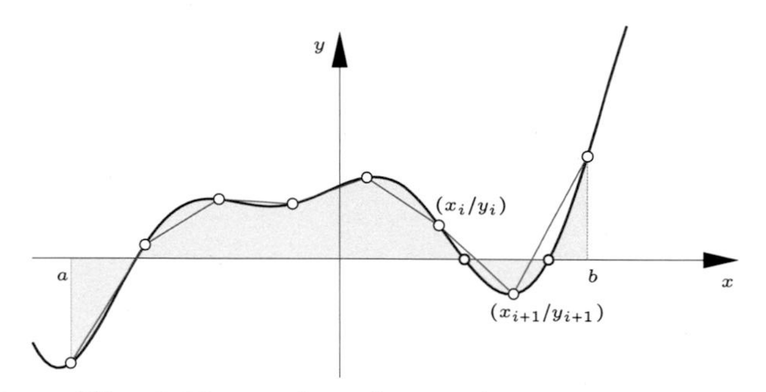


**Fig. 6.36** zeros by polynomial approximation

A given interval $[a, b]$ has to be specified in which the solutions are to be sought. In principle, for a sufficiently large number of equally distributed $x$-values $x_i \in [a, b]$, the function values $y_i = f(x_i)$ are initially calculated. If the signs of all $y_i$ are equal, then it is assumed that there is no zero in the interval $[a, b]$. In order to avoid missing any possible solution, one will – in the era of fast computers – test dozens or hundreds of values $x_i$ (see Application p. 42). If there is a change in the signs of the function values in the interval $[x_i, x_{i+1}]$ (Fig. 6.36, right), then a zero has to occur between $x_i$ and $x_{i+1}$. If the $x$-values are close enough, one can replace the curve by the straight line spanned by the points $(x_i/y_i)$ and $(x_{i+1}/y_{i+1})$ and quickly find an approximation $x_0$ for the zero:

$$\begin{pmatrix} x_i \\ y_i \end{pmatrix} + t \begin{pmatrix} x_{i+1} - x_i \\ y_{i+1} - y_i \end{pmatrix} = \begin{pmatrix} x_0 \\ 0 \end{pmatrix} \Rightarrow x_0 = x_i + t(x_{i+1} - x_i) \text{ with } t = \frac{-y_i}{y_{i+1} - y_i}.$$

Only in windfall will $y_0 = f(x_0)$ now be exactly zero. If $|y_0|$ is still too large, one has to repeat the process of the linear approximation to improve the approximation of the zero. Now, use $(x_0/y_0)$ as the first boundary of the newly calculated interval. The second boundary of the newly calculated interval is that point (either $(x_i/y_i)$ or $(x_{i+1}/y_{i+1})$) which is located on the other side of the abscissa compared to the first point.
If the function in the neighborhood of the root is "harmless" – for instance, it may not oscillate strongly –, then the following approximation method named after *Newton* converges very quickly: One chooses an initial value $x_0$, to which the point $(x_0/y_0)$ corresponds. In a first approximation, the

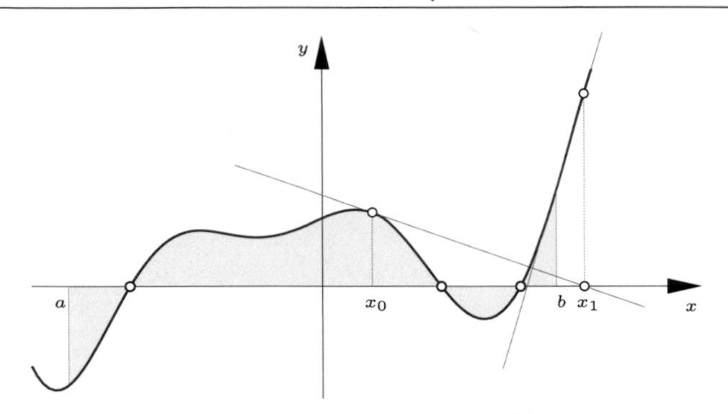


**Fig. 6.37** zeros according to *Newton*

curve is replaced there by its tangent whose slope is given by the derivative $f'(x_0) = y_0'$. If we intersect the tangent, instead of the curve, with the $x$-axis, we get a new value $x_1$:

$$\begin{pmatrix} x_0 \\ y_0 \end{pmatrix} + t \begin{pmatrix} 1 \\ y_0' \end{pmatrix} = \begin{pmatrix} x_1 \\ 0 \end{pmatrix} \Rightarrow x_1 = x_0 - \frac{y_0}{y_0'}.$$

Most of these new values will be better approximations for the zero and it takes two or three repetitions of the operation (iterations) to quickly approach the zero. Yet, it is also possible to have bad luck and leave the interval $[a, b]$ (Fig. 6.37). In this case, a different initial value has to be chosen, and one hopes – often rightly – that the process is so stable that it will already "stabilize" at the next iteration. Then again, if there are several zeros, it will not always be easy to find the right one.
In practice, one can mix the two methods successfully: By means of the polygon method (linear approximation), a zero can be detected quite accurately. In small intervals, the very efficient *Newton* method will find the right zero with high probability. If not, the interval is to be reduced further.

▸▸▸ **Application**: ***filling liquid*** (Fig. 6.38)
One liter of water is poured into a hollow hemispherical bowl with an inner diameter of 20 cm. How high is the water?
*Solution*:
The formula for the volume of a spherical cap

$$V = \frac{\pi h^2}{3}(3r - h) \tag{6.33}$$

will be derived in Application p. 411 by means of *Cavalieri*'s principle ($h$ is the height of the spherical segment). With $r = 10\,\text{cm}$ and $V = 1{,}000\,\text{cm}^3$, we have

$$1{,}000 = \frac{\pi h^2}{3}(30 - h) \Rightarrow f(h) = \pi h^3 - 30\pi h^2 + 3{,}000 = 0.$$

The equation has only one solution in the interval $[0, 10]$, namely $h = 6.355$. So, the water is about 6.4 cm high. ◂◂◂

**Fig. 6.38** spherical bowl with liquid

**Fig. 6.39** submersion of a sphere

▸▸▸ **Application**: ***submersion of a sphere*** (Fig. 6.39)
How deep does a ball plunge into water (density $\varrho < 1$)?

*Solution*:

This example is similar to Application p. 282. For the sake of simplicity, we can set the radius of the sphere to be 1. The immersion depth $t$ is given relative to the radius.

The volume of the sphere is, thus, $V = \dfrac{4\pi}{3}$, and the ball displaces the volume $V_W = \dfrac{\pi t^2}{3}(3-t)$.

Balancing buoyancy and weight force by equating them, we obtain

$$\varrho\frac{4\pi}{3} = \frac{\pi t^2}{3}(3-t).$$

Therefore, we get

$$f(t) = t^3 - 3t^2 + 4\varrho = 0.$$

⊕ *Remark*: With $\varrho = 1\frac{\mathrm{g}}{\mathrm{cm}^3}$, the trivial solution equals $t = 2$, because $f(2) = 0$ (the ball sinks completely). If $\varrho = 0.5$, obviously $f(1) = 0 \Rightarrow t = 1$. The immersion depth of the ball equals its radius, i.e. half the diameter. Fig. 6.39 illustrates the situation for $\varrho = 0.63$. In this case, $t \approx 1.175$ is the only solution in the interval $[0, 2]$, that is, the height of the protruded ball cap is about 5/6 of the sphere radius. ⊕ ◂◂◂

▸▸▸ **Application**: ***"folding" a circular sector into a cone of revolution***
From a circular sector of radius $s = 10$ cm, the "coat" of a cone of revolution is to be folded such that the volume of the cone equals $270\,\mathrm{cm}^3$. Calculate the central angle of the circular sector.

*Solution*:

Let $\alpha$ be half of the angle of aperture of the cone of revolution. Then, for the radius $r$, the height $h$, and the volume $V$ of the cone, we have

$$r = s\sin\alpha,\quad h = s\cos\alpha \Rightarrow V = \frac{\pi r^2 h}{3} = \frac{\pi}{3}s^3\sin^2\alpha\cos\alpha.$$

Specifically, we have, therefore,

$$\frac{\pi}{3}10^3\sin^2\alpha\cos\alpha = 270 \Rightarrow \sin^2\alpha\cos\alpha = 0.25783.$$

With $\sin^2\alpha + \cos^2\alpha = 1$, we have

$$\sin^2\alpha\sqrt{1-\sin^2\alpha} = 0.25783 \Rightarrow x\sqrt{1-x} = 0.257831 \text{ (with } x = \sin^2\alpha)$$

or, after squaring both sides,

$$x^2(1-x) - 0.257831^2 = 0 \Rightarrow f(x) = x^3 - x^2 + 0.0664768 = 0.$$

**Fig. 6.40** folding and unfolding ...

This equation has three solutions for $x = \sin^2\alpha$, of which only the positive ones are relevant:

$$x_1 = -0.232265 < 0, \quad x_2 = 0.310506 \quad x_3 = 0.921759.$$

We also have two real solutions for $\alpha$:

$$\sin^2\alpha_1 = 0.310506 \Rightarrow \alpha_1 = 34.18°, \quad \sin^2\alpha_2 = 0.921759 \Rightarrow \alpha_2 = 73.76°.$$

On the other hand, for the central angle $\omega$ (measured in *degrees*), the circular sector – according to Formula (4.18) – yields

$$\omega° = 360° \sin\alpha.$$

The two solutions for $\omega$ are, therefore,

$$\omega_1° = 202.2°, \quad \omega_2° = 345.6°.$$ ◂◂◂

▸▸▸ **Application**: ***precisely timed simulation of a planetary orbit***

According to the first two of *Kepler*'s laws, the orbit of a planet $P$ around the Sun $S$ is an ellipse with $S$ as a focal point. The orbit is traced with permanently changing speed, but "equal areas are covered in equal periods of time" (Fig. 6.41). As simple as this theory may sound, it is still impossible to describe the elliptical orbit by means of elementary functions such that it is "parametrized by time".

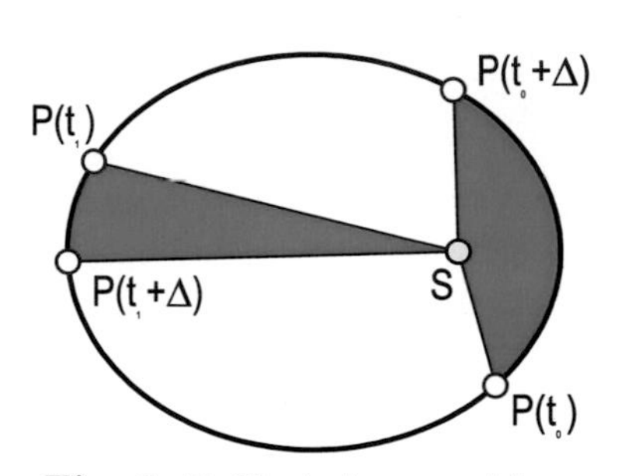

**Fig. 6.41** *Kepler*'s second law

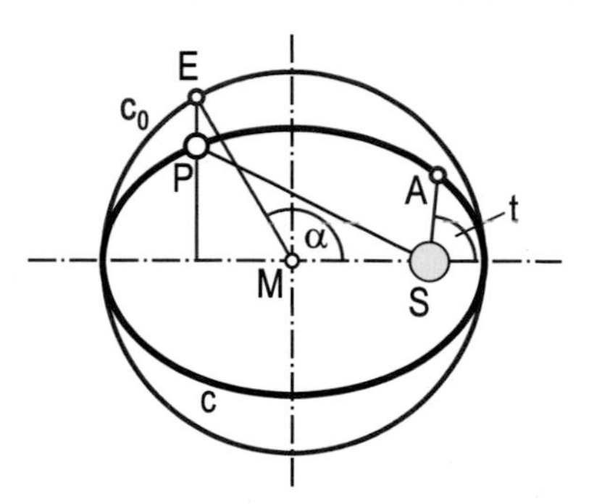


**Fig. 6.42** virtual planetary orbits

*Kepler* solved the problem by examining the two virtual planets besides the planet in question (Fig. 6.42): The "middle planet" $A$ and the "eccentric planet" $E$. $A$ moves on the same orbital ellipse $c$ as $P$ (center $M$, semi-axis lengths $a$, $b$), but with *constant angular velocity* – where $A$ and $P$ have the same total orbital time. The eccentric planet $E$ traces the circumcircle $c_0$ (center $M$, radius $a$) of the orbital ellipse $c$ as a continual companion of $P$ on a line parallel to the ellipse's minor axis (Fig. 6.42). Thus, $E$ results from $P$ by multiplying its distance to the principal axis by the constant factor $a/b$.

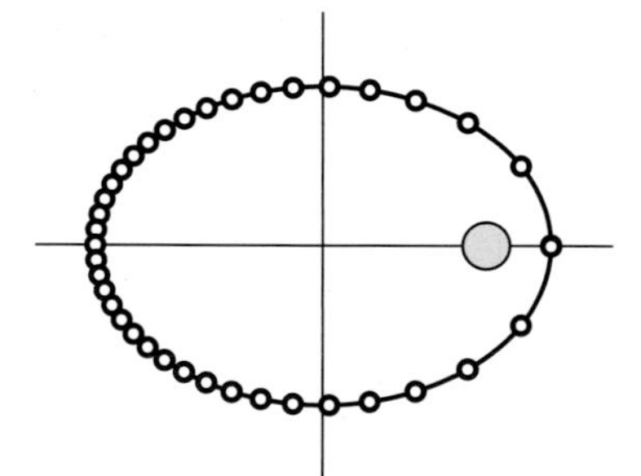

**Fig. 6.43** "stroboscope recording"

**Fig. 6.44** *Galilei* and his telescope

Obviously, it is sufficient to calculate the course angle $\alpha$ of $E$ and then multiply $P$'s distance by the reciprocal factor $b/a$. *Kepler* deduced a relation between the unknown angle $\alpha$ and the course angle $t$ of $A$ which is proportional to the elapsed time. "*Kepler*'s equation" reads

$$f(\alpha) = \alpha - \varepsilon \sin \alpha - t = 0 \quad \text{with} \quad \varepsilon = \sqrt{1 - \left(\frac{b}{a}\right)^2}. \tag{6.34}$$

The *numerical eccentricity* $\varepsilon$ that governs the shape of the ellipse is given in astronomic tables. When specifying the time $t$, $\alpha$ can be found with our method as the only real root of Formula (6.34).

⊕ *Remark*: Fig. 6.43 shows a number of positions of a planet for equal time intervals. The major axis of the orbital ellipse is the apse line. On this line, we find the *perihelion* (position nearest to the Sun) and the *aphelion* (position furthest away from the Sun). Higher orbital speeds can clearly be observed near the Sun. For our Earth, the numerical eccentricity $\varepsilon \approx \dfrac{1}{60}$ is pretty small and the elliptical orbit is almost circular. Nevertheless, the winter season in the northern hemisphere is at least six days shorter than the corresponding summer season. In the annual calendar, this fact is compensated by the following agreements:

1. February has only 28 days.
2. Both July and August have 31 days.
3. Autumn begins two days later than the other seasons. ⊕ ◂◂◂

### ▸▸▸ Application: *power function and exponential function*

When does the exponential function $y = 2^x$ "overtake" the power function $y = x^5$ (Fig. 6.14)?

*Solution*:
In the moment of "overtaking", the following holds:

$$2^x = x^5 \Rightarrow f(x) = 2^x - x^5 = 0.$$

The two solutions are $x_1 = 1.1773$ and $x_2 = 22.4400$. The value $(x_2)^5 = 2^{(x_2)}$ is roughly estimated via $2^{10} \approx 10^3$, and it is pretty large, namely $\approx 2^{2.4} \cdot 10^6 \approx 5 \cdot 10^6$.

⊕ *Remark*: We have already said that any exponential function can overtake even the "mightiest" power function. Looking for the "point of overtaking", one soon encounters numerical difficulties. For example, the method fails already in the computation of the zeros of $f(x) = 2^x - x^6 = 0$, though $f(x) = 2^x - x^{5.969} = 0$ is still solvable (the zero $x_2 \approx 29$, then, corresponds to the value $2^{29} \approx 0.5 \cdot 10^{10}$). ⊕ ◂◂◂

## 6.7 Further applications

▸▸▸ **Application**: ***the problem with lilies ...***

Water lilies, which grow in a pond, increasingly spread across its surface. Say, for example, the area that they occupy grows by 10% every day, and it will only take 30 days until the entire pond is covered. How large was the initial area, and when exactly was half of the pond covered?

*Solution*:

Let $S$ be the surface area covered by the water lilies on their first day and $A$ the surface of the entire pond. On the second day, the lilies would cover an area of $kS$ with $k = 1.1$ (increase of 10%), and on the $n$-th day, they would cover an area of $k^n S$. From $k^{30} S = A$, it follows that $S = A/k^{30} \approx A/17.5$. This means that, at the beginning, the pond was covered by less than 6%. Now we have to calculate the value of $n$ from $k^n S = 0.5A$:

$k^n S = 0.5 \cdot k^{30} S \Rightarrow k^{n-30} = 0.5 \Rightarrow n = 30 + \log 0.5 / \log k \approx 22.73$.

It would seem that the pond was covered exactly halfway on the 23rd day.

**Fig. 6.45** a near-closed area of reeds on the Neusiedlersee (Austria)

⊕ *Remark*: The problem of water lilies is typical of exponential growth phenomena in general, which, in nature, are manifested most dramatically in the reproduction of bacteria. Such growth is forced to collapse eventually in one form or another due to a limited amount of resources being available. In the case of the pond, it is especially easy to see why, as it makes no sense to cover it with another layer of water lilies (Fig. 6.45). ⊕ ◂◂◂

▸▸▸ **Application**: ***population statistics***

Let $f(x)$ be the proportion of the population of a country which is only $x$ years old ($f(x) \in [0, 1]$). This function is described by means of the illustration in Fig. 6.46.

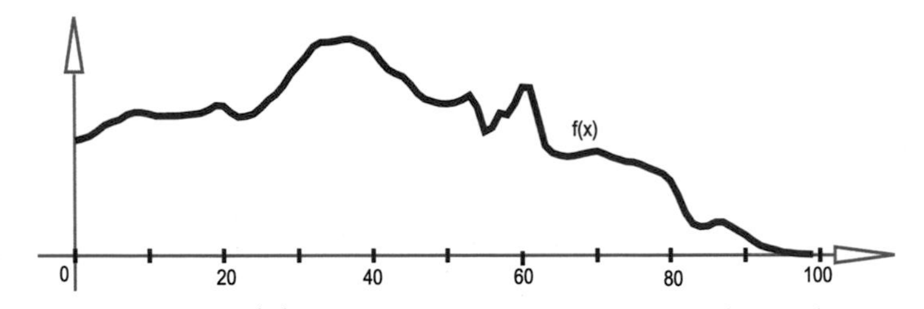


**Fig. 6.46** proportion $f(x)$ of the population at the age of $x$ in Austria (2003)

*Solution*:
If the age classification is only accurate up to one year, then $f(x)$ is not a smooth function, but a "staircase function" ("bar charts" with a bar width of 1). In "developed countries", the proportion of the population $f(x)$ decreases anywhere from 80 and onward: $f(100)$ is already almost zero – even in nations where "many" people are more than 100 years old. Therefore, one can limit the maximum human age to ≈ 100 years. It is striking that, in a typical European country such as Austria, the number of newborns decreases continuously. In a typical developing country, this is reversed. In these countries, however, the mortality – also and especially at a young age – is much higher. The proportion of people who are $i$ years old equals $f(i)$. Since each person has exactly one age, we have

$$\sum_{i=0}^{m} f(i) = 1.$$

The area under the function graph has, therefore, the value 1. In the diagram, the ordinate is extremely superelevated, because otherwise, we would not be able to accurately see anything and the diagram would be useless. ◂◂◂

### ▸▸▸ Application: *reconstruction of date and time*

The vector equation (5.52) for the light ray corresponding to the date $\alpha$ ($\alpha$ measures the number of days from March 21) at the time $z$ allows us to calculate $\alpha$ and $z$ if the (normalized) direction vector $\vec{s}_0(s_x, s_y, s_z)$ pointing towards the Sun is given together with the latitude $\varphi$. First, we calculate the auxiliary variables $\omega$ and $\mu$ as

$$\omega = \arctan \frac{s_x}{s_z \cos\varphi - s_y \sin\varphi}, \quad \mu = \arctan \sqrt{\frac{\sin^2 \omega}{s_x^2} - 1}. \tag{6.35}$$

Then, $z$ and $\alpha$ are as follows ($\alpha$ is ambiguous and corresponds to the angle $\alpha^\circ$):

$$z = 12 - \frac{1}{15}\omega^\circ, \quad \alpha_1^\circ = \arcsin \frac{\mu}{23.44^\circ}, \quad \alpha_2^\circ = 180^\circ - \alpha_1^\circ. \tag{6.36}$$

◂◂◂

▸▸▸ **Application**: ***monetary value at various times***

Three people are making an offer for a house: $A$ will immediately pay $200,000$ €, $B$ wants to pay $100,000$ € as a first instalment and procure $120,000$ € in 4 years, $C$ proposes $230,000$ € in 3 years. Which offer is best for the seller if we calculate an annual interest rate of 5% and 3.5% respectively?

*Solution*:

We relate all monetary values to the time "4 years ahead": A 5% interest rate will then yield the following amounts for the three respective offers (in $100,000$ €-units):

$A = 2.0 \cdot 1.05^4 \approx 2.43$, $B = 1.0 \cdot 1.05^4 + 1.2 = 2.415$, $C = 2.3 \cdot 1.05^1 = 2.41$.

The three deals are almost equal, but $A$ offers the most to the seller. With an interest rate of only 3.5%, the seller can expect to get:

$A = 2.0 \cdot 1.03^4 \approx 2.25$, $B = 1.0 \cdot 1.03^4 + 1.2 = 2.33$, $C = 2.3 \cdot 1.03^1 = 2.37$. This time, the margins are a bit larger and $C$ is the best bidder. ◂◂◂

▸▸▸ **Application**: ***cold coffee*** (Fig. 6.47)

If we are in a hurry in the morning, we may drink our coffee in one gulp immediately after pouring it into a cup. To avoid burning our mouths, we might add some cold milk from the fridge. However, if we decide to let several minutes pass between pouring the coffee into a cup and drinking the "melange" after adding the milk, the following question arises for connoisseurs who wish the mixture to be still as warm as possible: Assuming that we add the same amount of milk in any case, should we pour the milk into the coffee right away or just before drinking?

**Fig. 6.47** cold or warm milk? (right stereo image, see also Fig. 5.50)

*Solution*:

We have already discussed the cooling process in Application p. 251, where we have computed with temperature differences. Let $k$ be the temperature of the black coffee in the beginning, and let $m$ ($m < 0$) be the temperature of the milk. The quantity of milk is a fraction $q$ of the amount of coffee. According to the *law of energy conservation*, the temperature difference of the "mixture" is

$$\Delta = \frac{k + q\,m}{1 + q}.$$

The latter decreases after $t$ minutes by a factor of $\mu = \lambda^t$ (with $\lambda < 1 \Rightarrow \mu < 1$) to

$$\Delta_t = \mu\,\Delta = \mu\,\frac{k + q\,m}{1 + q}.$$

If we mix after $t$ minutes, then we have

$$\overline{\Delta_t} = \frac{(\mu\,k) + q\,m}{1 + q}.$$

Then, we form

$$\Delta_t - \overline{\Delta_t} = \frac{qm}{1+q}\,(\mu - 1).$$

Since $m < 0$ and $\mu - 1 < 0$, this expression is positive, which means that the "mixture" has cooled less. The difference is more noticeable the warmer the room (because then $|m|$ is greater) or the more milk is added. The term is obviously independent of the initial temperature of the coffee.

⊕ *Remark*: On the right in Fig. 6.47, one clearly sees the so-called focal curves. These are parts of curves that are enveloped by all reflected light rays. See the section on envelopes. ⊕ ◂◂◂

**▸▸▸ Application**: ***spline curves***

We want to interpolate $n+1$ points $P_1(u_1/v_1), \ldots, P_{n+1}(u_{n+1}/v_{n+1})$ (so-called "knots") by a "cubic spline". We are confronted with this task, for instance, when we want to design smooth curves ("free form curves") in a CAD system.

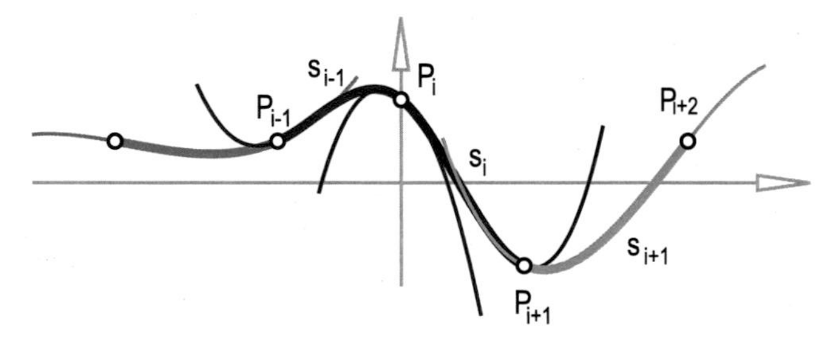


**Fig. 6.48** composition of a cubic spline from cubic parabolas

*Solution*:

In Application p. 33 in the section on systems of linear equations, we computed a parabola of degree $n$ on $n+1$ points. The higher the number of points to be interpolated, the higher the degree of the curve and the more oscillations occur (especially at the end points). For that purpose, the following procedure is frequently used in computer graphics (Fig. 6.48):

Consider four adjacent nodes $P_{i-1}$, $P_i$, $P_{i+1}$, and $P_{i+2}$ and the tangents at these points (initially arbitrarily selectable). Then, two neighboring points plus their tangents define a unique cubic parabola (since a tangent together with its point of contact counts for two points). Adjacent cubic arcs are

tangent to each other (this is frequently called $G^1$ *continuity*). We shall soon see that the tangents can be chosen such that the cubic parabolas have the same curvature at the transition points (this is called $G^2$ *continuity*). All $n$ parabolic arcs are then combined to form a cubic spline.
For the $i$-th parabolic arc $s_i$ from $P_i$ to $P_{i+1}$, we make the ansatz

$$y = a_i\,t^3 + b_i\,t^2 + c_i\,t + d_i \ (i = 1, \ldots, n).$$

For each parabola, this will yield four unknowns $a_i$, $b_i$, $c_i$, and $d_i$, that is, a total of $4\,n$ unknowns. The value $t = 0$ will correspond to $P_i$, which already yields $n$ indeterminates:

$$v_i = d_i \ (i = 1, \ldots, n).$$

That leaves $3\,n$ unknowns $a_i$, $b_i$, $c_i$, which can be calculated if one takes $3\,n$ linear equations independent from each other.
For the parameter $t = 1$, we will get the point $P_{i+1}$:

$$v_{i+1} = a_i + b_i + c_i + d_i \Rightarrow a_i + b_i + c_i = v_{i+1} - v_i \ (i = 1, \ldots, n) \tag{6.37}$$

($n$ linear equations).
Now, for the tangents: We compute the first derivative

$$\dot{y} = 3\,a_i\,t^2 + 2\,b_i\,t + c_i \ (i = 1, \ldots, n)$$

of the cubic parabolas. We need the tangent of a neighboring parabola, for example, the left $s_{i-1}$ that exists only for $i > 1$:

$$\dot{y} = 3\,a_{i-1}\,t^2 + 2\,b_{i-1}\,t + c_{i-1}.$$

There, the same value must be achieved for $t = 1$ on the $i$-th parabola $s_i$ for the parameter $t = 0$:

$$3\,a_{i-1} + 2\,b_{i-1} + c_{i-1} = c_i \ (i = 2, \ldots, n) \tag{6.38}$$

which yields a further $n - 1$ linear equations.
Now for the second derivatives, which, together with the first derivative, determine the curvature of the curve at $t = 0$:
The second derivative of the $i$-th parabola

$$\ddot{y} = 6\,a_i\,t + 2\,b_i$$

equals that of the $(i - 1)$-th parabola at $t = 1$. With the same considerations as before, we get

$$6\,a_{i-1} + 2\,b_{i-1} = 2\,b_i \ (i = 2, \ldots, n) \tag{6.39}$$

which are an additional $n - 1$ linear equations. Overall, we have now found $3\,n - 2$ linear equations. The last two can be found by prescribing the second derivative at the start and end point, and thus,

$$2\,b_1 = 0, \quad 6\,a_n + 2\,b_n = 0. \tag{6.40}$$

One can solve this $(3n, 3n)$ system of linear equations efficiently because of all the zeros in the determinants (the best method was invented by *Gauß*). It turns out that the computation time is only linearly (and not quadratically as in the general case) increasing with the number of nodes.

⊕ *Remark*: Normally, cubic splines look as we expect them to look (which is important in the design of free-form curves).
One should choose relatively equal distances between the control points. A disadvantage, however, is that any change of a control point affects the curve only locally. Therefore, other spline types (for example, *Bezier* splines) are offered by CAD programs, which we will not discuss here. ⊕ ◂◂◂

▸▸▸ **Application**: ***approximation of a point cloud through a straight line***
Show that the points $P_1(x_1/y_1), \ldots, P_n(x_n/y_n)$ in the plane uniquely define the linear fit $r:\ y = kx$ through their "center of gravity" $S(\overline{x}/\overline{y}) = \frac{1}{n}\sum\limits_{i=1}^{n} P_i$ with the coefficient $k = \sum_{i=1}^{n}(x_i - \overline{x})(y_i - \overline{y}) / \sum_{i=1}^{n}(x_i - \overline{x})^2$ that minimizes the sum of *squared* distances to the data points.

***Proof***: Let the straight line $r$ be given by $y = kx + d$ (Fig. 6.49). The obvious next step is to move $r$ through the "center of gravity" $S(\overline{x}/\overline{y})$ of the point cloud, thus $d = \overline{y} - k\,\overline{x}$.

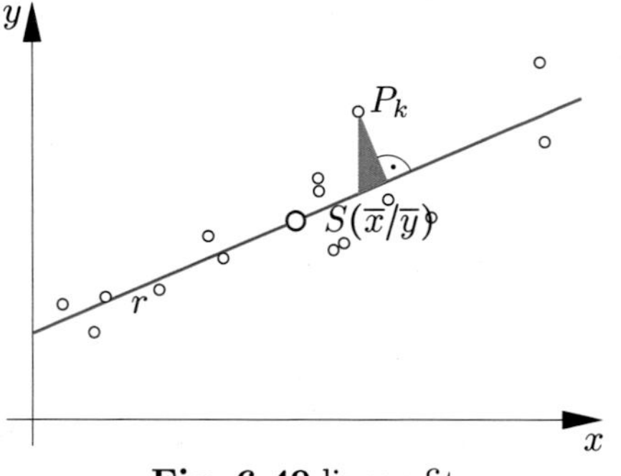


**Fig. 6.49** linear fit

The normal distance $\overline{Pr}$ of a point $P(x/y)$ from the line $r$ with the angle of inclination $\alpha = \arctan k$ (with respect to the $x$-axis) is proportional to the difference $\Delta y_i$ of all $y_i$-values of all points $P_i$ and the corresponding points on the line $r$ at $x_i$ $(\overline{Pr} = \Delta y_i \cdot \cos\alpha)$. So, it is sufficient to choose $k$ in

$$\sum(\Delta y_i)^2 = \sum[y_i - (kx_i + d)]^2 = \sum[y_i - (kx_i + \overline{y} - k\,\overline{x})]^2 =$$
$$= \sum[(y_i - \overline{y})^2 + k^2(x_i - \overline{x})^2 - 2k(y_i - \overline{y})(x_i - \overline{x})]$$

such that we obtain a minimum. For that purpose, we have to find the zeros of the derivative with respect to $k$.

$$\sum[2k(x_i - \overline{x})^2 - 2(y_i - \overline{y})(x_i - \overline{x})] = 0 \Rightarrow k\sum(x_i - \overline{x})^2 = \sum(y_i - \overline{y})(x_i - \overline{x}),$$

whereby the above assertion has been shown. ⊙

⊕ *Remark*: The linear fit plays an important role in statistics (Application p. 497): If the deviation of the points from the line is small, one can make predictions about events beyond the given data. ⊕ ◂◂◂

▸▸▸ **Application**: ***strain on an insect wing*** (Fig. 6.50)
Insects rapidly flutter their wings. Guess the maximum acceleration of the outer parts of the wings of the depicted hawk-moth (8 cm wingspan).

*Solution*:
The wing motion is ideally "harmonic". The path-time diagram of the wing-tips is, therefore, a sine curve.

**Fig. 6.50** hawk moth in flight

Let the wingtips have an amplitude of $s$. The frequency of flapping is $n$. The function $y = s\sin(2\pi\, n\, t)$ oscillates $n$ times up and down in the time interval $[0,1]$. So, it is sufficient to simulate the situation within one second. The acceleration is obtained by differentiating twice with respect to time: $\ddot{y} = -s(2\pi\, n)^2 \sin(2\pi\, n\, t)$. The maximum acceleration is, therefore, $4\pi^2\, s\, n^2 \approx 40\, s\, n^2$.
In fact, for the depicted butterfly, the roughly estimated values $s = 1\text{cm} = 0.01\text{m}$ and $n = 60$ result in a nearly 150-fold gravitational acceleration. Similar values also appear for the wings of hummingbirds. Bees have a frequency of about $n = 200$ and an amplitude of $s = 2\text{mm} = 0.002\text{m}$. The maximum acceleration for the outer wings is, then, about $300g$.

⊕ *Remark*: Insects can definitely cope with larger accelerations. During their rapid turning manoeuvres, flies and dragonflies can stand up to $30g$ with their whole bodies. The wing parts are much more resilient because they do not contain any vital organs. The smaller the insect, the greater the resilience: The cross-sectional areas responsible for the stability relative to the weight are substantially greater, as we have discussed in detail in Chapter 3. Short term accelerations of $300g$ are no problem for a bee's wing. ⊕ ◂◂◂

▸▸▸ **Application**: ***efficiency of a wing or fin stroke*** (Fig. 6.51)
The so-called Strouhal-number $S$ corresponds to the frequency of a wing beat or flapping multiplied by its amplitude and divided by the forward velocity.

The highest efficiency is achieved at values between 0.2 and 0.4. This number is used to describe the locomotion of flying and floating animals – regardless of their size – so, ranging from mosquitoes to birds, bats, and even baleen whales.[3]

**Fig. 6.51** different frequencies by huge and tiny warm-blooded animals

For the hawk-moth in Application p. 293, we obtain

$$S = \frac{\frac{60}{\mathrm{s}}\,0.01\mathrm{m}}{v} = 0.3 \Rightarrow v = 2\frac{\mathrm{m}}{\mathrm{s}},$$

i.e. the greatest efficiency at an airspeed of about 2 m/s. If an insect or bird is in the air (Fig. 6.51, right), the Strouhal-number assumes an infinitely large value, which means "zero efficiency" – at least in terms of overcoming distances. ◂◂◂

▸▸▸ **Application**: ***detailed analysis of a landing flight*** (Fig. 6.52)
Especially with smaller birds, the landing flight can happen in a flash, and it is difficult to make out any details with the naked eye. For this reason, such a landing flight shall be examined in close detail here by using biophysical methods of slow-motion analysis.[4]

**Fig. 6.52** landing flight of a blue tit

The image series Fig. 6.52 was originally shot at 250 frames per second (see also Fig. 3.69). Pictured here and highlighted in the graph Fig. 6.53 is every

[3] `http://www.nature.com/news/2003/031016/full/news031013-9.html`
[4] From Glaeser, Paulus, Nachtigall: *The Evolution of Flight*, Springer Nature, Heidelberg 2017.

third frame, that is, frames number 0 – 3 – 6 etc. The interval $\Delta t$ that has elapsed between one trio of frames and the next thus equals about 1/85 s.
Since it is not possible to determine a centroid, the centers of the eyes in subsequent images have been drawn over each other to create a distance-time graph $s(t)$. Between each trio of frames, the average landing speed $v = \Delta s/\Delta t$ has been calculated by difference analysis and is plotted as $v(t)$ both in meters per second and kilometers per hour. An approximate delay $-b = \Delta v/\Delta t(t)$ may also be calculated and plotted as $-b(t)$ in meters per square second and in units of Earth acceleration $g$.
This comparison is quite instructive (the numbers used in the evaluation are rounded to two decimal places, as is common in technology; the accuracy of the evaluation on the basis of the drawings is, of course, considerably lower).
The blue tit lands in a slightly wavy line, which is inclined downwards at an average angle of no more than 15°. Shortly before frame 0, the bird makes a braking flap with its wing, then two more on its way to the landing site, and a third weak flap upon touching down with its feet (frame 18, cf. drawing). One would think that the landing velocity is abruptly reduced with each braking flap of the wings. As the graphs show, this is the case to some extent. The braking flaps cause fluctuations in the $v(t)$-curve, but the fluctuations are not particularly strong. These flaps may produce some kinks in the curve, but in general, the landing velocity decreases at a constant pace. This is due to the inert mass of the bird, which puts a damper on any rapid changes.
Between frames 0 and 3, the blue tit is landing at a velocity of about 2.34 m/s or 8.42 km/h. Up to the moment when the bird touches down with its feet, that is, over a flight distance of 12.7 cm, which the tit crosses in 0.072 s, aerodynamic braking alone causes the velocity to be reduced to 1.21 m/s or 4.36 km/h, that is, by half.
In the remaining images until the bird comes to rest upon landing (frame 18 – final position), the work is done by the bird's legs, which are initially stretched out at an angle of 145° and are gradually folded in until they hang at an angle of about 45°. These legs are astonishingly effective as landing devices and shock absorbers.
The delay can only approximately be determined by means of "manual" differential calculations. Due to the braking flaps of the wings, it fluctuates considerably. Over a flight distance of 12.7 cm, that is, until the moment of the feet's touch-down, the tit slows down from 2.34 m/s to 1.21 m/s within 0.072 s. The average delay thus equals $-b = (2.34 - 1.21)/0.072 = 15.69\ \mathrm{ms}^{-2}$ or 1.60 $g$.
When comparing individual frames, however, the delay may vary considerably, namely between 22.10 $\mathrm{ms}^{-2}$ (which equals 2.25 $g$) within the first trio of frames and 41.65 $\mathrm{ms}^{-2}$ (which equals 4.25 $g$) within the final trio. So, over this short distance of braking flight, the delay changes by a factor of 1.89. During its landing, the bird even accelerates slightly, but this may not be the case in other pictures of a blue tit's landing flight.
Upon touching down with its legs, the bird slows down further until it comes to rest. The deceleration forces thereby generated must cushion the impact of the landing legs. It has already been mentioned that the legs are bent in by about 100° upon touch-down. The leg extension muscles, which are already activated and thus

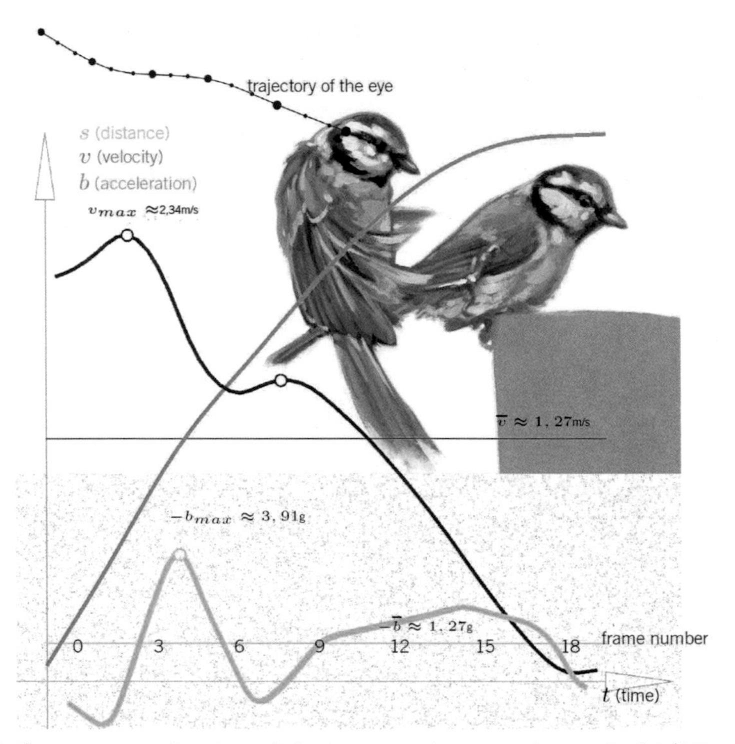


**Fig. 6.53** Computer evaluation of the image series at an interval of 1/250 s. Green: distance-time-diagram, blue: velocity-time diagram, orange: acceleration-time-diagram (mirrored on the abscissa). Every third frame (1/83 s) is numbered. Initially, only these frames were "manually" evaluated, but the final outcome already became apparent at this point.

prepared for the strain, as well as their tendons, are further stretched in the process, like a previously stretched tension spring, on which further weight is attached. At the same time, the bird tilts clockwise around its lateral axis and thrusts its head forward until it comes to rest in the position drawn with dashed lines. In the final phase, other muscles, especially those in the back, also help to alleviate the impact of the touch-down. The energy absorbed in the process is released as heat into the environment unless it is used to change the position of the bird's trunk. The deceleration in the final phase amounts to an average of 2 g. Good "shock absorbers" are needed to alleviate such an impact!

The seemingly effortless landing process does not just "happen by itself," but it must be precisely controlled and navigated, which is enabled by a complex interplay of sensory organs acting as sensors on the one hand and flight, leg, and trunk muscles acting as effectors on the other hand. And this must all take place within a timeframe of just a few milliseconds. ◂◂◂

# 7 Curves and surfaces

This chapter can be seen as a continuation of the chapter concerning vectors. It will introduce so-called matrices (especially rotation matrices), which can be used to provide elegant descriptions of rotations in the plane, as well as in three-dimensional space. Naturally, we will often speak of congruence motions, which are the subject of kinematics.

The fundamental theorem of planar kinematics states that two positions of one object can always be transformed into each other by means of a single rotation (or translation). The spatial analogon (the fundamental theorem of spatial kinematics) states that two congruent positions of an object can always be transformed into each other by means of a single helical motion (eventually a rotation or a translation). In the limiting case, we even speak of instantaneous rotations or instantaneous helical motions.

This chapter contains important examples of mathematically describable curves (such as conic sections) and surfaces. We will focus, in particular, on their generation via motions and on the various applications that can be found in nature and technology.

Rigid body motions often produce envelopes of curves and surfaces, which we will also study with differential calculus. Envelopes (curves or surfaces) play a fundamental role in the study of motions.


G. Glaeser, *Math Tools*, https://doi.org/10.1007/978-3-319-66960-1_7

## 7.1 Rigid body motions

In this section, we will focus more closely on translations, rotations, and helical motions. These are the "shape preserving" transformations (isometries) in the plane and in space which do not change the relative position of points on the moved object.
Vector calculus will be augmented by calculations involving rotational matrices.

### Translation

Let $P(p_x/p_y/p_z)$ be a point in space with the position vector $\vec{p}$. The parallel displacement (translation) of the point $P$ towards $P_1$ (with position vector $\vec{p_1}$) along the translation vector $\vec{t}$ is accomplished by a simple vector addition:

$$\vec{p_1} = \vec{p} + \vec{t} \quad \text{or, more detailed,} \quad \vec{p_1} = \begin{pmatrix} p_x + t_x \\ p_y + t_y \\ p_z + t_z \end{pmatrix}. \tag{7.1}$$

In $\mathbb{R}^2$, the third component is simply zero.

### Rotation in $\mathbb{E}^2$ about the coordinate origin

If the point $P(p_x/p_y)$ is rotated about the origin through the angle $\varphi$, then the new point $P_1$ has the position vector (according to Fig. 7.1)

$$\vec{p_1} = \begin{pmatrix} \cos\varphi\, p_x - \sin\varphi\, p_y \\ \sin\varphi\, p_x + \cos\varphi\, p_y \end{pmatrix}. \tag{7.2}$$

From now on, we will use the short notation

$$\vec{p_1} = \begin{pmatrix} \cos\varphi\, p_x - \sin\varphi\, p_y \\ \sin\varphi\, p_x + \cos\varphi\, p_y \end{pmatrix} = \begin{pmatrix} \cos\varphi & -\sin\varphi \\ \sin\varphi & \cos\varphi \end{pmatrix} \cdot \vec{p} = \mathbf{R}(\varphi) \cdot \vec{p}. \tag{7.3}$$

The number scheme

$$\mathbf{R}(\varphi) = \begin{pmatrix} \cos\varphi & -\sin\varphi \\ \sin\varphi & \cos\varphi \end{pmatrix} \tag{7.4}$$

is called a *rotation matrix*. In the next section, we will discuss matrices in much greater detail.

▸▸▸ **Application**: ***refracted light*** (Fig. 7.2)
According to *Snell*'s law of refraction (the equation is derived in Application p. 378), the following holds true for the angle of incidence $\alpha$ and the angle $\beta$ of the refracted ray:

$$\frac{\sin\alpha}{\sin\beta} = n = \text{const} \Rightarrow \beta = \arcsin\frac{\sin\alpha}{n} \tag{7.5}$$

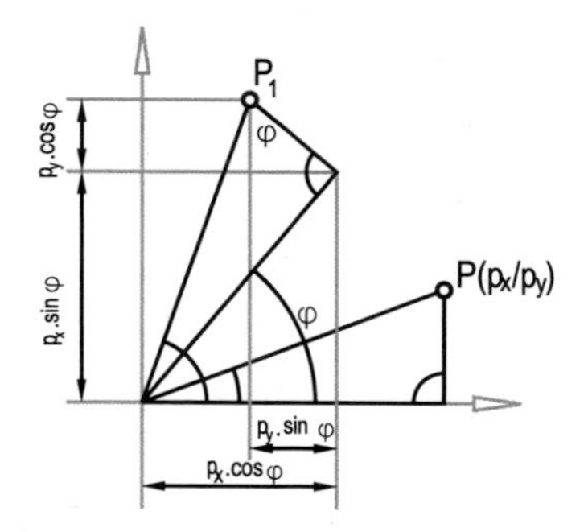

**Fig. 7.1** rotation about the origin

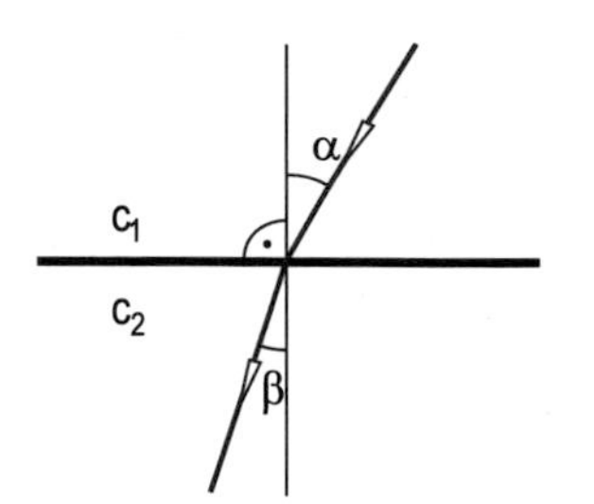

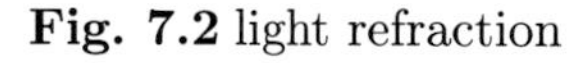
**Fig. 7.2** light refraction

where $n$ is the material-dependent refractive constant ($n = 4/3$ for the transition from air to water). The ray that hits the surface of the water has the direction $\vec{l} = (-1, -2)$. What is the direction of the refracted light ray $\vec{r}$ in the water?

*Solution*:

From $\alpha = \arctan 1/2 = 26.565°$, we infer $\beta = 19.597°$. The vector has to be rotated through the angle $\varphi = \alpha - \beta = 6.968°$. Since $\sin\varphi = 0.1213$ and $\cos\varphi = 0.9926$, we have

$$\vec{r} = \begin{pmatrix} 0.9926 \cdot (-1) - 0.1213 \cdot (-2) \\ 0.1213 \cdot (-1) + 0.9926 \cdot (-2) \end{pmatrix} = \begin{pmatrix} -0.750 \\ -2.107 \end{pmatrix}.$$ ◂◂◂

## Rotation in $\mathbb{E}^2$ about an arbitrary point

Let $C(c_x/c_y)$ be an arbitrary center of rotation (position vector $\vec{c}$), and let $\varphi$ be the angle of rotation. Then, we need three steps in order to describe the rotation of $P$ about $C$:

1. translation of $P$ by $-\vec{c}$
   ↦ auxiliary point $H$ with the position vector $\vec{h} = \vec{p} - \vec{c}$,
2. rotation of $H$ about the origin
   ↦ auxiliary point $H_1$ with the position vector $\vec{h}_1 = \mathbf{R}(\varphi) \cdot \vec{h}$, and
3. translation of $H_1$ by $\vec{c}$
   ↦ final result $P_1$ with the position vector $\vec{p_1} = \vec{h}_1 + \vec{c}$.

All in all, we could also have written:

$$\boxed{\vec{p_1} = \mathbf{R}(\varphi) \cdot (\vec{p} - \vec{c}) + \vec{c}.} \tag{7.6}$$

## Main theorem of planar kinematics

For planar kinematics, the following theorem, already mentioned in Application p. 195, is of great importance:

Any two positions of a rigid planar figure can always be transformed by rotation (or a translation) into each other.

In the course of a "rigid body motion", planar figures do not change their shape, and thus, the theorem is valid at every instant of the motion, regardless of the complexity of the figure. The instantaneous center of rotation is called the *instantaneous pole*. In a rotation, the orbits are always circles around the center of rotation, and therefore, the following theorem on the orbit's normals holds true:

At any instant, the path normals of a rigid body motion pass through a certain fixed point, namely the instantaneous pole.

In an instantaneous translation, the paths of all points are straight lines, and so, the path normals are parallel. Examples can be found in the following application and Application p. 318.

**▸▸▸ Application**: ***trochoidal motion (planetary motion)***

This motion is the superposition of two arbitrary rotations in the plane: A point $R$ rotates about a fixed point $F$ with constant angular velocity, and another point $C$ rotates about the point $R$ at the same time with the proportional angular velocity. Then, $C$ undergoes a so-called planetary motion. Show that the same motion can also be produced as the trace of a point on a circle rolling on another circle. Determine the radii of the circles.

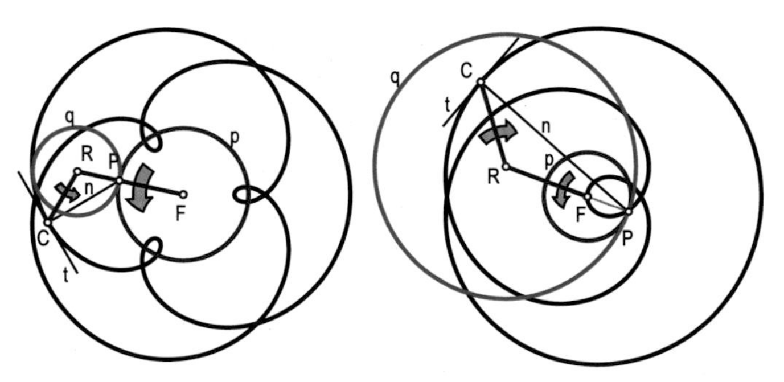


**Fig. 7.3** trochoids

*Solution*:

First of all, let us consider the points $F$ and $R$ as fixed and as the centers of two touching rubber-coated wheels $p$ and $q$ (Fig. 7.3). If one turns $p$, then $q$ automatically rotates (for example, $w$ times as fast). A point $C$ that is firmly attached to $q$ also rotates about $R$.

Now you hold $p$ steady and the axial connection $FR$ of fixed-length $s$ is turned instead. Then, $q$ rolls on $p$, $R$ rotates about $F$, and the fixed ratio of the radii ensures the proportionality of the rotations. The point $C$ describes the desired trochoid.

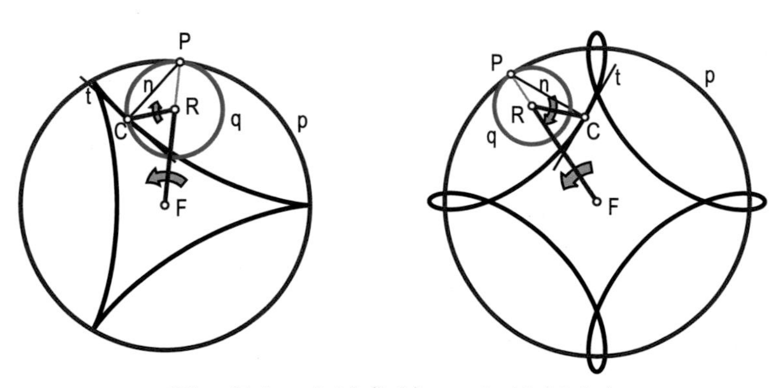


**Fig. 7.4** cycloid (left), trochoid (right)

The rubber coating on the wheels guarantees a rolling without gliding, and thus, equal arcs $r$ are transmitted. Let $r_p$ and $r_q$ be the circles' radii ($r_p + r_q = s$). Then we always get $r_p = w\,r_q$. From the two conditions, we deduce

$$r_q(w+1) = s \Rightarrow r_q = \frac{s}{w+1},\ r_p = \frac{s\,w}{w+1}.$$

The ratio of the radii is, thus, indirectly proportional to the ratio of the angular velocities ($r_q : r_p = 1 : w$). Admitting also negative radii results in so-called hypocycloids, i.e. curves that are traced by "rolling a circle" inside the fixed circle (Fig. 7.4). In any case, the instantaneous velocity is zero at the point of contact of the circles, and therefore, it is the instantaneous pole, through which all path normals $n$ are running (cf. the theorem on path normals p. 300). The path tangent $t$ of the point $C$ is, thus, perpendicular to its connection with $P$ (see Fig. 5.28).

⊕ *Remark*: Actually, the name planetary motion comes from astronomy: The planets move around the Sun. In addition to the motion we observe, there is also the Earth's own rotation. Thus, planet orbits appear more complex in the firmament than ordinary stellar orbits. ⊕

Among the trochoidal curves, we find ellipses for the special ratio $1 : -1$ (Application p. 316). Other special circumstances will yield a multitude of curves that play a role in some area of mathematics, such as the Cartesian quadrifolium in Application p. 43. ◂◂◂

In space, we do not rotate about points. We rotate about straight lines. The simplest case is the rotation about the $z$-axis or about parallel axes.

## Rotation about vertical axes

Such rotations can be traced back to the two-dimensional case. If a point $P(p_x, p_y, p_z)$ rotates about the $z$-axis through the angle $\varphi$, then the new point $P_1$, according to Formula (7.2), has the position vector

$$\vec{p_1} = \begin{pmatrix} \cos\varphi\, p_x - \sin\varphi\, p_y \\ \sin\varphi\, p_x + \cos\varphi\, p_y \\ p_z \end{pmatrix} = \begin{pmatrix} \cos\varphi\cdot p_x & - & \sin\varphi\cdot p_y & + & 0\cdot p_z \\ \sin\varphi\cdot p_x & + & \cos\varphi\cdot p_y & + & 0\cdot p_z \\ 0\cdot p_x & + & 0\cdot p_y & + & 1\cdot p_z \end{pmatrix}. \tag{7.7}$$

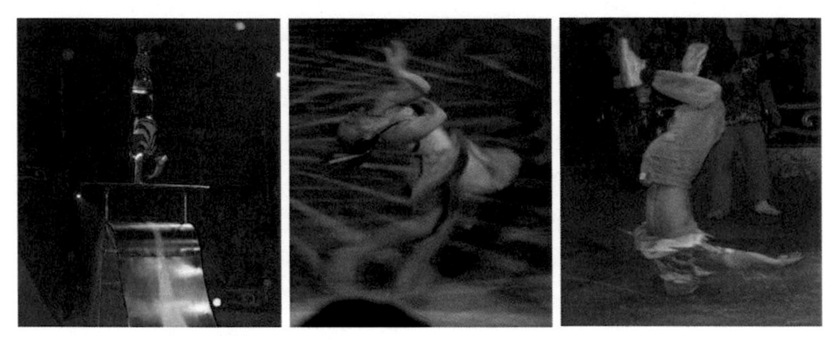

**Fig. 7.5** rotations about vertical axes

Again, we use an abbreviation in the form of a rotation matrix

$$\vec{p_1} = \begin{pmatrix} \cos\varphi & -\sin\varphi & 0 \\ \sin\varphi & \cos\varphi & 0 \\ 0 & 0 & 1 \end{pmatrix} \cdot \vec{p} = \mathbf{R}_z(\varphi)\cdot\vec{p}. \tag{7.8}$$

The rotation matrix $\mathbf{R}_z(\varphi)$ arises from the matrix $\mathbf{R}(\varphi)$ in Formula (7.4) by adding a column and a row of zeros with a 1 on the crossing point, i.e. in the matrix's diagonal:

$$\mathbf{R}_z(\varphi) = \begin{pmatrix} \cos\varphi & -\sin\varphi & 0 \\ \sin\varphi & \cos\varphi & 0 \\ 0 & 0 & 1 \end{pmatrix}. \tag{7.9}$$

The rotation about a common parallel $z$-axis is accomplished by translating the $z$-axis, applying the above rotation (matrix), and applying the inverse translation.

## Rotations about arbitrary axes

Swapping the coordinates yields the formulas for the rotation about the other two coordinate axes.
For example, the rotation matrix for the rotation about the $x$-axis reads

$$\mathbf{R}_x(\varphi) = \begin{pmatrix} 1 & 0 & 0 \\ 0 & \cos\varphi & -\sin\varphi \\ 0 & \sin\varphi & \cos\varphi \end{pmatrix}. \tag{7.10}$$

A rotation about an *arbitrary line $g$ through the origin by the angle $\varphi$* can always be interpreted as a combination of special rotations. As *Euler* has

**Fig. 7.6** rotations about arbitrary axes

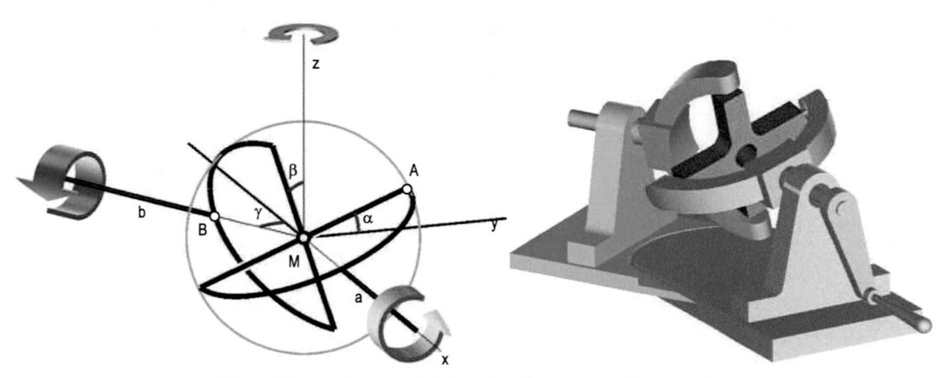


**Fig. 7.7** universal joint in theory and practice

shown, all rotations can be decomposed into a rotation about the $z$-axis, a subsequent rotation about the $x$-axis, and finally a rotation about the new $z$-axis. The corresponding rotation angles are called the *Euler* angles.

▸▸▸ **Application**: ***universal joint*** (Fig. 5.67)
We recall the universal joint for the transmission of the rotation about an axis $a$ – to a non-proportional rotation about an axis $b$ intersecting $a$. The point $A$ of the first fork is rotated through an angle $\alpha$ about the $x$-axis. The point $B$ of the second fork rotates initially about the $x$-axis by an angle $\beta = \arctan \frac{\tan\alpha}{\cos\gamma}$ (see Section 2, Formula (5.48)) and finally about the $z$-axis rotated through the angle $\gamma$.
Let $r$ be the radius of the forks. Then, the position vectors $\vec{a}$ and $\vec{b}$ of the points $A$ and $B$ can be given by means of the matrix notation as

$$\vec{a} = \mathbf{R}_x(\alpha)\cdot\begin{pmatrix} r \\ 0 \\ 0 \end{pmatrix}, \quad \vec{b} = \mathbf{R}_z(\gamma)\cdot\left[\mathbf{R}_x(\beta)\cdot\begin{pmatrix} 0 \\ 0 \\ r \end{pmatrix}\right].$$

◂◂◂

▸▸▸ **Application**: ***oblique views of an object***
Explain the "photography machine" in Fig. 7.8 or Fig. 7.9.
*Solution*:
In order to view or photograph an object from all possible directions, one

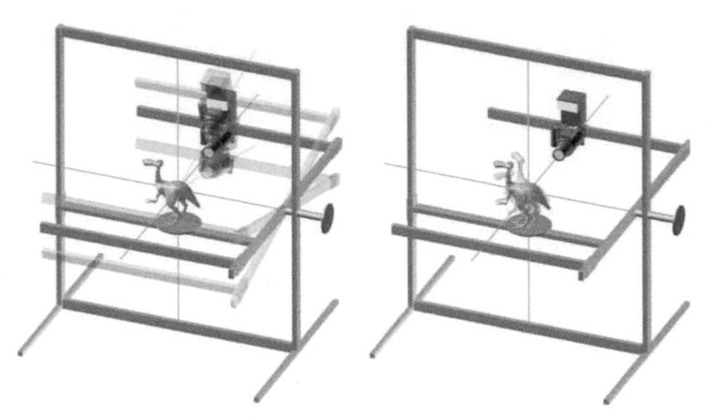

**Fig. 7.8** change of elevation angle and azimuth angle

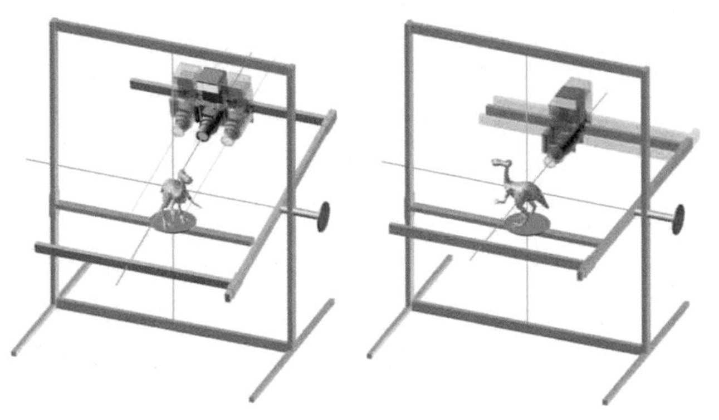

**Fig. 7.9** change of target point and distance

can proceed economically as follows: The camera can be rotated about a fixed horizontal axis, and the object can be independently rotated about a vertical axis. In this manner, you can set any line of view and thus create any orthogonal projection.
Unfortunately, photographs are central projections. The target point (the point in the center of the image) and the distance (from the lens center to the target point) are of importance for the image. ◂◂◂

▸▸▸ **Application**: ***rotation of the firmament in the course of a day***
In the course of one day, the Earth will remain roughly in the same position on its orbit around the Sun (actually it rotates about $1^\circ$ about the axis of the orbital plane). Relatively speaking, the Sun rotates (and of course, the positions of the "fixed stars" in the firmament) about the Earth's axis ("around the North Star"). The projection of this axis into the horizontal base plane $\pi$ points towards North and the axis itself encloses the angle $\varphi$ with $\pi$, where $\varphi$ equals the latitude of the current position (the lead encloses the complimentary angle $90^\circ - \varphi$ with the Earth's axis).

⊕ *Remark*: Currently, the Earth's axis points towards a star which we call the Pole Star (North Star). Over the course of approximately 25,700 years, the axis of the Earth rotates about a "gyro axis" perpendicular to the orbital plane (see for example:

`http://astro.wsu.edu/worthey/astro/html/lec-precession.html`). So, when the ancient Egyptians aligned their pyramids precisely with the north around 4,500 BC), our present North Star was certainly not seen as a star whose position in the night sky remains fixed. ⊕ ◂◂◂

▸▸▸ **Application**: ***Why does the sky rotate in the reverse direction on the southern hemisphere?***

This question is not trivial at all: The answer "Because on the southern hemisphere we stand kind of upside-down" is wrong!

*Solution*:

The correct answer is the following: On the northern hemisphere, we have to look towards the North Star to perceive the Earth's rotation. On the southern hemisphere, we must look in the opposite direction of the Earth's axis (i.e. towards the southern celestial pole). A rotation in space cannot be oriented if the viewing direction is not given.

⊕ *Remark*: If we just make a headstand, this does *not* change the orientation. It will only result in a twist of one and the same image by about 180°.
The Sun's rotation seems to have the opposite orientation than that of the other stars. This is so, because if we watch the sun, we turn off the corresponding celestial pole. On the northern hemisphere, we look southwards. Thus, the orientation changes, and the Sun seems to rotate clockwise. ⊕ ◂◂◂

## Rotations about arbitrary axes

Rotations with axes that do not run through the origin can be accomplished analogously to the two-dimensional case as follows: If $C$ is an arbitrary point on the axis of rotation, we first move the axis of rotation and the point to be rotated about $-\vec{c}$. Then, we rotate about the displaced axis, and translate everything back (translation vector $\vec{c}$).

▸▸▸ **Application**: ***rotoidal motion*** (Fig. 7.10)

A rotoidal motion is the composition of two proportional rotations about perpendicular skew axes. Such motions occur as relative motions at *worm gears* that transmit a rotation from one axis to another skew perpendicular axis.

A point is rotated about the $z$-axis (through an angle $\alpha$) and it simultaneously *rotates* about the axis $b$ which is also undergoing the first rotation. The rotation about $b$ is proportional to that about $a$, and thus, the second rotation angle is $\beta = n\,\alpha$ (Fig. 7.10a). The same result is achieved if we *first* rotate about the axis $(0/-r/0)$ parallel to the $x$-axis through the angle $\beta$ and, *after that*, about the $z$-axis through the angle $\alpha$. A point $P$ (position vector $\vec{p}$) is transformed to the position

$$\vec{p_1} = \mathbf{R}_z(\alpha)\cdot\left[\mathbf{R}_x(\beta)\cdot(\vec{p}-\begin{pmatrix}0\\-r\\0\end{pmatrix})+\begin{pmatrix}0\\-r\\0\end{pmatrix}\right].$$

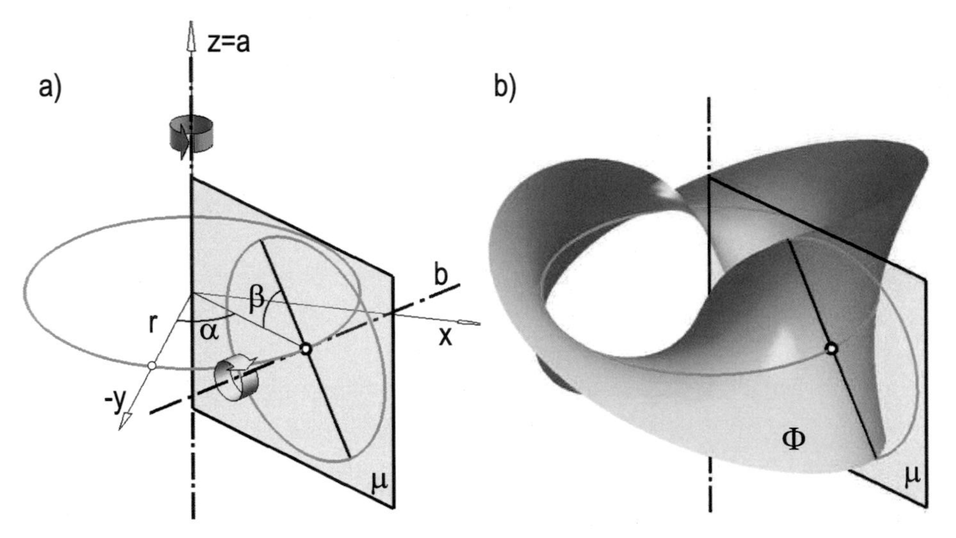


**Fig. 7.10** rotoidal motion

Fig. 7.10b shows which surface arises when, for example, all points lying on a straight line that intersects $b$ perpendicularly and meets the line $a$ are transformed. ◂◂◂

▸▸▸ **Application**: ***Everything revolves around the parabolic flight.***

**Fig. 7.11** Everything revolves around the parabolic flight.

In Fig. 7.11, different, and sometimes complicated, trajectories are depicted. Nevertheless, they have one thing in common: The paths of additional non-moving body parts – for example, in an ideal somersault – will be planar curves generated by rotations about horizontal axes through the center of gravity moving along the bomb trajectory, provided that the rotation is uniform. ◂◂◂

## Helical motion

If a point $P(p_x/p_y/p_z)$ is rotated about an axis through the angle $\varphi$ and simultaneously translated along this axis *proportionally* to the translation, then one speaks of a *helical motion*. The ratio **c** between the translation distance and the angle $\varphi$ of rotation is the *parameter* of the helical motion. If the helical axis coincides with the $z$-axis, then the new point $P_1$ has the position vector

$$\vec{p_1} = \mathbf{R}_z(\varphi) \cdot \vec{p} + \varphi \begin{pmatrix} 0 \\ 0 \\ \mathbf{c} \end{pmatrix} = \begin{pmatrix} \cos\varphi\, p_x - \sin\varphi\, p_y \\ \sin\varphi\, p_x + \sin\varphi\, p_y \\ p_z + \mathbf{c}\varphi \end{pmatrix}. \tag{7.11}$$

For fixed coordinates of $P$, the equation (7.11) already describes the trajectory of $P$ in the helical motion around the $z$-axis by parameter **c**. One speaks of the *parametric representation* of the helix. In the following section, we will develop a number of these parametric equations.

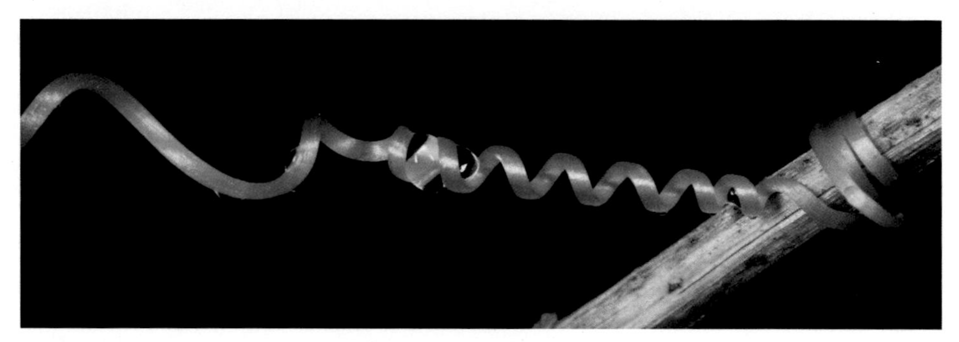

**Fig. 7.12** twining plant

▸▸▸ **Application**: ***twining*** (Fig. 7.12)
Creeper, climber, and twiner are names for plants that try to spread across surfaces like twigs. They combine a translation and a rotation in order to find new carrier objects and thus perform helical motions. Sometimes they obviously switch the orientation of the rotation (in the left part of the image). ◂◂◂

▸▸▸ **Application**: ***simple and effective damping*** (Fig. 7.15)
If you want to fasten your boat in the harbour, it is a good idea to have something between the rope and the fixed attachment that protects the rope. The depicted helical spring does the job efficiently. When the boat is jerkily teared off by waves, the two hooks compress the spring and, mathematically speaking, diminish the parameter of the helix. This allows for more distance between the boat and the harbor. The spring forces then "harmonically" tear the boat back. ◂◂◂

▸▸▸ **Application**: ***highway exit*** (Fig. 7.16)
A 6 m wide constantly inclined uphill road (appearing circular from the top view) needs to overcome a height difference of 12 m. A 270° rotation will take

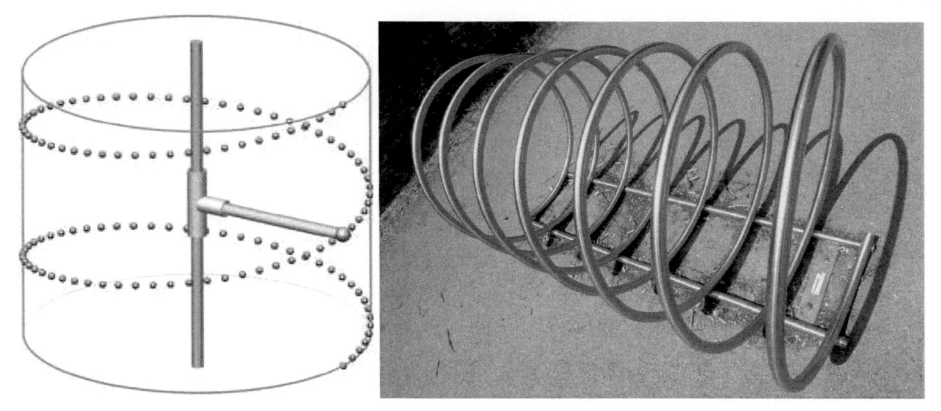

**Fig. 7.13** helical motion **Fig. 7.14** designer bike stand

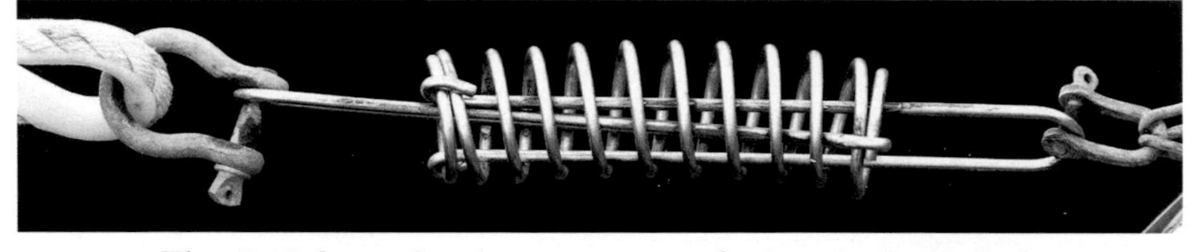

**Fig. 7.15** damped spring as a mount of a boat in the harbour

place, and along the center line, the slope equals 10%. How big is the inner radius $r_m$ and the slope along the inner border line?

**Fig. 7.16** motorway or equivalent ...

The top view of the center line is 120 m long, because of the 10% slope. A circular arc with radius $r_m$ and central angle of 270° has arc length $270\frac{\pi}{180}r_m = 4.712\,r_m$, and thus, $r_m = 120\,\mathrm{m}/4.712 \approx 25.5$ m. The inner side of the road has radius $r = r_m - 3\,\mathrm{m} = 22.5\,\mathrm{m}$. The arc length of its top view is $\approx 106.03\,\mathrm{m}$. Thus, the slope is $12\,\mathrm{m}/106.03\,\mathrm{m} \approx 11.32\%$.
We find the helical parameter

$$\mathbf{c} = \frac{12\,\mathrm{m}}{\frac{3\pi}{2}} = 2.546\,\mathrm{m}.$$

In the coordinate system shown in the figure, the helices with radius $r_0$ have the parametric representation

$$\vec{x}(\varphi) = \mathbf{R}_z(\varphi) \cdot \begin{pmatrix} 0 \\ -r_0 \\ 0 \end{pmatrix} + \begin{pmatrix} r_m \\ r_m \\ \mathbf{c}\,\varphi \end{pmatrix} \quad \left(0 \le \varphi \le \frac{3\pi}{2}\right).$$

This equation was used to generate Fig. 7.16. Furthermore, surfaces of constant slope are shown along the constantly inclined part of the street. These surfaces are *helical developables.* ◂◂◂

**Fig. 7.17** "elliptical" helical staircase (top/bottom view) and "ordinary" staircase

▸▸▸ **Application**: ***helical staircase*** (Fig. 7.17)
Helical staircases are usually formed by applying a helical motion to a "prototype" of a step. The smoothly polished bottom is called a "helicoid". The banister is a helix. In Fig. 7.17, the stairs have been subjected to a "compression" so that the top views of the helices are ellipses and not circles. Write the equation of the upper curve assuming that the banister's height is 1 m. The base is an ellipse with a semi-major axis of 2 m and a semi-minor axis of 1 m. One full turn of the helix covers a height difference of 4 m.
*Solution*:
Initially, we do not worry about the compression. The curve is a helix with radius $r_0 = 2$ and parameter $\mathbf{c} = \dfrac{4}{2\pi}$. Let the cylinder's axis be the $z$-axis and the first point of the banister have the coordinates $P(r_0/0/1)$. Then, the equation of the helix is

$$\vec{p_1} = \begin{pmatrix} r_0 \cos\varphi \\ r_0 \sin\varphi \\ 1 + \mathbf{c}\,\varphi \end{pmatrix}.$$

The compression is achieved by multiplying the $y$-coordinate by $1/2$. ◂◂◂

## The main theorem of spatial kinematics

The following theorem is of fundamental importance for the theory of spatial motions:

Any two positions of a rigid body can always be transformed into each other by a helical motion.

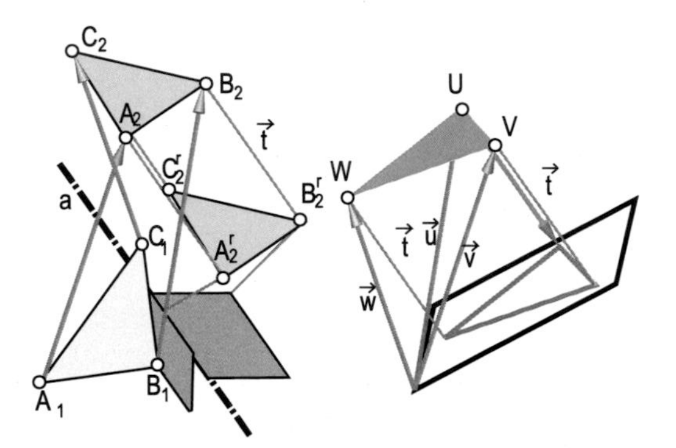


**Fig. 7.18** the main theorem of spatial kinematics

***Proof***: The two congruent bodies can be defined by the two associated (and thus, also congruent) triangles $\Delta_1 = A_1B_1C_1$ and $\Delta_2 = A_2B_2C_2$ (Fig. 7.18). Now, we assume that there is a helical displacement that transforms $\Delta_1$ into $\Delta_2$ and show that the axis and the parameter are uniquely determined.
A helical motion is a composition of a rotation and a proportional translation along the rotation axis $a$. Therefore, the vectors

$$\vec{u} = \overrightarrow{A_1A_2}, \quad \vec{v} = \overrightarrow{B_1B_2}, \quad \text{and } \vec{w} = \overrightarrow{C_1C_2}$$

can be decomposed into the translational component $\vec{t}$ parallel to $a$ and the rotational component normal to $a$. For all three vectors, $\vec{t}$ is the same, so that $\vec{u} - \vec{t}$, $\vec{v} - \vec{t}$, and $\vec{w} - \vec{t}$ are normal to $a$.
We interpret $\vec{u}$, $\vec{v}$, and $\vec{w}$ as position vectors of the three points $U$, $V$, and $W$ that form a plane $\varepsilon \perp a$ at the distance $s = |\vec{t}|$ from the origin, with $s$ being the length of the translation distance. The normal vector $\vec{a}$ of $\varepsilon$ is the direction vector of the helical axis, parallel to $\vec{t}$. We obtain it via

$$\vec{a} = \overrightarrow{UV} \times \overrightarrow{UW} = (\vec{v} - \vec{u}) \times (\vec{w} - \vec{u}).$$

We normalize the vector

$$\vec{a}_0 = \frac{1}{|\vec{a}|}\vec{a}$$

and obtain the equation of $\varepsilon$ as

$$\varepsilon: \ \vec{a}_0 \bullet x = \vec{a}_0 \bullet \vec{u} = c.$$

The normal distance $s$ from the coordinate origin to $\varepsilon$ is equal to the constant $s = c$ taken from the plane's equation in Formula (5.38). The translation vector is now $\vec{t} = s\vec{a}_0$. Now, we apply the translation $-\vec{t}$ to the triangle $\Delta_2$ and get the new triangle $\Delta_2^r$ which is joined with $\Delta_1$ by a rotation about an axis. This axis equals the desired helical axis and can be found as the intersection of the bisector planes of $A_1A_2^r$ and $B_1B_2^r$.

⊕ *Remark*: In any smooth spatial motion, two "neighboring positions" are congruent. The corresponding helical motion is then called the *instantaneous helical motion*, and the axis is called the *instantaneous helical axis.* ⊕ ⊙

▸▸▸ **Application**: ***determination of the helical motion***

Find the helical motion which moves the triangle $A_1(-3/0/0)$, $B_1(0/0/0)$, $C_1(0/2/0)$ into the congruent triangle $A_2(1/0/2)$, $B_2(1/-3/2)$, $C_2(3/-3/2)$.

*Solution*:

With the notations of the above proof, we have

$$\vec{u} = \overrightarrow{A_1A_2} = \begin{pmatrix}4\\0\\2\end{pmatrix}, \quad \vec{v} = \overrightarrow{B_1B_2} = \begin{pmatrix}1\\-3\\2\end{pmatrix}, \quad \vec{w} = \overrightarrow{C_1C_2} = \begin{pmatrix}3\\-5\\2\end{pmatrix}$$

and

$$\vec{a} = (\vec{v} - \vec{u}) \times (\vec{w} - \vec{u}) = \begin{pmatrix}-3\\-3\\0\end{pmatrix} \times \begin{pmatrix}-1\\-5\\0\end{pmatrix} = \begin{pmatrix}0\\0\\12\end{pmatrix}.$$

The vector is easy to normalize:

$$\vec{a_0} = \begin{pmatrix}0\\0\\1\end{pmatrix}.$$

The helical axis is parallel to the $z$-axis. The plane $\varepsilon$ has the equation

$$\begin{pmatrix}0\\0\\1\end{pmatrix} \bullet \vec{x} = \begin{pmatrix}0\\0\\1\end{pmatrix} \bullet \begin{pmatrix}4\\0\\2\end{pmatrix} = 2,$$

so that with $s = 2$, we get:

$$\vec{t} = s \cdot \vec{a}_0 = \begin{pmatrix}0\\0\\2\end{pmatrix}.$$

We subtract $\vec{t}$ from $A_2$, $B_2$, $C_2$ and obtain

$$A_2^r(1/0/0), \ B_2^r(1/-3/0), \ C_2^r(3/-3/0).$$

According to Formula (5.27), the planes of symmetry of the segments $A_1A_2^r$ and $B_1B_2^r$ are

$$\sigma_1: \begin{pmatrix}4\\0\\0\end{pmatrix} \bullet \vec{x} = \frac{1}{2}\left[\begin{pmatrix}1\\0\\0\end{pmatrix}^2 - \begin{pmatrix}-3\\0\\0\end{pmatrix}^2\right] = -4,$$

$$\sigma_2: \begin{pmatrix}1\\-3\\0\end{pmatrix} \bullet \vec{x} = \frac{1}{2}\left[\begin{pmatrix}1\\-3\\0\end{pmatrix}^2 - \begin{pmatrix}0\\0\\0\end{pmatrix}^2\right] = 5.$$

The intersection $a$ of these two vertical planes is the helical axis. It is parallel to the $z$-axis and passes through the point $(-1/-2/0)$. ◂◂◂

## 7.2 Matrix calculations and some applications

In the following, we will calculate with matrices. We need them for some applications, such as coordinate transformations, but also for tasks from business economics.

*An $(n,m)$-matrix* $\mathbf{A}$ *is nothing more than a rectangular array of numbers with n rows and m columns.* The *transposed matrix* $\mathbf{A}^T$ is an $(m,n)$-matrix produced by interchanging rows and columns.

Vectors are also matrices in this sense. An ordinary three-dimensional vector can be considered as a $(3,1)$-matrix, or, transposed to a row vector, as a $(1,3)$ matrix. Our rotation matrices from above are $(3,3)$-matrices if we calculate in three-dimensional space or $(2,2)$-matrices if we calculate in the plane.

Each row or each column of a matrix is called a *row vector* or a *column vector.* Matrices can be used – as we have covertly seen with the rotation matrices – to write confusing situations (with many indices) in a very clear way. In business economics, they are used to manage large amounts of interdependent data. The great advantage of writing in this way is that now you can perform tasks by means of so-called matrix multiplication that would otherwise almost reach the breaking point of being barely readable.

The definition of the *multiplication of two matrices* is as follows:

Let $\mathbf{A}$ be an $(m,n)$-matrix, and $\mathbf{B}$ an $(n,k)$-matrix. $\mathbf{A}$ must have as many columns as $\mathbf{B}$ has rows. We define the result $\mathbf{A}\bullet\mathbf{B}$ of the multiplication as an $(m,k)$-matrix $\mathbf{C}$. Its elements $c_{ij}$ are the respective dot product of the $i$-th row vector of $\mathbf{A}$ with the $j$-th column vector of $\mathbf{B}$.

Described in detail, the definition looks like this:

$$\mathbf{A} = \begin{pmatrix} a_{11} & a_{12} & \cdots & a_{1n} \\ a_{21} & a_{22} & \cdots & a_{2n} \\ \vdots & \vdots & \cdots & \vdots \\ a_{m1} & a_{m2} & \cdots & a_{mn} \end{pmatrix}, \quad \mathbf{B} = \begin{pmatrix} b_{11} & b_{12} & \cdots & b_{1k} \\ b_{21} & b_{22} & \cdots & b_{2k} \\ \vdots & \vdots & \cdots & \vdots \\ b_{n1} & b_{n2} & \cdots & b_{nk} \end{pmatrix}$$

$$\Rightarrow \mathbf{C} = \mathbf{A}\bullet\mathbf{B} = \begin{pmatrix} c_{11} & \cdots & \cdots \\ \vdots & \vdots & \vdots \\ \cdots & c_{ij} & \cdots \\ \vdots & \vdots & \vdots \\ \cdots & \cdots & c_{nk} \end{pmatrix} \text{ with } c_{ij} = a_{i1}b_{1j} + a_{i2}b_{2j} + \ldots + a_{in}b_{nj}. \quad (7.12)$$

▸▸▸ **Application**: ***multiplication of vector and matrix***

Multiply the vector $\overrightarrow{v} = \begin{pmatrix} 4 \\ -5 \end{pmatrix}$ once from “the left” and once from “the right” by the matrix $\mathbf{A} = \begin{pmatrix} 1 & -2 \\ 3 & -1 \end{pmatrix}$ by considering it once as a row vector and then as a column vector.

*Solution*:

$$\vec{v}^T \cdot \mathbf{A} = (4 \; -5) \cdot \begin{pmatrix} 1 & -2 \\ 3 & -1 \end{pmatrix} = (4 \cdot 1 + (-5) \cdot 3 \quad 4 \cdot (-2) + (-5) \cdot (-1)) = \begin{pmatrix} -11 \\ -3 \end{pmatrix}^T,$$

$$\mathbf{A} \cdot \vec{v} = \begin{pmatrix} 1 & -2 \\ 3 & -1 \end{pmatrix} \cdot \begin{pmatrix} 4 \\ -5 \end{pmatrix} = \begin{pmatrix} 1 \cdot 4 + (-2) \cdot (-5) \\ 3 \cdot 4 + (-1) \cdot (-5) \end{pmatrix} = \begin{pmatrix} 14 \\ 17 \end{pmatrix}.$$

⊕ *Remark*: The example shows that the order of multiplication matters. A row vector will yield another row vector, and a column vector will result in another column vector. ⊕ ◂◂◂

## Change of coordinate systems

We can consider a coordinate system as a *spatial tripod*. It consists of mutually orthogonal unit vectors $\vec{e}$, $\vec{f}$, $\vec{g}$, and the position vector $\vec{t}$ towards the origin $T$. For the "basis vectors", we have

$$\vec{e}^2 = \vec{f}^2 = \vec{g}^2 = 1, \; \vec{e} \bullet \vec{f} = \vec{e} \bullet \vec{g} = \vec{f} \bullet \vec{g} = 0. \tag{7.13}$$

We denote the two matrices that can be defined by the triple $\vec{e}$, $\vec{f}$, $\vec{g}$ by $\mathbf{K}$ and $\mathbf{K}^T$, where

$$\mathbf{K} = \begin{pmatrix} e_x & e_y & e_z \\ f_x & f_y & f_z \\ g_x & g_y & g_z \end{pmatrix}, \quad \mathbf{K}^T = \begin{pmatrix} e_x & f_x & g_x \\ e_y & f_y & g_y \\ e_z & f_z & g_z \end{pmatrix}. \tag{7.14}$$

Via Formula (7.13), it can be seen that the product of $\mathbf{K}$ and $\mathbf{K}^T$ yields the *identity matrix* $\mathbf{E}$. The entries of the product matrix are defined as the dot products of the corresponding row and column vectors. One may even swap matrices (this is a little harder to show), so that finally, the following holds:

$$\mathbf{K} \cdot \mathbf{K}^T = \mathbf{K}^T \cdot \mathbf{K} = \mathbf{E} = \begin{pmatrix} 1 & 0 & 0 \\ 0 & 1 & 0 \\ 0 & 0 & 1 \end{pmatrix}. \tag{7.15}$$

⊕ *Remark*: Matrices with this property are called *orthogonal matrices*. ⊕

Now, we move a *spatial tripod* from one position to another, taking a point $P$ which is firmly connected to the system (Fig. 7.19) with it. Based on the current position of the tripod, the point has constant coordinates or a constant position vector $\vec{p}$. What position vector does the point $P^*$ have relative to the original coordinate system? Conversely, a point in the original system has the position vector $\vec{p}^{\,*}$, what is its position vector in the new system?

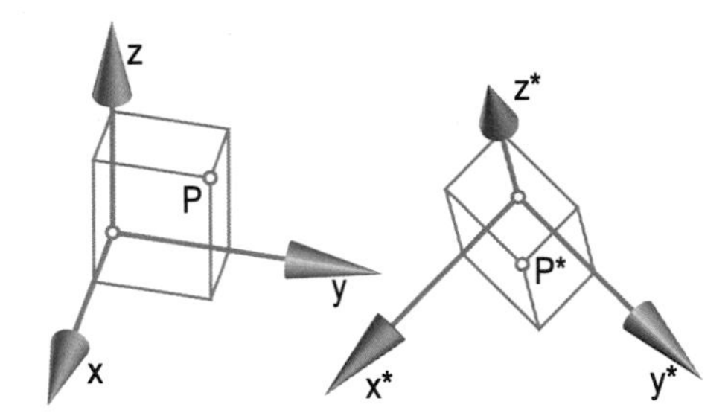


**Fig. 7.19** motion of the coordinate systems

Linear algebra provides the clear formulas

$$\boxed{\overrightarrow{p^*} = \overrightarrow{t^*} + \mathbf{K}^T \cdot \overrightarrow{p}, \quad \overrightarrow{p} = \mathbf{K} \cdot (\overrightarrow{p^*} - \overrightarrow{t^*}).} \tag{7.16}$$

***Proof***: Initially, we have: $\overrightarrow{e} \bullet \overrightarrow{p} = p_x$, $\overrightarrow{f} \bullet \overrightarrow{p} = p_y$, $\overrightarrow{g} \bullet \overrightarrow{p} = p_z$, because if $\varepsilon$ is the angle between $\overrightarrow{e}$ and $\overrightarrow{p}$, then

$$\cos\varepsilon = \frac{\overrightarrow{e} \bullet \overrightarrow{p}}{|\overrightarrow{e}| \cdot |\overrightarrow{p}|} = \frac{\overrightarrow{e} \bullet \overrightarrow{p}}{|\overrightarrow{p}|} \Rightarrow \overrightarrow{e} \bullet \overrightarrow{p} = |\overrightarrow{p}| \cdot \cos\varepsilon = p_x, \text{ etc.}$$

Furthermore, the following holds:

$$\overrightarrow{p} = \overrightarrow{p^*} - \overrightarrow{t^*}.$$

Therefore, according to Formula (7.12), we get,

$$\mathbf{K} \cdot (\overrightarrow{p}^* - \overrightarrow{t}^*) = \begin{pmatrix} e_x & e_y & e_z \\ f_x & f_y & f_z \\ g_x & g_y & g_z \end{pmatrix} \cdot \begin{pmatrix} p_x \\ p_y \\ p_z \end{pmatrix} = \begin{pmatrix} \overrightarrow{e} \bullet \overrightarrow{p} \\ \overrightarrow{f} \bullet \overrightarrow{p} \\ \overrightarrow{g} \bullet \overrightarrow{p} \end{pmatrix} = \begin{pmatrix} p_x \\ p_y \\ p_z \end{pmatrix} = \overrightarrow{p}.$$

This proves the right-hand side of the formula. Now, we multiply the right equation on the left by $\mathbf{K}^T$ and find

$$\mathbf{K}^T \cdot \mathbf{K} \cdot (\overrightarrow{p}^* - \overrightarrow{t}^*) = \mathbf{K}^T \cdot \overrightarrow{p}.$$

Further, we have $\mathbf{K}^T \cdot \mathbf{K} = \mathbf{E}$, cf. Formula (7.15). “Nothing happens” if we multiply a vector by $\mathbf{E}$. After addition of $\vec{t}^*$ to both sides, we face the left-hand side of the equation in Formula (7.16). ⊙

## 7.3 Parametrization of curves

### Parametric representation of a circle in $\mathbb{R}^2$

A circle "in principal position" (center $M(0/0)$, radius $r$) can immediately be parametrized. It arises by rotating the point $(r/0)$ about the origin (the rotation angle $u$ is given in radians for the purpose of calculating with a computer):

$$\vec{x}_0 = \begin{pmatrix} r\cos u \\ r\sin u \end{pmatrix} \quad (0 \le u \le 2\pi). \tag{7.17}$$

Now, the midpoint (center of rotation) shall be the point $M(m_x/m_y)$. Then, we have

$$\vec{x} = \vec{x}_0 + \vec{m} = \begin{pmatrix} r\cos u + m_x \\ r\sin u + m_y \end{pmatrix} \quad (0 \le u \le 2\pi). \tag{7.18}$$

### Intersection of a circle with a straight line

Now, let $\vec{n}\cdot\vec{x} = c$ be the implicit equation of a line $g$. The intersection with an arbitrary circle then satisfies the condition

$$\vec{n}\cdot\begin{pmatrix} r\cos u + m_x \\ r\sin u + m_y \end{pmatrix} = c.$$

This leads to an equation of the type $P\cos u + Q\sin u + R = 0$:

$$\underbrace{n_x r}_{P}\ \cos u + \underbrace{n_y r}_{Q}\ \sin u + \underbrace{n_x m_x + n_y m_y - c}_{R} = 0,$$

thus, a quadratic equation in $u$. Real solutions are present if $\overline{gM} \le r$.

▸▸▸ **Application**: ***satellite in the Earth's shadow, lunar eclipse*** (Fig. 7.20)
A satellite orbits the Earth at an altitude of 500 km (by the way, this enforces an orbital period of about 1.5 hours). How long is the satellite in the Earth's shadow?

*Solution*:
Essentially, the problem is two-dimensional. That is, we can consider the circular satellite orbit in the $xy$-plane:

$$\vec{x} = \begin{pmatrix} r\cos u \\ r\sin u \end{pmatrix} \quad \text{with} \quad r = R + 500\text{km},\ R = 6{,}370\,\text{km}.$$

The streak of light to the Earth's contour is the line of intersection

$$\begin{pmatrix} 0 \\ 1 \end{pmatrix}\cdot\vec{x} = \begin{pmatrix} 0 \\ 1 \end{pmatrix}\cdot\begin{pmatrix} 0 \\ R \end{pmatrix} = R.$$

**Fig. 7.20** satellite

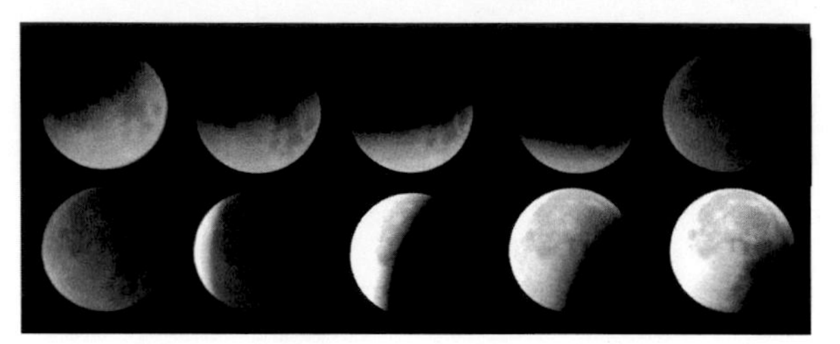

**Fig. 7.21** the Moon in the Earth's shadow ($3\frac{1}{2}$ h)

Now, we intersect the straight line and the circle:

$$\begin{pmatrix}0\\1\end{pmatrix} \cdot \begin{pmatrix}r\cos u\\ r\sin u\end{pmatrix} = R \Rightarrow \sin u = \frac{R}{R+500} = 0.9272 \Rightarrow u = 68°.$$

The second solution $u = 180° - 68°$ is not relevant. The satellite is in the Earth's shadow for $-68° < u < 68°$, thus, for $\frac{2\cdot 68}{360}\,1.5\,\mathrm{h} \approx 34$ min.
In practice, this time is even shorter: The streak of light is refracted in the Earth's atmosphere towards the lead direction, where the lower-energy red portion of the light is further refracted. This phenomenon occurs during the lunar eclipse, where, instead of the satellite, the Moon disappears in the Earth's shadow. The refracted red light *always* hits the Moon because of its great distance. That is why the Moon becomes "blood red", but not totally dark (Fig. 7.21). ◀◀◀

▸▸▸ **Application**: ***special ellipse motion*** (Fig. 7.22)
Let us assume that we are given two perpendicularly intersecting lines on which the two points $A$ and $B$ move. The distance $s = \overline{AB}$ is constant. Then, each point $C$ on the line $AB$ traces an ellipse. Prove this. How can the traced curve be determined pointwise?

***Proof***: If one chooses $A$ on the first line $a$, then one obtains $B$ on the second line $b$ as the intersection with a circle around $A$ of radius $s$.
Let $M = a \cap b$. Let us consider the rectangle $MAPB$ and its circumcircle $q$ (midpoint $R$). The diagonals $MP$ and $AB$ enclose the same angle $\alpha$ with the axis $a$. Let us now consider the points $K$ and $N$ on $MP$ and $b$ (or on the line parallel to $a$ through $C$). The distance $\overline{MK}$ equals $\overline{AC}$, and therefore, it is constant. Thus, $k$ moves along a circle around $M$. Furthermore, $\overline{NK}$ is always proportional to $\overline{NC}$, because it is constant since $t = \overline{RC}$ = constant, and thus,

$$\frac{\overline{NC}}{\overline{NK}} = \frac{(s/2+t)\cos\alpha}{(s/2-t)\cos\alpha} = \frac{s/2+t}{s/2-t}.$$

So, the locus of $C$ is obtained by stretching $K$'s circular orbit and is, hence, an ellipse. ⊙

Since $\overline{MP} = \overline{AB} = s$, the circle $q$ has the constant radius $s/2$. For each position of $AB$, the point $P$ lies on the circle $p$ centered at $M$ with the double radius

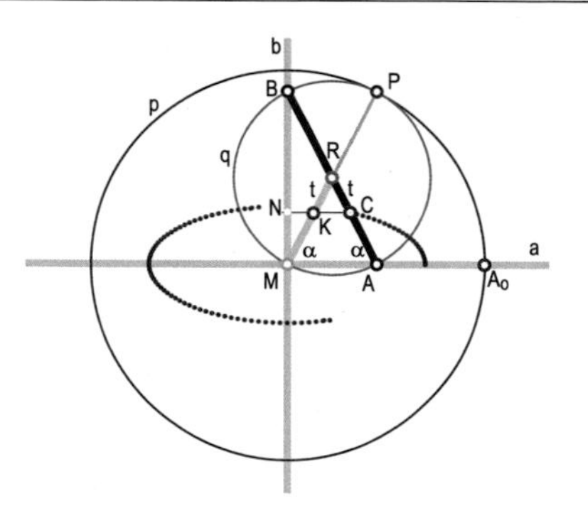


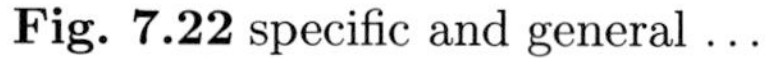

**Fig. 7.22** specific and general ...

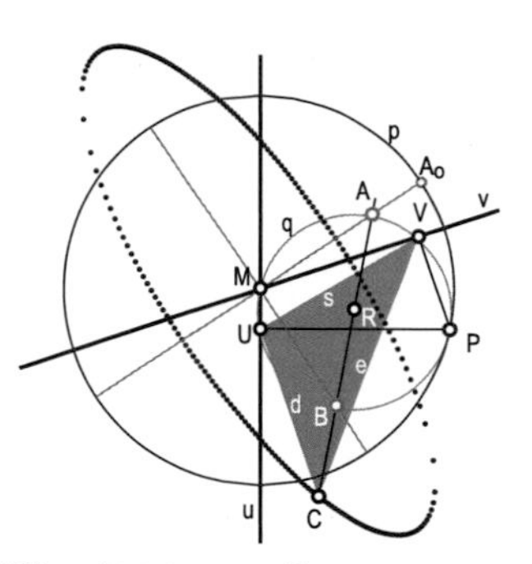


**Fig. 7.23** ... ellipse motion

$r$. Since the point $P$ is the intersection of the path normals of $A$ and $B$, it is the instantaneous pole of the motion (theorem on path normals p. 300). The circle $q$ can be viewed as the isoptic circle of the segment $AP$ (see more Fig. 4.52). According to the inscribed angle theorem, $\angle ARP = 2\angle AMP$. Therefore, the circular arc $A_0P$ (measured on $p$) is of the same length as the arc $AP$ (measured on $q$) for any position of $P$. So, we can say: The circle $q$ rolls without gliding on the circle $p$ that is twice the size. Therefore, an ellipse motion can also be generated by rolling circles (ratio of radii 2 : 1). These are special types of hypotrochoids (Application p. 300). ◂◂◂

## Intersecting two circles in the plane

The intersection of two circles can be traced back to the intersection of a circle with a straight line (radical line): The straight line joining the points of intersection is orthogonal to the connection of the two centers.

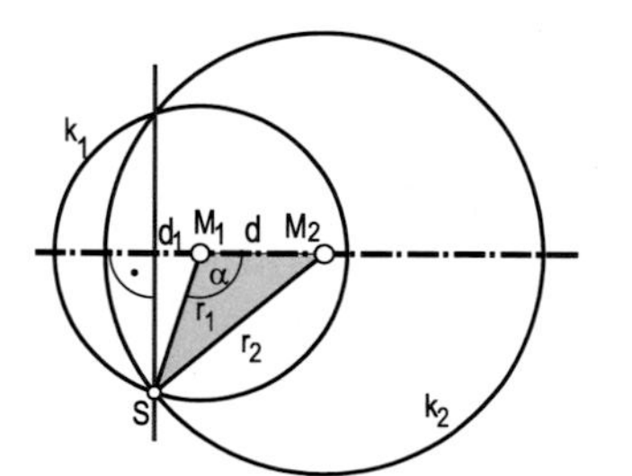


**Fig. 7.24** intersection of two circles

Its distance $d_1$ from the center of the first circle results from the *Law of Cosines* (with the notations from the adjacent figure):

$$r_2^2 = d^2 + r_1^2 - 2d\underbrace{r_1 \cos\alpha}_{d_1}$$

$$\Rightarrow d_1 = \frac{d^2 + r_1^2 - r_2^2}{2d}.$$

▸▸▸ **Application**: ***general ellipse motion*** (Fig. 7.23)

Let there be two intersecting lines $u$ and $v$ which form an arbitrary angle. On these lines, two points $U$ and $V$ move. The distance $s = \overline{UV}$ must be kept constant. Then, each point $C$ rigidly connected to the line $UV$ traces an ellipse (with $\overline{UC} = d$ and $\overline{VC} = e$). How do we find the point $C$ if we are given the position of $U$?

*Solution*:
The circle about $U$ with radius $s$ meets $v$ in two possible positions of $V$. The circles about $U$ of radius $d$ and the circle about $V$ with radius $e$ deliver the point $C$ (two solutions).
The intersection of $M = u \cap v$ is the midpoint of the orbit. The circle $q$ through $U$, $V$, and $M$ (center $R$) is called the pitch circle. It can be interpreted as the isoptic circle of the segment $UV$, because the angle $\angle UMV$ is constant (see Fig. 4.52). Thus, $q$ is independent of the choice of $U$. According to Thales's theorem, the point $P$ opposite to $M$ on the pitch circle is the intersection of the path normals of $U$ and $V$. Thus, it is the instantaneous pole (theorem of path normals on p. 300), and the radius of the circle $p$ about $M$ through $P$ is also constant. The ratio of the radii of $p$ and $q$ equals $2:1$, and $q$ always touches $p$ from the inside.
Now consider the path of the point $A$ on the pitch circle opposite to $C$. According to the theorem of inscribed angles, the central angle $\angle PRA$ is twice the contact angle $\angle PMA$. Thus, the length of the arc $PA_0$ on $p$ is equal to the length of the arc $PA$ on $q$. Since this is true for any position of $C$, there exists a real rolling of $q$ in $p$. Therefore, the two arcs are always of the same length, and there is a real rolling of circles in the ellipse motion previously described. Here, $A$ travels obviously on the line $MA_0$. The same applies to the point $B$ opposite to $A$ on the pitch circle. The two orbital lines of $A$ and $B$ are perpendicular. The segment $AB$ is of constant length. Thus, one can also achieve the same motion when $A$ moves on the fixed line $MA$ and $B$ on the fixed perpendicular line $MB$. For further information on the proof, see the previous example. ◂◂◂

**▸▸▸ Application**: ***four bar mechanisms***
The following mechanisms are used very often in technical applications: First a rod $LA$ rotates about its fixed bearing point $L$, and a second rod $MB$ rotates about its fixed bearing point $M$. The motion of the rods is coupled by a fixed orbit length $AB$. Each additional point $C$ rigidly attached to the rod $AB$ traces a so-called *coupler curve*.

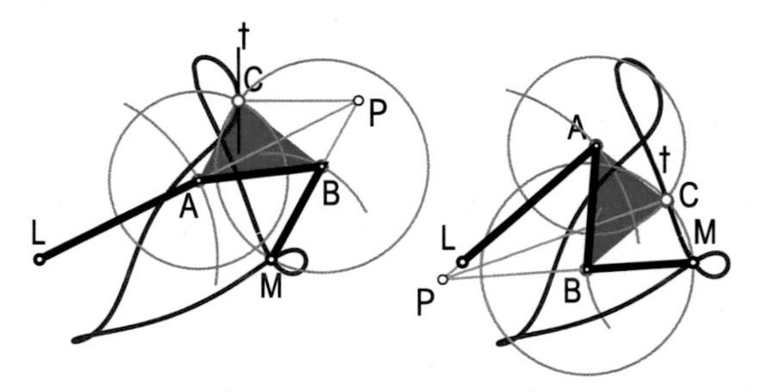


**Fig. 7.25** a typical coupler curve (two snapshots)

How is the position of $C$ calculated provided that the first rod $LA$ is given? Determine the path tangent and instantaneous velocity of $C$.

*Solution*:
Everything leads to the intersection of two circles. First, $B$ is determined through the intersection of the circle around $A$ with radius $\overline{AB}$ with the circle centered at $M$ with radius $\overline{MB}$ (two solutions). Then, $C$ lies in the intersection of the circle about $A$ with radius $\overline{AC}$ with the circle about $B$ with radius $\overline{BC}$. This time, because of the sense of revolution of $ABC$, only one point of intersection is permitted.
The path tangent of $C$ can easily be determined: We know the orbital circles of $A$ and $B$. The associated path normals are the lines of the rods $LA$ and $MB$. They intersect each other according to the theorem on path normals (p. 300) at the instantaneous pole $P$. The path tangent at $C$ is, then, perpendicular to the polar radius $PC$. The distance $\overline{PC}$ to the instantaneous pole (center of rotation) is also proportional to the instantaneous velocity of $C$. ◂◂◂

▸▸▸ **Application**: ***crane clearing the cargo ship***
In what direction does the cable hook of the crane located on the pier (Fig. 7.26) move?

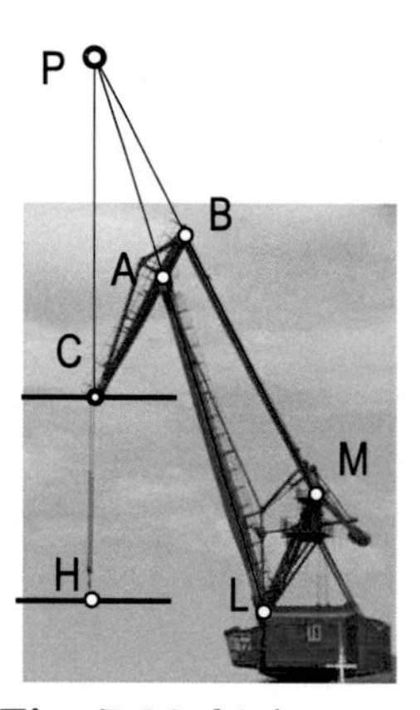


**Fig. 7.26** ship's crane

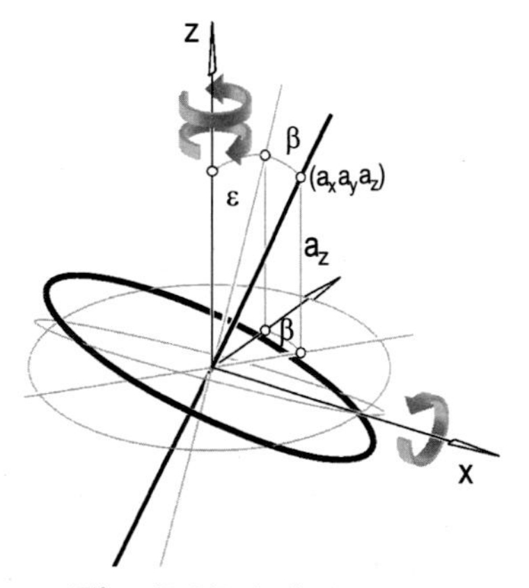


**Fig. 7.27** circle in space

*Solution*:
One recognizes the quadrangle $LABM$ with the fixed points $L$ and $M$ about which the points $A$ and $B$ rotate. This results in the instantaneous pole $P$ and the path tangent at the point $C$ which are – as expected – horizontal. The hook $H$ is always right below $C$ and also has a horizontal tangent provided that the pitch is constant. ◂◂◂

## Parameter representation of a circle in $\mathbb{R}^3$

In three-dimensional space, the situation is generally much more complicated. First, we consider the special case in which the center equals the origin and the circles' axis is the $z$-axis:

$$\vec{x}_0 = \begin{pmatrix} r \cos u \\ r \sin u \\ 0 \end{pmatrix} \quad (0 \le u \le 2\pi). \tag{7.19}$$

Now, we want to create a circle by rotation of a point about an arbitrary axis.
We let

$$\vec{a} = \begin{pmatrix} a_x \\ a_y \\ a_z \end{pmatrix} \quad \text{with} \quad \sqrt{a_x^2 + a_y^2 + a_z^2} = 1$$

be the *normalized* direction vector of the rotation axis. The axis may contain the origin.
The following considerations shall be supported by a look at Fig. 7.27: We find the direction of the axis if we rotate the $z$-axis through the angle $-\varepsilon$ about the $x$-axis and then rotate through the angle $-\beta$ about the $z$-axis. The following relations are valid:

$$\cos\varepsilon = a_z, \quad \sin\varepsilon = \sqrt{a_x^2 + a_y^2} = s, \quad \text{or} \quad \cos\beta = \frac{a_x}{s}, \quad \sin\beta = \frac{a_y}{s}. \tag{7.20}$$

The rotation of the points on the circle (7.19) about the $x$-axis (rotation angle -$\varepsilon$) yields

$$\vec{x}_1 = \mathbf{R}_\mathrm{x}(-\varepsilon) \cdot \vec{x}_0 = \begin{pmatrix} r \cos u \\ r\, a_z \sin u \\ r\, s \sin u \end{pmatrix}.$$

The subsequent rotation about the $z$-axis finally yields

$$\vec{x}_2 = \mathbf{R}_\mathrm{z}(-\beta) \cdot \vec{x}_1 = \begin{pmatrix} \frac{r}{s}(a_x \cos u + a_y\, a_z \sin u) \\ \frac{r}{s}(a_y \cos u - a_x a_z \sin u) \\ r\, s \sin u \end{pmatrix}.$$

Now, we can even prescribe an arbitrary center $M(m_x/m_y/m_z)$ and find the general parametric representation of a circle in three-dimensional space, which, though not simple, is often quite useful:

$$\vec{x} = \begin{pmatrix} m_x + \frac{r}{s}(a_x \cos u + a_y\, a_z \sin u) \\ m_y + \frac{r}{s}(a_y \cos u - a_x\, a_z \sin u) \\ m_z + r\, s \sin u \end{pmatrix} \quad (s = \sqrt{a_x^2 + a_y^2},\ 0 \le u < 2\pi). \tag{7.21}$$

▸▸▸ **Application**: ***boundary of the shadow on the globe***
The day-night boundary on the Earth is a great circle in a plane perpendicular to the direction of the light ray. On which points does the Sun rise above or below the latitude $\varphi$?

**Fig. 7.28** and yet it moves ...

*Solution*:

The (normalized) vector of the light ray is the direction vector $\vec{a}$ of the axis of the Earth's circular terminator. The radius $R$ of the terminator equals that of the Earth.
We obtain the desired points on the given latitude by the intersection of the shadow's boundary with the plane $z = R\sin\varphi$. Consequently, we have

$$Rs\,\sin u = R\sin\varphi \Rightarrow u = \arcsin\frac{\sin\varphi}{s}.$$

For $\sin\varphi > s$, there is no real solution: On the corresponding circle of latitude, there is either a 24 h day or a 24 h night. ◂◂◂

**Fig. 7.29** The sphere is not developable.

▸▸▸ **Application**: ***The sphere is not developable.***
At this point, we shall have a brief excursion to cartography. As is known, it is not possible to flatten the surface of a sphere into a plane without distortion. Fig. 7.29 illustrates that the shell of a reasonably spherical mandarin still remains curved although it is "torn" in several places (mandarins are easy to peel, because they have a poorly developed inner pericarp layer).
There are now countless methods that somehow allow us to transform the spherical surface into the plane. None of them can preserve lengths, areas, and angles simultaneously. The simplest method is to identify the longitude

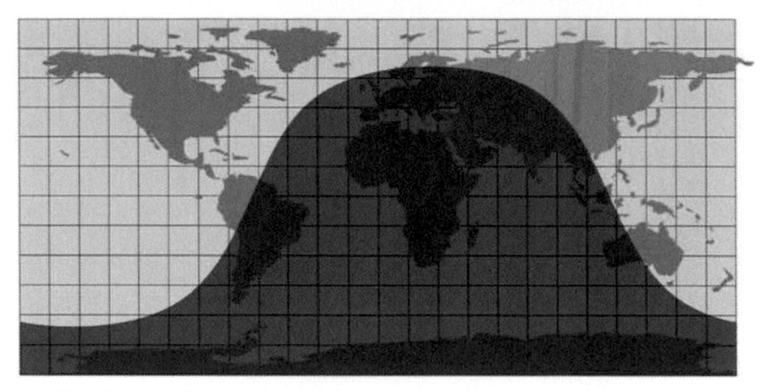

**Fig. 7.30** the shadow boundary in a rectangular map

and latitude of a point with Cartesian coordinates. In this coordinate frame, the sphere is mapped to a rectangle. Fig. 7.30 shows what the shadow boundary from Fig. 7.28 looks like in a rectangular map. The intensity of this deformation of the spherical surface can best be seen close to the North and South Pole. The shadow curve is obviously not a sine curve, as one might assume (also see Fig. 3.38). The Earth's rotation corresponds to a left shift in the coordinate grid. ◂◂◂

▸▸▸ **Application**: ***shortest flight route*** (Fig. 7.91)
The shortest route from $P$ to $Q$ is a great circle, i.e. a circle with the center of the Earth $M$ as its center. The circles' axis is the normal of the plane $PQM$ and the circle has the radius $R = 6{,}370\,\text{km}$ plus the altitude (aircraft $10\,\text{km}$, satellite between $150\,\text{km}$ and $36{,}000\,\text{km}$). ◂◂◂

## Intersection of a circle with a plane

Let $\vec{n}\cdot\vec{x} = c$ be the equation of a plane $\varepsilon$, and let $k$ be an arbitrary circle with radius $r$. We insert $\vec{x}$ from the equation (7.21) into the equation of the plane, and thus, we obtain one more equation of the type

$$P \sin u + Q \cos u + R = 0$$

(see p. 143). It has two real solutions if the intersection line of $\varepsilon$ with the circle's plane has a distance $\leq r$ from the axis of the circle.

## Intersection of two circles in space

In general, two circles $c_1$ and $c_2$ (centers $M_i$, radii $r_i$) in space have no point of intersection. If they lie on one sphere (center $K$, radius $\varrho$), then they may have common points. In such a case, the axes must intersect at the point $K$ (center of the sphere) on the one hand, and on the other hand, the following has to hold for both circles:

$$\overline{KM_i}^2 + r_i^2 = \varrho^2.$$

▸▸▸ **Application**: **Monge*'s "sphere technique"*** (Fig. 7.31)
In order to determine the intersection curve of two surfaces of revolution with

intersecting axes, we choose an auxiliary sphere around the intersection of the axes with variable radius. Any such sphere meets the surfaces of revolution along one or more circles. Any two circles on either surface have potential intersections belonging to the intersection curve. The radii of the auxiliary spheres may not be too small or too large. Otherwise, the desired intersection circles may not be real. ◂◂◂

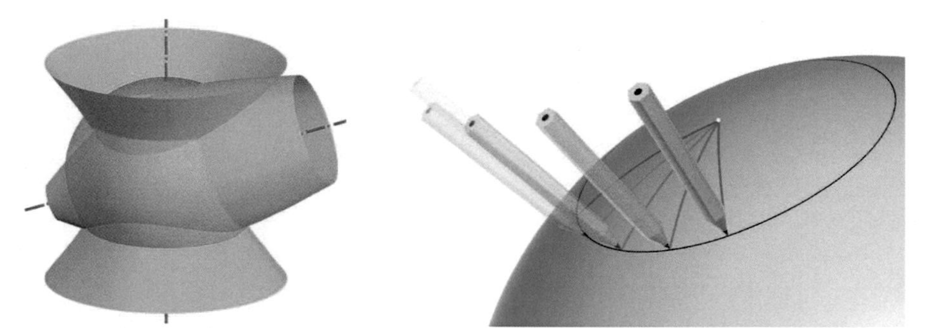

**Fig. 7.31** *Monge*'s technique **Fig. 7.32** distance circle on a sphere

▸▸▸ **Application**: ***distance circles on the sphere (spherical trigonometry)***
Compute those points on 15° E that are 2,000 km away from London (on the prime meridian, $\varphi = 51.5°$ N). Does the Sun rise at these points an hour earlier?

*Solution*:
A distance of 2,000 km (measured along a great circle on the sphere) corresponds to a central angle of 18° (radius $R = 6{,}370\,\mathrm{km}$, circumference $40{,}000\,\mathrm{km}$). The radius $r$ of the distance circle is, then,

$$r = R \sin 72° = 1{,}968\,\mathrm{km}.$$

The distance of the center of the circle from the origin is

$$a = \sqrt{R^2 - r^2}.$$

The plane carrying the distance circle has the equation

$$\begin{pmatrix} 0 \\ -\cos\varphi \\ \sin\varphi \end{pmatrix} \bullet \vec{x} = \begin{pmatrix} 0 \\ -\cos\varphi \\ \sin\varphi \end{pmatrix} \bullet \begin{pmatrix} 0 \\ -a\cos\varphi \\ a\sin\varphi \end{pmatrix}.$$

Therefore,

$$-\cos\varphi\, y + \sin\varphi\, z = a(\cos^2\varphi + \sin^2\varphi) = a.$$

The carrier plane with a longitude of 15° ($\lambda = 15°$) has the equation

$$-\cos\lambda\, x + \sin\lambda\, y = 0.$$

For the intersection of the two planes, we immediately find a parametric representation

$$x = \tan\lambda t,\ y = t,\ z = (a + \cos\varphi t)/\sin\varphi.$$

For us, the latitudes of the two intersections are of interest. One equals 37.5° N (southern coast of Sicily). The other one equals 68° N (above the Arctic Circle on the Norwegian Arctic coast). The two points are about 3,500 km apart from each other. Berlin is almost exactly halfway between the two points. On the same latitude as London, the Sun will rise on 15° E an hour earlier throughout the year. In Sicily, the lengths of the days (shorter in summer, longer in winter) are more balanced during the year than in London. The Sun rises more than an hour earlier in winter and in less than an hour earlier in summer. The opposite is true for northern Scandinavia. The aforementioned point on 68° N even lies above the Arctic Circle, and on top of that, it has a polar night in the second half of December! This could also be formulated as follows: Both sunrise and sunset take place at 12h noon. ◂◂◂

▸▸▸ **Application**: *distance circles on curved surfaces*

Surprisingly, a two-dimensional being on a surface can find out whether and to what extent "its carrier surface" is curved without leaving the surface and moving to three-dimensional space: It only needs to draw a circle of radius $r$ and measure its circumference (Fig. 7.33). If this circumference equals $2\pi \cdot r$, the being lives on a parabolic surface (specifically, in a plane). If the circumference is larger, then it lives on a hyperbolic surface. On an elliptic surfaces – like the globe – the circumference is smaller than expected.

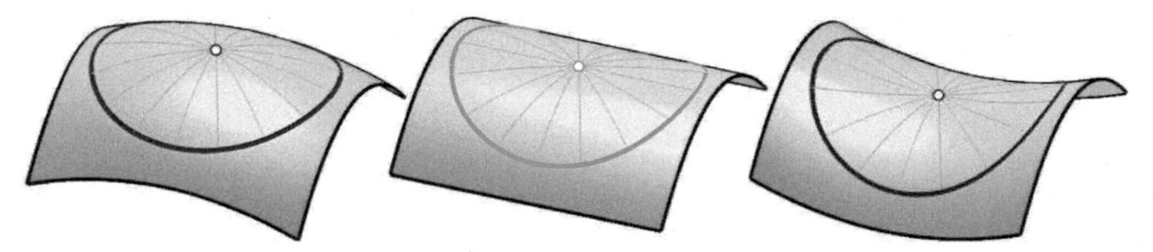

**Fig. 7.33** distance circles on a variety of curved surfaces

For example, if we draw a circle around the North Pole of the globe with a radius of $r = 10,000$ km (which equals one quarter of the circumference of the Earth), all points on the circle will lie on the equator. Its circumference is 40,000 km. However, we compute $2\pi \cdot r \approx 62,800$ km. Our carrier surface is, therefore, strongly elliptically curved. ◂◂◂

## Different representations of curves

A curve in a plane can be represented in four ways:

1. *Explicit representation*
   If the curve has no tangents parallel to the $y$-axis, it can be interpreted as a function graph and the $y$-value can be *explicitly* given:

$$y = f(x), \quad x_1 \le x \le x_2.$$

   In vector notation, a point on the curve has the coordinates $(x/f(x))$. The associated tangent is parallel to the direction vector $(1/f'(x))$. Typical examples are the basic functions discussed in the previous chapter $y = x^n$, $y = \sin x$, $y = \cos x$, $y = \tan x$, and $y = e^x$ (all these functions have no restriction on $x$), as well as their inverse functions (with restrictions on certain intervals).

2. *Parameter representation*
   If the curve also has tangents parallel to the $y$-axis, one may use the parameter representation

$$\vec{x} = \vec{x}(u) = \begin{pmatrix} x(u) \\ y(u) \end{pmatrix}, \quad u_1 \le u \le u_2.$$

   The tangent that corresponds to the curve point is determined by the direction vector $(\dot{x}(u)/\dot{y}(u))$. A classical example is the circle

$$\vec{x} = \begin{pmatrix} \cos u \\ \sin u \end{pmatrix}, \quad 0 \le u \le 2\pi.$$

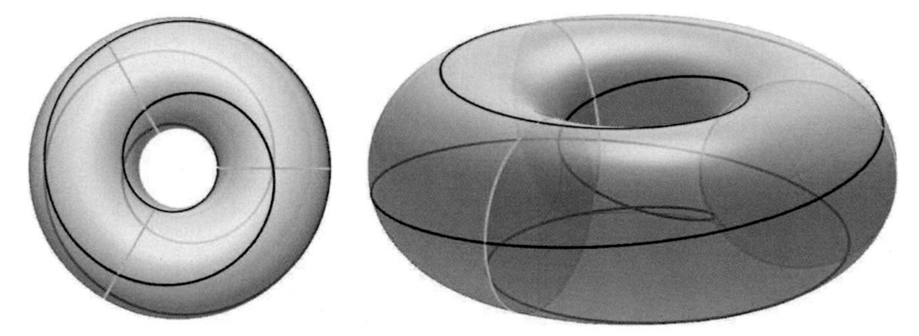

**Fig. 7.34** a space curve using parametric representation

In "the computer age", this description of curves is most widely used, because it is the most general. It also works well for curves in space, where we just need to get a third equation for $z$. In order to illustrate this, we look at the equation of a rotoid (Fig. 7.34, left: top view, right: general view) that winds along a torus (see also Fig. 7.71):

$$\vec{x} = \begin{pmatrix} (a - b \cos n\,u) \cos u \\ (a - b \cos n\,u) \sin u \\ b \sin(n\,u) \end{pmatrix} \quad (0 \le u \le 6\pi, \; n = -\frac{1}{3}).$$

3. *Representation in polar coordinates* (Fig. 7.45)
Sometimes, it is convenient to use so-called *polar coordinates.* Here, the distance $r$ from the origin is a function of the polar angle $\varphi$:

$$r = r(\varphi), \quad \varphi_1 \leq \varphi \leq \varphi_2.$$

The tangent that corresponds to a generic point on the curve and the polar ray subtend the angle

$$\psi = \arctan \frac{r(\varphi)}{\dot{r}(\varphi)} \tag{7.22}$$

(proof in Application p. 327). Classic examples of the description of curves by means of polar coordinates are *logarithmic spirals* (Fig. 7.37)

$$r = r_0 \, e^{k\varphi} \quad (k = \text{constant}, \ \varphi \in \mathbb{R}) \tag{7.23}$$

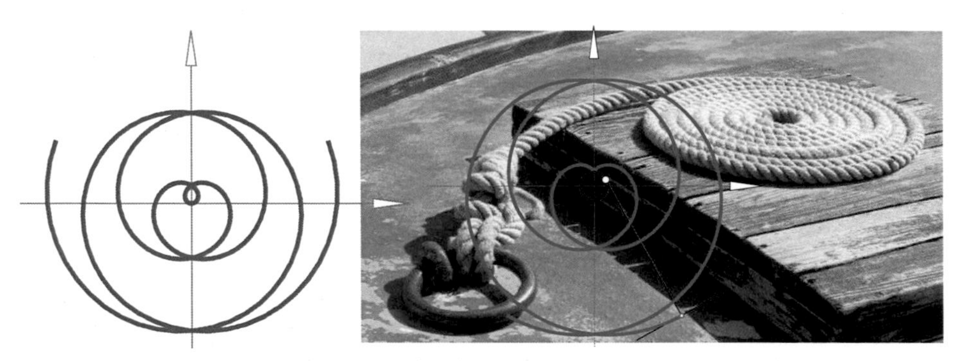

**Fig. 7.35** An Archimedean spiral (left) and a practical approach (right). Strictly speaking, the center line of the cable is called an *involute*, which arises when one "uncoils" a thread on a cylinder of revolution (shown in red).

and – in comparison – the *Archimedean spiral* (Fig. 7.35, left)

$$r = k\varphi \quad (k = \text{constant}, \ \varphi \in \mathbb{R}). \tag{7.24}$$

For the latter, the radius increases by the constant value $2\pi k$ "per revolution", which is nicely shown on the right in Fig. 7.35.

Polar coordinates can be easily converted into Cartesian coordinates and vice versa (p. 331) so that the representation of the curve is equivalent to a parametric representation in polar coordinates.

4. *Implicit representation*
We have occasionally seen this type of curve representation. There is no coordinate explicitly shown in this representation. The curve's tangent is obtained by *implicit differentiation*, but will not be discussed in detail. As an example, we use once again the circle (radius $r$, center $M(m_x/m_y)$)

$$(x - m_x)^2 + (y - m_y)^2 = r^2$$

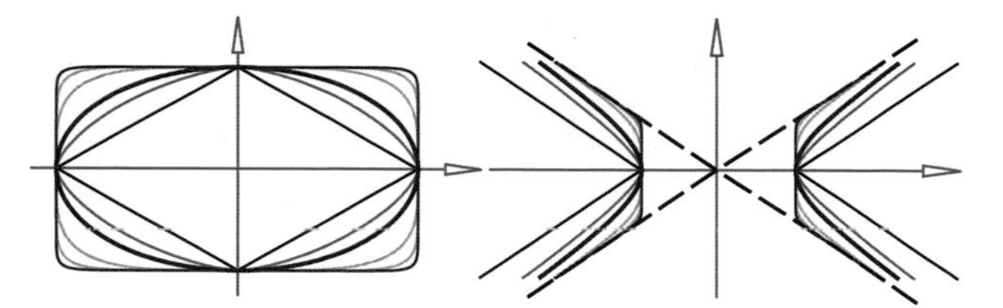

**Fig. 7.36** superellipses with respect to superhyperbolas

or – as a generalization – the so-called superellipses

$$(x/a)^n + (y/b)^n = 1.$$

Among them we find, for $n = 2$, an ordinary ellipse with axis lengths $2a$ and $2b$. On the left-hand side of Fig. 7.36, such curves are displayed for $n = 1, 1.5, 2, 4, 8, 16$. If we modify the equation to $(x/a)^n - (y/b)^n = 1$, we obtain "superhyperbolas" (Fig. 7.36, right).

▸▸▸ **Application**: ***course angle of a logarithmic spiral***

Prove Formula (7.22) and show the following: *The "course angle" of the logarithmic spiral is constant* (Fig. 7.37).

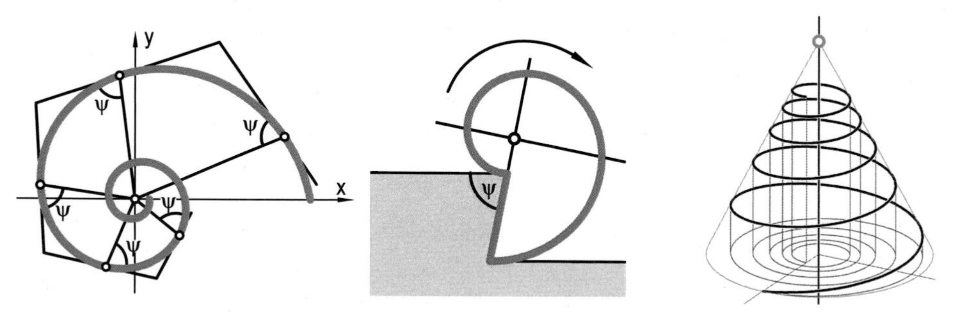


**Fig. 7.37** constant course angle of a logarithmic spiral (right: in space)

***Proof***: We show Formula (7.22) using Fig. 7.38, middle: For $d\varphi \to 0$, we obviously have

$$\tan\psi = \frac{r\,d\varphi}{dr} = \frac{r}{\dfrac{dr}{d\varphi}} = \frac{r(\varphi)}{\dot r(\varphi)}.$$

⊙

For the logarithmic spiral given by Formula (7.23), the following holds

$$\dot r(\varphi) = k\,r_0\,e^{k\,\varphi} = k\,r(\varphi) \Rightarrow \psi = \arctan\frac{1}{k} = \text{constant}.$$

⊕ *Remark*: During the day, butterflies navigate with the Sun. At night, they use the Moon for orientation. In both cases, they point one of their *ommatidia* (conical segment of the compound eye) towards a light source "at infinity" (Fig. 7.39, left). The light rays are parallel to each other so that butterflies "only need to flutter" in

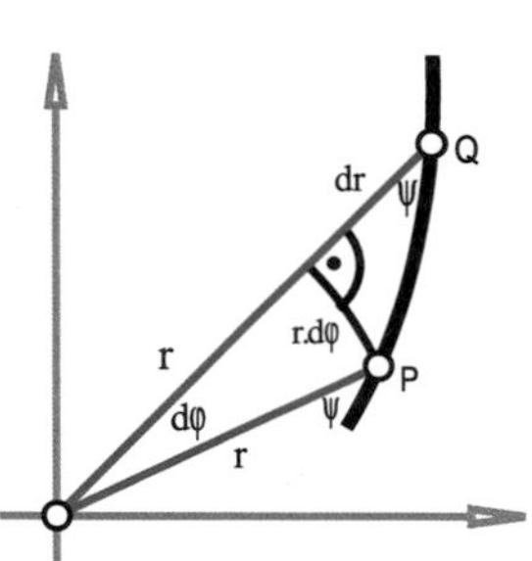

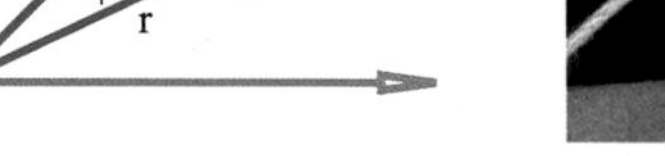
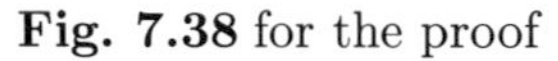

**Fig. 7.38** for the proof

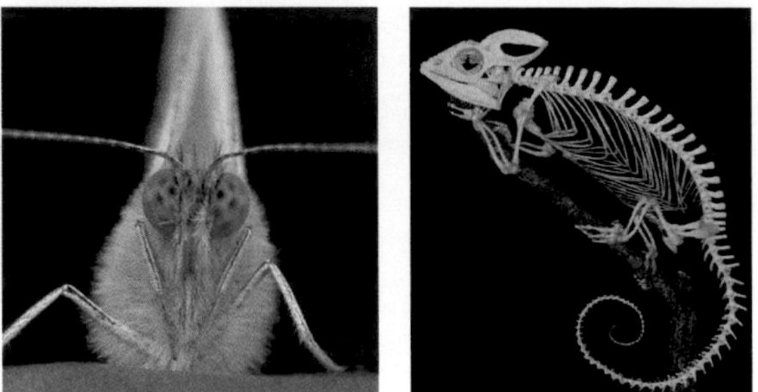

**Fig. 7.39** a butterfly and a chameleon

order to fly straight.
If moths meet an artificial light source (perhaps even fire), then – as they attempt to pass straight – they fly towards the light source on a trajectory that approximates a logarithmic spiral due to the constant course angle.
It is also due to the constant course angle that the chameleon rolls its tail (Fig. 7.39, right). The vortex lengths behave in a reasonably "exponential" manner. A snake, on the other hand, winds itself into an Archimedean spiral, because its (reasonably constant) body diameter is added up with every turn (see Fig. 7.35). Viewed in this manner, a record groove is also an Archimedean spiral. ⊕ ◂◂◂

**Fig. 7.40** Different spirals: an Archimedian spiral (left), a logarithmic spiral (middle), and several logarithmic spirals that produce a visual effect when the image is rotated (right).

▸▸▸ **Application**: ***How to distinguish spirals?*** (Fig. 7.40)
Spirals are fascinating for humans. Therefore, they often appear in daily life. Although there are many different kinds of spirals, the two "classics" appear more frequently. How can we see the difference quickly?

*Solution*:
The following criterion allows us to distinguish between the Archemedian spiral and the logarithmic spiral: The former has constant distance between subsequent windings. In the latter, the distance between subsequent windings grows exponentially.

⊕ *Remark*: If we rotate the circular disc in Fig. 7.40 (right), we will immediately have the impression that we are either "kicked out of" or "sucked into" the image, depending on the way we rotate (clockwise or counter-clockwise). ⊕ ◀◀◀

▸▸▸ **Application**: ***snail shell*** (Fig. 7.41, Fig. 7.42)
Why are the contours (seen from the the top view) of snail shells (mussel shells, nautilus shells) logarithmic spirals?

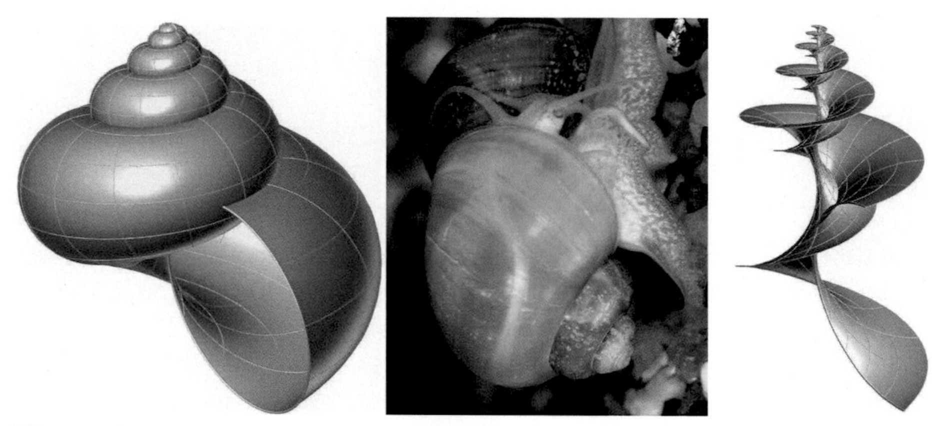

**Fig. 7.41** middle: fresh water snails mating; left and right: digital model. (The interior of a snail shell can be quite complicated.)

**Fig. 7.42** clam and a digitalized model

*Solution*:
Obviously, snails have the following "growth strategy":

- The animal grows "around an axis" (apparently to remain compact).

- The magnification $dr$ of the distance to the axis is directly proportional to the previous maximum distance $r$ (proportionality factor $k$).

- The increase $d\varphi$ of the rotation angle is proportional to the increase in distance $dr$.

**Fig. 7.43** 200 million years old ammonites: spinning to the left or to the right?

"Viewed from above", i.e. in the direction of the spiral axis (Fig. 7.37, right), we get the contour of the snail shell and thus obtain the relation

$$dr = k\, r\, d\varphi.$$

We have just noted (in Application p. 326) the following property of the logarithmic spiral:

$$\frac{r}{\dot{r}} = \frac{1}{k} \Rightarrow \frac{\dot{r}}{r} = k \Rightarrow \frac{dr}{r} = k\, d\varphi.$$

If the growth follows the mentioned principles, intermediate irregularities can occur without changing the overall geometric shape. If the snail finds more food during the season, it will grow faster. Otherwise, it can take a break in between growth phases, without this becoming visible at a later stage.

⊕ *Remark*: We have just solved a "differential equation" by means of a comparison The example could, therefore, be given in the next chapter, where we provide general solution strategies for simple differential equations. ⊕

⊕ *Remark*: Snail shells or mussel shells that "spread to three dimensions" can be either left- or right-handed. Virtually, almost all snail shells wind in the clockwise direction (mathematically negative) when viewed from above. This is not the case with the "snail king". It "winds to the right". For the ammonites in Fig. 7.43 – at least at first glance – there is no above and below. Therefore, you cannot say in which direction the shell spirals. ⊕ ◂◂◂

### ▸▸▸ Application: *Stay the course!*

What happens if you travel long distances by plane or boat with a constant course angle (see also Application p. 547)?

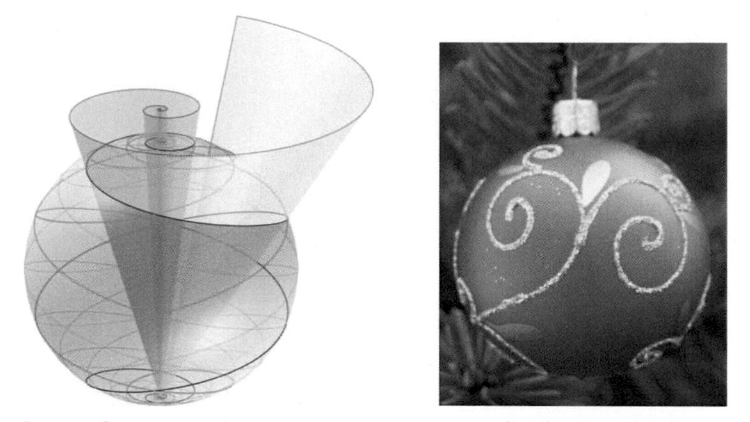

**Fig. 7.44** odyssey on the sphere: with compass (left) and without (right)

*Solution*:
One does not travel on a great circle (which would be the shortest route from start to finish, Application p. 361), but starts – like a moth cruising around a light source – around one of the poles in spiral form. The exact solution is obtained by *stereographic projection* from the South Pole to the tangent plane at the North Pole (Fig. 7.44, left). This projection is conformal, as is shown, for example, in *Geometry and its Applications in Art, Nature and Technology*. The meridians of the sphere are mapped to straight rays through the North Pole in the tangent plane. The constant rate angle produces a logarithmic spiral. Projected back onto the sphere, one obtains a *loxodrome* on the sphere.

⊕ *Remark*: Exceptions: If you drive precisely towards the north or south, you move on a meridian. If you drive east or west, then you move on a circle of latitude. In polar regions, the latter means a pretty big detour. You are probably familiar with the following riddle: A bear marches 10 km to the south, 10 km to the west, and finally 10 km to the north. At the end of its journey, the bear realizes that it has arrived at the starting point. What color is the bear? For the solution, see Fig. 2.16. ⊕ ◂◂◂

## Converting Cartesian coordinates into polar coordinates and back

1. Let the point $P(x/y)$ (Fig. 7.45) be given. Then, its distance from the origin is

$$r = \sqrt{x^2 + y^2}$$

and the course angle is computed from the coordinates, with their signs taken into account, as

$$\tan\varphi = \frac{y}{x} \Rightarrow \varphi = \begin{cases} \operatorname{sign} x \cdot \arctan\frac{y}{x}, \text{if } x \neq 0, \\ \operatorname{sign} y \cdot \frac{\pi}{2}, \text{ if } x = 0. \end{cases}$$

2. Let the point $P$ be given in polar coordinates $P(r;\varphi)$. Then, its Cartesian coordinates are
$$x = r\cos\varphi, \quad y = r\sin\varphi.$$

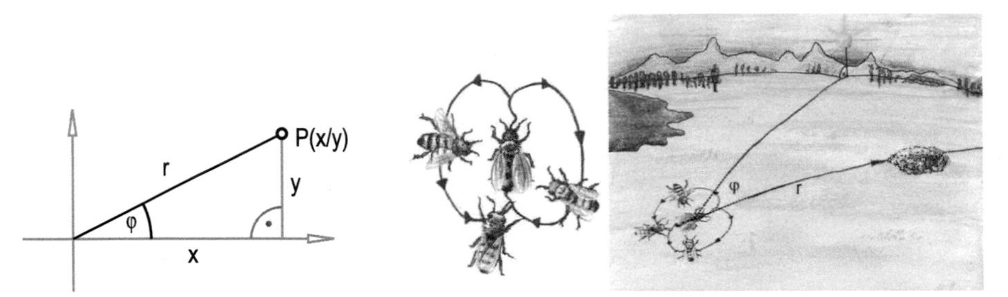

**Fig. 7.45** polar coordinates

**Fig. 7.46** bees and polar coordinates

⊕ *Remark*: *Bees and polar coordinates*: Through their performance, dancing bees communicate the location of a food source to their sisters (Fig. 7.46). The central axis of their dance shows the direction (the polar angle $\varphi$ to the base point of the Sun) and the frequency of the waggle motion along these lines serves as a measure for the distance $r$ of the food source. ⊕

## Conic Sections

In addition to straight lines and circles, the most important curves are conics. These are curves of degree 2, i.e. they have a maximum of two points of intersection with a generic straight line. If the intersection with a tangent line is counted twice, and if you further admit "complex" solutions (Section B), a conic *always has exactly two points of intersections with a straight line.*

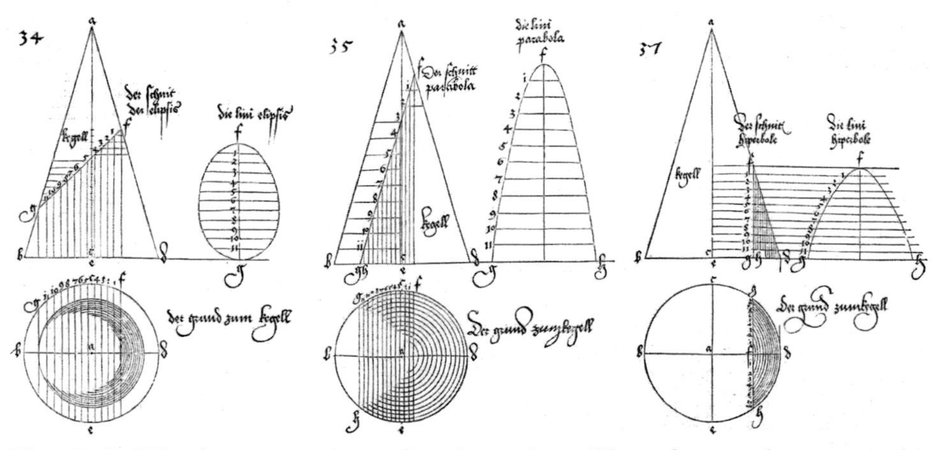

**Fig. 7.47** *Dürer*'s constructions of conic sections. These famous drawings in his book *Underweysung der Messung, mit dem Zirckel und Richtscheyt, in Linien, Ebenen unnd gantzen corporen*, from 1525 lack a little bit of symmetry.

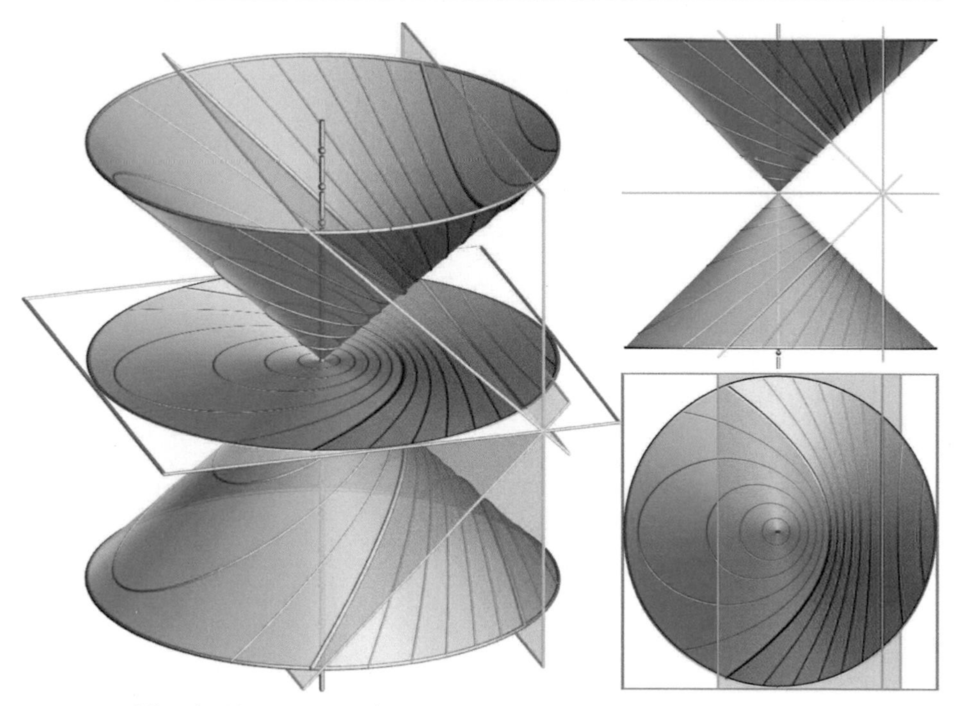

**Fig. 7.48** conics as planar intersections of cones of revolution

The name *conic* already gives a hint: It is the planar intersection curve of a cone of revolution. These curves include the *ellipse* (the circle is a special case of it), the *parabola*, and the *hyperbola*. The decisive factor is the location of the intersecting plane. Let us imagine the cone of revolution with a vertical axis so that all generators (and also all tangent planes) form the same angle of inclination with the horizontal base plane $\beta$ (Fig. 7.48). The cone is "infinitely long" in both directions. Then, all planes $\sigma$ that are less inclined than the tangent planes intersect along ellipses, and all steeper planes intersect along hyperbolas. In the limit, i.e. between ellipses and hyperbolas, we will find parabolas. Since antiquity, conic sections have been of great interest in mathematics and related disciplines. Therefore, one could give a whole lecture on them. Here, we will cite only some important geometrical properties, especially those which are common to all conics.

- Ellipses and hyperbolas have two focal points, whereas parabolas have only one. These focal points are the points of contact with those two spheres that are inscribed into the cone and touch the plane $\sigma$. The latter spheres are called the *Dandelin* spheres after their discoverer (*G.P. Dandelin*, 1794–1847). The *Dandelin* spheres enable us to show that, for the points on an ellipse/hyperbola, the sum/difference (absolute value) of the distances to the focal points is constant. We will take a closer look at the special case of the parabola because of its great technical importance (Application p. 342).

- The *parallel projection* of a conic to a fixed plane is again a conic of the same affine type. That is, an ellipse is mapped to an ellipse, a hyperbola is mapped to a hyperbola, a parabola is mapped to a parabola with their known focal properties. Accordingly, the "top view" – the view from above or below – is again a conic of the same type. The top view of the cone's tip is one of the two foci.

  For the sake of simplicity, let us think about the cone whose generators/tangent planes are inclined under 45° (as shown in Fig. 7.48). We are only interested in the top view. The generators $e$ have the parametric representation $\vec{x} = r\,(\cos\varphi, \sin\varphi, -1)^T$. Here, $r$ is the radial distance from the vertical axis, and $\varphi$ is the rotation angle. The plane of intersection with the inclination angle $\beta$ through the point $(0/0/-s)$ has the equation $\sigma:\ \varepsilon\,x - z = s$ (with $\varepsilon = \tan\beta$). The intersection $e \cap \sigma$ results in the relation

$$r = \frac{s}{1 + \varepsilon\,\cos\varphi}, \tag{7.25}$$

  which is a representation of the intersection curve in *polar coordinates*. The characteristic value $\varepsilon = \tan\beta$ is called the *numerical eccentricity*. For $\varepsilon = 0$, there is obviously a circle of radius $r_0$. For $0 < |\varepsilon| < 1$, the denominator will never be 0. So, we are dealing with an ellipse. For $|\varepsilon| = 1$, we get a parabola. For $|\varepsilon| > 1$, we get a hyperbola.

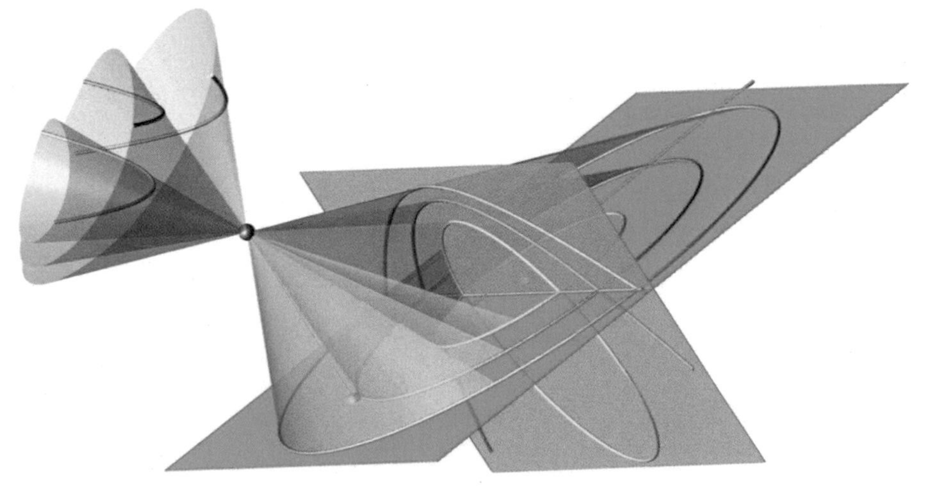

**Fig. 7.49** Any kind of conic can be the perspective image of a circle.

- The *central projection* (projection from a fixed point to a fixed plane) of a conic may change the type, but the result is always a conic. For example, an ellipse (the special case of a circle) can be mapped to a hyperbola. This occurs in photography over and over again, because there is a central projection from the optical center of the lens onto a photosensitive film plane.

### ▸▸▸ Application: *vertices and foci of conics*

Find the coordinates of the principal vertices, the center, and the focal points of the conic $r(\varphi) = \dfrac{r_0}{1+\varepsilon\cos\varphi}$. Further, calculate $\varepsilon$ and $r_0$ for a given distance $2a$ of the principal vertices or the distance $2e$ between the focal points.

*Solution*:

The first focal point $F_1$ is the origin of the coordinate system. The principal vertices occur for $\varphi = 0$ and $\varphi = \pi$. We obtain $r(0) = \dfrac{r_0}{1+\varepsilon}$. Thus, the first principal vertex has the coordinates $A\left(\dfrac{r_0}{1+\varepsilon}\Big/0\right)$, and the second principal vertex has the coordinates $B\left(\dfrac{-r_0}{1-\varepsilon}\Big/0\right)$. Therefore, the length of the major axis is

$$\overline{AB} = 2a = \frac{r_0}{1+\varepsilon} + \frac{r_0}{1-\varepsilon} = \frac{2r_0}{1-\varepsilon^2}.$$

The center is given by

$$M\left(\frac{1}{2}\left[\frac{r_0}{1+\varepsilon} + \frac{-r_0}{1-\varepsilon}\right]\Big/0\right) = M\left(\frac{r_0\,\varepsilon}{\varepsilon^2-1}\Big/0\right).$$

Its $x$-coordinate $e = \overline{MF_1} = \left|\dfrac{r_0\,\varepsilon}{1-\varepsilon^2}\right|$ is referred to as the *linear eccentricity.* The second focal point $F_2$ is symmetric to $F_1$ with respect to $M$, and thus, $F_2\left(\dfrac{2r_0\,\varepsilon}{\varepsilon^2-1}\Big/0\right)$.

Now let $2a$ and $2e$ be given. Then, we find

$$\frac{e}{a} = \left|\frac{\dfrac{r_0\,\varepsilon}{\varepsilon^2-1}}{\dfrac{r_0}{1-\varepsilon^2}}\right| = \varepsilon, \quad \text{and subsequently,} \quad r_0 = a(1-\varepsilon^2).$$

◂◂◂

### ▸▸▸ Application: *locating a sound source*

At three locations $F_1$, $F_2$, and $F_3$ whose relative positions are known, one hears an explosion at the respective times $t_1$, $t_2$, and $t_3$. Where did the explosion take place?

*Solution*:

Let $S$ be the position at which the explosion has occurred, and let $s_1 = \overline{SF_1}$ and $s_2 = \overline{SF_2}$. Let $c$ be the speed of sound at the given time, then, $s_2 - s_1 = c\,(t_2-t_1)$. Thus, $S$ is located on a hyperbola with foci $F_1$ and $F_2$ ($2e = 2\overline{F_1F_2}$) and $2a = c\,(t_2-t_1)$. Its equation reads

$$r(\varphi) = \frac{r_0}{1+\varepsilon\cos\varphi} \quad \text{with } \varepsilon = \frac{e}{a} \quad \text{and} \quad r_0 = a(1+\varepsilon^2).$$

Analogously, $S$ lies on a (homofocal) hyperbola with foci $F_1$ and $F_3$ and the equation

$$r(\varphi) = \frac{r_0}{1+\varepsilon\cos(\varphi+\varepsilon)} \quad \text{with adjusted } a \text{ or } e.$$

Here, $\varepsilon$ is the angle at $F_1$ in the triangle $F_1F_2F_3$. The intersection of these homofocal conics leads to a quadratic equation (Application p. 144) for $\cos\varphi$. So, there are two solutions for $\varphi$. The right solution can be seen by testing the values.

⊕ *Remark*: A slightly modified method – namely by introducing spherical coordinates – is used to determine the epicenter of an earthquake. Before the introduction of GPS, it was also the prevalent method for the determination of a ship's position at sea. ⊕ ◂◂◂

Among the remarkable physical properties of conic sections, we will mention only *Kepler's First Law*:

> The relative orbits of planets or asteroids and comets around the Sun are ellipses, and the Sun is one of the two foci.

The proof of this is cumbersome and it took the genius of *Newton* to give. It is given in detail in G. Glaeser, H. Stachel, B. Odehnal: *The Universe of Conics. From the ancient Greeks to 21st century developments.* Springer Spektrum, 2016, pp. 62–76.

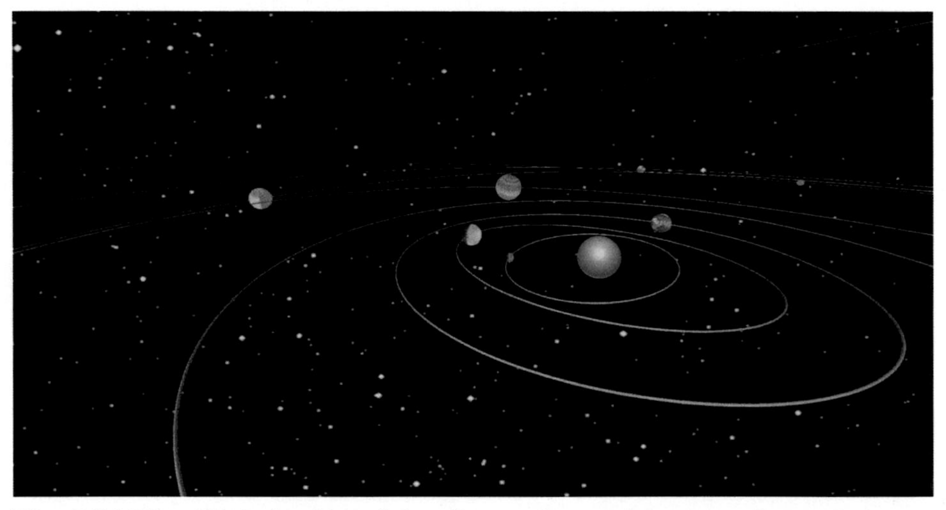

**Fig. 7.50** The elliptical orbits of the planets of our solar system lie nearly in one plane. Unfortunately, this also holds true for asteroids (small rocks) and comets.

▸▸▸ **Application**: ***The comet is coming!*** (Fig. 7.50)
For millennia, mankind has trembled at the thought of a comet hitting their planet, which could destroy all life. In fact, 65 million years ago, a comet caused an appalling devastation, wiping out the formerly successful kingdom of dinosaurs. There are always potentially dangerous situations even if the probability of the occurrence of a disaster in the "short term" – that is, approximately in the next 10,000 years – is quite low.

The Earth's elliptical orbit is given by the equation $r = \dfrac{r_1}{1 + \frac{1}{60}\cos\varphi}$. A comet moves in the same plane on the elliptical orbit $r = \dfrac{r_2}{1 + 0.7\cos(\varphi + \delta)}$. Here, $\delta$ is the constant twisting angle of both ellipse's principal axes. Where are the points of intersection and hence the potential collision points of both conics?

*Solution*:

By equating the radii, we obtain

$$\frac{r_1}{1 + \frac{1}{60}\cos\varphi} = \frac{r_2}{1 + 0.7\cos(\varphi + \delta)}.$$

We have already solved this equation in Application p. 144.

⊕ *Remark*: The planetary orbits do not intersect each other. So, no collision of planets can ever occur. ⊕ ◂◂◂

⊕ *Remark*: Shooting stars (meteorites) are completely harmless, only haven a diameter of between 1 to 10 mm. Their appearance varies according to the time of day and the time of year:

First of all, the greatest amounts of meteorites can be seen just before sunrise. If we are on the Earth, we move on a circle of latitude because of the Earth's rotation. At the same time, the Earth moves in a direction that is perpendicular to the direction of the Sun, because the orbit is almost circular. Now, let us assume that the Earth travels into a swarm of meteorites. The points whose latitude coincides with the breaking of dawn will be most heavily bombarded by the "shower of particles". Points on the opposite side are shielded. They are only caught by particles which are about to "overtake" the Earth.

Furthermore, the Earth always collides with a heap of meteorites in late July/early August: In this period, the Earth crosses the meteor shower of the Perseids, which is a tube-like elliptical zone containing a swarm of small particles. The particles come mainly from the direction that runs opposite to the direction of the Earth's trajectory, and according to the rules of perspective, many shooting stars will appear to exit from the same point in the firmament (the common ideal point of the proceeding direction). ⊕

## The theory of space curves

Planar curves have tangents and normals. If the curve does not lie in a plane, then the whole issue becomes more demanding. The term tangent does not change its meaning. The curve normal, however, is no longer uniquely defined: Each normal to the tangent through the point of contact is a *curve normal*, and lies in the *normal plane* $\nu$ of the curve tangent.

In space, we can also consider a curve point $P$ and *two* neighboring points $Q$ and $R$. If the three points do not lie on a straight line, they define a circle. Now, we let $Q$ and $R$ move towards $P$. The circumcircle of the three points $P$, $Q$, and $R$ reaches a *limit position*, called the *osculating circle* of

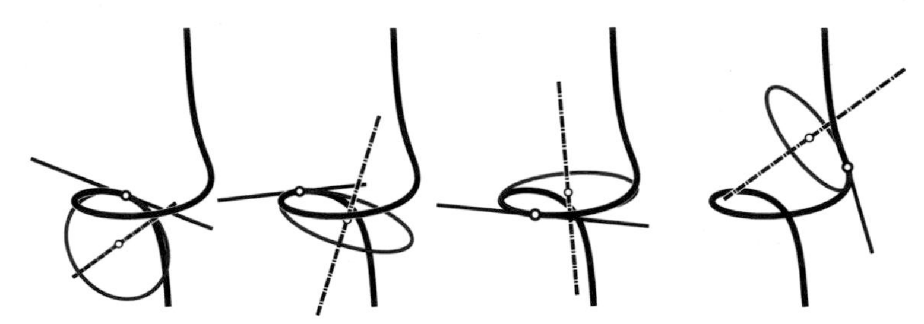

**Fig. 7.51** osculating circles of a space curve

the curve at $P$. Its carrier plane is called the *osculating plane*. We say: The osculating plane shares three (infinitely close) points with the curve at $P$. In Application p. 339, it is shown how to determine the osculating plane of a parametrized curve $\vec{x}(u)$ via the cross product $\dot{\vec{x}}(u) \times \ddot{\vec{x}}(u)$ of the first and the second derivative of the curve.

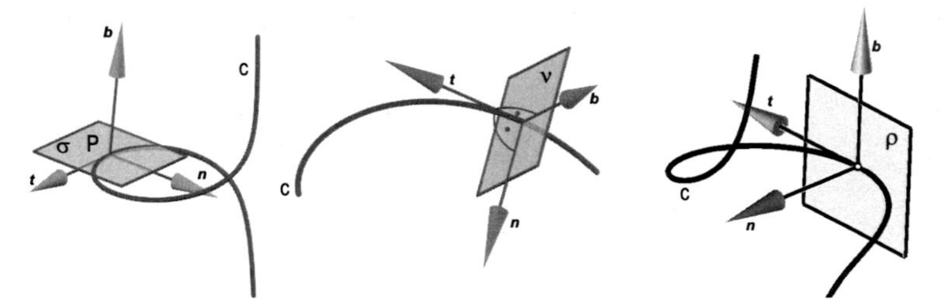


**Fig. 7.52** accompanying tripod (also called *Frenet frame*) of a space curve

Among all normals of the curve at the point $P$, we can now highlight two normals in particular: The first one lies in the osculating plane and is called the *principal normal.* It carries the center of the osculating circle. The second one is perpendicular to the osculating plane and is, therefore, located in the normal plane. It is called the *binormal* of the curve at $P$. The tangent, the normal, and the binormal form the curve's *accompanying tripod.* The plane spanned by the tangent and the binormal is called the *rectifying plane.*
The curve's orthogonal projection onto the osculating plane $\sigma$ at $P$ is a planar curve that shares the osculating circle at $P' = P$ with the space curve (Fig. 7.53, left).
The curve's orthogonal projection onto the rectifying plane $\varrho$ has (in general) a point of inflection at $P'' = P$ (the osculating circle appears as a straight line, Fig. 7.53, middle). If $P$ is a so-called *handle point* – which may be found in a symmetry plane of the curve – a *flat point* on the image curve shows up. When we finally project orthogonally onto the normal plane $\nu$ (Fig. 7.53, right), a cusp $P''' = P$ occurs because the curve tangent appears as a point. Next to the curvature, geometry also employs the concept of *torsion* of a curve (Fig. 7.54).

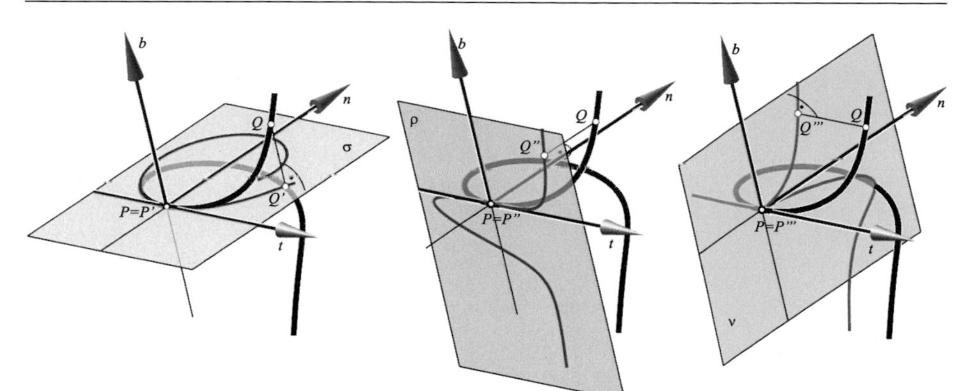


**Fig. 7.53** special projections of a curve

**Fig. 7.54** torsion of a space curve and something akin to an accompanying tripod

Circles (viewed as space curves) have constant curvature and torsion. However, the torsion equals zero all over the curve. Helices as "generalized circles" also have constant curvature and constant non-zero torsion.
The torsion at a curve point is a measure of the osculating plane's winding about the curve tangent (by analogy, the curvature of a planar curve which measures the rotation of the tangent about the center of the osculating circle). A curve with zero torsion lies in a fixed plane which is its osculating plane (by analogy, a planar curve without curvature stays in the direction of a straight line, which is its tangent).

### ▸▸▸ Application: *osculating plane of a space curve*

In Application p. 358, we will need to determine the rectifying plane of a point of a space curve. It contains the tangent of the curve and is perpendicular to the osculating plane. Prove that if $\vec{x}(u)$ is a parametrized curve, the osculating plane $\sigma(u_0)$ in $u = u_0$ is spanned by the derivation vectors $\dot{\vec{x}}_0$ and $\ddot{\vec{x}}_0$. Thus, its normal is given by $\dot{\vec{x}}(u_0) \times \ddot{\vec{x}}(u_0)$.

***Proof***: Let $\vec{x}(u)$ be a parametrization of a curve whose coordinate functions can be expanded in Taylor series. The osculating plane $\sigma$ at a point $\vec{x}_0 = \vec{x}(u_0)$ is by definition spanned by the point and two "neighboring points":

$$\vec{x}_0^+ = \vec{x}(u_0 + k) = \vec{x}_0 + k\dot{\vec{x}}_0 + \frac{k^2}{2}\ddot{\vec{x}}_0 + T_3,$$
$$\vec{x}_0^- = \vec{x}(u_0 - h) = \vec{x}_0 - h\dot{\vec{x}}_0 + \frac{h^2}{2}\ddot{\vec{x}}_0 - T_3$$

where we have used the Taylor expansion and $T_3$ indicates terms of degree 3 or higher. The normal $n$ of the plane $\sigma$ spanned by $\vec{x}_0$, $\vec{x}_0^+$, and $\vec{x}_0^-$ is parallel to

$$\vec{n} = (\vec{x}_0^+ - \vec{x}_0) \times (\vec{x}_0^- - \vec{x}_0) = \frac{1}{2}hk(h+k)\dot{\vec{x}}_0 \times \ddot{\vec{x}}_0 + T_4$$

with $T_4$ indicating terms of degree 4 and higher. Obviously, the normal $n$ is parallel to $\dot{\vec{x}}_0 \times \ddot{\vec{x}}_0$.

In order to perform the limit procedure, we cancel the maximum power of $h$ and $k$, i.e., the factor $\frac{1}{2}hk(h+k)$. Then, after $h \to 0$ and $k \to 0$, we see that the limit of the normal $n$ is parallel to $\dot{\vec{x}} \times \ddot{\vec{x}}$. Hence, the limit of the plane through three sufficiently close points on the curve contains both the first and the second derivative of the curve. ⊙ ◀◀◀

▸▸▸ **Application**: ***more surfaces determined by a space curve***

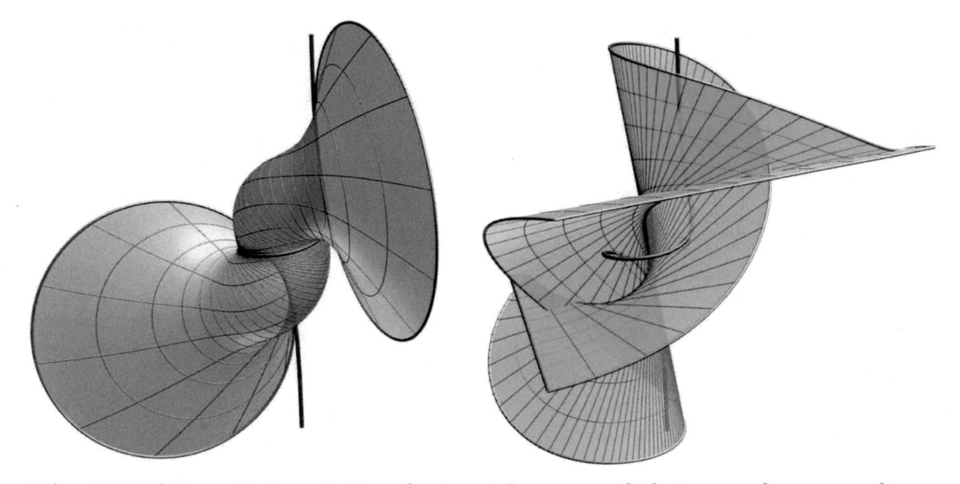

**Fig. 7.55** All osculating circles of a spatial curve and their axes form a surface.

There are more "accompanying surfaces" like the rectifying developable (Application p. 358) which are determined solely by the space curve. E.g., one can consider the circular surface that consists of all the osculating circles (Fig. 7.55, left). Furthermore, all the axes of these circles form a developable surface called the *polar developable*, see Fig. 7.55, right. ◀◀◀

## 7.4 Envelopes

If a straight line $g$ moves, it envelopes a curve. The simplest case is that of a line rotating about a point $M$. Then, it envelopes a circle with center $M$ and radius $r = \overline{Mg}$. In general, the following theorem holds:

Assume that we are given the equation of a line dependent on some parameter $t$ (for example, time): $a(t)\,x + b(t)\,y = c(t)$. Then, one obtains the point of contact with the envelope $c$ by intersecting the straight line with its "derivative line" $\dot{a}(t)\,x + \dot{b}(t)\,y = \dot{c}(t)$.

***Proof***: Let $\vec{c}\,(t) = \begin{pmatrix} u(t) \\ v(t) \end{pmatrix}$ be a parametrization of the enveloped curve $c$. Then, its tangent at the point $C(u(t)/v(t))$ reads

$$\vec{x}\,(t) = \begin{pmatrix} u(t) \\ v(t) \end{pmatrix} + \lambda \begin{pmatrix} \dot{u}(t) \\ \dot{v}(t) \end{pmatrix}.$$

Multiplication by the normal vector $\begin{pmatrix} \dot{v}(t) \\ -\dot{u}(t) \end{pmatrix}$ yields the non-parametric representation

$$\dot{v}\,x - \dot{u}\,y = \dot{v}\,u - \dot{u}\,v.$$

The equation of the tangent describes the line $g$. We differentiate the equation with respect to $t$ and obtain again a straight line (the "derivative of the line" $\dot{g}$):

$$\ddot{v}\,x - \ddot{u}\,y = \ddot{v}\,u + \dot{v}\,\dot{u} - (\ddot{u}\,v + \dot{u}\,\dot{v}) = \ddot{v}\,u - \ddot{u}\,v.$$

Apparently, $x = u$ and $y = v$ satisfy both linear equations, and therefore, $C(u/v)$ is the point of intersection of the lines. ⨀

An envelope of lines is often visible in specular cups and is called a *focal curve* (Fig. 7.56; Fig. 6.47, right). If the cup is a cylinder of revolution, a pure reflection can be observed in the top view (Fig. 4.55).

**Fig. 7.56** envelopes in a cooking pot (two symmetric light sources)

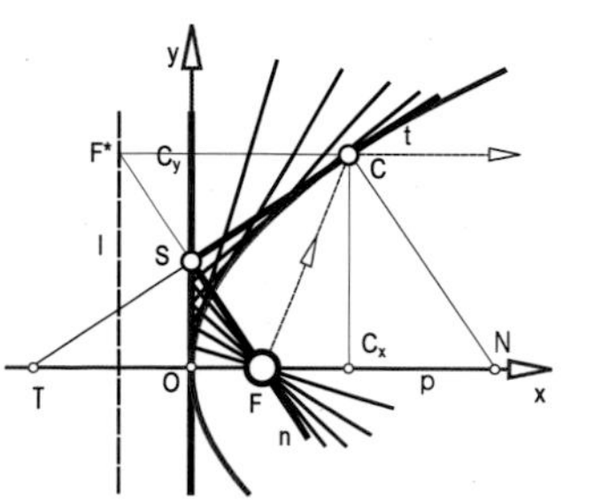


**Fig. 7.57** a parabola as an envelope of a right-angle hook

▸▸▸ **Application**: ***a parabola as an envelope*** (Fig. 7.57)
A right-angle hook is placed in such a way that the apex lies on the $y$-axis and one of the legs goes through a fixed point $F(a/0)$. Calculate the equation of the envelope of the second leg.

*Solution*:

Let $S(0/t)$ be the apex of the right-angle hook. Then, $\vec{n} = \overrightarrow{SF} = \begin{pmatrix} a \\ -t \end{pmatrix}$ is the normal vector of the line $g$ of the second leg. Consequently,

$$g: \begin{pmatrix} a \\ -t \end{pmatrix} \vec{x} = \begin{pmatrix} a \\ -t \end{pmatrix} \cdot \begin{pmatrix} 0 \\ t \end{pmatrix} \Rightarrow a\,x - t\,y = -t^2.$$

By differentiating with respect to $t$, we obtain the derivative of the line

$$\dot{g}: \ -y = -2\,t \Rightarrow y = 2\,t.$$

We intersect the two lines $g$ and $\dot{g}$ by substituting $y = 2\,t$ in the equation of $g$:

$$a\,x - t\,(2\,t) = -t^2 \Rightarrow x = \frac{t^2}{a}.$$

A parametrization and an equation of the envelope are, thus,

$$x = \frac{t^2}{a}, \ y = 2\,t \quad \text{or} \quad y^2 = 4a\,x.$$

They describe a parabola. The $y$-axis is tangent at the vertex of the parabola which is the origin $O$ of the coordinate system. $F$ is the focal point.

**Fig. 7.58** parabolic headlights **Fig. 7.59** parabolic cylinder

Let $C_x$ be the projection of the parabola's point $C$ onto the $x$-axis (Fig. 7.57) and $C_y$ the projection onto the $y$-axis. The main properties of the parabola can directly be derived from $y = 2t \Rightarrow \overline{OS} = \overline{SC_y}$:

1. If one reflects $F$ in the tangent, the "counter-point" $F^*$ is always on $x = -a$ which is parallel to the $y$-axis (the "directrix" $l$). Because of the mirror property given by the "classical parabola definition", we clearly see:

$$\overline{CF} = \overline{CF^*}.$$

Thus, the parabola is the locus of all points equidistant to a fixed point $F$ and a fixed line $l$.

2. The focal rays $FC$ are reflected in the parabola and emerge parallel to the axis (also as a result of reflection). This lovely property is utilized for headlights (Fig. 7.58). Conversely, incoming rays that are parallel to the axis are reflected to the focal point, which, for example, is utilized in radio telescopes and "satellite receivers" (Fig. 7.59).

3. The subnormal $\overline{C_x N}$ has a constant length of $2a$.

4. The subtangent $\overline{C_x T}$ is halved by the vertex. Similarly, the tangent section $\overline{CT}$ is bisected by the vertex tangent ($\overline{TS} = \overline{CS}$).

◂◂◂

▸▸▸ **Application**: ***parabolic headlights and sun collectors***
By positioning a point light source at the focal point of a reflecting paraboloid of revolution, it can be ensured that all light rays that hit the mirror will be reflected parallel to the axis. Such lights are called "long distance beams". In practice, rod-shaped light sources are often used.

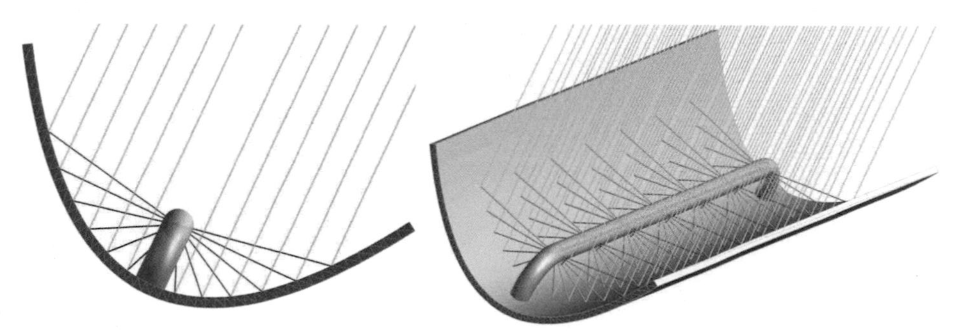

**Fig. 7.60** parabolic screen in an east-west-direction

Parabolic cylinders are utilized for such cases, which are also useful for focusing sunlight on a pipe. This pipe may contain liquid (e.g., thermo oil) which is first heated and then diverted. The screen is oriented in an east-west direction and is constantly moved to ensure that the sun stays in its symmetry plane. ◂◂◂

▸▸▸ **Application**: ***the envelope of a straight line in an ellipse motion*** (Fig. 7.61)
A rod $XY$ with constant length is guided so that $X$ moves along the $x$-axis and $Y$ along the $y$-axis. Calculate the curve enveloped by the rod (demo program `astroid.exe`).

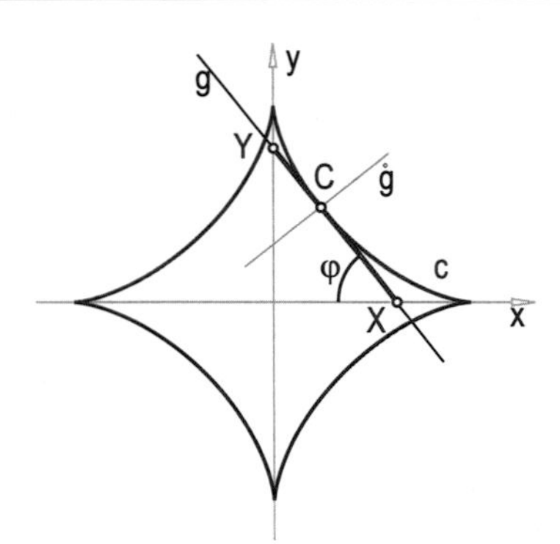


**Fig. 7.61** an astroid as an envelope

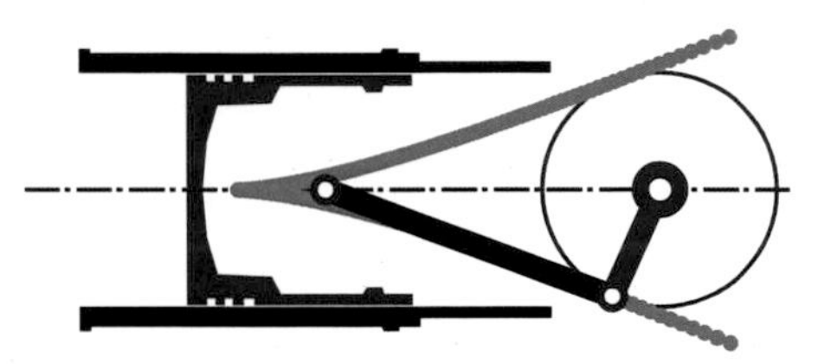

**Fig. 7.62** space requirement of the connecting rod

*Solution*:
Let $a = \overline{XY}$ be the length of the rod and $\varphi$ the angle enclosed with the $x$-axis. Then, we get $X(a\cos\varphi/0)$ and $Y(0/a\sin\varphi)$. Substituting these coordinate pairs proves that the carrier line $g = XY$ is described by

$$\frac{x}{a\cos\varphi} + \frac{y}{a\sin\varphi} = 1.$$

We multiply by the common denominator and arrive at

$$g:\ \sin\varphi\, x + \cos\varphi\, y = a\sin\varphi\cos\varphi.$$

This time, of course, $\varphi$ is the parameter. If we differentiate with respect to $\varphi$, we obtain the derivative of the line

$$\dot{g}:\ \cos\varphi\, x - \sin\varphi\, y = a[\cos\varphi\cos\varphi + \sin\varphi\,(-\sin\varphi)].$$

In order to calculate the intersection point $C = g \cap \dot{g}$ – coincidentally, the two direction vectors of $g$ and $\dot{g}$ are always perpendicular to each other because their dot product always disappears (Fig. 7.61) – we multiply the equation of $g$ by $\sin\varphi$ and that of $\dot{g}$ by $\cos\varphi$ and add both equations (thus, $y$ is eliminated):

$$(\sin^2\varphi + \cos^2\varphi)\,x = a[\sin^2\varphi\,\cos\varphi + \cos^3\varphi - \sin^2\varphi\,\cos\varphi].$$

Since $\sin^2\varphi + \cos^2\varphi = 1$, we obtain $x = a\cos^3\varphi$. Analogously, we find $y = a\sin^3\varphi$. The thus described curve is known as an *astroid.*

⊕ *Remark*: From $\cos\varphi = \sqrt[3]{\frac{x}{a}}$ or $\sin\varphi = \sqrt[3]{\frac{y}{a}}$ we can eliminate $\varphi$ because $\sin^2\varphi + \cos^2\varphi = 1$, and we get

$$\sqrt[3]{x^2} + \sqrt[3]{y^2} = \sqrt[3]{a^2} = \text{constant}.$$

Cubing twice will then show that the astroid is a curve of degree 6. ⊕ ◂◂◂

The envelopes of straight lines are frequently needed in order to determine space requirements of mechanisms (Fig. 7.62). Nowadays, computer animations are used for the development of transmissions. However, when using such animations, a helpful parameter representation of the moving line $g(t)$ is rarely available. In such a case, one can try the approximation with two "neighboring lines" $g(t-\varepsilon)$ and $g(t+\varepsilon)$ – with a "very small $\varepsilon$" (with regard to numerical instabilities, the chosen $\varepsilon$ should not be *too* small; experience shows that $\varepsilon = 10^{-4}$ is a good value) – and so we get the following statement:

> The intersection of two "neighboring straight lines" also provides an approximation of the point of contact with the envelope.

***Proof***: The statement is obviously true for an ordinary rotation around a fixed point. Following the main theorem of planar kinematics, even the most complicated motion can be interpreted as an infinitesimal rotation (or translation) at each instant. ⨀
The statement above applies not only to envelopes of straight lines, but quite generally to any envelope of curves (contact is only related to the tangent and not to the curvature). So, you only have to intersect two "neighboring curves", and this will yield the points of the envelope.

▸▸▸ **Application**: ***envelope of circles*** (Fig. 7.63)
Without using differential calculus, we will investigate the envelope(s) of a circle $k$ whose center $M$ moves along a curve $m$ (for example, an ellipse).
*Solution*:
One intersects pairs of "neighboring circles", which always provides two points on the normal to $m$ at $M$. Properly connected, this results in two curves. If $m$ is an ellipse, the outer branch is always oval and approximately elliptical, the inner branch – with a sufficiently large radius of $k$ – may have cusps.
Fig. 7.63 can be interpreted spatially as the silhouette of a torus under normal projection. ◂◂◂

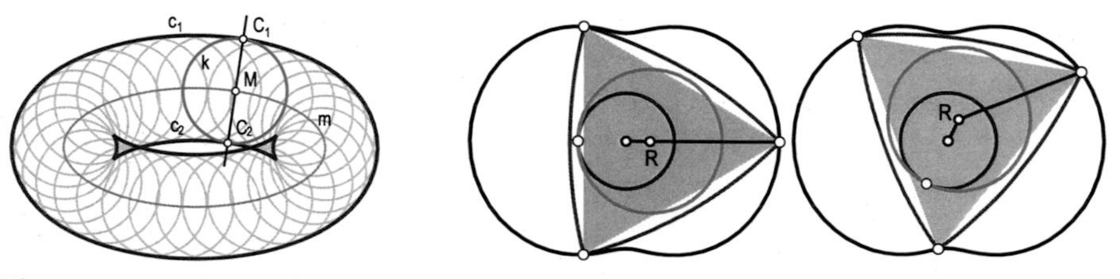


**Fig. 7.63** envelope of circles

**Fig. 7.64** Wankel engine

▸▸▸ **Application**: ***envelope in the Wankel engine*** (Fig. 7.64)
One practical application of an envelope shows up in the Wankel engine. It is basically an equilateral triangle (the core of the piston) that moves through

gears (and is, thereby, turned around within the piston chamber outlining a trochoid).
The enveloped curves of the triangle's sides fit without difficulty into the piston chamber. For the purpose of high-compression (and simultaneously better sealing), the center piece can be "thickened" as long as the envelope has just enough space in the piston chamber. In Fig. 7.64, this is done by means of circular arcs (in practice, one has opted for a slightly deviating form in order to create a more compressing profile). ◂◂◂

▸▸▸ **Application**: ***stability of a ship*** (Fig. 7.67)
A ship is sinking into the water due to its own weight. One might be tempted to think that the exclusive criterion for the stability of the ship in a heavy swell is that the center of gravity should be as low as possible. A low center of gravity is desirable, but the decisive factor is the position of the so-called "metacenter" $M$ – which is the cusp of all the "buoyancy lines" of the envelope (the buoyancy line is the vertical line through the center of gravity of the displaced water; relative to the pitching and tossing ship this line is, then, of course no longer vertical). The location of the metacenter $M$ is decisively influenced by the hull's shape. The higher $M$ is above the center of gravity $G$ of the ship's profile, the more stability it provides to the ship. Explain why this is the case.

**Fig. 7.65** the inventor himself – Eureka!

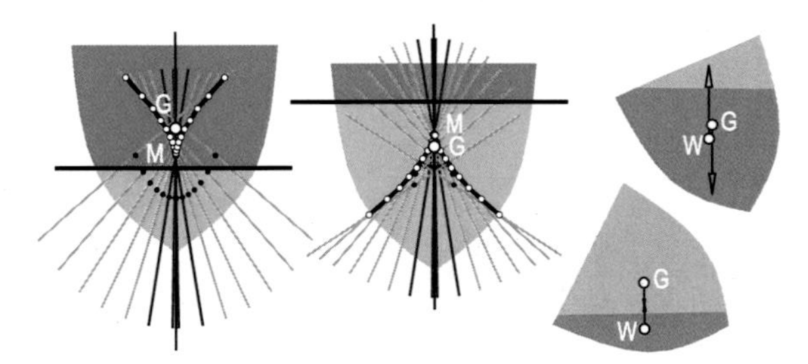


**Fig. 7.66** metacenter for an unfavorable cross-section – the center of gravity must then be lower

*Solution*:
In Application p. 218, we can see an inclined position of the hull. For this position, the center of gravity of the displaced water can be calculated. The buoyant force acts upwards in this case. In order for the occurring torque to raise the ship, the "buoyancy line" $a$ must pass the center of gravity $G$ underneath. The uplifting torque is proportional to the normal distance $\overline{aG}$. Now, we consider a series of successive slanting positions and plot the buoyancy line in the "cross-section". Then, we get a series of straight lines that envelope a curve. This "metacurve" has an apex on the line of symmetry of the cross-section. Since the buoyancy line in a non-inclined position coincides with the line of symmetry, the curve has a cusp (the metacenter $M$) at this point.

According to the aforementioned considerations, $M$ has to lie above $G$. In fact, it should lie as far above G as possible so that for all curve tangents – up to a maximum tilt angle – $G$ is always on the right side.

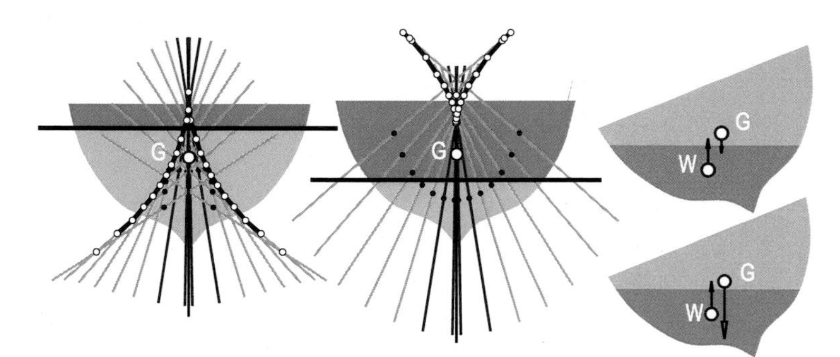


**Fig. 7.67** the metacenter of a ship in a typical situation

⊕ *Remark*: Nowadays, sailing vessels are constructed so that they set up right on their own. This is achieved by ballasting the bottom with plumb. ⊕ ◂◂◂

▸▸▸ **Application**: ***hazardous profiles*** (Fig. 7.66)
If the metacenter (Application p. 346) is under the center of gravity, then the ship sinks under unfavorable conditions (especially if high waves bring more and more water into the hull).

**Fig. 7.68** The metacenter changes due to water penetration.

⊕ *Remark*: There are countless reports on the sinking of supposedly unsinkable ships. One of them concerns the ferry "Estonia", which capsized near the Polish coast on September 28th, 1994, which led to the drowning of 852 people in icy water. By means of a lifelike replica model, the shipyard wanted to prove that the sinking was caused by poor maintenance and later modifications. However, this is not completely true because the simulation of real conditions at sea is still difficult. On the website that accompanies this book you will find the demo program `stormy_ocean.exe`. You can select different profiles and simulate a stormy sea. It will forecast whether the selected object will sink or not. This is performed by means of a relatively complex calculation of metacenters. ⊕ ◂◂◂

## Evolutes

Differential geometry deals with the limits of functions. For instance, if one considers two points $P$ and $Q$ of a planar curve, then its connection line forms a *secant* (lat. *secare* = to intersect). If $Q$ now moves increasingly closer to $P$, then there exists an unambiguous limiting position which is called *tangent* to the curve (lat. *tangere* = to touch). One simply says: At the point of contact $P$, the tangent of a curve and the curve share two points which are infinitely close together. The normal to the tangent at $P$ is called the *curve normal.* If the curve is a circle, all curve normals pass through the center. However, in general, the totality of normals to the curve envelops a curve – the so-called *evolute.*

**Fig. 7.69** tangent, normal, evolute of different curves

We can also consider *two* neighboring points $Q$ and $R$ to the curve point $P$. If the three points do not lie on a straight line, they define a circle. We now let $Q$ and $R$ move closer to $P$. This leads to a well-defined limiting position of the circle defined by three points – the *osculating circle* at the point $P$ (a circle can also degenerate into a straight line or a point). The radius of the osculating circle is called the radius of curvature. In the vicinity of the curve point, it approximates the curve much better than the tangent. The osculating circle at $P$ is said to share three points with the curve.

| The locus of the midpoints of all osculating circles is the evolute enveloped by the curve normals. |
|---|

***Proof***: The osculating circle is determined by three neighboring points $P$, $Q$, and $R$. Its center is, therefore, located at the intersection of the perpendicular bisectors of $\overline{PQ}$ and $\overline{PR}$. Let $Q^*$ be the midpoint of $PQ$ and $R^*$ that of $PR$. The perpendicular bisectors then converge towards the curve normal at $Q^*$ and $R^*$. The circumcenter of $PQR$, thus, converges towards the intersection of two adjacent curve normals. ⊙

### ▸▸▸ Application: *a circle as an evolute*

Fig. 7.69 shows curves with their evolutes. One can see that evolutes have cusps whenever the curve has an apex – a point whose radius of curvature is stationary. The right curve is especially notable. It is constructed in such a way that its evolute is a circle. It is, thus, called the *circle's involute.* ◂◂◂

▸▸▸ **Application**: ***A curve can be its own evolute.***
In Fig. 7.70, one can see the logarithmic spirals with their evolutes. Since the normals of the original spiral are tangents of the evolutes, the evolute also intersects a pencil of rays at a constant angle $\psi$. Its equation in polar coordinates $(r, \varphi)$ is $r = e^{p \cdot \varphi}$. The constant parameter $p = \cot \psi$ determines how fast the logarithmic spiral wraps around its asymptotic point. The evolute of such a curve is another logarithmic spiral that is congruent to the initial curve if $p$ is chosen appropriately. In special cases, it can even be equal to the initial curve (right image). ◂◂◂

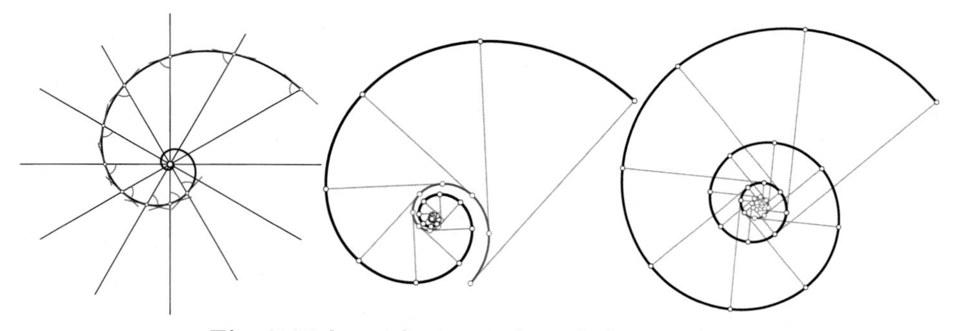

**Fig. 7.70** logarithmic spirals with their evolutes

## 7.5 Surfaces

**Fig. 7.71** A regular surface (computer-generated), and photos from a corresponding 3D-model, which was created by a 3D-Printer. A "mathematical surface" (given by parametric equations) has, of course, no thickness.

When a curve $c$ moves in space, it traces a surface that is generally "curved twice". This also occurs when the "generating curve" is a straight line. In this case, one speaks of a *ruled surface* (Fig. 7.71, left: "rotoidal helicoid", see also Fig. 7.10). The generating curve may change its shape during the course of the motion (Fig. 7.72: "*Escher* torus", Fig. 7.73).

**Fig. 7.72** an "*Escher* torus" in two different views

The generating curves $c$ may be circles, parts of circles, or other curves. Fig. 7.73 shows such surfaces as parts of snail shells.
If all circles are of equal size and it is also possible to roll a sphere such that it envelopes the same surface that is generated by the congruent circles, we obtain a *pipe surface* (Fig. 7.74 "feng shui spiral" and sculpture in Berlin). Among the pipe surfaces, we find the well-known example of the cylinder of

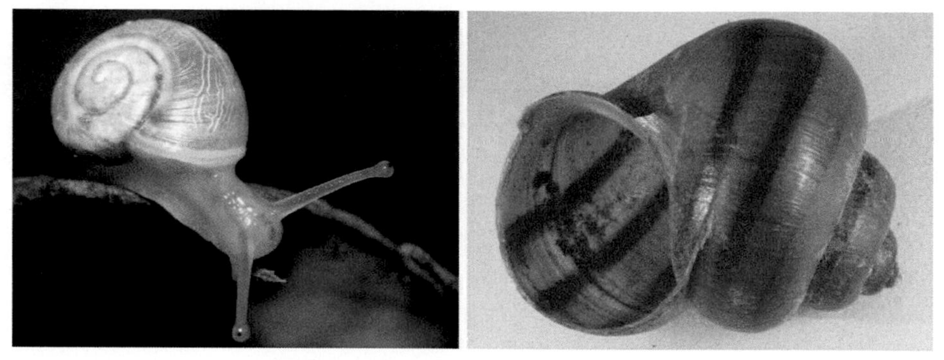

**Fig. 7.73** spiral surfaces in nature

revolution, the sphere, and the torus (Fig. 8.40). Toric parts occur very often in practice (Fig. 7.75, Application p. 351).

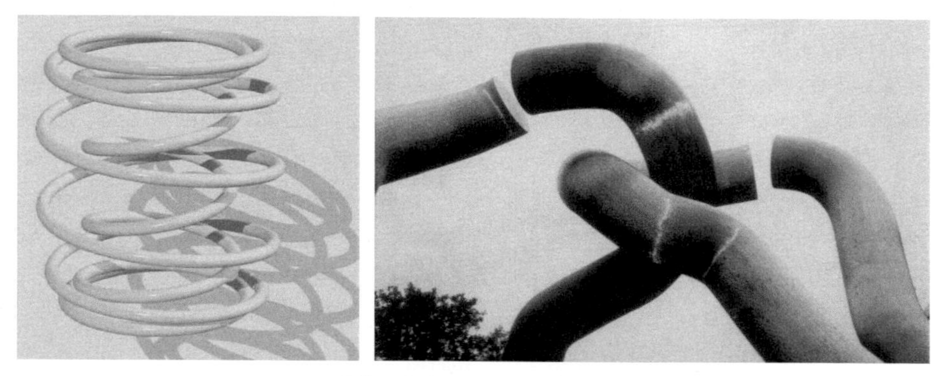

**Fig. 7.74** pipe surfaces

### ▸▸▸ Application: *boundary surfaces of technical objects*

Describe the boundary surfaces of the object depicted in Fig. 7.75.

*Solution*:

The bearing retainer is composed of the following parts:

- The main part consists of two hollow cylinders $Z_1$ and $Z_2$, which are joined by a quarter of a torus $T$ (a section is cut out for a better view into the interior).
- The clamps are made of prismatic parts $P$ (i.e. the parts are bounded by planes) which are joined by a half-cylinder $Z_3$ and a quarter-cylinder $Z_4$.

The universal transmits a rotation about one axis to another axis in a *uniform* manner (this is an improvement compared to the universal, see Application p. 219 and Application p. 303). It is composed of the following parts:

- Two axes $a_1$ and $a_2$, materialized by solid cylinders which are rigidly connected with two spherical parts $\Sigma_1$ and $\Sigma_2$. In the figure, the axis $a_1$

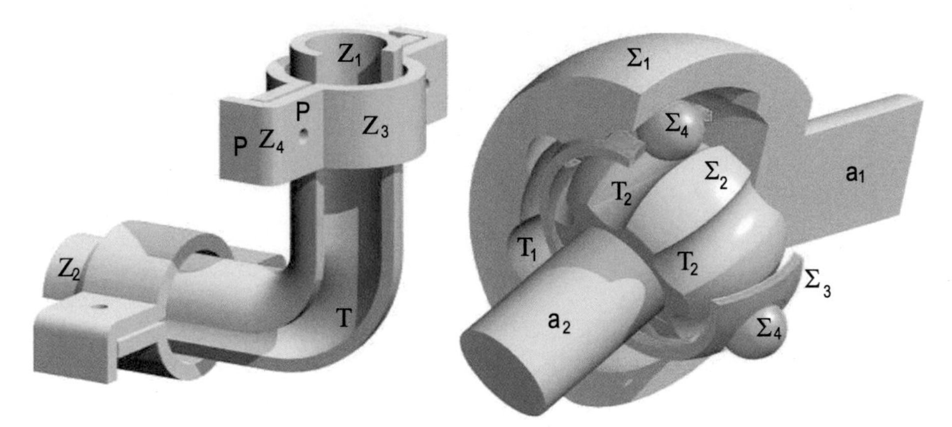


**Fig. 7.75** torus at a bearing retainer and a universal joint

and the part of the attached hollow sphere $\Sigma_1$ are cut in half, giving us a view of the interior of the joint.

- Between the outer hollow sphere $\Sigma_1$ and the inner spherical segment $\Sigma_2$, there is a thin spherical layer $\Sigma_3$ that is mounted and fixed on $\Sigma_1$. It has circular holes in which the spheres $\Sigma_4$ can roll.
- These rolling spheres can move both on the inner wall of $\Sigma_1$ and on the outer wall of $\Sigma_2$ in toroidal tracks $T_1$ and $T_2$. This enables a constant transmission of rotation at any adjustable angle. ◂◂◂

*Surfaces of revolution* are formed when a curve $c$ rotates about a fixed axis. The curve does not need to lie on a "meridian plane" through the axis, although this would make it easier to imagine the shape of the surface. However, one can rotate the points of $c$ into such a meridian plane. The resulting curve $c_0$ generates the same surface (Fig. 7.78) during rotation.

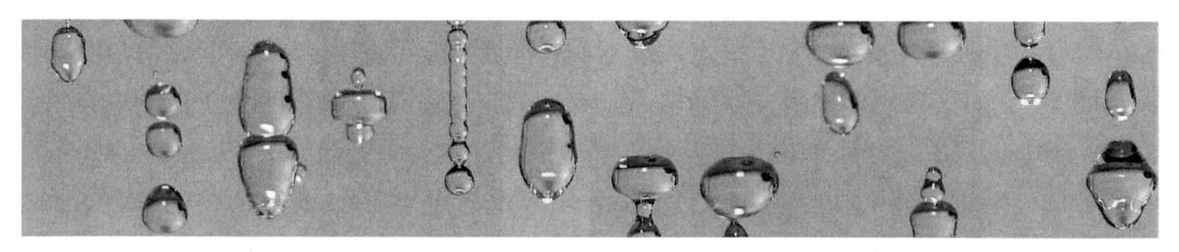

**Fig. 7.76** when a jet of water "severs" ...

When a jet of water flows from a faucet, it finally "breaks away" due to the increasing rate of the descending water (see Application p. 368) and forms rotation-surface-like drops within fractions of a millisecond. The only shape that you will look for in vein is the classic drop or teardrop shape that tapers downwards while pointing upwards.

*Helical surfaces* arise when a curve undergoes a helical motion about an axis. Again, there are, of course, different types of generation of these surfaces. Intuitively, the surface is best understood when a *profile section* (perpendicular

**Fig. 7.77** dew drops descending slowly from a tree bark

to the axis) and/or a *meridian section* (planar intersection through the axis) is known. Fig. 7.79 shows a helical pipe surface. On the left, we see how it is generated by a circle. The image on the right shows its profile sections.

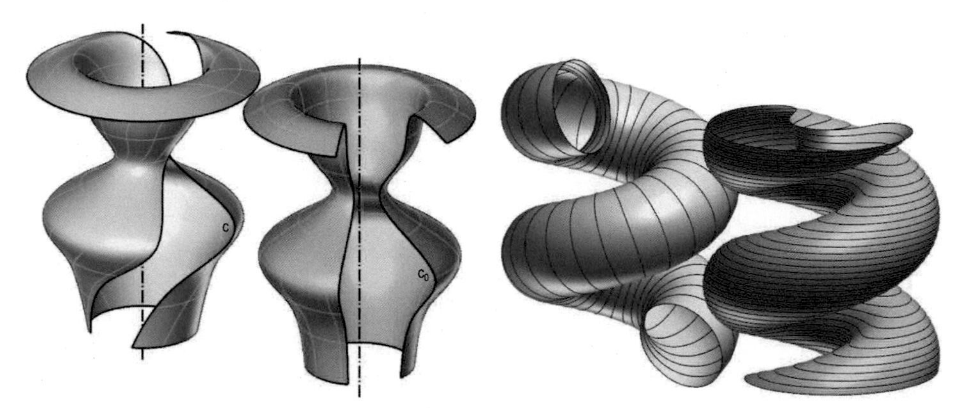


**Fig. 7.78** surface of revolution generated in two different ways

**Fig. 7.79** helical surface: two generations of one and the same surface

The various positions and the trajectories of the generating curves are called *parameter lines*. In order to distinguish between the different curves, we call them $v$-lines and $u$-lines: When the $v$-lines of the motion parameter is set to $v = v_0$, we obtain the position of the curve $c$ to the parameter $v_0$. Similarly, the paths of the $u$-lines remain constant on the points of $c$ in which the curve parameter is $u = u_0$.
The $v$-lines on ruled surfaces, surfaces of revolution, helical surfaces, and spiral surfaces are lines, parallel circles, helices, and cylindro-conical spirals. Surfaces can simultaneously belong to more than one of the surface classes discussed above. So, a cylinder of revolution, a sphere, and a torus are both surfaces of revolution and pipe surfaces. If a surface is formed by screwing a sphere ("helical pipe surface"), then it is both a helical surface and a pipe surface. In the Baroque period, columns were often made from such surfaces. In technology, they are used as chutes (Fig. 4.33).

**Fig. 7.80** Pipe surfaces are common in nature.

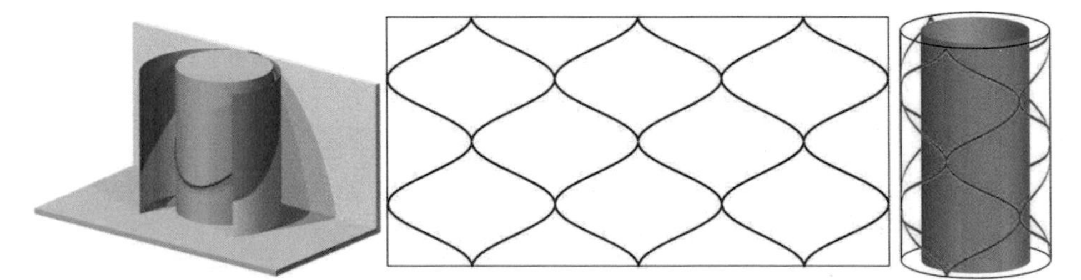

**Fig. 7.81** the unfolding, or rather the folding, of a cylinder of revolution (left side), "tessellation" (right)

If the generating curve is a straight line $c$ and if any two "neighboring positions" intersect in a point (which may also be at infinity), then the surface swept by $c$ is *developable.* This means it can be unfolded into the plane without deformation (Fig. 7.81). There are a few developable surfaces: the cylinder (even if it is not a cylinder of revolution), the cone (even if it is not a cone of revolution), and the tangent surfaces of space curves.

▸▸▸ **Application**: ***from space curves to famous planar curves*** (Fig. B.5)
Consider 2D-curves with the parametric equation

$$x = a \sin(t + \alpha), \; y = b \sin(nt + \beta).$$

Among them are ellipses ($n = 1$), eight-loops ($n = 2$ and $n = 1/2$), cubic parabolas ($n = 3$ and $n = 1/3$ with $\alpha = \beta = 0$), etc. The curves can be interpreted as normal projections of space curves which are generated by "drawing sine curves on vertical cylinders" (with horizontal axis). For $n = 1$, we have the special case that the whole space curve lies in a plane and is, therefore, an ellipse (Fig. 7.81, left).
A rotation of the carrier cylinder induces a phase shift.

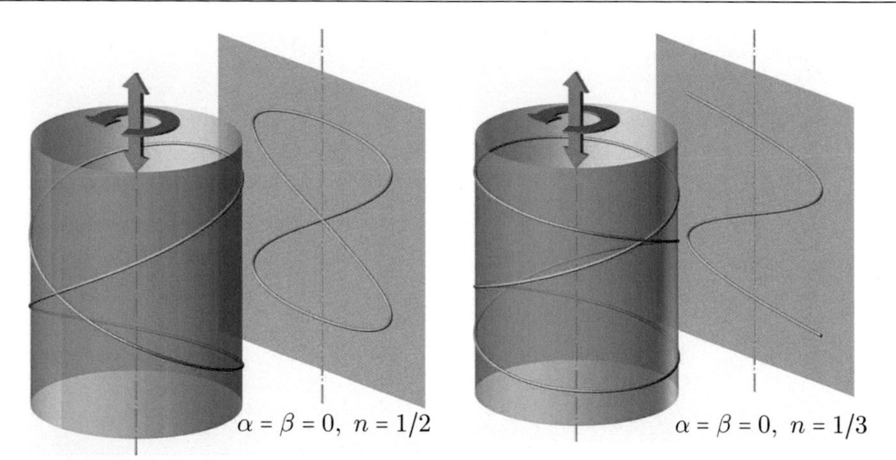


**Fig. 7.82** Sine curves coiled around a cylinder are projected onto *Lissajous* curves.

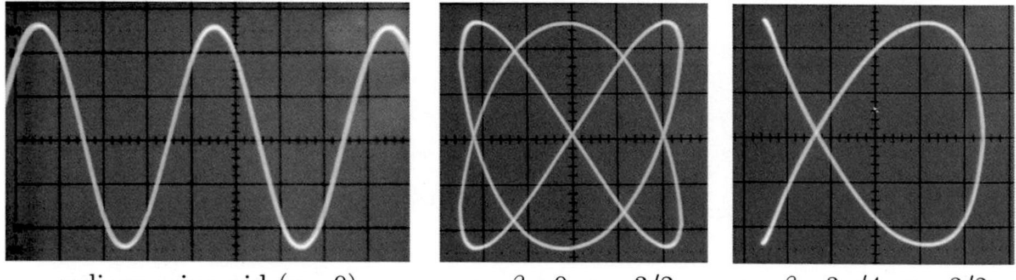


**Fig. 7.83** *Lissajous* curves on an oscilloscope

All electrical engineers have seen such curves on an oscilloscope (Fig. 7.83), when a harmonically alternating voltage is chosen for the entrance of both the $x$- and the $y$-deviation. In fact, a continuous phase shift creates this exact same illusion on the oscilloscope, as if a cylinder of revolution with coiled sine curve is rotating. In special cases, the curves are tracked twice for symmetry reasons, and they seem to end abruptly because the direction keeps changing. ◂◂◂

▸▸▸ **Application**: **Möbius *strip*** (Fig. 7.84)
In 1865, *A.F. Möbius* described a simple method for the creation of a surface with only one boundary line and one face. He suggested to take a rectangular strip and glue together opposite edges. Ever since, all surfaces of this kind have been called "Möbius strips". The strip inspired *M.C. Escher*'s famous image with the scuttling ants. ◂◂◂

▸▸▸ **Application**: ***another non-orientable surface*** (Fig. 7.85)
Besides the Möbius strip, the Klein bottle is another famous example of a non-orientable surface. It has neither an interior nor an exterior. There are two Möbius strips on the surface, and along these strips, one can move

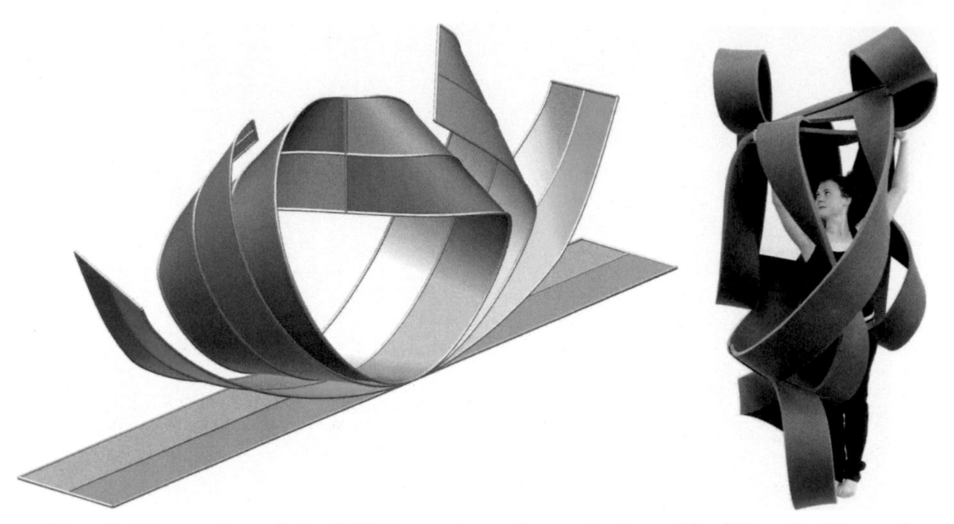

**Fig. 7.84** creation of the Möbius strip and a variation of it (Klaudia Kozma)

**Fig. 7.85** non-orientable surfaces with self-intersections: the Klein bottle and an artwork that was probably inspired by it ("Ferryman" by *Tony Cragg*)

continuously in both directions to each point of the surface. On the right-hand side of Fig. 7.85, part of such a strip is shown (in green). ◂◂◂

▸▸▸ **Application**: ***mathematical crochet work*** (Fig. 7.86, Fig. 7.87)
Another remarkable surface with self-intersections (in 3-space) was introduced in 1901 by Werner Boy. Although it possesses no singularities, it is very hard to understand its "inner life".

*Solution*:
One way to try to solve the mystery is to go through the surface slice by

**Fig. 7.86** Can you comprehend this surface?

slice. This technique is proposed in the paper cited below,[1] which, in turn, inspired the artist Lilian Boloney.

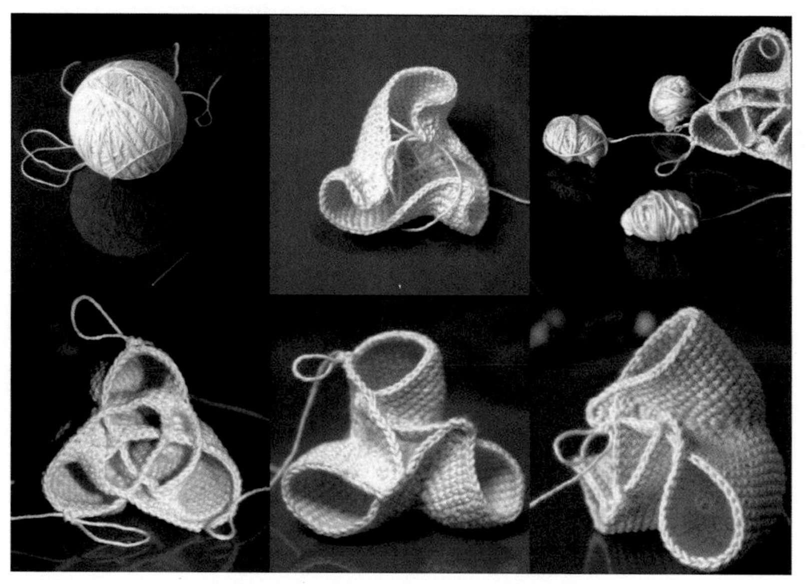

**Fig. 7.87** the artist's approach ...

Using her crocheting technique, she discovered a "geometric slide rule" that can be used to understand arbitrary complicated surfaces. With a soft yarn texture, any curvature can be obtained. By stringing together the whole surface mesh by mesh, she approximated its final shape. In all likelihood, this makes her one of only a few people with a deep understanding of Boy's surface! ◂◂◂

[1] O. Karpenko, W. Li, N. J. Mitra, M. Agrawala: *Exploded View Diagrams of Mathematical Surfaces* IEEE Transactions on Visualization and Computer Graphics archive, Vol. 16 Issue 6, Nov. 2010, Pages 1311–1318.

## 7.6 Further applications

▸▸▸ **Application**: ***How to create a developable strip?*** (Fig. 7.88)
We know that cylinders and cones can be developed, i.e., unfolded into the plane without distortions. Let us take a rectangular strip of paper and attach the narrow edges to each other. The generated ribbon is, of course, developable, since we cannot stretch or compress the paper without tearing it. If we do not twist the paper strip, then we get a cylinder. The midline of the strip is, thus, its normal section. Twisting produces a developable with the midline being a curve in space. The question is now: Is there a "paper strip" for an arbitrary space curve $c$ so that $c$ is its midline? This would allow us to create paper strips deliberately.

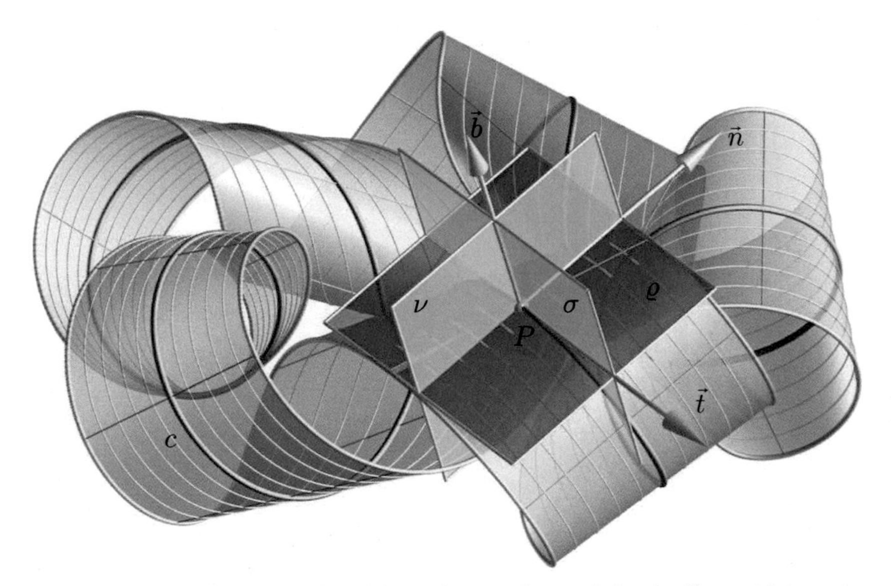


**Fig. 7.88** The rectifying developable – the envelope of the (red) rectifying plane $\varrho$ – of a space curve can be parametrized so that the whole surface develops into a rectangular strip with the midline as the rectified curve. One immediately suspects that the surface is developable, since all contour lines are straight.

*Solution*:
A theorem in Differential Geometry says: A surface that is generated as the envelope of a moving plane is developable. Let us consider three developables that are connected to the space curve: the *tangential surface* as the envelope of all osculating planes, the *normal developable* as the envelope of all normal planes, and the *rectifying developable* as the envelope of all rectifying planes (planes orthogonal to the principal normal).

The midline $c$ becomes a straight line when spread out in the plane and is, thus, a *geodesic curve* on the developable. It is known that the osculating planes of these are perpendicular to the tangent plane at any point. The tangential surface consisting of the osculating planes of the contact curve is, thus, not qualified for our purpose. We can see from Fig. 7.55 on the right that the normal developable does not even carry the curve. This only leaves the rectifying developable, for which all the necessary conditions are met: The curve lies on the developable and is also its geodesic line.
Despite the fact that the development preserves angles, the generatrices of the developable do not meet the midline orthogonally. Quite on the contrary: The only straight line in the rectifying plane which is orthogonal to the tangent of $c$ is the *binormal*, and the locus of all binormals is not a developable. Thus, the method of intersecting two neighboring normal planes of the curve does not lead to the solution. In this case, it results in the binormal developable which is irrelevant here. ◂◂◂

▸▸▸ **Application**: ***cutting a pipe connection***
Two water pipes made of stainless steel are shown in (Fig. 7.89), intersecting each other. Find a parametric representation of the intersection curve and thus the blank of the thin tube.

⊕ *Remark*: The thickness of the steel sheet is to be neglected. Any calculations are correct with thin sheets (for example, 2 mm thickness) up to about one millimeter, which is sufficient in practice (welded pipes). ⊕

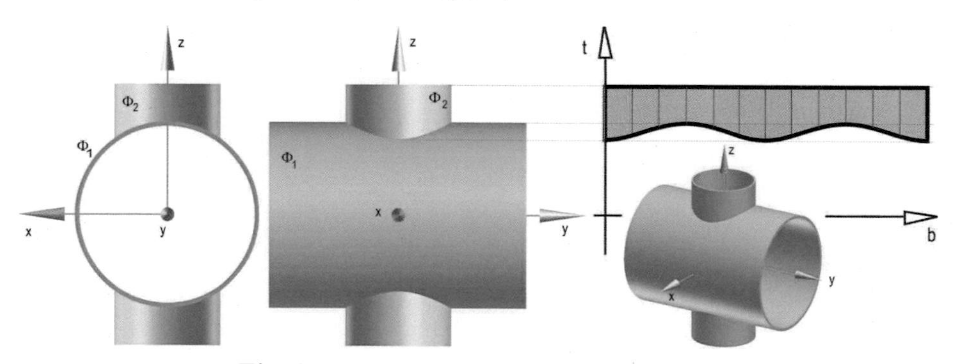


**Fig. 7.89** pipe joint made of stainless steel

*Solution*:
We identify the axes of two cylinders of revolution $\Phi_1$ and $\Phi_1$ with the $y$-axis and the $z$-axis. The radii of the cylinders are $r_1$ and $r_2$.
The generator $e$ lying in the $yz$-plane rotates about the $z$-axis (rotation angle $u$) and generates the cylinder $\Phi_2$. The line $e$ is intersected with $\Phi_1$:

$$e:\ \vec{x}\cdot\begin{pmatrix} r_2\cos u \\ r_2\sin u \\ t \end{pmatrix} = 0,\ \Phi_1:\ x^2+z^2=r_1^2 \ \Rightarrow\ e\cap\Phi_1:\ (r_2\cos u)^2+t^2=r_1^2.$$

Thus, we have $t(u) = \pm\sqrt{r_1^2 - r_2^2 \cos u^2}$ and a parameter representation $\vec{x}(u)$ of the intersection curve consisting of two branches in general.
In the normal projections in the direction of the cylinders' axes, the two curve parts appear as circles or circular arcs. In the normal projection onto the plane spanned by the axes (the $yz$-plane), the intersection appears as an equilateral hyperbola

$$z^2 - y^2 = (r_1^2 - r_2^2 \cos u^2) - r_2^2 \sin u^2 = r_1^2 - r_2^2,$$

"in the front view", see the middle of (Fig. 7.89).

When creating the cutting (development or gradation) of $\Phi_2$, a joint of the edge has the arc length $b = r_2 u$ for the abscissa and the $t$-value for the ordinate. In the Cartesian $(b, t)$-coordinate system, the bipartite curve of intersection (symmetric with respect to the $xy$-plane) is determined by the parametric representation $b = r_2 u,\ t = \pm\sqrt{r_1^2 - r_2^2 \cos^2 u}$ $(0 \le u \le 2\pi)$.

⊕ *Remark*: If $r_1$ is much bigger than $r_2$, the development of the intersection curve looks like a sine curve, but differs in the curvatures in the upper and lower vertices. The $t$-values vary between $\sqrt{r_1^2 - r_2^2}$ and $r_1$. For symmetry reasons, only a part of the curve is interesting.
For $r_1 = r_2$, the curve splits into two ellipses, and the developable curve into two sine curves (Fig. 8.64). ⊕ ◂◂◂

▸▸▸ **Application**: ***a non-trivial "tessellation" of a triangular rod***
Referring to Fig. 7.90, one takes a sine curve with vertices $C$, $A$, $D$ – temporarily located in one plane – and reflects them in $CD$ $(A \mapsto B)$. The two flat sine curves are now projected onto a sinusoidal cylinder with horizontal generators parallel to $AB$. The points $C$ and $D$ remain fixed and the vertices $A$ and $B$ are "shifted back". This creates a simply curved papule-leaf-like patch that is suitable for "tessellating" a three-sided rod:

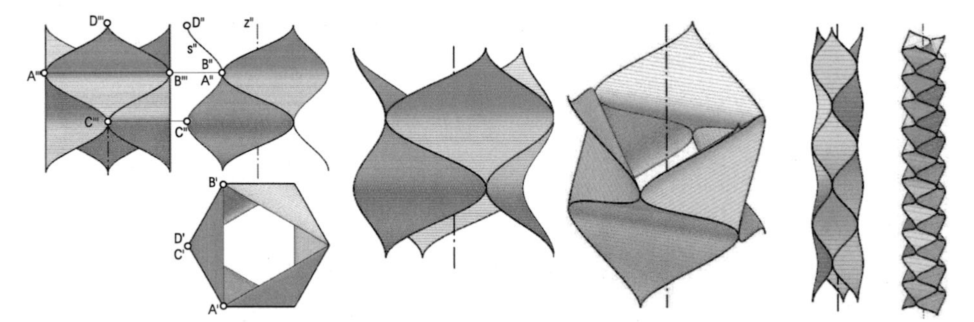


**Fig. 7.90** cylinder generated by sinusoids (Talia Nadermann & Sebastian Gomez)

Reflecting the straight line $CD$ in the shifted back line $AB$, one obtains the line $z$ which shall serve as the axis of a rotation. By rotating the "papule leaf" $ABCD$ twice by $\pm 120°$, it is possible to make the (blue marked) "leaves" touch each other, provided that a suitable choice of $\overline{AB}$ is given.

After rotating this "wreath" by $\pm 60°$ about the axis and translating it along the axis by the distance $CD/2$, the screwed wreath (marked in red) fits seamlessly together with the initial wreath. Through repeated application of this helical motion, both aesthetic and very stable structures, like the ones shown on the right in Fig. 7.90, can be produced. For representations with a computer, parametric descriptions for the leaves' edges are advantageous. Derive the parametrizations and show that these curves are (ordinary planar) sinusoids.

*Solution*:

Let $a$, $b$, and $h$ be the dimensions of the circumscribed cuboid as in Fig. 7.90 on the left. In order to bring reflected cylinder parts into contact, $b/2 : a/2 = \tan 60° = \sqrt{3}$ ($\Rightarrow b = \sqrt{3}\,a$) has to hold. A possible parameter representation of the symmetric sections of the cylinders is now

$$\vec{x} = \begin{pmatrix} x \\ y \\ z \end{pmatrix} = \begin{pmatrix} \pm\frac{\sqrt{3}a}{4}\,(1+\cos u) \\ \frac{a}{8}\,(3+\cos u) \\ h\,\frac{u}{2\pi} \end{pmatrix}.$$

For $0 \le u \le 2\pi$, one obtains the section depicted on the left in Fig. 7.90. The curves lie in $z$-parallel planes because $x/\sqrt{3} \mp 2y =$ constant holds (pictured bottom left). Thus, the paper edges do not only appear as sinusoids in the projection. They are also sinusoidal in three-dimensional space.

⊕ *Remark*: At a constant ratio $h : a$, we get a similar structure. Varying $h : a$, one obtains affinely distorted paper wreaths (see the variants pictured right in Fig. 7.90). To calculate the blank (developable) of a "poplar leaf", we need integral calculus. ⊕

◂◂◂

▸▸▸ **Application**: ***analysis of the shortest route***

We want to travel along the shortest route with a cruising speed of 800 km/h from Vienna (16° E, 48° N) to Los Angeles (135° W, 35° N). (In practice, not only the distance, but also the wind directions at an altitude of 10 – 11 km have to be taken into account. In this specific example, the aircraft uses the so-called "jet stream" over the Azores on its return flight.) How long is the flight path? Where is the airplane after $n$ hours? Where and when does it reach the northernmost point? In which direction does it start (apart from the predetermined direction of the runway)?

*Solution*:

1. The shortest path on a sphere is a great circle, i.e. a circle whose center is the center of the sphere. We calculate the Cartesian coordinates of Vienna ($P$) and Los Angeles ($Q$). $P$ and $Q$ together with the Earth's center $M$ (the origin) span the plane of the circle. The plane's normal vector $\vec{a}$ specifies the parametric representation of the circular orbit.

2. We determine the parameter values $u_P$ and $u_Q$ corresponding to $P$ and $Q$. The flight distance $f$ and the total flight time are

$$f = (u_Q - u_P)\, R \qquad (R = 6{,}380\,\text{km}), \quad \text{or} \quad T = \frac{f}{800} h.$$

3. We fly at 800 km per hour, or $\frac{800}{40{,}000} = \frac{1}{50}$ of the entire circumference of the Earth. This corresponds to a central angle of $\Delta u = \frac{2\pi}{50}$. After $n$ hours, we are at the position $u = u_P + n \cdot \Delta u$.
4. At the northernmost point $z = r\,s\,\sin u$ is maximal, thus, $u_0 = \pi/2$. This point is reached after $\frac{u_0 - u_P}{\Delta u}$ hours.
5. The "starting tangent" $\vec{t}$ lies in the circle's plane ($\Rightarrow\ \vec{t} \perp \vec{a}$) and is perpendicular to the position vector of the starting point $P$ ($\Rightarrow\ \vec{t} \perp \vec{p}$). So, we get $\vec{t} = \vec{a} \times \vec{p}$. This East-West direction $\vec{w}$ in the tangential plane of $P$ equals the normal vector of the meridian circle's plane $AMN$, where $N(0/0/R)$ is the North Pole. The northern direction follows from $\vec{n} = \vec{a} \times \vec{w}$. Therefore, we have the departure angle $\varepsilon$ with

$$\cos\varepsilon = \frac{\vec{t} \cdot \vec{n}}{|\vec{t}| \cdot |\vec{n}|}. \qquad ◂◂◂$$

▸▸▸ **Application**: ***intersection of two flight routes*** (Fig. 7.91)
An airplane flies along a great circle from $A$ to $B$, and another airplane flies along a different great circle from $C$ to $D$ at a constant altitude 10 km. Where can they potentially meet?

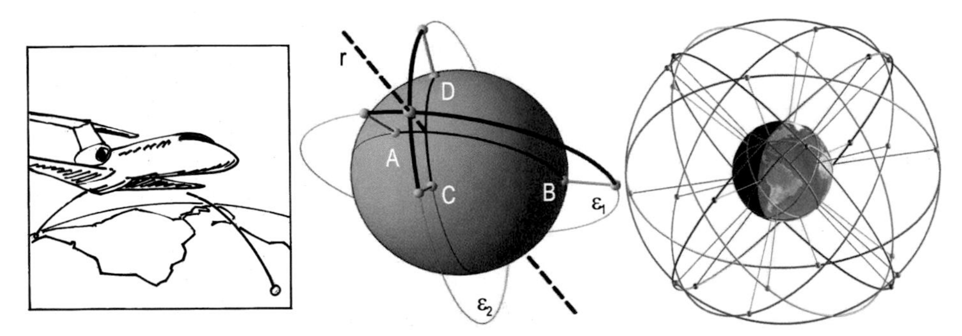


**Fig. 7.91** Center: two trajectories and their potential meeting. Right: 24 GPS satellites at an altitude of more than 20,000 km, evenly distributed into groups of four across six planes, each plane being inclined towards the equator plane by 55°. Through rotation about the Earth's axis by 60°, the trajectories of the satellites overlap.

*Solution*:
Let $M$ be the center of the Earth. Then, the trajectories are in the great circle planes $\varepsilon_1 = ABM$ and $\varepsilon_2 = CDM$. The potential intersection is, therefore, on the intersection of the two planes, at an altitude of 10 km.
We determine the normal vector of the planes:

$$\varepsilon_1 : \vec{n_1} = \overrightarrow{AM} \times \overrightarrow{BM}, \quad \varepsilon_2 : \vec{n_2} = \overrightarrow{CM} \times \overrightarrow{DM}.$$

The direction vector $\vec{r}$ of the intersection is perpendicular to both normal vectors:

$$\vec{r} = \vec{n_1} \times \vec{n_2}.$$

We normalize $(\vec{r_0} - \vec{r}/|\vec{r}|)$ and scale in order to get the direction vector $\vec{s}$ of the intersection point:

$$\vec{s} = \pm(6,370 + 10)\,\text{km} \cdot \vec{r_0}.$$

The relevant sign depends on the specific situation. ◂◂◂

▸▸▸ **Application**: ***determining the position of an aircraft*** (Fig. 7.92)
How can an aircraft determine its position by measuring the distances to three fixed points $A$, $B$, and $C$ (e.g. by means of a laser or radar)?

*Solution*:
Let $a$, $b$, and $c$ denote the distances of the aircraft position $P$ relative to the three fixed points $A$, $B$, and $C$. Then, $P$ is contained in the intersection of the three spheres $\Sigma_A$, $\Sigma_B$, and $\Sigma_C$ centered at these points with radii $a$, $b$, and $c$. Any two spheres intersect each other along a circle $k_{AB}$, $k_{AC}$, and $k_{BC}$ (see Application p. 147). The carrier planes of two such intersections – for instance, $\varepsilon_{AB}$, $\varepsilon_{AC}$ – intersect each other along a straight line $s$. This line is to be intersected with one of the three spheres (resulting in two possible positions of $P$, of which only "the upper" one is possible). ◂◂◂

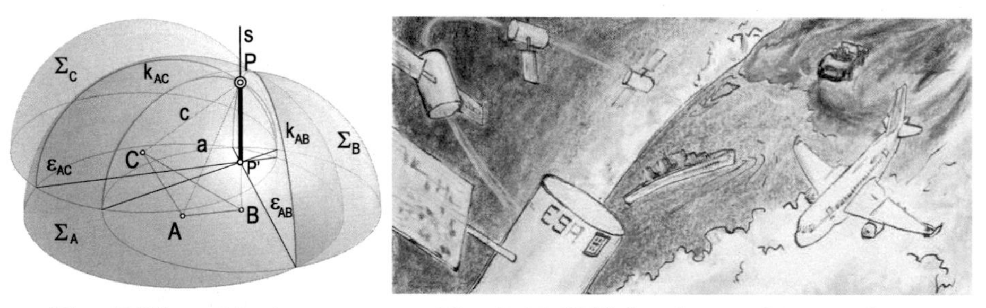


**Fig. 7.92** positioning **Fig. 7.93** GPS for planes, ships, and cars

▸▸▸ **Application**: ***Global Positioning System – GPS*** (Fig. 7.93)
The GPS has brought a revolution in the determination of locations. By means of handy devices, one can know his/her exact position at any time with an accuracy of up to a few meters (!). This is made possible by the fact that a relatively small number (24) of strategically distributed satellites (Fig. 7.91, right) constantly transmits signals containing their current position and the corresponding exact time. How can the device determine its own position?

*Solution*:
In principle, the GPS operates as described in Application p. 363: The current positions $A_1$, $B_1$, and $C_1$ of (at least) three satellites are obtained by radio.

For a first calculation, a position $Q$ close to the actual position $P$ is chosen (e.g., the last calculated position).
This yields the distances $a_1 = \overline{A_1Q}$, $b_1 = \overline{B_1Q}$, and $c_1 = \overline{C_1Q}$. Now, we intersect these three spheres centered at $A_1$, $B_1$, and $C_1$ (as in Application p. 363) and obtain a solution. However, the new positions $A_2$, $B_2$, and $C_2$ of the quickly moving satellites are assumed and the old distances are calculated as a result. This results in a "mixed position" $Q_{12}$. This, of course, does not agree with $Q$ or $P$ for $P \neq Q$. The distance $d = \overline{QQ_{12}}$ is a measure of the error: the smaller $d$, the closer the actual position $P$ to $Q$.
Now, we do the following: We repeatedly carry out the above-described "calculation using a sample" by systematically varying the coordinates of $Q(q_x/q_y/q_z)$ and observing the direction in which the distance $d$ decreases: Then, for example, the values of $d$ are obtained only by varying the $x$-value. Next, we can find a better $q_x$-value through relatively coarse interpolation. Now, we change the $q_y$-value and then the $q_z$-value to get the result. In the meantime, we will have to include new satellite positions into the calculation in order to account for possible changes in position.

⊕ *Remark*: Optimizing the step-by-step (*iterative*) search for the solution is a very challenging task, but we will not discuss this here. ⊕ ◂◂◂

**▸▸▸ Application**: ***position determination in earlier times***
GPS is an invaluable asset, especially in shipping: One can almost immediately get lost out in the ocean. How was the position determined in earlier times while travelling on the high seas?

**Fig. 7.94** Portuguese galley

**Fig. 7.95** The GPS fails – what now?

*Solution*:
The determination of the latitude $\varphi$ is relatively simple: On the northern hemisphere, we see the Pole Star under an inclination angle that equals the latitude $\varphi$ (on the southern hemisphere, we use the *Southern Cross*). It was also possible – using tables – to draw conclusions about $\varphi$ given the Sun's highest position (the Vikings presumably did so).

⊕ *Remark*: It was much more difficult to determine the longitude out on the ocean (the well-known Greenwich Observatory in the east of the city of London has been internationally recognized since 1885 as lying on the zero meridian). In order to determine the current longitude, it was necessary to have an absolutely precise clock on board, which was not technically feasible until the 19th century. The problem was indeed serious, and it was not uncommon for a captain to "lose his nerves" during an intermediate storm on a week-long voyage from the mainland to some remote island and to return on the right circle of latitude in the wrong direction because he believed he had missed his destination. ⊕ ◂◂◂

▸▸▸ **Application**: ***the "wrongly tilted" crescent Moon*** (Fig. 7.96)
The following phenomenon is well-known and appears over and over again as an unanswered question in the literature and on internet forums: In most cases, the bisector of the crescent Moon does not seem to be precisely directed at the Sun. Particularly at sunset, when you would expect the bisector to be horizontal, it mostly points upwards. How large is this "deviation angle"?

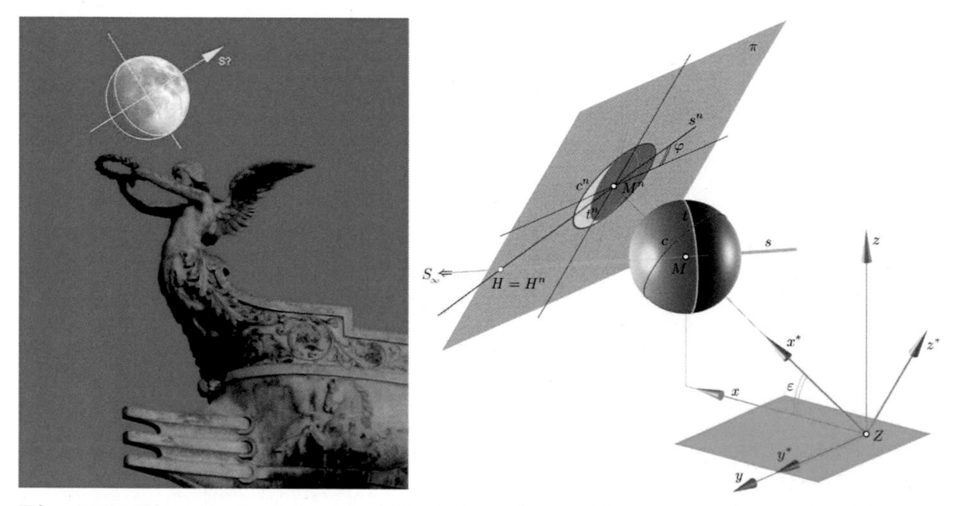


**Fig. 7.96** The seemingly wrong tilt of the crescent Moon at sundown: The bisector of the terminator points upwards!

We deal with close-ups of the Moon, which can only be produced by means of an efficient telephoto lens and deliver a good approximation of a normal projection where the theorem of the right angles (page 188) holds.
The path of the light from the Sun to the Moon is straight in any case. The outline of the Moon's sphere is circular. The terminator (a great circle on the Moon) appears as a half-ellipse. The center of this ellipse is the projection of the Moon's center. The vertices of the image ellipse lie diametrically opposed on the sphere's outline: The connecting line of these points in space has "principal position". Therefore, it is parallel to the sensor plane. The minor axis is orthogonal to the connection of the two vertices, and according to the theorem of right angles, it coincides with the normal projection of the circle's axis.
Now, we want to derive a formula for the tilt of the minor axis or the "bulbousness" of the ellipse. Imagine that the lens center is the origin of a Cartesian

coordinate system. For the sake of simplicity, we define the optical axis as the $x$-axis, the horizontal direction of the sensor plane as the $y$-direction, and the line of steepest slope of the sensor plane as the $z$-direction (Fig. 7.96). The direction to the Moon is also given by $\vec{m} = (1,0,0)$. The direction to the Sun is defined by the direction vector $\vec{s} = (s_x, s_y, s_z)$. Its normal projection onto the sensor plane has the components $\vec{s}_n = (0, s_y, s_z)$. We measure the angle $\varphi$ to the $y$-axis by normalizing $\vec{s}_n$ and multiplying by $\vec{y} = (0,1,0)$:

$$\cos\varphi = \vec{s}_n \bullet \vec{y} = s_y / \sqrt{s_y^2 + s_z^2 + s_z^2}. \tag{7.26}$$

The coordinates of the Moon and those of the Sun (in horizontal polar coordinates given through their azimuth angles $\alpha$, $\alpha^*$ or their difference $\delta = \alpha - \alpha^*$ and elevation angles $\varepsilon$ and $\varepsilon^*$) may be at our disposal (for example, by means of relevant software). Let us look at the Cartesian coordinates in a coordinate system which has the same $y$-axis, but whose $z$-axis is vertical: There, the direction vector to the Sun has the components

$$\vec{s}^* = (s_x^*, s_y^*, s_z^*) = (\cos\varepsilon^* \cos\delta, \cos\varepsilon^* \sin\delta, \sin\varepsilon^*). \tag{7.27}$$

Relative to the coordinate system that is twisted through the elevation angle, the following applies:

$$\vec{s} = (s_x^* \cos\varepsilon + s_z^* \sin\varepsilon, s_y^*, -s_x^* \sin\varepsilon + s_z^* \cos\varepsilon). \tag{7.28}$$

If we again insert (7.27) in (7.28), $\varphi$ can be calculated directly by inserting

$$s_y = \sin\delta \cos\varepsilon^*, \quad s_z = -\cos\delta \cos\varepsilon^* \sin\varepsilon + \sin\varepsilon^* \cos\varepsilon$$

in (7.26). The tilt $\psi$ of the sun rays to the picture plane is equivalent to the complementary angle to the negative $x$-axis $\vec{m}$ determined by

$$cos\psi = -\vec{s} \bullet \vec{m} = -s_x / \sqrt{s_y^2 + s_z^2} \text{ (with } \quad s_x = \cos\delta \cos\varepsilon^* \cos\varepsilon + \sin\varepsilon^* \sin\varepsilon).$$

This cosine value is also a measure of the thickness of the image of the terminator. ◂◂◂

# 8 Infinitesimal calculus

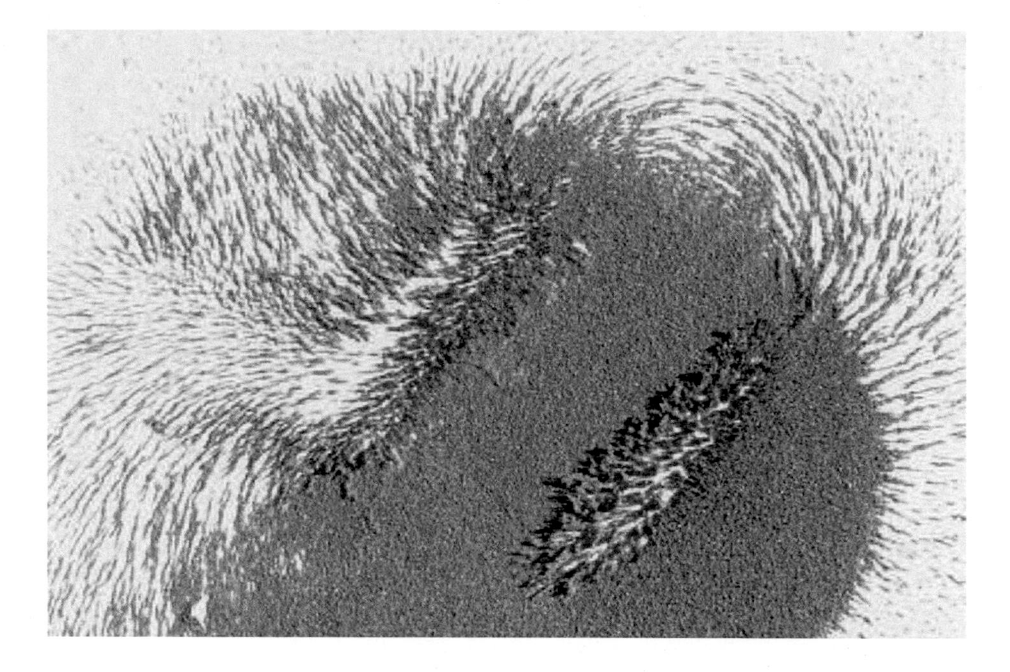

From Chapter 6, we know that differentiation yields information about the *instantaneous growth of a function.* This growth may be interpreted geometrically as well as physically (inclination of the tangent, instantaneous velocity, instantaneous acceleration, etc.). In this chapter, we will discover additional fields of application, such as examples involving minima and maxima, and the progression of power series.
Integral calculus can be described as the "inversion" of differential calculus. Broadly speaking, we get information about the *area* under a function graph by performing integration. This property may be interpreted in several ways. Geometrically speaking, it becomes possible to calculate areas, volumes, arc lengths, or a necessary amount of energy. Probabilities, life expectancies (and much more) likewise fall into the realm of integral calculus. The practical applications are so diverse that we can only scratch their surface.
The integration of functions is often taxing to novice mathematicians. Sometimes it can even be impossible. In any case, integrals are now being increasingly evaluated through numerical means by computers.


G. Glaeser, *Math Tools*, https://doi.org/10.1007/978-3-319-66960-1_8

## 8.1 Calculation with infinitesimal quantities

Differential and integral calculus are not fundamentally different. Both are concerned with "infinitely small" (infinitesimal) quantities – in both cases, we speak of *infinitesimal calculus.* In general, it can be stated that the calculation steps for non-linear problems often require "infinitesimal deliberations". This new method of thinking may require some getting used to, but is ultimately quite easy to learn. At first, the most essential part is to imagine the relations as "very small", but not yet infinitesimally (or "infinitely") so. Then, one may attempt a *transition* into the infinitesimal way of thinking. When analysed infinitesimally, the relations are often much easier to understand than the rather complicated result would lead us to believe.

▸▸▸ **Application**: ***tapering of a jet of water*** (Fig. 8.1)
When a jet of water exits a faucet, it produces a surface of revolution that tapers down towards its lower end. How is its meridian curve defined?

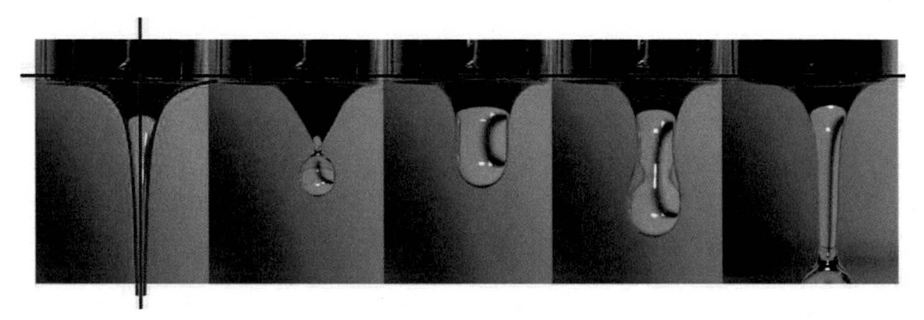

**Fig. 8.1** continuous and interrupted jet of water

*Solution*:

Let $Q_0$ be the cross-section at the height where the jet of water exits the tap. The exit velocity is $v_0$. Over time $dt$ the amount of water $dV = Q_0\, v_0\, dt$ emerges from the tap. After $t$ seconds, the emerging water is accelerated to the speed $v = v_0 + g\,t$, and the cross-section $Q = Q(t)$ equals

$$Q_0\, v_0\, dt = Q\, v\, dt \;\Rightarrow\; Q(t) = \frac{Q_0\, v_0}{v_0 + g\,t}.$$

Considering an idealized circular cross-section, the following relation holds for the radius $x$ of the cross-section:

$$x(t) = \frac{1}{\sqrt{\pi}}\sqrt{\frac{Q_0\, v_0}{v_0 + g\,t}}.$$

The following is true for the distance passed by the water, see Formula (2.28):

$$s(t) = v_0\,t + \frac{g}{2}\,t^2.$$

Thus, the meridian curve is already given by the parameter representation.

⊕ *Remark*: In practice, the jet will keep changing its shape due to the non-uniform flow of water. When the intensity of the water stream is reduced, droplets eventually start to form (see Fig. 7.76). ⊕ ◂◂◂

We have already made a few infinitesimal deliberations. First, let us consider that we have interpreted the tangent at the point $P$ of a curve as a straight line connecting $P$ with a very close "adjacent point" $Q$.
If a transition to the limit $Q \to P$ is performed, then the straight line $PQ$ becomes exactly the tangent. This transition can be performed geometrically by differentiation resulting in the inclination of the tangent value.
The first derivative of a function is a function measuring the change of its values. In general, a function may be differentiated an arbitrary number of times. The second derivative yields a measure of the change of the tangent inclination, and so on. In the following section, we will have a closer look at this.
To encourage a more careful manner of thinking about infinitesimal relations, let us now cite the following famous paradox:

**▸▸▸ Application: Zeno's *paradox of the tortoise and* Achilles**

A tortoise is moving at a constant speed of $1\,\frac{\mathrm{m}}{\mathrm{s}}$ (this number may be a little high for such a slow animal, but that is not relevant here), and it has an advantage of 10 m. A fast runner – in *Zeno*'s example, it is the mythological figure of *Achilles* (≈ 450 BC, Fig. 8.2) – moves 10 times faster. When does the Greek hero catch up with the slower reptile?

**Fig. 8.2** *Achilles* and the tortoise

*Solution*:
The following line of reasoning contains an "error":
When the runner has reached the starting position of the tortoise, it has

already moved on by 1 m. Once this new position is reached by *Achilles*, the animal is already 0.1 m farther. When he reaches once again the tortoise's position, it has already moved away from the preceding position by 0.01 m – and so it continues. This step can be repeated an infinite number of times, with the tortoise always remaining in front. It would seem that *Archilles* can never catch up with the tortoise ...
The error lies in the conclusion: This potentially infinite number of steps describes only a small interval of time: $1\,\mathrm{s} + 0.1\,\mathrm{s} + 0.01\,\mathrm{s} + \ldots = 1.1111\ldots\,\mathrm{s}$. What happens *after* this interval is an altogether different question.
Let us now turn to the "true" solution to the problem:
For the trajectories of the runners, the following must hold true: $s_2 = s_1 + 10\,\mathrm{m}$. At the time of overtaking, the time $T$ has elapsed. Then,

$$10\,\frac{\mathrm{m}}{\mathrm{s}}\,T = 1\,\frac{\mathrm{m}}{\mathrm{s}}\,T + 10\,\mathrm{m} \Rightarrow T = 1.\dot{1}\,\mathrm{s} \Rightarrow s = 1.\dot{1}\,\mathrm{m}.$$ ◂◂◂

▸▸▸ **Application**: ***sum of fractions – the joy of proof***
Prove the formula

$$\sum_{k=1}^{\infty} \frac{1}{4^k} = \frac{1}{4} + \frac{1}{4^2} + \frac{1}{4^3} + \ldots = \frac{1}{3}$$

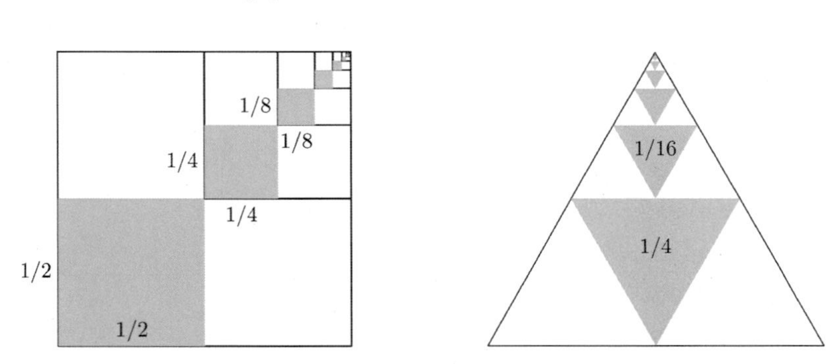


**Fig. 8.3** proofs without words

*Solution*:
Instead of just using well-known formulas, we give two geometrical proofs:[1]
According to Fig. 8.3 (left), the squares on the upper left- and lower right-hand are left as is, but the upper right-hand square is itself subdivided into four squares, and we repeat recursively the process begun with the initial square. Of the area that has been "processed" (three-fourths of the total), one-third has been colored. At the next step, we again color one-third of the area, and so on. The amount of the area that has been processed converges to 1, and so the colored area converges to one-third.
The second proof (Fig. 8.3, right) is left to the reader. ◂◂◂

[1] *Roger B. Nelsen*: Proofs without Words: Exercises in Visual Thinking (Classroom Resource Materials).

## 8.2 Curve sketching

Drawing graphs of ever more complex functions has become easy in the age of computers. It is often not common points of the curve that are of practical interest, but points with horizontal tangents, which are known as *minima* and *maxima* respectively. Occasionally, the so-called *inflection points* are also important. These are the points where a curve passes from a "right curve" to a "left curve" (and vice versa).
We have the fundamental theorem:

> The local extremal values of a curve $y = f(x)$ must satisfy the condition $y'(x) = 0$. The inflection point must satisfy the condition $y''(x) = 0$.

***Proof***: The first part is trivial: At a minimum or maximum (that is, at an extreme value), the tangent of the curve has to be horizontal, that is, $y' = 0$.
Now, consider the part of the graph $y(x)$ that is curved to the right. If we consider the tangents with increasing $x$, their slopes $y'(x)$ decrease continuously. The increase of $y'$ is, therefore, negative: $\frac{dy'}{dx} = y'' < 0$. Similarly, the preceding argument applies to "left curves" where $y'' > 0$ holds. Therefore, the transition point (inflection point) has to satisfy $y'' = 0$. ⊙

⊕ *Remark*: The above-mentioned conditions are necessary but not sufficient for the existence of a local extremum or point of inflection:
If $y'(x_0) = 0$ and $y''(x_0) = 0$ at a specific place $x_0$, then $(x_0/y(x_0))$ is a *saddle point* (provided that $y'''(x_0) \neq 0$).
If both $y''(x_0) = 0$ and $y'''(x_0) = 0$ at $x_0$, then a *higher-order point of inflection* or a *flat point* is present (provided that either an even-ordered or an odd-ordered derivative is the first to be not equal to zero).
In practice, however, you first draw the graph with a computer and then "optically" check such subtleties. ⊕

▸▸▸ **Application**: **Gauss*ian bell curve*** (Fig. 8.4)
Consider the "normal distribution" curve introduced by *Gauß*

$$y = \frac{1}{\sqrt{2\pi}} e^{-\frac{x^2}{2}}. \tag{8.1}$$

Where does the curve have extreme values or inflection points? Calculate the slope of the "inflection tangents".

*Solution*:
Since the argument $x$ occurs only in an even power, we have $y(x) = y(-x)$. Therefore, the curve is symmetric with respect to the $y$-axis. Like the expo-

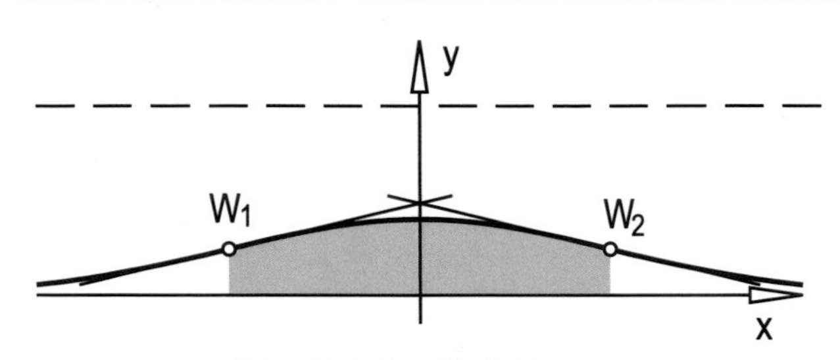


**Fig. 8.4** *Gauß*'s bell curve

nential function, it does not have a root but approaches the $x$-axis asymptotically on both sides.

We compute the first derivative and use the abbreviation $c = \frac{1}{\sqrt{2\pi}}$, which is a constant factor. According to the chain rule, we have

$$y' = c \cdot e^{-\frac{x^2}{2}} \cdot (-x) = -x\,y.$$

Since $y \neq 0$, we have $y' = 0$ (extreme value) only for $x = 0$. There is obviously a maximum. In order to calculate the points of inflection, we need the second derivative. The product rule yields

$$y'' = -(1 \cdot y + x \cdot y') = -(y - x^2 \cdot y) = (x^2 - 1)\,y.$$

As might be expected, $y''(0) < 0$, which confirms that a maximum is present at $x = 0$. Inflection points generally occur at $y'' = 0$. Since $y \neq 0$ always holds, $x^2 - 1 = 0$ gives $x_w = \mp 1$. The corresponding $y$-values are both equal because of the symmetry of the curve: $y_w = c\,e^{-1/2} = \frac{1}{\sqrt{2\pi\,e}}$. The inflection tangents have the slope $y'_w = \mp(-x_w\,y_w) = \pm y_w$.

⊕ *Remark*: The bell curve or normal distribution curve has a fundamental significance in probability theory. We will meet it again when it comes to the calculation of areas. The inflection points of the curve have to do with the so-called "variance of the distribution". ⊕ ◂◂◂

### ▸▸▸ Application: *maximal circle of curvature of a parabola*

What radius of curvature $\varrho$ does the basic parabola $y = a\,x^2$ have at the point $x = 0$? For this, we use the formula

$$\varrho = \frac{(1 + y'^2)^{\frac{3}{2}}}{y''}. \tag{8.2}$$

*Solution*:
With $y' = 2a\,x$ and $y'' = 2a$ and using Formula (8.2), we get

$$\varrho(x) = \frac{\left(1 + (2a\,x)^2\right)^{\frac{3}{2}}}{2a} \Rightarrow \varrho(0) = \frac{1}{2a}.$$

◂◂◂

**▸▸▸ Application**: ***osculating circles of the sine curve***

Show that the zeros and the points of inflection of the sine curve $y = a\sin(b\,x + c)$ are identical. Calculate the minimal osculating circle.

*Solution*:

Let $y' = a\,b\cos(b\,x + c)$ and $y'' = -a\,b^2\sin(b\,x + c) = -b^2\,y$. The conditions for the zero ($y = 0$) and the point of inflection ($y'' = 0$) are, therefore, identical. If $\sin(b\,x + c) = 0$, then $\cos(b\,x + c) = \pm 1$. Therefore, we have the tangent of inflection with the slope $\pm a\,b$. A minimum $x = x_s$ occurs if $y'(x_s) = 0$ and $y(x_s) = \pm a$. Therefore, it follows that $y''(x_s) = \mp a\,b^2$.

With Formula (8.2), we get $\varrho(x_s) = \mp\dfrac{1}{a\,b^2}$.

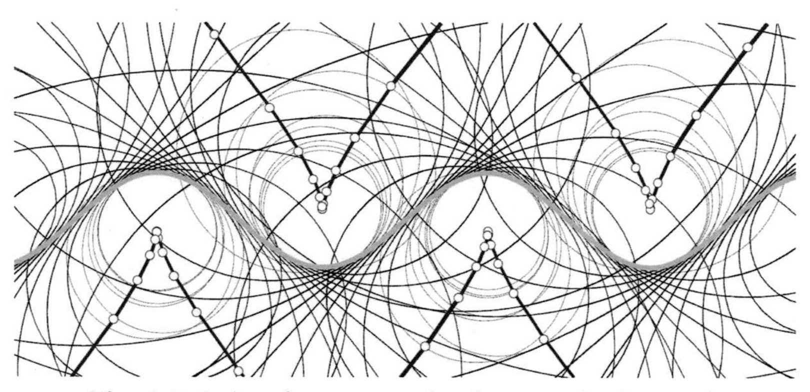

**Fig. 8.5** circles of curvature of a sine curve in theory ...

**Fig. 8.6** ... and in practice

For the prototype of the sine curve $y = \sin x$, we have $a = b = 1$ and $c = 0$. The tangents are, then, inclined by 45°, and the maximal osculating circle has the radius $\varrho(x_s) = 1$. Fig. 8.5 shows a general sine curve with a number of osculating circles. The position of the midpoints of the circles of curvature (light blue) is called the *evolute*. This curve has its points at infinity in the directions normal to the tangents of inflection. ◂◂◂

## 8.3 Optimization problems

So far, we have determined the extremal values of a function $y = f(x)$ by computing the zeros of its first derivative. This is a very common technique in applied mathematics. The most striking task in optimization problems is to find the *target function* $f(x)$. The variables need not be called $x$ and $y$. We will give some examples in order to illustrate how to deal with so-called "optimization problems".

▸▸▸ **Application**: ***cardboard box of maximal volume***

A rectangular cardboard sheet is to be formed in such a way that the paper box with the largest possible volume is produced.

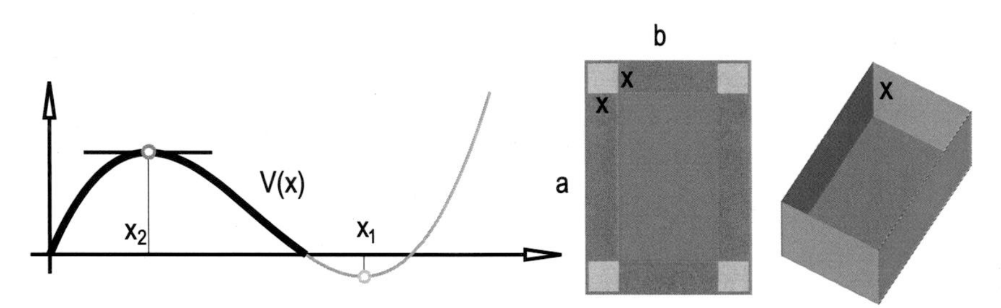


**Fig. 8.7** box made from a rectangle

*Solution*:

With the conditions of Fig. 8.7, the dimensions $a$ and $b$ of the sheet are given. We now cut small squares of side length $x$ at the four corners. When we then fold and join the resulting rectangles, we have constructed a cardboard box with the dimensions $(a-2x)\cdot(b-2x)\cdot x$. Its volume is $V(x) = (a-2x)(b-2x)x$, and it depends on $x$, which is to be determined such that $V(x)$ attains a maximum. $V(x)$ is our target function. The condition for a maximum is $V'(x) = 0$. We expand

$$V(x) = (ab - 2b\,x - 2a\,x + 4x^2)x = 4\,x^3 - 2(a+b)x^2 + ab\,x$$

and, therefore,

$$V'(x) = 12\,x^2 - 4(a+b)x + ab.$$

In order to find the zeros of $V'(x) = 0$, we have to solve a quadratic equation:

$$x_{1,2} = \frac{4(a+b) \pm \sqrt{16(a+b)^2 - 4\cdot 12\cdot (ab)}}{24}.$$

We divide by 4 to simplify the equation and get

$$x_{1,2} = \frac{(a+b) \pm \sqrt{(a+b)^2 - 3ab}}{6} = \frac{(a+b) \pm \sqrt{a^2 - ab + b^2}}{6}.$$

Which of the two solutions is the right one? We can recognize the correct solution either from the computer drawing, or we find the second derivative – which is justifiable in the present situation:

$$V''(x) = 24\,x - 4(a+b).$$

We substitute $x_{1,2}$ and obtain

$$V''(x_{1,2}) = 24\,\frac{(a+b) \pm \sqrt{a^2 - ab + b^2}}{6} - 4(a+b) = \pm 4\sqrt{a^2 - ab + b^2}.$$

Therefore, the minimum is $V(x_1)$ since $V''(x_1) > 0$, while $V''(x_2) < 0$ yields the maximum because $V''(x_2) < 0$. ◀◀◀

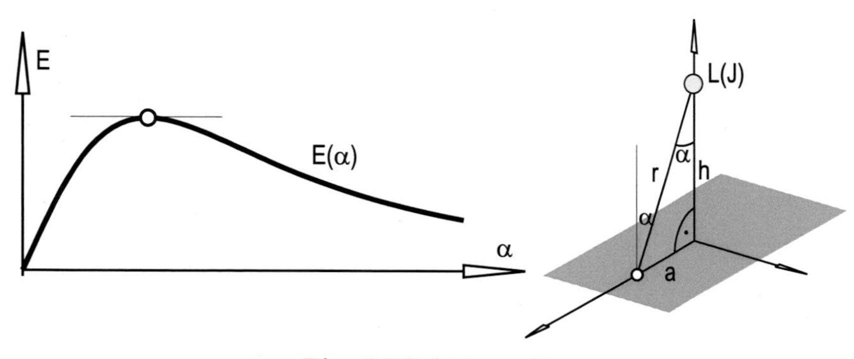


**Fig. 8.8** light intensity

▸▸▸ **Application**: ***optimal lighting*** (Fig. 8.8)
How high does a light source have to be placed above a horizontal plane so that a certain point $P$ at a fixed lateral distance $a$ is as well illuminated by the source as possible? For the illumination intensity $E$ as a function of the distance $r$, the angle of incidence $\alpha$, and the (constant) intensity $J$ of the light source (light intensity), we have the equation

$$E = \frac{J}{r^2}\cos\alpha.$$

*Solution*:
The so-called target function $E$ depends on two variables, namely $r$ and $\alpha$. However, the two variables are not independent of each other, because $\frac{a}{r} = \sin\alpha$, and consequently, $r^2 = \frac{a^2}{\sin^2\alpha}$. Therefore, we have

$$E(\alpha) = \frac{J}{a^2}\sin^2\alpha\,\cos\alpha.$$

The positive factor $\frac{J}{a^2}$ only causes a scaling of the target function. As a result, the simplified function

$$\widetilde{E}(\alpha) = \sin^2\alpha\,\cos\alpha$$

will have the same extreme values for $\alpha$. For these extreme values, the following must apply: $\widetilde{E}' = 0$. Of course, the prime $'$ indicates the derivative with respect to the variable $\alpha$. We apply the product rule and get

$$\widetilde{E}'(\alpha) = \frac{d\widetilde{E}(\alpha)}{d\alpha} = 2\sin\alpha\cos\alpha\cos\alpha + \sin^2\alpha(-\sin\alpha) = 0.$$

We can assume $\sin\alpha \neq 0$, and thus, $\alpha \neq 0$. This is only possible if the light source is directly above $P$. Thus, cancelling $\sin\alpha$ is permitted, and we find

$$2\cos^2\alpha - \sin^2\alpha = 0 \Rightarrow 2\cos^2\alpha = \sin^2\alpha \Rightarrow \tan^2\alpha = 2 \Rightarrow \tan\alpha = \pm\sqrt{2}.$$

Therefore, for optimal illumination, the light source has to be at a height of

$$h = \frac{a}{\tan\alpha} = \frac{a}{\sqrt{2}}.$$ ◂◂◂

▸▸▸ **Application**: ***beams with maximal load bearing capacity*** (Fig. 8.9)
A beam of greatest bearing capacity $W$ is to be cut out from a cylindrical tree trunk of a given diameter $d$. What is the ratio *width* : *height* ($b : h$) when the following function should achieve a maximum:

$$W = c\,b\,h^2 \quad (c > 0 \text{ material constant})?$$

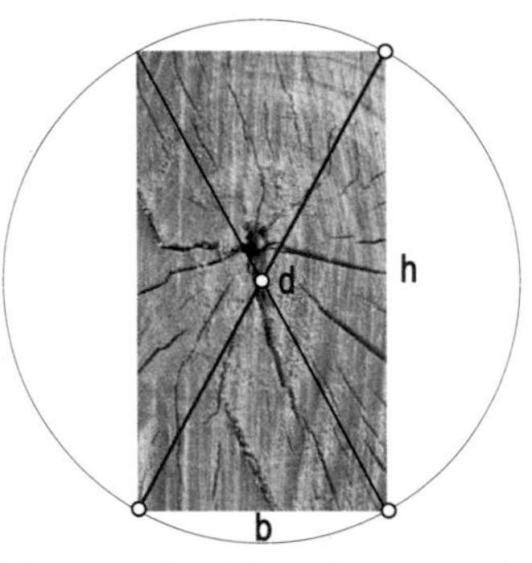


**Fig. 8.9** load bearing capacity in theory ...

**Fig. 8.10** ... and in practice

*Solution*:
$W$ depends on $b$ and $h$. On the other hand, $b$ and $h$ depend on each other by the relation $b^2 + h^2 = d^2$. Thus,

$$W(b) = c\,b(d^2 - b^2) = c\,(d^2\,b - b^3).$$

$W(b)$ is maximal (or minimal) if the derivative with respect to $b$ vanishes:

$$\frac{dW}{db} = c\,(d^2 - 3\,b^2) = 0.$$

Therefore, one obtains $b$ and, as a further consequence, $h$ from

$$\Rightarrow b = (\pm)\frac{d}{\sqrt{3}} \Rightarrow h = \frac{d\sqrt{2}}{\sqrt{3}} \Rightarrow b : h : d = 1 : \sqrt{2} : \sqrt{3}.$$

⊕ *Remark*: Fig. 8.10 shows that beams with this cross-section are, in fact, used in practice. The reasons behind the collapse of a building (such as the former Reichsbrücke in Vienna, which cracked in 1976) depends on a variety of factors. Naturally, one reason is that the load bearing capacity increases only with the square of the scale, while the dead weight of objects increases by the power of 3. This means that the load-bearing beams of a model of a bridge can hold significantly more than the beams of the real bridge (see Chapter 3). ⊕ ◂◂◂

▸▸▸ **Application**: ***optimal roof inclination***

How large should the inclination angle $\alpha$ of a roof's surface be for the rainwater to flow off as quickly as possible?

*Solution*:

The time $t$ required for rainwater to run down shall be given as a function of the inclination angle $\alpha$. Let $\Delta x$ be a small piece in the horizontal direction. The corresponding rafter length is, then,

$$\Delta s = \frac{\Delta x}{\cos\alpha}.$$

The height difference is $\Delta h = \Delta x \tan\alpha$. The increase in speed, i.e. the run-off acceleration, is thus

$$\Delta v = \sqrt{2g\,\Delta h} = \sqrt{2g\,\tan\alpha\,\Delta x}.$$

The increase of elapsed time $\Delta t$ when running off by $\Delta s$ equals

$$\Delta t = \frac{\Delta s}{\Delta v} = \frac{\frac{\Delta x}{\cos\alpha}}{\sqrt{2g\,\tan\alpha\,\Delta x}}.$$

Thus, the square of the increase of elapsed time reads

$$(\Delta t)^2 = \frac{\frac{(\Delta x)^2}{\cos^2\alpha}}{2g\,\tan\alpha\,\Delta x} = \frac{1}{2g}\frac{\Delta x}{\cos^2\alpha\,\tan\alpha} = \frac{1}{g}\frac{\Delta x}{2\sin\alpha\cos\alpha} = \frac{1}{g}\frac{\Delta x}{\sin 2\alpha}.$$

If the increase of time should be minimal, then so should its square. However, the reciprocal of this is then *maximal*. Thus, we can say, $\Delta t$ becomes minimal if the expression $\sin 2\alpha$ becomes maximal, thus

$$\frac{d}{d\alpha}\sin 2\alpha = 0 \Rightarrow 2\cos 2\alpha = 0 \Rightarrow 2\alpha = 90^\circ \Rightarrow \alpha = 45^\circ.$$

⊕ *Remark*: With an inclination angle > 45°, the run-off path that is required to overcome the horizontal distance $\Delta x$ already increases to such an extent that more time is required for draining the rainwater. Therefore, the angle 45° represents the best "compromise" between the brevity of the run-off path $\Delta x$ and the acceleration caused by the slope. ⊕ ◂◂◂

▸▸▸ **Application**: **Snell's *law of refraction*** (Fig. 8.11)
We consider two optical media $I$ and $II$ with different optical densities in which light propagates at respective speeds $c_1$ and $c_2$. The two media are separated by a single layer. Experiments have given consistent results and confirmed that the law of refraction comes into effect at the transition point (kink). Let the respective angles $\alpha$ and $\beta$ be enclosed by the incoming and the outgoing ray and the layer's normal. Then, *Snell*'s law reads

$$\frac{\sin\alpha}{\sin\beta} = \frac{c_1}{c_2}.$$

Show that this law, named after *Snell*, is based on *Fermat*'s principle, according to which light "looks for" that path from $A$ to $B$ where it reaches $B$ in the shortest possible *time*.

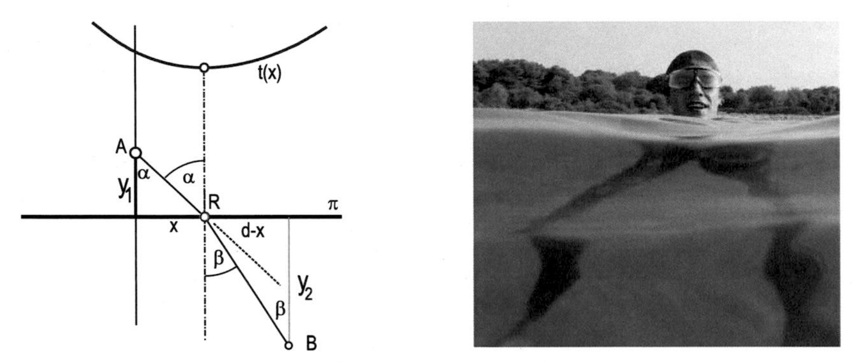


**Fig. 8.11** total time function during refraction and imaging

*Solution*:
The problem can be considered in the plane $\varepsilon$, which contains $A$ and $B$ and is perpendicular to the planar interface $\pi$. We identify the intersection line $\varepsilon \cap \pi$ with the $x$-axis and assign the coordinates $A(0/y_1)$ and $B(d/y_2)$ to the two points. The kink shall have the coordinates $R(x/0)$. The following holds: *time* = *distance* / *speed.* Therefore, the "total time function" has the form

$$t(x) = \frac{\overline{AR}}{c_1} + \frac{\overline{RB}}{c_2} = \frac{\sqrt{y_1^2 + x^2}}{c_1} + \frac{\sqrt{y_2^2 + (d-x)^2}}{c_2}.$$

If the total time is minimal, then $t'(x) = 0$ has to hold:

$$t'(x) = \frac{1}{c_1}\frac{2x}{2\sqrt{y_1^2 + x^2}} + \frac{1}{c_2}\frac{2(d-x)(-1)}{2\sqrt{y_2^2 + (d-x)^2}} = 0$$

or, equivalently,

$$t'(x) = \frac{1}{c_1}\frac{x}{\overline{AR}} - \frac{1}{c_2}\frac{d-x}{\overline{RB}} = 0.$$

With $\dfrac{x}{\overline{AR}} = \sin\alpha$ and $\dfrac{d-x}{\overline{RB}} = \sin\beta$, we simplify the equation to

$$\frac{1}{c_1}\sin\alpha - \frac{1}{c_2}\sin\beta = 0.$$

From this, we find the law of refraction

$$\frac{\sin\alpha}{\sin\beta} = \frac{c_1}{c_2}.$$

⊕ *Remark*: If we consider all the rays which pass through $B$, and apply the law of refraction to them, the outgoing rays in their extension to the side of $B$ form a curve called the *diacaustic* (Fig. 8.11, right). An observer at $A$ can now only guess the position of $B$ and assume it to be at the contact point with the diacaustic $B^*$. The vertical bar through $B$ appears strangely "uplifted". ⊕ ◂◂◂

▸▸▸ **Application**: ***the most voluminous conical cocktail glass*** (Fig. 8.12)
At which opening angle $2\alpha$ does the conical cocktail glass have the largest volume with a given slant height $s$ (cone)?

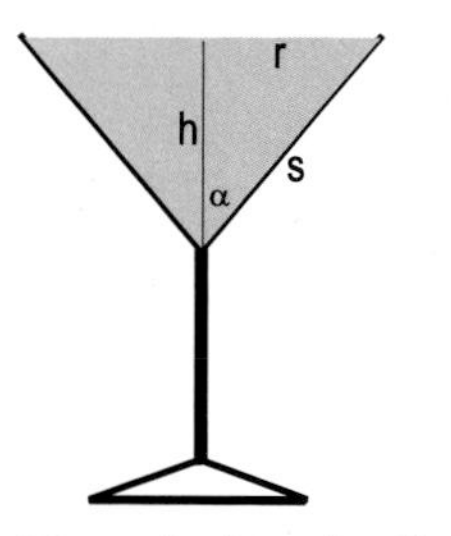


**Fig. 8.12** conical cocktail glass

**Fig. 8.13** optimization

*Solution*:
The volume of the cone is

$$V = \frac{1}{3}\pi r^2 h = \frac{1}{3}\pi\,(s\,\sin\alpha)^2\,s\,\cos\alpha = \frac{1}{3}\pi\,s^3\,\sin^2\alpha\,\cos\alpha.$$

It becomes maximal if the expression $\widetilde{V}(\alpha) = \sin^2\alpha\,\cos\alpha$ becomes maximal, i.e. if $\dfrac{d}{d\alpha}\widetilde{V} = 0$:

$$2\sin\alpha\cos\alpha\cos\alpha + \sin^2\alpha(-\sin\alpha) = 0 \Rightarrow 2\sin\alpha\cos^2\alpha = \sin\alpha\sin^2\alpha.$$

The case $\sin\alpha = 0 \Rightarrow \alpha = 0$ leads to a minimum, so that we can reduce by $\sin\alpha \neq 0$ and obtain

$$2\cos^2\alpha = \sin^2\alpha \Rightarrow \tan^2\alpha = 2 \Rightarrow \tan\alpha = \sqrt{2} \Rightarrow \alpha \approx 55°.$$

Thus, the opening angle of the cone is $2\alpha \approx 110°$. ◂◂◂

▸▸▸ **Application**: ***maximum trajectory length on a hillside*** (Fig. 8.14)
What is the angle $\alpha$ at which a projectile has to be launched to get as far as possible in a terrain with inclination angle $\varepsilon$?

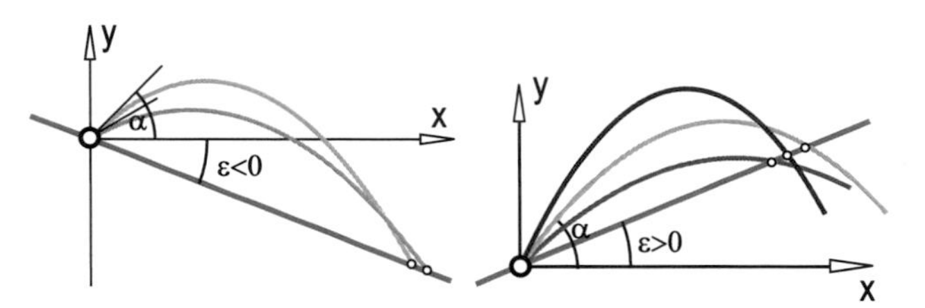


**Fig. 8.14** trajectory on a hillside

*Solution*:
We have often worked with the ballistic trajectory (parabola):

$$(1) \quad x = v_0 t \cos\alpha, \quad (2) \quad y = v_0 t \sin\alpha - \frac{g}{2}t^2.$$

Now, we intersect it with the inclined terrain

$$(3) \quad y = x \tan\varepsilon = k\,x.$$

From (1) and (3), we conclude

$$y = k\,x = k\,v_0\,t\cos\alpha \Rightarrow y\tan\alpha = k\,v_0 t\sin\alpha.$$

Substituting this into (2) yields

$$y = \frac{y\tan\alpha}{k} - \frac{g}{2}\cdot\frac{y^2}{k^2 v_0^2\cos^2\alpha}.$$

We may reduce by $y$ if we exclude the trivial solution $y = 0$:

$$\frac{g}{2}\cdot\frac{y}{k^2 v_0^2\cos^2\alpha} = \frac{\tan\alpha}{k} - 1 \Rightarrow y = \frac{2}{g}k^2 v_0^2\cos^2\alpha\left(\frac{\sin\alpha}{k\cos\alpha} - 1\right)$$

$$\Rightarrow y = \frac{2}{g}k v_0^2(\sin\alpha\cos\alpha - k\cos^2\alpha).$$

The value $y$ is an extremum if the expression

$$f(\alpha) = \sin\alpha\cos\alpha - k\cos^2\alpha = \frac{\sin 2\alpha}{2} - k\cos^2\alpha$$

is an extremum, which happens for $f'(\alpha) = 0$:

$$f'(\alpha) = \cos 2\alpha + 2k\cos\alpha\sin\alpha = \cos 2\alpha + k\sin 2\alpha = 0 \Rightarrow \tan 2\alpha = -\frac{1}{k}.$$

*Numerical example*:
$\varepsilon = -45°$ (on a downward terrain):

$$k = \tan\varepsilon = -1 \Rightarrow \tan 2\alpha = 1 \Rightarrow 2\alpha = 45° \Rightarrow \alpha = 22.5°,$$

$\varepsilon = 45°$ (on an upward terrain):

$$k = \tan\varepsilon = 1 \Rightarrow \tan 2\alpha = -1 \Rightarrow 2\alpha = 135° \text{ (!)} \Rightarrow \alpha = 67.5°.$$

◂◂◂

## 8.4 Series expansion

Up until a few decades ago calculation with trigonometric functions, logarithms, and exponential functions was a very tedious task. If one wanted to work with precision, one had to constantly refer to tables and then interpolate between the table values. In applied mathematics, one was satisfied – justifiably – with less precision and worked with the "slide rule".
With the advent of electronic pocket calculators, sine values or logarithms appeared on the "flashy" display (in less than a second). Today, an ordinary PC needs only (a fraction of) microseconds to accomplish the same task. How is this possible?
The solution to the problem is that non-algebraic (transcendental) functions can be approximated by algebraic power functions which are called *power series*. For the latter functions, one "only" needs to be able to multiply and divide if one knows the derivatives of the functions.
With certain limitations, *Taylor*'s fundamental formula holds:

$$\boxed{f(x) = \sum_{k=0}^{\infty} \frac{f^{(k)}(x_0)}{k!}(x - x_0)^k.} \tag{8.3}$$

The value $x_0$ is called the *knot*. The number $k! = k\cdot(k-1)\cdot(k-2)\cdot\ldots\cdot3\cdot2\cdot1$ is called the $k$ factorial. It proves to be convenient to define $0! = 1$.
The prerequisite for this series is that $|x - x_0|$ has to be "small enough". In many cases, $|x - x_0| < 1$ is sufficient.
***Proof***: Written in detail, we obtain

$$f(x) = f(x_0) + \frac{f'(x_0)}{1!}(x - x_0) + \frac{f''(x_0)}{2!}(x - x_0)^2 + \frac{f'''(x_0)}{3!}(x - x_0)^3 + \ldots.$$

We make the ansatz

$$f(x) = a_0 + a_1(x - x_0) + a_2(x - x_0)^2 + a_3(x - x_0)^3 + \ldots$$

and attempt to determine the coefficients $a_k$. This is done by repeated differentiation:

$$f'(x) = 1\cdot a_1 + 2\cdot a_2(x - x_0) + 3\cdot a_3(x - x_0)^2 + 4\cdot a_4(x - x_0)^3 + \ldots,$$
$$f''(x) = 2\cdot1\cdot a_2 + 3\cdot2\cdot a_3(x - x_0) + 4\cdot3\cdot a_4(x - x_0)^2 + \ldots,$$
$$f'''(x) = 3\cdot2\cdot1\cdot a_3 + 4\cdot3\cdot2\cdot a_4(x - x_0) + 5\cdot4\cdot3\cdot a_5(x - x_0)^2 + \ldots.$$

Assuming $x = x_0$ in all these equations, we obtain

$$f(x_0) = a_0 \Rightarrow a_0 = f(x_0) = \frac{f(x_0)}{0!} \text{ (since } 0! = 1),$$
$$f'(x_0) = 1\cdot a_1 \Rightarrow a_1 = f'(x_0) = \frac{f'(x_0)}{1!} \text{ (since } 1! = 1)$$
$$f''(x_0) = 2\cdot1\cdot a_2 \Rightarrow a_2 = \frac{f''(x_0)}{2!},$$
$$f'''(x_0) = 3\cdot2\cdot1\cdot a_3 \Rightarrow a_3 = \frac{f'''(x_0)}{3!},$$
$$\vdots$$

$$f^{(k)}(x_0) = k \cdot (k-1) \cdot \ldots \cdot 3 \cdot 2 \cdot 1 \cdot a_k \Rightarrow a_k = \frac{f^{(k)}(x_0)}{k!}. \qquad \odot$$

For the knot $x_0 = 0$, the formula was already given before *Taylor* by *Maclaurin*:

$$\boxed{f(x) = f(0) + \frac{f'(0)}{1!}x + \frac{f''(0)}{2!}x^2 + \frac{f'''(0)}{3!}x^3 + \ldots.} \tag{8.4}$$

Again $|x - x_0|$ has to be small enough. Formula (8.4) allows us to calculate most of the basic transcendental functions.

►►► **Application**: ***series expansion of the exponential function***
Show that the exponential function is given by

$$e^x = 1 + x + \frac{x^2}{2!} + \frac{x^3}{3!} + \frac{x^4}{4!} + \ldots. \tag{8.5}$$

*Solution*:
We simply "prove" this by direct calculation: The Euler function can be differentiated as often as desired and does not change:

$$(e^x)' = (e^x)'' = \ldots = e^{(k)}(x) = e^x.$$

Therefore, $e^{(k)}(0) = e^0 = 1$, and thus we have already verified Formula (8.5). ◄◄◄

It is barely any more difficult to calculate the series expansion of the sine or cosine function:

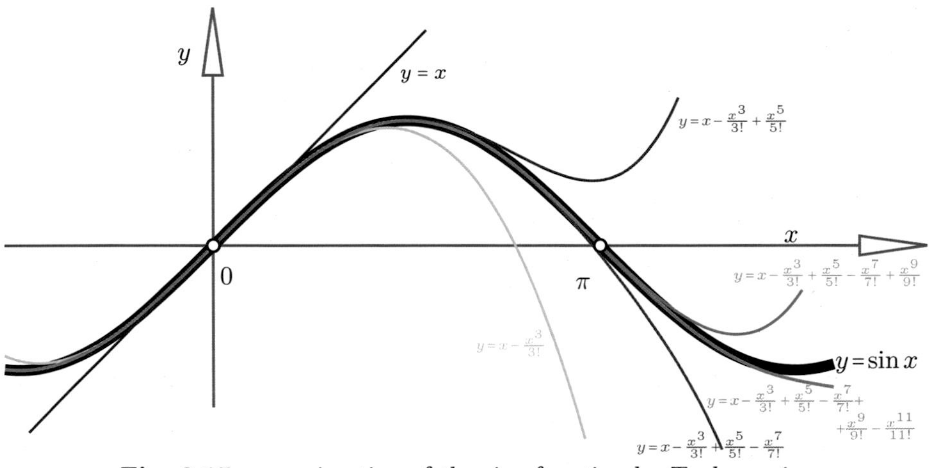


**Fig. 8.15** approximation of the sine function by Taylor series

►►► **Application**: ***series expansion of the sine function*** (Fig. 8.15)
Show that

$$\sin x = x - \frac{x^3}{3!} + \frac{x^5}{5!} - \frac{x^7}{7!} + \ldots, \tag{8.6}$$

$$\cos x = 1 - \frac{x^2}{2!} + \frac{x^4}{4!} - \frac{x^6}{6!} + \ldots. \tag{8.7}$$

***Proof***: We have $(\sin x)' = \cos\ x$, $(\sin x)'' = -\sin(x)$, $(\sin x)''' = -\cos x$. The fourth derivative is again the function itself ($(\sin x)^{(4)} = \sin x$). So, the derivatives repeat themselves periodically. Therefore, $\sin 0 = 0$, $\sin' 0 = 1$, $\sin'' 0 = 0$, $\sin''' 0 = -1$, and thus we have the series for the sine function in Formula (8.6). The development of the cosine function is done analogously. ⊙ ◂◂◂

▸▸▸ **Application**: ***useful approximations of trigonometric functions***
Show the following frequently used approximations "for small $x$" and test the quality of the approximation for different $x < 0.5$:

$$\boxed{\sin x \approx \tan x \approx \sqrt{2(1-\cos x)} \approx x.}$$

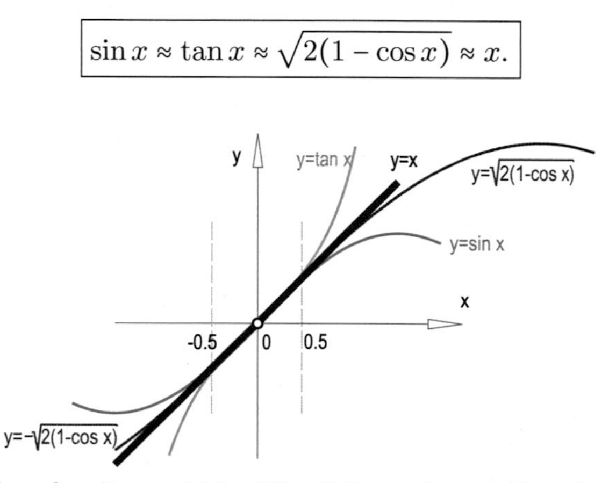


**Fig. 8.16** four functions which differ little in the neighbourhood of $x = 0$

| deg. $\frac{180}{\pi}x$ | rad. $x$ | error [%] for $\sin x$ | error [%] for $\tan x$ | error [%] for $\sqrt{2(1-\cos x)}$ |
|---|---|---|---|---|
| 5 | 0.09 | 0.1 | 0.3 | 0.03 |
| 10 | 0.17 | 0.5 | 1.0 | 0.13 |
| 15 | 0.26 | 1.1 | 2.4 | 0.29 |
| 20 | 0.35 | 2.0 | 4.3 | 0.59 |
| 25 | 0.44 | 3.1 | 6.9 | 0.71 |
| 30 | 0.52 | 4.5 | 10.3 | 1.14 |

***Proof***: The approximation of $\sin x \approx x$ follows directly from $\sin x = x - \frac{x^3}{3!} + \ldots \approx x + x^3$, because $x^3$ is already *very* small for small $x$. Further,

$$\tan x = \frac{\sin x}{\cos x} \approx \frac{x}{1 - \frac{x^2}{2!} + \ldots} \approx x.$$

Finally,

$$\sqrt{2(1-\cos x)} = \sqrt{2\left(1 - \left(1 - \frac{x^2}{2!} + \ldots\right)\right)} \approx x.$$

For negative $x$, the negative expression $-\sqrt{2(1-\cos x)}$ is to be used. ⊙

⊕ *Remark*: From the table, one can see: The error made by the approximation of the sine function for $x^\circ < 20^\circ$ and of tangent function for $x^\circ < 15^\circ$ is less than

2%. The best approximation is given by $\sqrt{2(1-\cos x)}$, where the error is only worth mentioning for $x > 30°$ (which can also be seen in Fig. 8.16)! We need the given approximations, for example, in order to derive the oscillation duration of a pendulum (Application p. 420). ⊕ ◂◂◂

We need *Taylor*'s Formula (8.3) for the series expansion of the logarithm:

▸▸▸ **Application**: ***series expansion for*** $\ln x$ (Fig. 8.18)
Show that the natural logarithm can be approximated by

$$\ln x = (x-1) - \frac{(x-1)^2}{2} + \frac{(x-1)^3}{3} - \frac{(x-1)^4}{4} + \dots \tag{8.8}$$

*Solution*:
The derivatives of the function $f(x) = \ln x$ are $f' = \frac{1}{x} = x^{-1}$, $f'' = (-1)x^{-2}$, $f''' = (-1)(-2)x^{-3}$, $f^{(4)} = (-1)(-2)(-3)x^{-4}$, ....
Since we cannot divide by 0, we expand at the knot $x_0 = 1$. There we have:
$f(1) = 0,\ f'(1) = 1,\ f''(1) = -1,$
$f'''(1) = (-1)(-2),\ f^{(4)}(1) = (-1)(-2)(-3),\ \dots.$
Obviously, the derivatives at the knots are

$$\frac{f^{(k)}(1)}{k!} = \frac{(-1)^{k-1}}{k}$$

and Formula (8.8) is verified. ◂◂◂

The number of terms to which the series must be expanded in order for the result to be accurate enough depends on the respective series and also on how much the function value $x$ differs from the knot $x_0$. Usually, the expansions are stopped somewhere between the 11th and 15th power.

▸▸▸ **Application**: ***approximation formula for body extension*** (Application p. 386)
If the temperature of a body of length $L_1$ is increased by $\Delta t$, its length increases to $L_2$ according to the formula

$$L_2 = L_1 \sqrt[3]{1 + \gamma\,\Delta t}.$$

There, $\gamma$ is the material constant. Show that in the case of a small expansion in length, one can do well with the approximation formula

$$L_2 = L_1 \left(1 + \frac{\gamma}{3}\,\Delta t\right).$$

*Solution*:
We set $x = \gamma\,\Delta t$ and expand the function

$$f(x) = \sqrt[3]{1+x}$$

in a power series at the knot 0:

$$f(x) = (1+x)^{\frac{1}{3}} \Rightarrow f'(x) = \frac{1}{3}(1+x)^{-\frac{2}{3}} \Rightarrow f''(x) = \frac{1}{3}\left(-\frac{2}{3}\right)(1+x)^{-\frac{5}{3}}\ \text{, etc.}$$

At the knot $x_0 = 0$, we have $(1+x)^u = 1$ for each exponent $u$, and thus, we obtain

$$f(x) = 1 + \frac{1}{3}x - \frac{2}{9 \cdot 2!}x^2 + \dots.$$

For small $x$, $x^2/9$ is still much smaller (e.g., $x = 0.001 \Rightarrow x^2/9 \approx 0.0000001$) and can be neglected, because length measurement is limited in any case. ◂◂◂

▸▸▸ **Application**: *conversion of fisheye photographs*

If you want to photograph as much of the interior of a room as possible, you have to use a wide-angle lens. The spiral staircase in Fig. 8.17 could only be captured in its entirety with a so-called fisheye lens. Such images can be transformed by means of a power series expansion into ultra wide-angle images (Fig. 8.17, right).

**Fig. 8.17** left: fisheye image, right: wide angle view

*Solution*:

An ordinary photograph is a "perspective", which is a projection from the center of the lens system onto the photosensitive layer. This mapping is "line preserving": Lines in space are mapped to lines in the image plane. Let $P^c$ be the perspective image of a point $P$ in space. Its distance from the center of the image (the principal point of the perspective) equals $r$.

Let us distort a perspective image so that all points are stretched or compressed from the center by a scaling factor $x = f(r)$ that is dependent on the distance $r$. This scaling function $f(r)$ can be any continuous function.

Fisheye images actually arise as follows: Image points far from the principal point are "shifted" towards the center. For a given fisheye image ($x$ is known), "only" the inverse function $g(x) = f^{-1}(x)$ that shifts back the points has to be found. Irrespective of what the function $g(x)$ may look like, it has a power series expansion at the knot $x = 0$ of the form $g(x) = \sum_{k=0}^{\infty} a_k\, x^k$. Since the distortion is close to zero near the image center, we have $g(x) = x$ for a small $x$. Therefore, $a_0 = 0$ and $a_1 = 1$. The series expansion, thus, has the form

$$g(x) \approx x + a_2\, x^2 + a_3\, x^3 + a_4\, x^4 + a_5\, x^5 + \ldots.$$

In the computer age, the coefficients can be found through interactive testing. Thereby, one must try to straighten curved images of rectilinear edges (Fig. 8.17, right). Once a fisheye lens has been calibrated, each additional image is quickly rectified. ◂◂◂

▸▸▸ **Application**: ***convergence problems of the sine function***

Up until this point, we have assumed the convergence condition $|x - x_0|$ as being "small enough". It is convenient to take $|x - x_0| < 1$, because $(x - x_0)^k$ goes towards zero for large $k$. What does the computer do when, for example, $\sin 7.3$ is to be calculated?

*Solution*:

First of all, arbitrary multiples of $2\pi$ can be added or subtracted to $x$ without changing the sine value. Thus, $\sin 7.3 = \sin(7.3 - 2\pi) = \sin 1.01681$. This value does not satisfy the condition $|x| < 1$ either. However, we can use the formula $\sin x = \cos(\pi/2 - x)$. Therefore, $\sin 1.01681 = \cos 0.55398$ where the cosine series converges.

The method initially described appears to be quite complicated. That is why it took such a long time for the sine value to become available at the "touch of a button": It requires the execution of a computer program covering all possible cases. This program is now usually "hard-wired in a chip". That is to say, it is implemented into the hardware and thus works quite fast. ◂◂◂

▸▸▸ **Application**: ***convergence problems of the*** $e$***-function or the*** $\ln$***-function***

Calculate $e^{-2.34}$ or $\ln 16.98$ through series expansion.

*Solution*:

We can restrict the domain of $x$ to positive values because $e^{-x} = \dfrac{1}{e^x}$. Yet, 2.34 does not fit into the desired range of $x < 1$ either (due to rapid convergence). If, however, $e = 2.71828\ldots$ is stored, then $e^{2.34} = e \cdot e \cdot e^{0.34}$, and for $x = 0.34$, the series expansion converges very well.

For the calculation of $\ln x$, the convergence interval is to be found in the neighbourhood of 1 (Fig. 8.18). Numbers outside the interval $[1, e]$ will be divided by $e$ or multiplied by $e$, until the number lies in the desired interval. Each division/multiplication decreases/increases the logarithm by 1. If the number is less than $\sqrt{e} = 1.64872\ldots$, then we can start the series expansion. Otherwise, we divide by $\sqrt{e}$, decreasing the logarithm by 0.5, and then, we expand. In the specific case, 16.98 can be divided twice by $e$ and once by $\sqrt{e}$. The logarithm of the quotient $1.3938\ldots$ can then be calculated $(0.33204\ldots)$.

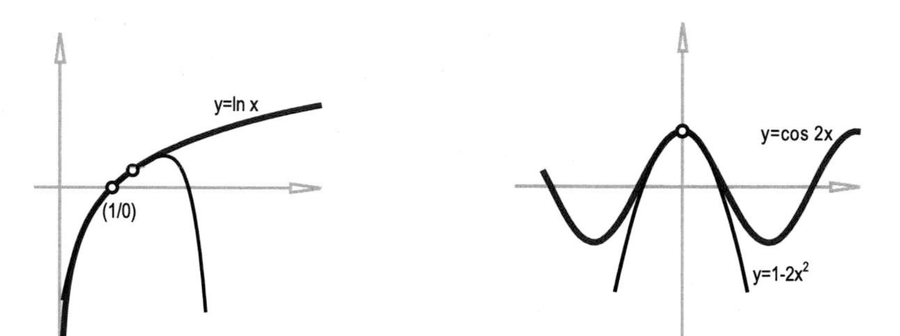


**Fig. 8.18** $y = \ln x$, polynomial of deg. 6 **Fig. 8.19** osculating parabola of deg. 2

We have to add 2.5 to this result. Fig. 8.18 shows the poor approximation of the *Taylor* function if $x$ is too far from 1. In the case of the logarithm, even a development to very high powers results only in a small improvement. ◂◂◂

▸▸▸ **Application**: ***osculating parabola*** (Fig. 8.19)
The first three terms of the Taylor series indicate the equation of an "osculating parabola" of the function at the knot. Such a parabola approximates the function graph there very well. Find such an osculating parabola of $f(x) = \cos 2x$ at the point $x_0 = 0$.
*Solution*:
The series expansion of $f(x)$ is given by Formula (8.6)

$$f(x) = 1 - \frac{(2x)^2}{2!} + \frac{(2x)^4}{4!} - \dots .$$

So, $a_0 = 1$, $a_1 = 0$, and $a_2 = -2$. The osculating parabola has the equation

$$y = 1 - 2x^2.$$

Fig. 8.19 illustrates how well the parabola approximates the curve. ◂◂◂

▸▸▸ **Application**: ***three types of surface points*** (Fig. 8.20)
One can extend the considerations of Application p. 387 to three-dimensional space. We consider an arbitrary point $P$ of an arbitrary doubly curved surface and impose a Cartesian coordinate system such that $P$ is the origin and the surface normal at $P$ is the $z$-axis. Each planar intersection of the surface through the $z$-axis yields a (planar) curve which has an osculating parabola at $P$. It is possible to show that all these osculating parabolas form a (hyperbolic or elliptic) paraboloid or a parabolic cylinder (Fig. 8.20). Consequently, one speaks of an elliptic, hyperbolic, or parabolic surface point $P$.
During the investigation of even the most complicated surfaces, it is now possible to relate to these three types of surfaces, and no further cases need to be distinguished.
In particular, the following important theorem holds:

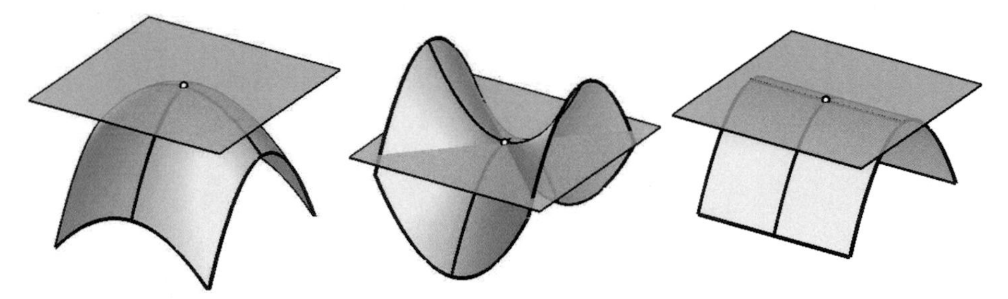

**Fig. 8.20** elliptic, hyperbolic, and parabolic point on a surface

A surface is developable if, and only if, it consists of parabolic points exclusively. A necessary (but by no means sufficient) condition for the developability is that the surface consists of straight lines. Visually, a developable surface can be recognized by the fact that all parts of the contour are rectilinear in any viewing direction.

This shows that doubly curved surfaces such as a sphere, a torus, a hyperboloid etc. cannot be developable. Arbitrary cones, cylinders, and tangent surfaces of space curves (frequently called *developables*, rarely called torses) can be developed into a plane.

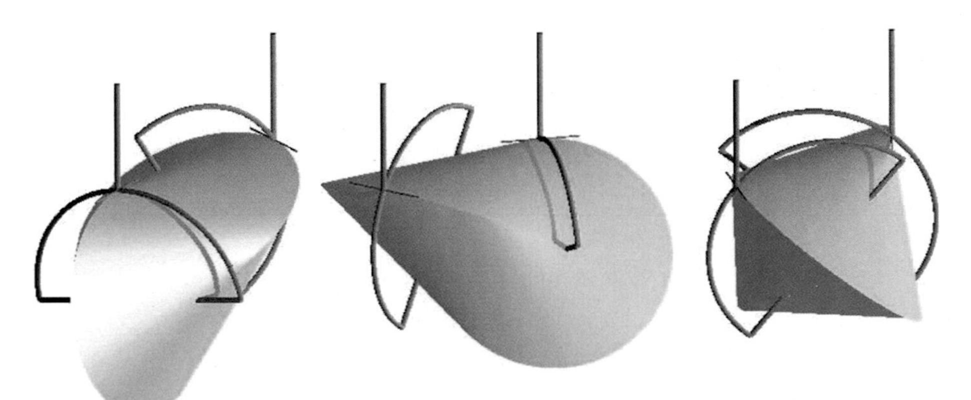

**Fig. 8.21** The oloid belongs to the class of developable surfaces. All of its generators are of equal length. Its body can skillfully be turned over by means of two modified universal joints which is exploited technologically (for example, for the efficient mixing of pond systems).

⊕ *Remark*: If a surface is developable, then there exist cones or cylinders touching the surface along each generator (straight line). Consequently, *the contour of the surface consists only of parts of straight lines.* This is a very good distinctive feature of a developable surface. Fig. 8.21 shows the rotation of an *oloid* (*wobbler*) by means of two Cardan joints. In each situation, the object has rectilinear contours. It can be developed. See the `wobbler.exe` demo program. ⊕ ◂◂◂

## 8.5 Integration as the inverse of differentiation

We can already differentiate functions. The question arises if a prescribed derivative $f'(x)$ can already determine the function $f(x)$.
The answer is: In principle, yes. If we have the derivative $f'(x)$ for every $x$-value, then we also know the tangent direction. If a point $P_0$ of the curve with the $x$-value $x_0$ is known, then we go a little bit along the tangent to a new point $P_1$ with the $x$-value $x_1 = x_0 + \Delta x$. There, we again know the tangent direction and arrive in an analogous way at a neighboring point $P_2$ with the $x$-value $x_2 = x_1 + \Delta x$, etc. In this way, we obtain an "approximating polygon" which will approximate the function graph of $f(x)$ more accurately for a smaller step size $\Delta x$. The polygonal curve converges to the function graph $f(x)$ for $\Delta x \to 0$.
A small problem still remains: We must know the starting point $P_0$, i.e. the function value of $f(x_0)$. Do we really have to know the starting point $P_0$? If we choose an arbitrary point with the $x$-value $x_0$ as the starting point, we still get a function whose derivative is always $f'(x)$ according to our construction. This curve differs only by parallel displacement along the $y$-axis from the original solution.
In the following, we call a function $G(x)$ the *anti-derivative* of the function $g(x)$ if $G'(x) = g(x)$. Clearly, $f(x)$ is an anti-derivative of $f'(x)$. Then, the following theorem is valid:

| Each function has infinitely many anti-derivatives which can be made to agree up to an additive constant through parallel displacement. The graphs of all anti-derivatives are, thus, congruent to one prototype. |
|---|

The construction described above is called *integrating* the function $g(x)$ and the anti-derivative $G(x)$ is the *indefinite integral* of the function (indefinite, because parallel shifts along the $y$-axis are possible). Symbolically, we write

$$G(x) = \int g(x)\, dx.$$

In particular, the following holds

$$f(x) = \int f'(x)\, dx.$$

▸▸▸ **Application**: ***graphical integration***
The *Coradi* integraph (Fig. 8.22) draws an anti-derivative for a given input function. The graph of the function to be integrated (the upper curve) has to be traced with a stylus.
The angle between the small cutting wheel and the horizontal axis is adjusted by a complicated linkage proportional to the function value (point on the

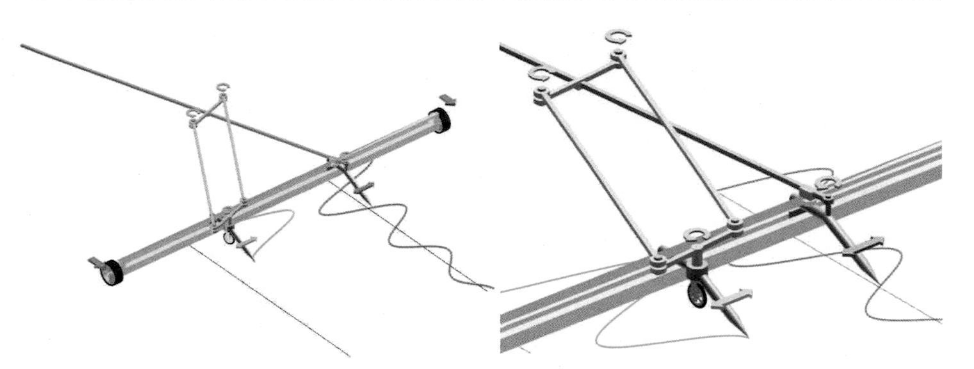

**Fig. 8.22** Coradi integraph (right: enlarged)

curve). As a result, the lower curve is drawn by a pencil. At each point of this curve, the function value of the input curve is the slope of the tangent. ◂◂◂

Differentiation and integration are somehow inverse operations, such as multiplication and division, squaring and finding the square root, evaluating sine and arcsine, or $e^x$ and $\ln x$.
Thus, we can already prove the following important rule:

$$\boxed{\int x^n\,dx = \frac{x^{n+1}}{n+1} + C \quad (n \in \mathbb{R},\ n \neq -1).} \tag{8.9}$$

Here, $C \in \mathbb{R}$ is called the *constant of integration.*
***Proof***: We only need to show that if the right-hand side is differentiated, the left-hand side of the equation emerges. In fact, according to the already known differentiation rules, we have

$$\left(\frac{x^{n+1}}{n+1} + C\right)' = \left(\frac{1}{n+1} \cdot x^{n+1}\right)' = \frac{1}{n+1} \cdot (n+1) \cdot x^{n+1-1} = x^n. \qquad \odot$$

In Formula (8.9), we have to exclude $n = -1$, because we cannot divide by zero. We also know the result of $\int x^{-1}\,dx = \int \frac{1}{x}\,dx$:

$$\boxed{\int \frac{1}{x}\,dx = \ln|x| + C.} \tag{8.10}$$

Taking the absolute value $|x|$ of $x$ is necessary, because the logarithm of a negative number is not defined within real numbers.
***Proof***: We distinguish two cases and differentiate the right-hand side again:

$$x > 0 \Rightarrow |x| = x: \quad (\ln x + C)' = \frac{1}{x},$$
$$x < 0 \Rightarrow |x| = -x: \quad (\ln(-x) + C)' = -\frac{1}{-x} = \frac{1}{x}. \quad \odot$$

Furthermore, the following can be easily proven by differentiating:

$$\boxed{\int \sin x\,dx = -\cos x + C, \quad \int \cos x\,dx = \sin x + C, \quad \int e^x\,dx = e^x + C.} \quad (8.11)$$

In practice, the following theorem in which $G(x)$ is the anti-derivative of $g(x)$, i.e. $G'(x) = g(x)$, is often used:

$$\boxed{\int g(a\,x+b)\,dx = \frac{1}{a}G(a\,x+b)+C.} \quad (8.12)$$

***Proof***: We apply the chain rule

$$\left(\frac{1}{a}G(a\,x+b)+C\right)' = \frac{1}{a}\cdot G'(a\,x+b)\,a = G'(a\,x+b) = g(a\,x+b). \quad \odot$$

▸▸▸ **Application**: ***Calculate the following anti-derivatives:***

$$\int \sin(a\,x+b)\,dx, \quad \int (2x-3)^6\,dx, \quad \int e^{\frac{x+1}{3}}\,dx, \quad \int \frac{dx}{1-3x}.$$

*Solution*:
Using Formula (8.11) and Formula (8.12), we get

$$\int \sin(a\,x+b)\,dx = \frac{1}{a}(-\cos(a\,x+b)) + C;$$

using Formula (8.9) and Formula (8.12), we get

$$\int (2x-3)^6\,dx = \frac{1}{2}\frac{(2x-3)^7}{7} + C;$$

using Formula (8.11) and Formula (8.12), we get

$$\int e^{\frac{x+1}{3}}\,dx = \frac{1}{\frac{1}{3}}\cdot e^{\frac{x+1}{3}} + C = 3\,e^{\frac{x+1}{3}} + C;$$

using Formula (8.10) and Formula (8.12), we get

$$\int \frac{dx}{1-3x} = \frac{1}{-3}\ln|1-3x| + C.$$

⊕ *Remark*: Of course, the notation of the variables does not matter. It is also correct to write $\int \sin\alpha\,d\alpha = -\cos\alpha + C$ or $\int dt = t + C$, etc. ⊕ ◂◂◂

More generally, the following formula is valid:

$$\boxed{\int g[h(x)]\,h'(x)\,dx = G[h(x)] + C.} \quad (8.13)$$

***Proof***: Analogously we use the chain rule. $\odot$
Thus, in the case of nested functions, the derivative of the inner function has to show up "somehow".

▸▸▸ **Application**: ***Calculate the following antiderivatives:***

$$\int \sin^2 x\cos x\,dx, \quad \int 2xe^{x^2}\,dx, \quad \int \frac{3}{x}\ln x\,dx, \quad \int x(3x^2-1)^6\,dx.$$

*Solution*:
The following solutions shall be checked by means of differentiation.
$\int \sin^2 x \cos x \, dx = \int [\sin x]^2 \cos x \, dx = ?$
We set $g(x) = x^2$ and $h(x) = \sin x$. Then, $h'(x) = \cos x$. Thus, the derivative of the inner function is already there, and we can write

$$\int [\sin x]^2 \cos x \, dx = \frac{\sin^3 x}{3} + C.$$

$\int 2xe^{x^2} \, dx = e^{x^2} + C$, since $2x$ is the derivative of $x^2$.
$\int \frac{3}{x} \ln x \, dx = 3 \cdot \int \frac{1}{x} \ln x \, dx = 3\frac{(\ln x)^2}{2} + C$, since $\frac{1}{x}$ is the derivative of $\ln x$.
$\int x(3x^2 - 1)^6 \, dx = \frac{1}{6} \int 6x(3x^2 - 1)^6 \, dx = \frac{1}{6}\frac{(3x^2 - 1)^7}{7} + C$, since $6x$ is the derivative of $3x^2$. ◂◂◂

If the derivative of the inner function does not occur, one must try to rewrite the integral with all possible tricks. We will learn about this in the next section. Sometimes even a clever transformation of the integral does not lead to success. Thus, for example, the integral which is significant in probability theory

$$\int e^{-x^2} \, dx$$

can only be evaluated numerically, to the chagrin of the user.
Without admitting subtleties of the integral calculus, we will consider some important applications in the next section. For this, we need the fundamental fact that one can determine surfaces with antiderivatives.
To conclude this section, let us look at a more practical example:

▸▸▸ **Application**: ***optimal lawn sprinkler***
A lawn sprinkler moves in such a way that it evenly irrigates a rectangular strip of lawn. What does this motion look like?

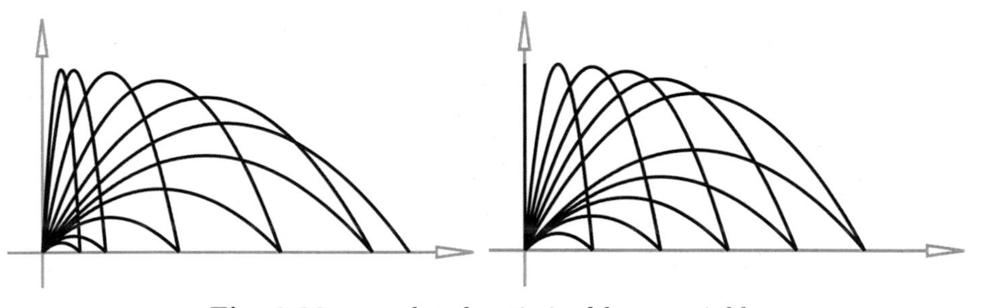

**Fig. 8.23** normal and optimized lawn sprinkler

*Solution*:
At any moment, water leaves the lawn sprinkler at point $A$ with the initial

speed $v_0$ and moves on a parabolic trajectory until it hits the lawn. The "throwing distance" $w$ is calculated in Application p. 142 as

$$w(\alpha) = \frac{1}{g} v_0^2 \sin 2\alpha.$$

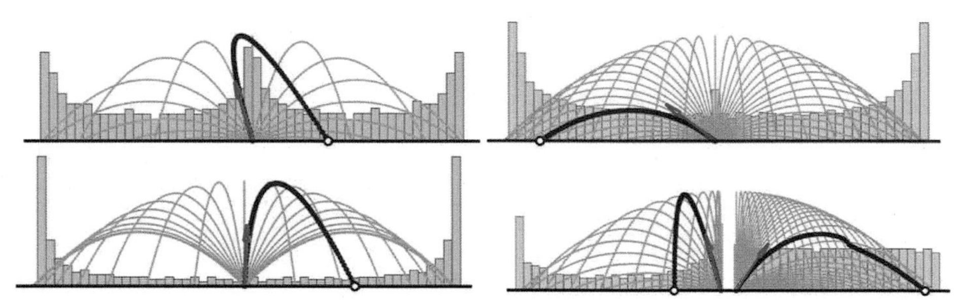

**Fig. 8.24** lawn-sprinkling variations and the corresponding results of a simulation

If the angle $\alpha$ is continuously changed with the time $t$

$$\alpha = \alpha(t),$$

the impact point of the water $B$ travels back and forth. Its distance from the point $A$ is, then, $w(\alpha(t))$. At $\alpha = 45°$, the water reaches the maximum distance. We obtain the instantaneous "initial velocity" $v_B$ of $B$ by differentiating with respect to time:

$$v_B = \frac{dw}{dt} = \text{(chain rule)} = \frac{dw}{d\alpha} \cdot \frac{d\alpha}{dt} = \frac{2}{g} v_0^2 \cos 2\alpha \cdot \frac{d\alpha}{dt}.$$

For optimal irrigation, $v_B$ has to be constant ($c_0$ is constant) such that

$$\frac{d\alpha}{dt} = \frac{1}{c_0 \cos 2\alpha}.$$

This equation is called a *differential equation.* If we want to solve it, we have to separate the variables. Then, we obtain

$$dt = c_0 \cos 2\alpha \, d\alpha.$$

Now, we can integrate both sides

$$\int dt = \int c_0 \cos 2\alpha \, d\alpha$$

and obtain

$$t = \frac{c_0}{2} \sin 2\alpha.$$

From this, $\alpha(t)$ can be calculated:

$$\alpha = \frac{1}{2} \arcsin \frac{2t}{c_0} = \frac{1}{2} \arcsin k\, t.$$

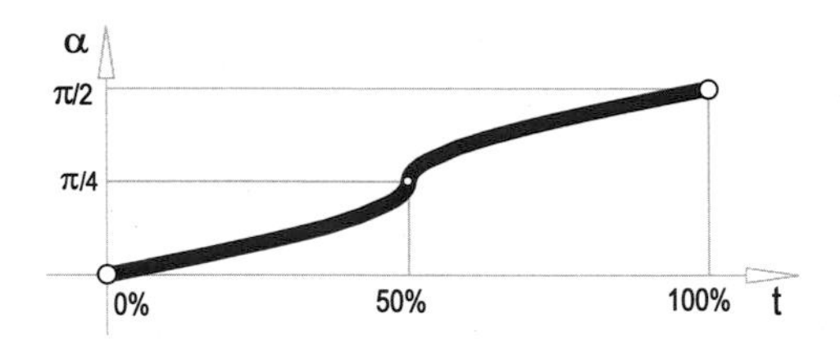


**Fig. 8.25** optimized function $\alpha(t)$ for the lawn sprinkler

The solution is by no means trivial. Since the arcsin is defined only for arguments with value $\leq 1$, one has to use the formula

$$\alpha = \frac{\pi}{2} + \frac{1}{2}\arcsin(kt - 2)$$

for $kt > 1$. Considering the graph of the described function (Fig. 8.25), we see that, in practice, the lawn sprinkler should be realized either only for angles up to 40° or for angles over 50°: If $\alpha = 45°$, the motion is too jerky due to the vertical tangent. In the restricted intervals, we can then limit ourselves to a linear growth of $\alpha$ with increasing time. In any case, the – quite common – lawn sprinklers with a "harmonic" sway are not optimal, as one recognizes from the illustration on the left of Fig. 8.23. ◂◂◂

# 8.6 Interpretations of the definite integrals

The following theorem is of great importance to us:

Let $G(x)$ be the antiderivative of $g(x)$ so that $G'(x) = g(x)$. Then, the definite integral of the function $g(x)$

$$\int_a^b g(x)\,dx = G(x)\Big|_a^b = G(b) - G(a) \tag{8.14}$$

gives the measure of the area underneath the curve $g(x)$ in the interval $[a, b]$.

⊕ *Remark*: Usually, we are not so nit-picky when using certain expressions. One could simply say *area* instead of the *measure of the area*. We do not want to fix certain physical dimensions. If the abscissa, for instance, is considered as the attractive force and the ordinate is considered as the distance (Application p. 403), then the measure of the area is the work (or potential energy) performed in the time interval $[a, b]$, since *work* $=$ *force* $\times$ *distance.* ⊕

***Proof***: We decompose the area between the $x$-axis and the curve $g(x)$ into small rectangular strips of width $dx$. The area starts at zero and increases by the value $g(x) \cdot dx$ with each rectangular strip.
Now, consider the function $G(x)$. It can have any value $G(a)$ at the point $a$ and increases by a very small $dx$ in the same way as its tangent does, hence by $G'(x) \cdot dx = g(x) \cdot dx$. At $b$, it reaches the value $G(b)$ and is, thus, increased by the value $G(b) - G(a)$.
The increase of the area and of the antiderivative are identical. Therefore, $G(b) - G(a)$ gives the total area. ⊙

⊕ *Remark*: With a variable upper boundary $x$, we have

$$\int_a^x g(u)\,du = G(x) - G(a) = G(x) + C,$$

and so, we obtain the *relation between the definite and indefinite integral.* ⊕

**▸▸▸ Application**: ***area underneath the sine curve***
Compute the area that is enclosed by a hunch of the sine curve and the $x$-axis.

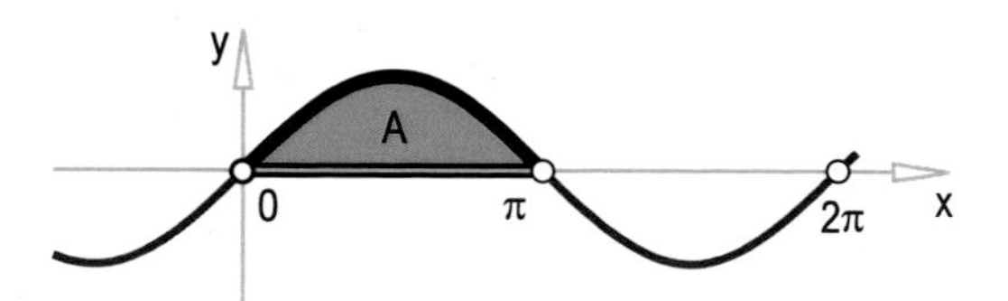


**Fig. 8.26** area underneath the sine curve

*Solution*:
Two adjacent zeros of the sine curve are $a = 0$ and $b = \pi$. In between, the sine curve is positive; hence, the area $A$ is also positive, and we have

$$A = \int_0^{\pi} \sin x\, dx = -\cos x\Big|_0^{\pi} = -\cos\pi - (-\cos 0) = -(-1) - (-1) = 2.$$

In the final step, a series of minus signs appeared and it is easy to make mistakes. In such a case, one should use the immediately obvious formula

$$-G(x)\Big|_a^b = +G(x)\Big|_b^a.$$

Here, we have

$$-\cos x\Big|_0^{\pi} = +\cos x\Big|_{\pi}^{0} = \cos 0 - \cos\pi = 1 - (-1) = 2.$$

The definite integral "responds" to a sign change and then returns a negative area. In fact,

$$\int_0^{2\pi} \sin x\, dx = -\cos x\Big|_0^{2\pi} = +\cos x\Big|_{2\pi}^{0} = \cos 0 - \cos 2\pi = 1 - 1 = 0$$

has not calculated the total area in the usual sense, but in the best case, one has proven the trivial fact that the two signed surfaces "cancel" each other (Fig. 8.29). ◂◂◂

▸▸▸ **Application**: ***area of the circle and the ellipse*** (Fig. 8.27, Fig. 8.28)
Calculate the formulas for the areas of the circle and the ellipse by means of integration.
*Solution*:
Today, everyone knows the formula for the area of a circle. The derivation of this formula requires integral calculus.
A circle around the origin with radius $r$ is given according to *Pythagoras* by the implicit equation $x^2 + y^2 = r^2$. Explicitly, $y$ can only be defined ambiguously by $y = \pm\sqrt{r^2 - x^2}$. Therefore, we have to exclude the lower semicircle and calculate the area under the upper semicircle. This means

$$\frac{A}{2} = \int_{-r}^{+r} \sqrt{r^2 - x^2}\, dx. \tag{8.15}$$

For a "practiced integrator", this integral may be easily computed. However, it requires a lot of computation and the solution is by no means visible to the "untrained eye".
Thus, we propose a different approach: We divide the circle into "infinitely many" sectors with the opening angle $d\varphi$. The area $dA$ of such an "elementary

sector" is equal to the area of an isosceles triangle with height $r$ and base line $r\,d\varphi$

$$dA = \frac{r^2}{2}\,d\varphi.$$

Now, we have

$$A = \int_0^{2\pi} dA = \int_0^{2\pi} \frac{r^2}{2}\,d\varphi = \frac{r^2}{2}\int_0^{2\pi} d\varphi = \frac{r^2}{2}\,\varphi\Big|_0^{2\pi} = \pi\,r^2.$$

Hence, the formula for the area of the circle is derived. We also know the result of the integral (8.15):

$$\int_{-r}^{+r} \sqrt{r^2 - x^2}\,dx = \frac{\pi}{2}\,r^2. \tag{8.16}$$

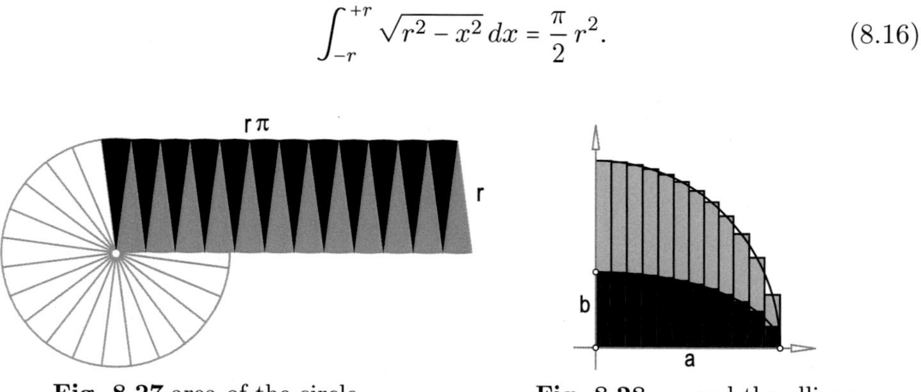


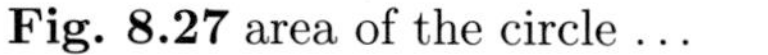

**Fig. 8.27** area of the circle ...

**Fig. 8.28** ... and the ellipse

⊕ *Remark*: In a purely heuristic way – even without integral calculus – we get an even simpler result: We consider the circle divided into $n$ equally large sectors (Fig. 8.27). Now, we put these sectors together so that the radii coincide and the arcs are alternatingly facing up and down as shown in Fig. 8.27. For $n \to \infty$, a rectangle with side lengths $r$ and $\pi\,r$ (half of the circumference) and area $\pi\,r^2$ is formed. ⊕
An ellipse "in principal position" with major axis length $2a$ and minor axis length $2b$ (Fig. 8.28) is described by the implicit equation

$$\frac{x^2}{a^2} + \frac{y^2}{b^2} = 1,$$

i.e. explicitly by

$$y = \pm\frac{b}{a}\sqrt{a^2 - x^2}.$$

Using Formula (8.16), we immediately get the area of the ellipse

$$\frac{A}{2} = \int_{-a}^{+a} \frac{b}{a}\sqrt{a^2 - x^2}\,dx = \frac{b}{a}\,\frac{\pi}{2}\,a^2 \quad \Rightarrow A = ab\pi. \tag{8.17}$$

◀◀◀

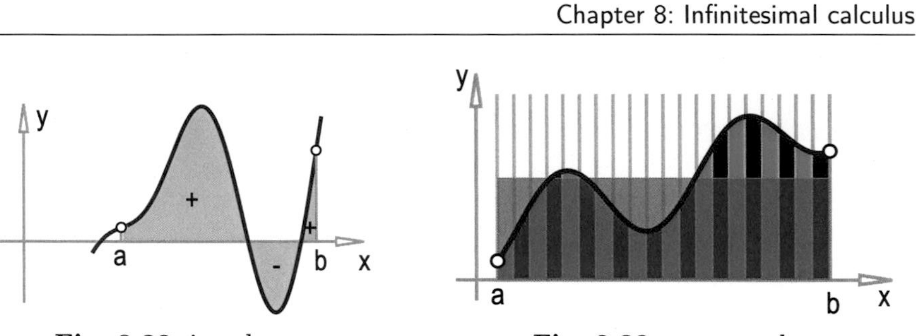


**Fig. 8.29** signed area

**Fig. 8.30** average values

If we know the area under a function graph $f(x)$, we can calculate the *average function value* $\overline{f}$. We have

$$\boxed{\overline{f} = \overline{f(x)} = \frac{1}{b-a}\int_a^b f(x)\,dx.} \tag{8.18}$$

***Proof***: We will give a quite heuristic (and, therefore, hopefully more intelligible) proof: Let us imagine a thin-walled aquarium with many narrow vertical lamellas (Fig. 8.30). Now we fill water into each lamella until it is as high as the function graph on the glass disc. The amount of water filled in can be interpreted as a measure of the area $\int_a^b f(x)\,dx$ under the graph. Now, we pull out the slats. Then, the water will form a thin parallelepiped with the same volume (and the same cross-sectional area). Its length is $b-a$. The height was the average water height $\overline{f(x)}$ in the individual lamella. The above formula follows from the area equality. ⊙

▸▸▸ **Application**: ***average value of the parabola*** $y = x^n$ ***in*** $[0, 1]$

*Solution*:

The area of the parabola is

$$A = \int_0^1 x^n\,dx = \frac{x^{n+1}}{n+1}\Big|_0^1 = \frac{1}{n+1}.$$

Thus, the average value is

$$\overline{y} = \frac{\frac{1}{n+1}}{1-0} = \frac{1}{n+1}.$$

Therefore, the parabola $y = x^n$ divides the area of the unit square in the ratio $1 : n$. In the special case $n = 2$, this was proven by *Archimedes* in the 3rd century BC using elementary methods! ◂◂◂

▸▸▸ **Application**: ***centroid of the semicircle***

Calculate the average positive sine value and the centroid of the semicircle.

*Solution*:

With Application p. 395 and Formula (8.18) we find

$$\overline{|\sin x|} = \frac{2}{\pi}. \tag{8.19}$$

◂◂◂

Now consider a semicircle with radius $r$ and divide it into many small sectors which can be approximated by triangles (Fig. 8.31). A triangle's centroid $S_i$ has the polar coordinates $x$ and $\frac{2}{3}r$ because the centroid divides the median in the ratio $2:1$.
All triangles have the same area. So, they are "equally significant". The position vector to the total center of gravity is, therefore, the arithmetic mean of the position vectors to the individual centers of gravity. The average abscissa is, of course, $s_x = 0$. The average ordinate is the average of $\frac{2}{3}r \sin x$ because of the uniform distribution of the individual centroids, i.e. in the limiting case

$$s_y = \frac{2}{3} r \overline{\sin x} = \frac{2}{3} r \frac{2}{\pi} = \frac{4}{3\pi} r \approx 0.424\, r.$$

The formula can be generalized in order to find the centroid of an arbitrary circle sector with central angle $\alpha$. We assume that the sector is placed symmetrically to the $y$-axis and then use the same considerations to calculate the average sine value $\overline{s}$ in the interval $\left[\frac{\pi}{2} - \frac{\alpha}{2}, \frac{\pi}{2} + \frac{\alpha}{2}\right]$, which, according to Formula (8.18), gives

$$s = \frac{1}{\alpha} \int\limits_{\frac{\pi}{2}-\frac{\alpha}{2}}^{\frac{\pi}{2}+\frac{\alpha}{2}} \sin x \, dx = \frac{1}{\alpha} \left[ \cos\left(\frac{\pi}{2} - \frac{\alpha}{2}\right) - \cos\left(\frac{\pi}{2} + \frac{\alpha}{2}\right) \right] = \frac{2}{\alpha} \sin \frac{\alpha}{2}.$$

Then, we have

$$y_s = \frac{2}{3} r\, s = \frac{4}{3\alpha} r \sin \frac{\alpha}{2}.$$

For $\alpha = \pi$, we have the above formula.

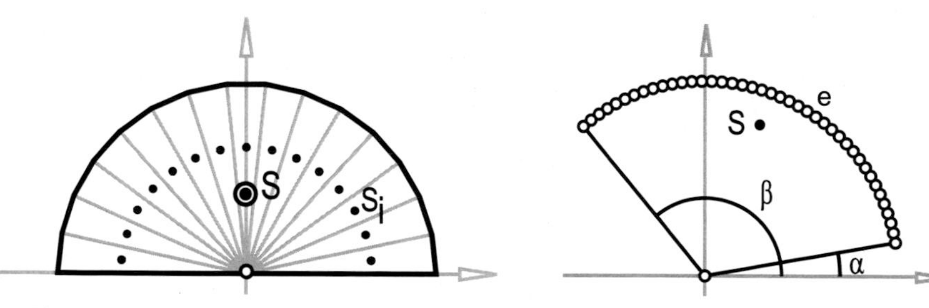


**Fig. 8.31** centroid of the semicircle

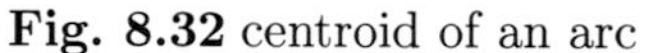

**Fig. 8.32** centroid of an arc

▸▸▸ **Application**: ***line centroid and areal centroid of a circular arc***
Calculate the line centroid of the arc $e$ in Fig. 8.32.
*Solution*:
The points of the arc are parametrized by

$$(r \cos\varphi / r \sin\varphi), \quad \varphi \in [\alpha, \beta].$$

Then, the line centroid of the curve has the average cosine and sine value multiplied by $r$ for its coordinates:

$$\overline{\cos\varphi} = \frac{1}{\beta-\alpha}\int_{\alpha}^{\beta}\cos\varphi\,d\varphi = \frac{1}{\beta-\alpha}\sin\varphi\Big|_{\alpha}^{\beta} = \frac{\sin\beta-\sin\alpha}{\beta-\alpha},$$

$$\overline{\sin\varphi} = \frac{1}{\beta-\alpha}\int_{\alpha}^{\beta}\sin\varphi\,d\varphi = -\frac{1}{\beta-\alpha}\cos\varphi\Big|_{\alpha}^{\beta} = \frac{\cos\alpha-\cos\beta}{\beta-\alpha}.$$

⊕ *Remark*: The centroid of the semicircle (Fig. 8.31) can obviously also be interpreted as the centroid of the concentric semicircle formed by the centroids of the individual triangles. ⊕ ◂◂◂

**Fig. 8.33** conversion of muscle energy into kinetic energy

▸▸▸ **Application**: ***kinetic energy as a definite integral*** (Fig. 8.34)
Derive the formula

$$W = \frac{m\,v^2}{2} \tag{8.20}$$

for the kinetic energy in a translatory motion.

*Solution*:
A body that moves "has" kinetic energy. In order to accelerate a body from a speed $v_1$ to a speed $v_2$, energy (work) is required.
In general, the infinitesimal energy increase $dW$ at each moment is proportional to the instantaneously applied force $F$ and the infinitesimal distance $ds$ because of the relation *work* = *force* ×*displacement*:

$$dW = F\,ds. \tag{8.21}$$

Furthermore, *force* = *mass* × *acceleration* holds true where the acceleration $a$ is the first derivative of the instantaneous velocity $v$ at the time $t$. So,

$$F = m\,a = m\,\frac{dv}{dt}.$$

Applying the chain rule, we find

$$F = m\,\frac{dv}{ds}\,\frac{ds}{dt}.$$

However, now $\frac{ds}{dt} = v$ is the instantaneous velocity, and therefore,

$$F = m\,v\,\frac{dv}{ds}.$$

So, together with Formula (8.21)

$$dW = F\,ds = m\,v\frac{dv}{ds}ds = m\,v\,dv,$$

and thus,

$$W = \int_{v_1}^{v_2} m\,v\,dv = \frac{1}{2}m\,v_2^2 - \frac{1}{2}m\,v_1^2.$$

This energy must be generated in order to accelerate a body from $v_1$ to $v_2$. If the initial velocity $v_1 = 0$, Formula (8.20) for the kinetic energy holds. ◂◂◂

**Fig. 8.34** kinetic energy

**Fig. 8.35** potential energy

▸▸▸ **Application**: ***potential energy***

Deduce the formula for the potential energy

$$W = m\,g\,h \tag{8.22}$$

by means of integral calculus.

*Solution*:

Again, *work* = *force* × *displacement*. The force is the weight $F = m\,g$. The energy needed during lifting by the small displacement $ds$ is, thus,

$$dW = F\,ds = m\,g\,ds.$$

We lift the body from height $h_1$ to height $h_2$ and have

$$W = \int_{h_1}^{h_2} m\,g\,ds = m\,g\,s\Big|_{h_1}^{h_2} = m\,g\,h_2 - m\,g\,h_1.$$

For $h = h_2$ and $h_1 = 0$, we get Formula (8.22). ◂◂◂

▸▸▸ **Application**: ***potential energy with a very high difference in height***
Calculate the energy that is necessary to transport a body of mass $m$ from the surface of the Earth to a *very great* height.

*Solution*:
The force which is to be overcome equals the weight $F = m\,g$. Actually, this is the attraction force of the Earth onto our body. For small differences in height, the acceleration $g$ of gravity only negligibly changes. Yet, "already" at a height $R = 6{,}370\,\text{km}$, (where $R$ is the radius of the Earth; thus, we have doubled the distance to the center of the Earth), only one fourth of the attraction force acts according to the findings of *Newton*: The adjusted weight formula is, therefore,

$$F(x) = \frac{m\,g}{x^2},$$

where $x$ is our current altitude measured in Earth's radii from the Earth's center. Let us represent the height $x$ on the abscissa.
Now, we think "infinitesimally" as we have frequently done: We lift the body a little bit by $dx$. In the small range of $dx$, the current acceleration due to gravity does not change. (This is what we mean when we say: "On a small scale," things are often easier.) In order to raise the body a little bit by $dx$, we have to perform the work $dW = \dfrac{m\,g}{x^2} \cdot dx$ (*work* = *force* × *displacement*). The tiny amount of work $dW$ can be interpreted as the area of the narrow rectangle $F(x) \times dx$. The sum of all the work is the measure of the area underneath the curve:

$$W(x)_{x_1}^{x_2} = \int_{x_1}^{x_2} \frac{m\,g}{x^2}\,dx = m\,g \int_{x_1}^{x_2} x^{-2}\,dx.$$

The present integral is easy to solve:

$$\int_{x_1}^{x_2} x^{-2}\,dx = \left.\frac{x^{-1}}{-1}\right|_{x_1}^{x_2} = -\left.\frac{1}{x}\right|_{x_1}^{x_2}.$$

So, we have

$$W = m\,g\left(\frac{1}{x_1} - \frac{1}{x_2}\right).$$

If we set $x_1 = 1$, that is, if we start from the Earth's surface and let $x_2$ move towards infinity ($1/x_2 \to 0$), we get the measure for the work in our adapted coordinate system:

$$W_{x=1}^{\infty} = m\,g.$$

A unit corresponds to the radius of the Earth. If we want to proceed with the international measurement system, we set $R = 6{,}370{,}000\,\text{m}$, and thus,

$$W_R^{\infty} = m\,g\,R = F\,R. \tag{8.23}$$

Thus, it is easy to multiply the force $F$ by the radius $R$ of the Earth. Whatever this value is, it is not infinitely large, and that has a tremendous importance for all mankind. We do not need an infinite amount of energy to leave the Earth's gravity field! ◂◂◂

▸▸▸ **Application**: *escape speed from the Earth*

Although this example does not require integral calculus, it must be mentioned here: Use Formula (8.23) to calculate the velocity that a rocket needs in order to leave the Earth's gravitational field.

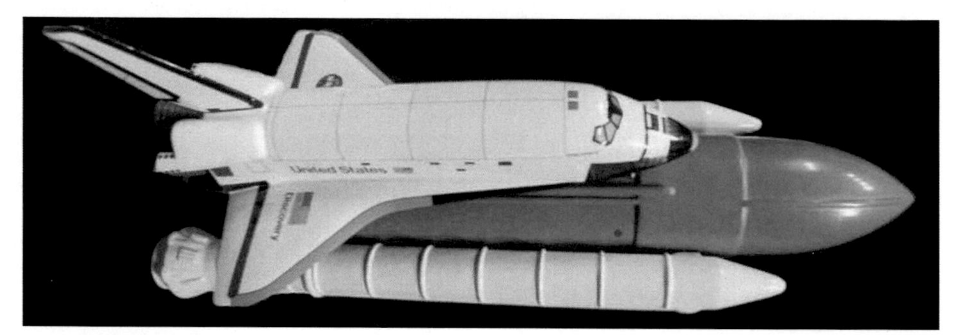

**Fig. 8.36** Space shuttle with carrier rockets: This object is not intended to leave the gravitational field!

*Solution*:

The work to be performed $m\,g\,R$ is converted into potential energy in the physical sense. If the object (the space capsule) is dropped again, this form of the energy changes into kinetic energy $m\,v^2/2$. The theoretical impact velocity (the air resistance only plays a role in the very last phase of the entrance into the dense atmospheric layers) can be calculated as:

$$m\,g\,R = \frac{m\,v^2}{2} \Rightarrow 2\,g\,R = v^2 \Rightarrow v = \sqrt{2\,g\,R}$$

$$\Rightarrow v \approx \sqrt{20\frac{m}{s^2} \cdot 6.37 \cdot 10^6 m} \approx 11.2\frac{\text{km}}{\text{s}} \approx 40,000\frac{\text{km}}{\text{h}}.$$

As in "usual" free fall, $v$ does not depend on the mass. Conversely, $v$ is the critical velocity that the rocket must reach so that it does not become "trapped" by the Earth.

⊕ *Remark*: Of course, "one has to keep some fuel in reserve", in order to be able to change the course in space at a later point. For flights to the Moon, the space capsule is also being "captured" by the gravitational field of the Moon. ⊕ ◂◂◂

## Volume and surface area

The following practical and often applicable theorem is about 400 years old and comes from *Bonaventura Cavalieri*:

> Cavalieri*'s principle*: If the areas of all parallel planar cross-sections of two bodies are equal, then the bodies' volumes are equal.

***Proof***: Consider a "stack" of thin cylindrical disks of arbitrary cross-section (Fig. 8.38). The volume does not change when the horizontal position of the panes are interchanged. Through refinement (limit process), this fact also applies to "infinitely thin" disks (one speaks of "elementary disks"). ⊙

This can analogously be applied to planar figures:
*If the chords of the intersections of two figures with straight lines are of the same length and if this is true for all chords on parallel lines, then these two figures have equal areas.*

The following modification of *Cavalieri*'s principle is immediately obvious: *If two bodies can be clamped in between two parallel planes so that their cross-sections are of the same area on average, then the two bodies have equal volumes.*

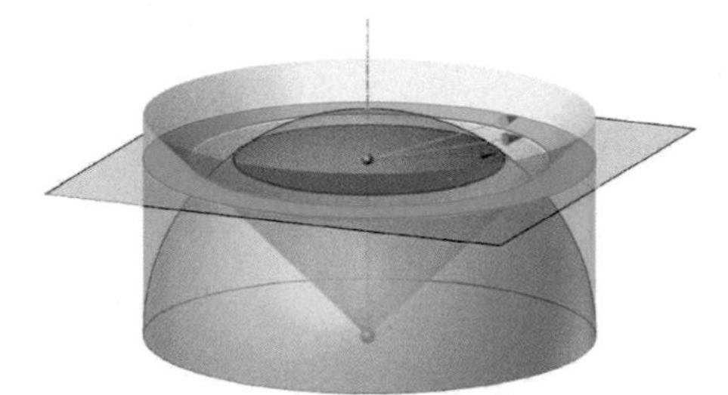

**Fig. 8.37** *Archimedes*'s method

**Fig. 8.38** *Cavalieri*'s principle

### ▸▸▸ Application: *oblique prisms and pyramids*

Following the above, oblique prisms and pyramids have the same volume $V$ as straight prisms and pyramids:

$$V = \text{base area} \times \text{height}, \quad V = \frac{1}{3} \times \text{base area} \times \text{height}. \tag{8.24}$$

By refinement, the theorem also applies to cylinders and cones with arbitrary cross-sections. ◂◂◂

### ▸▸▸ Application: *volume of a cone*

Show that the volume of an arbitrary cone (pyramid) is one third of the volume of a corresponding cylinder (prism) (Formula (8.24)).

***Proof***: It is sufficient to consider the respective prototype of the cone with a height of 1. Each similar cone results from the prototype by means of scaling, i.e. a dilation from a center.

Let $G$ be the measure of the base of the prototype. Let us consider the cross-sectional area $G_x$ of the cone at the distance $x \in [0,1]$ from the tip. It changes quadratically with $x$:

$$G_x = x^2 \cdot G.$$

According to Application p. 398, the function $y = x^2$ has the average value $\overline{y} = \frac{1}{3}$ in the interval $[0,1]$. Thus, the average cross-sectional area is one third of the base area $G$. The volume of the cone, therefore, corresponds to that of a cylinder with the same height and a similar base but reduced to a third, or the volume of a cylinder with the same base and a third of the height, see also Application p. 425. ⊙ ◂◂◂

▸▸▸ **Application**: ***volume and average cross-section of a sphere***

Show that the volume of a sphere with radius $r$ is equal to the volume of a cylinder of revolution with radius $r_z = \sqrt{2/3}\,r$ and height $2r$, and use this to calculate the "average cross-section of the sphere".

*Solution*:

The cylinder's radius $r_z$ is obtained by equating the volumes of sphere and cylinder:

$$\frac{4\pi}{3} r^3 = \pi\, r_z^2\, 2r \Rightarrow r_z = \sqrt{\frac{2}{3}}\, r.$$

Then, the average cross-section is obtained by reversing *Cavalieri*'s principle with $Q = \pi\, r_z^2 = \frac{2}{3}\pi\, r^2$ and is, therefore, 2/3 of the area of a large circle.
The volume of a cylinder of revolution circumscribed to a hemisphere is, thus, $\frac{3}{2}$ times as large as that of the hemisphere. Subtracting a cone of revolution from this volume, as in Fig. 8.37 (with one third of the cylinder volume, or half the volume of the sphere), the new structure has the same volume as the hemisphere. This discovery was made by *Archimedes*, who was so proud of it that he had it inscribed onto his gravestone. The proof, which anticipates the principle of *Cavalieri*(!): In the layer plane of height $z$, the sphere has a cross-section of area $\pi\,(r^2 - z^2)$. The circular ring cut out of the cone-shaped cylinder of revolution has outer radius $r$ and inner radius $z$ (the cone is inclined at 45°) and is, thus, the same surface. ◂◂◂

▸▸▸ **Application**: ***centroid of a hemisphere***

Show that the "auxiliary solid" in Fig. 8.37 (the cylinder with a conical drill hole) has the same centroid as the hemisphere of equal volume. Compute it.

*Solution*:

We divide both solids into arbitrarily thin layers parallel to the base circle. According to the preceding considerations, each layer has the same cross-section, and thus, the same volume or the same mass which we may think of as concentrated in points at the same height on the perpendicular axis of rotation. Instead of layers, we can, thus, think of particles of equal mass at the same height, which together have a common centroid. We calculate this centroid on the "auxiliary solid", which is produced by drilling out a cone of revolution (negative mass!) from the cylinder of revolution. The cylinder of revolution has mass $m_1$ and the cone of revolution has the negative mass $m_2 = -m_1/3$. The centroid of the cylinder lies at height $r/2$. For the cone, it is at height $3r/4$ (the centroid divides the height in the ratio $1:3$, Application p. 186). Using 5.15 we now get the height of the total centroid

$$s = \frac{1}{m_1 - \frac{m_1}{3}}\left(m_1 \cdot \frac{r}{2} - \frac{m_1}{3} \cdot \frac{3r}{4}\right) = \frac{3r}{8}.$$

◀◀◀

▸▸▸ **Application**: ***surface area of a sphere***
Calculate the surface of the sphere from its volume.

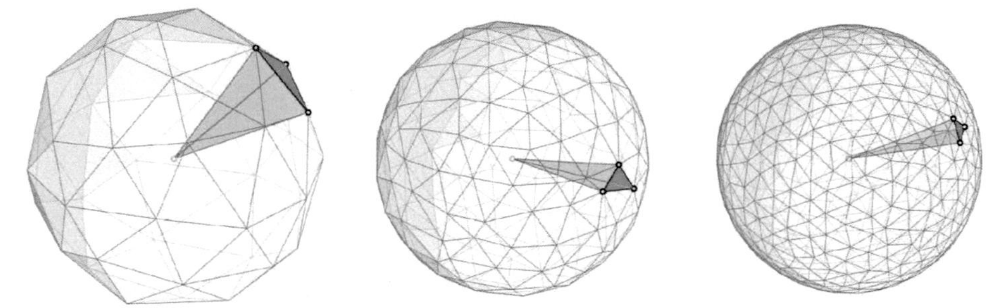

**Fig. 8.39** How to calculate the surface of the sphere from the volume.

*Solution*:
We choose many arbitrary points on the sphere and connect them, forming a triangular mesh (Fig. 8.39). Each triangle, together with the sphere's center, forms an arbitrary tetrahedron, to which the volume formula applies:

$$V_{\text{tetrahedron}} = \frac{1}{3}\text{base} \times \text{height}.$$

Approximately, the volume of the sphere is equal to the total sum of the volumes of the tetrahedra. If we further refine the triangular mesh on the sphere, then the height of all the tetrahedra converges towards the radius of the sphere and the sum of the base areas towards the spherical surface $S$. Thus,

$$V_{\text{sphere}} = \sum V_{\text{tetrahedra}} \Rightarrow \frac{4\pi}{3}r^3 = S \cdot \frac{r}{3} \Rightarrow S = 4\pi r^2.$$

Accordingly, the surface area of a hemisphere is twice as large as the surface area of its base circle or it is just as big as the surface area of the circumscribed cylinder of revolution. ◀◀◀

▸▸▸ **Application**: ***volume of a torus*** (Fig. 8.40)
Show that the volume of a ring torus with radii $a$ and $b$ equals the volume of a cylinder of revolution with radius $b$ and height $2\pi a$.

*Solution*:
The torus can be clamped between its "double tangent planes" on which it rests. A section parallel to these planes at the arbitrary height $z$ yields an annulus with radii $r_{1,2} = a \pm \varrho$ (with $\varrho = \sqrt{b^2 - z^2}$) whose area equals the area of a rectangle with sides $2a\pi$ and $2\varrho$:

$$A = \pi\left(r_1^2 - r_2^2\right) = \pi\left((a+\varrho)^2 - (a-\varrho)^2\right) = 4a\pi\,\varrho.$$

**Fig. 8.40** circular layers of a torus

The length of the meridian's chord at height $z$ is precisely $2\varrho$. This rectangle is interpreted as the planar section of a cylinder of revolution with a plane parallel to the axis. ◀◀◀

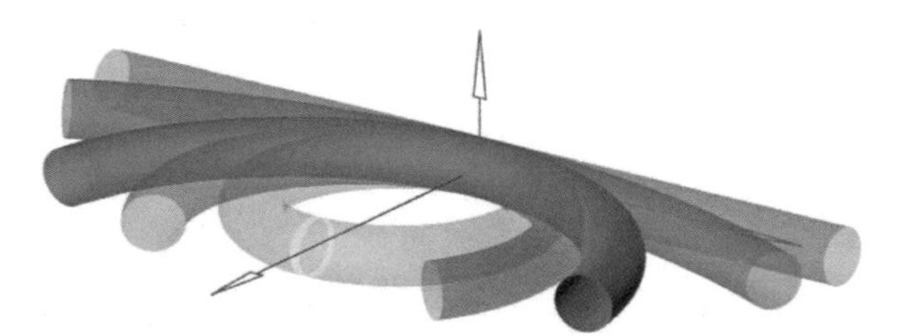

**Fig. 8.41** bending the torus into the cylinder of revolution

Thus, according to Application p. 406, a torus can be "straightened" (Fig. 8.41) without changing its volume. (It is only an imaginary bending. In practice, cracks and deformations would occur.) In this case, the circular path of the center of the meridian circle is stretched into a straight line segment of equal length. Even the surface of the torus coincides with that of the cylinder of revolution because the sum of "surface elements" of the torus corresponds to the symmetric surface elements of the cylinder at the same height.
This line of thought can be generalized: The rotating meridian can have any shape. The centroid of the meridian surface takes the role of the center of the circle. The following useful formulas, proved by Paul *Guldin* around 1600, were documented almost contemporaneously with the principle of *Cavalieri*. However, these rules can already be found, although without conclusive evidence, in the "Collectiones" of the Greek mathematician *Pappos of Alexandria*, who lived 1,300 years prior!

Guldin*'s first rule: The area of a surface of revolution is equal to the product of the length of the meridian curve* $m$ *lying on one side of the axis of rotation and the length of the trace of the* line centroid *of* $m$ *during a full rotation.*

Guldin*'s second rule: The volume of a solid of revolution is equal to the surface area of the planar meridian lying on one side of the axis of rotation and the length of the trace of the* surface centroid *of the meridian during a full rotation.*

Thus, it is possible to "bend" each solid of rotation into a cylinder without changing its surface or its volume. In this case, the circular path of the line centroid or the areal centroid of $m$ is stretched into an equally long straight line segment. The exact proof by means of integral calculus will be omitted here.

▸▸▸ **Application**: *surface area and volume of a torus*

One computes the formulas for the surface area and the volume of the ring torus ($a > b$) using *Guldin*'s rules.

*Solution*:

The meridian $m$ is a circle with radius $b$. For $a > b$, $m$ lies entirely on one side of the axis of rotation. The line centroid and the areal centroid of the meridian circle coincide with the center of the meridian. The path of the centroid in both cases is of length $2\pi\, a$. In other words,

$$S = 2\pi\, b \cdot 2\pi\, a, \quad V = \pi\, b^2 \cdot 2\pi\, a.$$

◂◂◂

▸▸▸ **Application**: *volume of a sphere*

Use *Guldin*'s second rule in order to verify the formula for the volume of the sphere, which has already been used several times.

*Solution*:

The sphere is generated by the rotation of a semicircle $m$ ("circle of longitude"). The surface of the semicircle is $\frac{\pi}{2} r^2$. In Application p. 398, we have calculated the centroid of the semicircle. Its distance from the axis of rotation is $\frac{4}{3\pi} r$. The distance of the centroids is, thus, $2\pi \frac{4}{3\pi} r = \frac{8}{3} r$, and the volume of the sphere, according to *Guldin*, is

$$V = \frac{8}{3} r \frac{\pi}{2} r^2 = \frac{4\pi}{3} r^3. \tag{8.25}$$

◂◂◂

▸▸▸ **Application**: *surface area of a spherical segment*

Calculate the surface of a spherical segment (and, as a special case, the total surface of the sphere) with *Guldin*'s first rule.

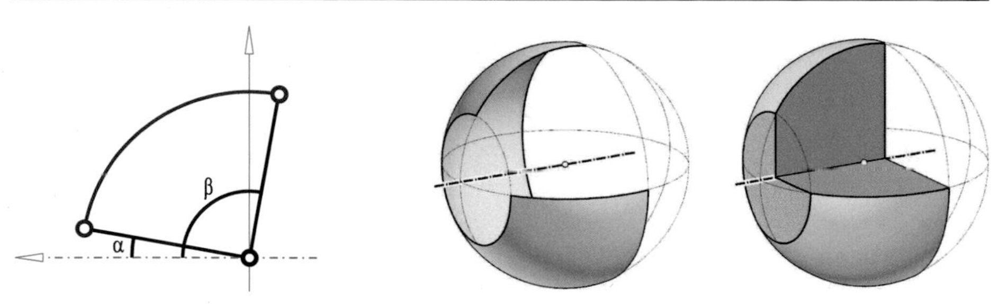


**Fig. 8.42** surface area and volume of a spherical segment.

*Solution*:
A spherical segment is created by the rotation of the circular arc

$$(r\cos\varphi / r\sin\varphi),\quad \varphi\in[\alpha,\,\beta]$$

about the $y$-axis. The distance from the circular arc's *line centroid* to the axis of rotation is given in Application p. 399:

$$r\,\frac{\sin\beta-\sin\alpha}{\beta-\alpha}.$$

The path of the centroid is $2\pi$ times as long. The length of the arc is $r\,(\beta-\alpha)$. Thus, according to *Guldin*'s first rule, we have

$$S = 2\pi\,r\,\frac{\sin\beta-\sin\alpha}{\beta-\alpha}\cdot r\,(\beta-\alpha) = 2\pi\,r^2(\sin\beta-\sin\alpha). \tag{8.26}$$

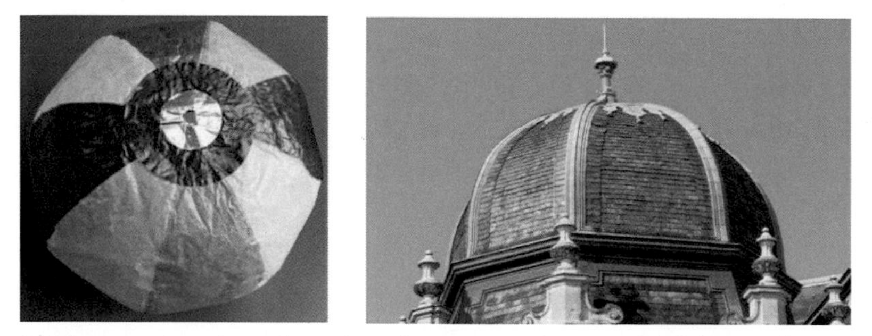

**Fig. 8.43** approximation of a sphere by means of parts of cylinders of revolution

An interesting application of this formula is given in Application p. 425.
The special case $\alpha=-\dfrac{\pi}{2}$, $\beta=\dfrac{\pi}{2}$ yields the well-known formula for the surface of the entire sphere:

$$S = 4\pi\,r^2. \tag{8.27}$$

⊕ *Remark*: This formula can also be found heuristically as follows: The sphere is approximated by parts of $n$ half cylinders of revolution (radius $r$). These intersect each other along half-ellipses. In the limit $n\to\infty$, one obtains a smooth spherical surface. The parts of the cylinders of revolution can be developed and transformed into lens-shaped structures: The elliptic oblique sections show up as sinusoids. The length of such a lens is half the circumference of the sphere ($l = \pi r$). The width is – for a large $n$ – an $n$-th of the circumference of the sphere ($b = 2\pi r/n$). According to Formula (8.19), we can convert the lens into a rectangle of width $\frac{2}{\pi} b = \frac{4r}{n}$. The $n$ subsections have a total area of $n \cdot \pi r \cdot \frac{4r}{n} = 4\pi r^2$. ⊕ ◂◂◂

Conversely, *Guldin*'s rules can be used for the computation of *areal centroids*. The distance $s$ of the centroid of the semicircle from the axis of rotation can be calculated from the volume $V$ of the sphere and the area $A$ of the semicircle (see Application p. 408):

$$s = \frac{1}{2\pi} \frac{V}{A}.$$

Thus, a two-dimensional problem is solved by "escaping" into the third dimension. We have already made use of this as we have interpreted the curves of degree 2 (which are planar anyhow) as planar intersections of a cone of revolution in three-dimensional space.
Up until now, we have included integral calculus in the calculation of volumes and areas insofar as we have used the principles of *Cavalieri* and *Guldin*'s theorems, whose exact proofs require integral calculus. It is also possible to calculate the volume of surfaces of revolution or their surface area directly:

If a function graph $y = f(x)$ ($x \in [a, b]$) rotates about the $x$-axis, then the surface area $S$ and the volume $V$ of the emerging surface and solid of revolution are

$$S = 2\pi \int_a^b y \sqrt{1 + y'^2}\, dx, \quad V = \pi \int_a^b y^2\, dx. \tag{8.28}$$

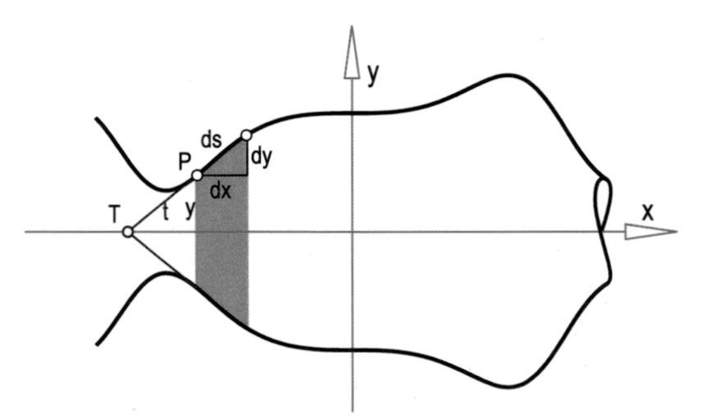


**Fig. 8.44** area of a surface of revolution

***Proof***: We cut the surfaces of revolution into thin slices (Fig. 8.44). These are truncated cones with height $dx$. Such an "elementary frustrum of a cone" has the surface area $dS$. In order to calculate it, we need the "arc element" $ds$ of $f(x)$. According to *Pythagoras*'s theorem, we have

$$ds = \sqrt{dx^2 + dy^2} = \sqrt{1 + \left(\frac{dy}{dx}\right)^2}\, dx = \sqrt{1 + y'^2}\, dx. \tag{8.29}$$

The elongated chord (in the limit the tangent) from the curve point $P$ to the $x$-axis (intersection point $T$) is a generator of the cone. The tip of the frustrum of the cone is denoted by $T$. The distance $\overline{TP}$ is denoted by $t$. Then, we have $dS = \pi[(y + dy)(t + ds) - yt] = \pi[y\,ds + t\,dy + dy\,ds]$. Since $y : t = dy : ds$, the equation $t\,dy = y\,ds$ holds, and we have $dS = \pi[2y\,ds + dy\,ds]$. The limit procedure with $dx \to 0$, and consequently, $dy \to 0$, $ds \to 0$, causes the product $dy\,ds$ "to be of higher order smaller" than $dy$ and $ds$, and can, therefore, be neglected. Now, we have $dS = 2\pi\,y\sqrt{1 + y'^2}\,dx$, and integration provides the above formula for $S$.
It is much easier to derive the formula for the volume: The volume $dV$ of the elementary frustrum of a cone for an "infinitely small" $dx$ is not to be distinguished from the volume of an "elementary cylinder", and thus,

$$dV = \pi\,y^2\,dx$$

from which the formula already follows by integration. ⊙

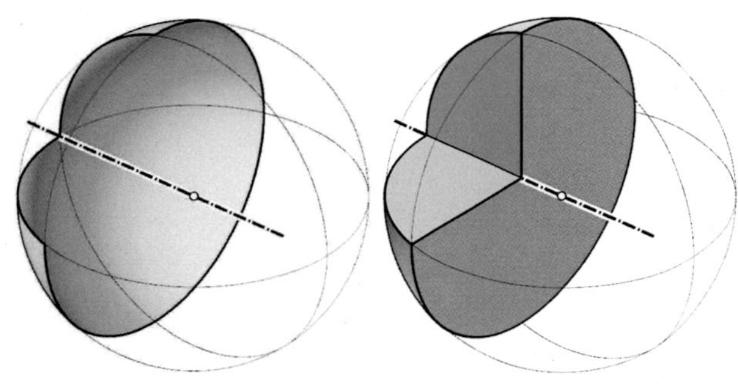

**Fig. 8.45** spherical cap

▸▸▸ **Application**: ***volume and area of a spherical cap*** (Fig. 8.45)
Calculate the volume and the area of a spherical cap with height $h$.

*Solution*:
For the meridian of the sphere, we have $x^2 + y^2 = r^2$, and thus, $y = \pm\sqrt{r^2 - x^2}$.
The volume is computed using Formula (8.28):

$$V = \pi \int_{r-h}^{r} y^2\,dx = \pi \int_{r-h}^{r} (r^2 - x^2)\,dx = \pi\Big[r^2\,x - \frac{x^3}{3}\Big]_{r-h}^{r} =$$

$$= \pi\Big[r^3 - \frac{r^3}{3} - r^2(r-h) + \frac{(r-h)^3}{3}\Big] = \frac{\pi}{3}h^2(3r - h).$$

Furthermore, the surface is obtained from

$$S = 2\pi \int_{r-h}^{r} y\sqrt{1 + y'^2}\,dx.$$

From the equation of the meridian, we get $y' = \dfrac{-2x}{2\sqrt{r^2 - x^2}} = -\dfrac{x}{y}$, and thus,

$$S = 2\pi \int_{r-h}^{r} y\sqrt{1 + \Big(\frac{x}{y}\Big)^2}\,dx = 2\pi \int_{r-h}^{r} y\sqrt{\frac{x^2 + y^2}{y^2}}\,dx =$$

$$= 2\pi \int_{r-h}^{r} r\,dx = 2\pi r x\Big|_{r-h}^{r} = 2\pi r h.$$

We compare this result with that of Application p. 408: There, we had the special case with $\beta = \dfrac{\pi}{2}$ and

$$S = 2\pi\, r^2(1 - \sin\alpha).$$

In fact, for $h = r - r\sin\alpha$, the formulas agree. ◂◂◂

**▸▸▸ Application**: ***capacity of a drinking glass***

A drinking glass has the shape of a surface of revolution which is generated by rotating a sinusoidal curve. Calculate the maximum volume of the drinking glass for $a = 3.2\,\text{cm}$, $b = 0.8\,\text{cm}$, and $h = 12\,\text{cm}$ (Fig. 8.46, left).

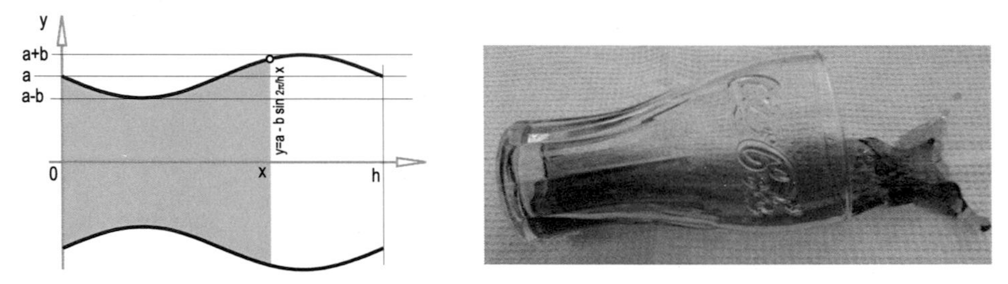


**Fig. 8.46** rotation of a sinusoidal curve (drinking glass)

*Solution*:

The sinusoidal curve can be described by $y = a - b\sin\dfrac{2\pi}{h}x$. For the volume, we obtain

$$V = \pi\int_0^h y^2\,dx = \pi\int_0^h \left(a - b\sin\frac{2\pi}{h}x\right)^2 dx =$$

$$= \pi\left[a^2\int_0^h dx - 2ab\int_0^h \sin\frac{2\pi}{h}x\,dx + b^2\int_0^h \sin^2\frac{2\pi}{h}x\,dx\right] =$$

$$= \pi\left[a^2h + 2ab\underbrace{\frac{h}{2\pi}\cos\frac{2\pi}{h}x\Big|_0^h}_{0} + b^2\int_0^h \underbrace{\frac{1}{2}\left(1-\cos 2\frac{2\pi}{h}x\right)}_{\text{according to Formula (4.15)}}\,dx\right] =$$

$$= \pi\left[a^2h + 0 + b^2\frac{h}{2} - \underbrace{\frac{b^2}{2}\int_0^h \cos\frac{4\pi}{h}x\,dx}_{0}\right] = \pi h\left[a^2 + \frac{b^2}{2}\right].$$

The interim calculation requires one transition to the double angle. For the given values, $V \approx 400\,\text{cm}^3 = 0.4\,l$.

⊕ *Remark*: Calculating the integral was quite tedious. We will soon learn an approximation method with which we can calculate the value as accurately as possible with a computer. Then, the obvious question can be answered: Where should the marking for $0.25\,l$ be placed? ⊕ ◂◂◂

## 8.7 Numerical integration

We have seen that integral calculus plays an important role in applied mathematics. There is often a problem in practice: The inability to express an antiderivative in terms of elementary functions.
However, each definite integral can be evaluated. For this, *Kepler* provided a decisive contribution with his so-called "Fassregel" ("barrel-rule"), sometimes also referred to as *Simpson*'s rule:

Kepler's rule: $$\int_a^b f(x)\,dx \approx \frac{b-a}{6}\left[f(a) + 4\,f\left(\frac{a+b}{2}\right) + f(b)\right].$$

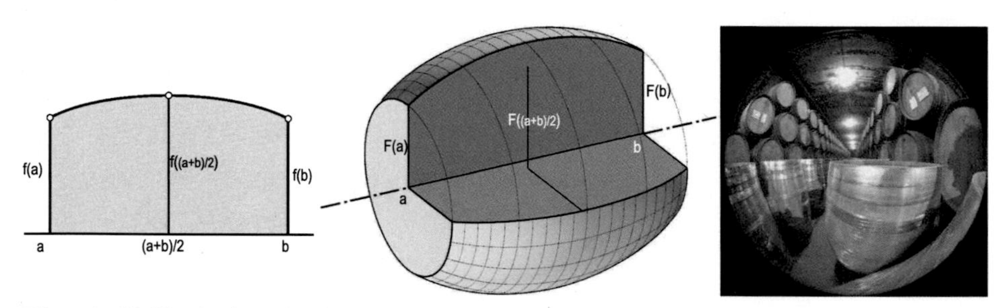


**Fig. 8.47** Kepler's rule: area underneath the curve and corresponding volume of the solid of rotation

***Proof***: The proof is merely sketched: We replace the meridian curve by a parabola with the equation $y = a_2\,x^2 + a_1\,x + a_0$ (see Application p. 33). The deviation will remain small if the curve looks parabolic in any case – like the meridian of a barrel (Fig. 8.47).
In order to shorten the calculation somewhat, we restrict ourselves to the parabola $y = x^2 + c$, which can always be achieved by shifting and scaling.
The average value of this prototype equals

$$\overline{f(x)} = \frac{1}{b-a}\int_a^b (x^2 + c)\,dx = \frac{1}{b-a}\Big[\frac{b^3 - a^3}{3} + c(b-a)\Big] =$$
$$= \frac{b^2 + a\,b + a^2}{3} + c = \frac{a^2 + b^2 + (a+b)^2 + 6c}{6}.$$

Now, we evaluate the meridian and get

$$f(a) = a^2 + c,\ f(b) = b^2 + c,\ f\big(\frac{a+b}{2}\big) = \frac{(a+b)^2}{4} + c$$

$$\Rightarrow a^2 = f(a) - c,\ b^2 = f(b) - c,\ (a+b)^2 = 4\,f\big(\frac{a+b}{2}\big) - 4c.$$

With

$$\overline{f(x)} = \frac{1}{6}\big[f(a) + 4\,f\big(\frac{a+b}{2}\big) + f(b)\big],$$

we find the above rule because $A = (b-a)\overline{f}$. ⊙

At first, *Kepler*'s rule was only used to calculate surfaces. However, *Kepler* computed the volume of the barrel by replacing $f(a)$, $f(b)$, and $f((a+b)/2)$ with the areas $F(a)$, $F(b)$, and $F((a+b)/2)$ of the cross-sections. Then, the rule also works for the function $F(x) = \pi f^2(x)$ and yields the volume as a result – only the function $F(x)$ has a different meaning, and its graph does not correspond to that of the meridian (Fig. 8.47).

According to what has just been said, Kepler's rule actually works for any – hence also non-rotationally-symmetric – solid, provided that the areas of the cross-sections on the boundaries and in the middle are known.

**Fig. 8.48** a huge wine barrel

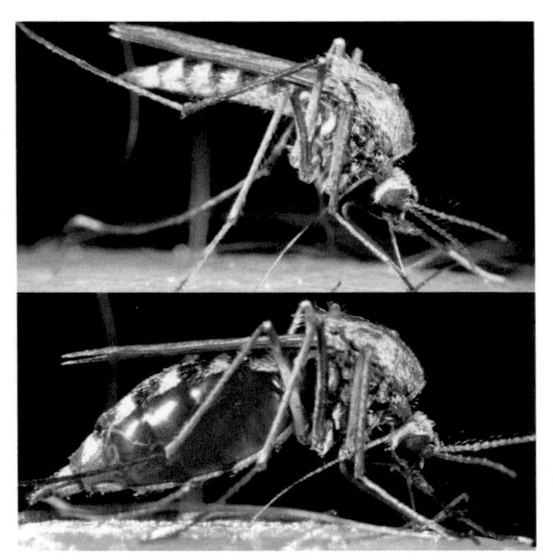

**Fig. 8.49** "an empty and a full" mosquito

### ▸▸▸ Application: *giant wine barrels*

Approximate the length of the huge wine barrel (Fig. 8.48) in the Croatian winery Kutjevo near Velika.

*Solution*:

We do not need to estimate the volume, because the barrel clearly shows the content: 53,520 liters, i.e. about 54 m$^3$. The two ladies in front of the barrel are 165 cm. Since the front diameter of the barrel is about twice as large, we have a cross-sectional area of about $8-9$ m$^2$ at the front and rear. The average cross-section should then be $10-11$ m$^2$. Thus, we have the following approximation (excessively exact computations would be "non-mathematical" because the chain is as strong as its weakest link):

$$54 \text{ m}^3 = \frac{b-a}{6}[8.5 + 4 \cdot 10.5 + 8.5]\text{m}^2 \Rightarrow b - a \approx 5.4 \text{ m}.$$

⊕ *Remark*: The largest wine barrel in the world is in Heidelberg and has four times as much capacity. The lengths are accordingly multiplied by $\sqrt[3]{4} \approx 1.6$. ⊕ ◂◂◂

▸▸▸ **Application**: ***How much blood fits into a mosquito?***
A mosquito has just refreshed itself on my left hand (Fig. 8.49, such experiments should only be made in areas free of malaria). How much blood was drained from me?

The shape of the "blood tank" (Fig. 8.49 below) is ideally suited for the *Kepler* formula. If we take the body as a symmetrical spindle, the cross-sections at the two ends are zero. The length $b-a$ is about 10 mm. The central cross-section with a diameter of 3 mm is about 7 square millimeters. Thus, the volume is

$$V = \frac{10 \text{ mm}}{6}[0+4\cdot 7+0]\text{mm}^2 \approx 46 \text{ mm}^3.$$

After 20 pricks, 1 $\text{cm}^3$ (1 milliliter) of blood has been removed.

⊕ *Remark*: In comparison with the initial phase of the prick (Fig. 8.49 above), the final blood-fattened mosquito is somewhat grotesque. However, it is not because of the loss of blood that mosquitoes are so unpleasant. It is rather because of the anti-coagulant that the mosquito injects and the diseases that are transmitted: Mosquito bites are painful and a danger for humans and animals. 1.5 million people die each year from malaria. ⊕ ◂◂◂

▸▸▸ **Application**: ***How useful is Kepler's rule?***
Compute the error that is made by *Kepler*'s rule when computing the volume of a sphere, a cone, and a cylinder.
*Solution*:
For the sphere, we want to estimate $\int_{-r}^{r} \pi(r^2-x^2)dx$. *Kepler*'s rule yields

$$V = \frac{2r}{6}\,(0+4\pi\, r^2+0) = \frac{4\pi}{3}\,r^3.$$

That is the exact formula (Application p. 408)!
For the cylinder of revolution, we have

$$V = \frac{h}{6}\,(\pi\, r^2 + 4\pi\, r^2 + \pi\, r^2) = \pi\, r^2\, h,$$

and for the cone of revolution, *Kepler*'s rule yields

$$V = \frac{h}{6}\,\Big(0+4\,\pi\Big(\frac{r}{2}\Big)^2 + \pi\, r^2\Big) = \frac{\pi}{3}\,r^2\, h,$$

again no deviation from the exact result (Application p. 404)!
Additional task:
Motivated by these results, we will test two more solids: The spherical cap gives the approximation

$$V = \frac{h}{6}\Big[0+4\pi\big(r^2-(r-\frac{h}{2})^2\big)+\pi\big(r^2-(r-h)^2\big)\Big] = \frac{\pi\, h^2}{3}\,(3r-h).$$

Again, this coincides with the exact formula (Application p. 411)!
For the torus, the exact formula is $V = 2\pi^2 a\, b^2$ (p. 408). The approximation gives

$$V \approx \frac{2b}{6}\Big[0 + 4\pi\left((a+b)^2 - (a-b)^2\right) + 0\Big] = \frac{16\pi}{3}\, a\, b^2.$$

This time, there is a deviation of about 18%. ◂◂◂

*Kepler*'s rule uses only three curve points (or cross-sections) and is, as we have just seen, often a sufficient approximation, especially for the calculation of the volume of simple solids of revolution. When calculating areas in the plane, it is also surprisingly accurate for parabola-like functions.
For more complicated curves or solids, $2m+1$ interpolation points are selected equidistantly with a distance of $h = \frac{b-a}{2m}$ to their neighbors, and one obtains $m$ adjacent approximating parabolas. The corresponding partial areas are evaluated with *Kepler*'s rule and summed up in order to get the total area. In this case, the boundary values $f(a)$ and $f(b)$ occur only once. The "odd vertices" occur four times. The "even vertices" but two times (because there are two parabolic arcs there). The corresponding formula is used by computers:

Simpson's approximation formula:

$$\int_a^b f(x)\,dx \approx \frac{h}{3}\big[f(a) + 4\,f(a+h) + 2\,f(a+2h) + 4\,f(a+3h) + \ldots$$

$$\ldots + 4\,f(b-3h) + 2\,f(b-2h) + 4\,f(b-h) + f(b)\big] \quad \text{with } h = \frac{b-a}{2m}.$$

▸▸▸ **Application**: ***numerical integration***

Calculate $\int\limits_0^1 \frac{dx}{1+x^2}$ once exactly and once by means of the approximation formula at five supporting points.

*Solution*:

The exact solution

$$\int\limits_0^1 \frac{dx}{1+x^2} = \arctan 1 - \arctan 0 = \frac{\pi}{4} = 0.785398\ldots$$

is known in this case. Now, we use the five evenly distributed knots:

$$h = \frac{1}{4} \implies (0/1),\ \left(\frac{1}{4}/\frac{16}{17}\right),\ \left(\frac{1}{2}/\frac{4}{5}\right),\left(\frac{3}{4}/\frac{16}{25}\right),\left(1/\frac{1}{2}\right).$$

With *Simpson*'s rule, we obtain:

$$\int_0^1 \frac{dx}{1+x^2} \approx \frac{1}{12}\left[1 + 4\,\frac{16}{17} + 2\,\frac{4}{5} + 4\,\frac{16}{25} + \frac{1}{2}\right] = 0.785392\ldots.$$

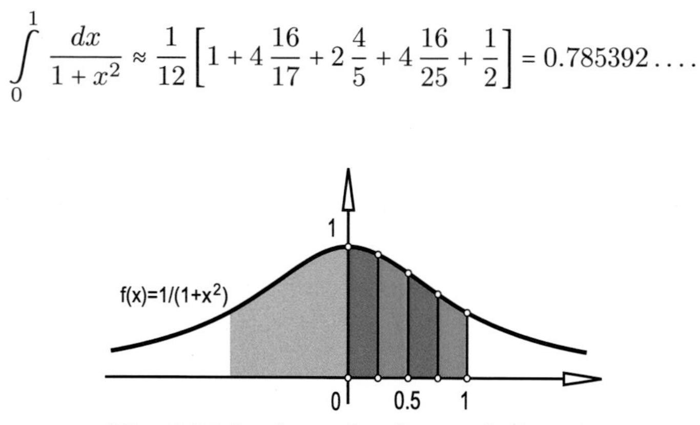


**Fig. 8.50** five knots for *Simpson*'s formula

We see that this "harmless" function already gives good results with only a few supporting points.

Using $\pi/4 = \arctan 1 \Rightarrow \pi = 4\arctan 1$, the number $\pi$ can be computed very precisely on a computer ($\pi$ is not predefined in every programming environment!). The computer calculates the arctan by means of expansion into a Taylor series (power series expansion). ◀◀◀

▸▸▸ **Application**: ***the area underneath the* Gaussian *curve***
(Application p. 371)

This area plays a fundamental role in probability theory (Chapter 8):

$$\Phi(t) = \frac{1}{\sqrt{2\pi}} \int_{-\infty}^{t} e^{-\frac{x^2}{2}}\,dx.$$

The only problem is that this particular integral can only be evaluated numerically. Moreover, it is an indefinite integral where even thousands of support points may deliver inaccurate results. One finds a way out of this dilemma.

*Solution*:

The curve $e^{-\frac{x^2}{2}}$ optically resembles the function $\dfrac{1}{1+x^2}$ (Fig. 8.4). It has $x = 0$ as a maximum and is "almost zero" for $|x| > 5$: $f(-5) = f(5) \approx 3{\cdot}10^{-6}$. Between $-\infty$ and $-5$ the area is $\approx 3 \cdot 10^{-7}$ and thus it is negligible – the same applies to the area between 5 and $\infty$. Practically, the entire area (0.9999994 of 1) is in the interval $[-5, 5]$, and there, only a few support points are sufficient.

A completely different solution is obtained as follows: According to Formula (8.5), the power series expansion of the function $y = e^{-x^2}$ is as follows:

$$e^{-x^2} = 1 - x^2 + \frac{x^4}{2!} - \frac{x^6}{3!} + \frac{x^8}{4!} - \ldots.$$

We need to compute

$$\int_a^b e^{-x^2}\,dx = \int_a^b \left(1 - x^2 + \frac{x^4}{2!} - \frac{x^6}{3!} + \frac{x^8}{4!} - \dots\right) dx =$$
$$= \left[x - \frac{x^3}{3} + \frac{x^5}{5\cdot 2!} - \frac{x^7}{7\cdot 3!} + \frac{x^9}{9\cdot 4!} - \dots\right]_a^b .$$

Now, we can prove that this series expansion converges for every $x$, although it converges *very* slowly for large $|x|$. For $|x| < 1$, the function converges very quickly, and one can also calculate probabilities with it. ◂◂◂

▸▸▸ **Application**: ***clothoid or* Cornu*'s spiral*** (Fig. 8.51)
In road construction, you need curves which continually increase or decrease the curvature, so that you can turn the steering wheel uniformly when going through the curve. These curves are given without proof in a parametric representation (parameter $t$) by the following integrals, which cannot be computed by means of elementary functions in a closed form:

$$x(t) = \int_0^t \cos\frac{x^2}{2A^2}\,dx, \quad y(t) = \int_0^t \sin\frac{y^2}{2A^2}\,dy. \tag{8.30}$$

The integrals can be calculated using *Simpson*'s approximation formula.

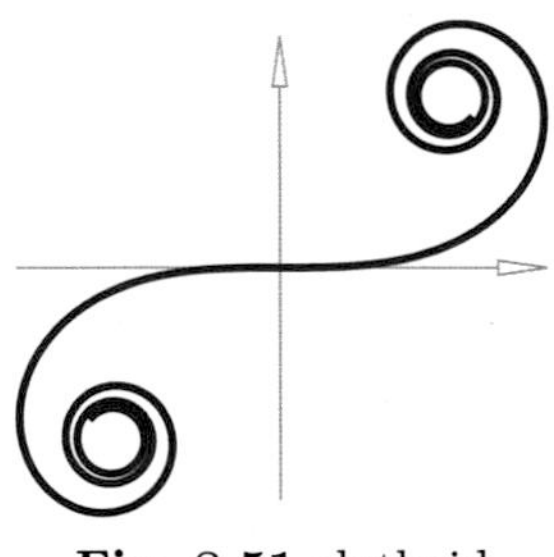

**Fig. 8.51** clothoid

**Fig. 8.52** turbulence over the Atlantic

⊕ *Remark*: Curves that almost resemble the form of clothoids are also found in nature. Fig. 8.52 shows that sometimes such forms are found even on weather maps, as if turbulence "evenly twists the steering wheel". ⊕ ◂◂◂

Obviously, *Simpson*'s formula also works if the function to be integrated is not given by mathematical formulas, but only by an odd number of evenly distributed measured values. If the measured values are unevenly distributed, the curve is approximated by spline curves in order to achieve an even distribution.

▸▸▸ **Application**: ***average dive depth, air consumption*** (Fig. 8.53)
Dive computers record the current depth of the diver at regular intervals and

calculate, among other things, the air consumption or the nitrogen content in the diver's blood. For a given "time-dive-depth diagram" $d = d(t)$, we give the average dive depth $D$ and deduce a formula for the consumption of air.

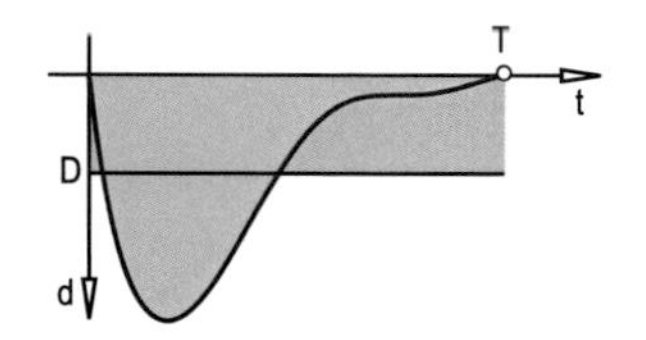


**Fig. 8.53** fast descent, slow ascent, safety stop

*Solution*:
Fig. 8.53 shows a typical "dive profile": The diver searches as quickly as possible for the maximum depth and remains there for a certain time ("bottom time"). Then he/she ascends not too fast (maximum 15–20 m per minute!) to about 5 m depth, and remains there for several minutes before finally leaving the water.
Let $T$ be the total dive time. The average dive depth, according to Formula (8.18), is

$$D = \frac{1}{T}\int_0^T d(t)\,dt. \tag{8.31}$$

Since the external pressure increases by 1 bar every 10 m, a pressure of $1 + \frac{d}{10}$ bar prevails at a depth of $d$ m. The valve of the tank with compressed air must push the same amount of compressed air into the lungs so that they do not collapse due to external pressure. Thus, air consumption is proportional to external pressure. Within time $T$, $V_0 = cT$ is the amount of air consumed on the surface. In each small time interval $dt$, $dV_0 = c\,dt$ of air is consumed. In the same time interval of $d$ m,

$$dV = c\left(1 + \frac{d(t)}{10}\right)dt$$

amount of air is required. Thus, the total air consumption equals

$$V = c\int_0^T \left(1 + \frac{d(t)}{10}\right)dt = c\int_0^T dt + \frac{c}{10}\int_0^T d(t)\,dt,$$

and therefore, using Formula (8.31),

$$V = cT + \frac{c}{10}T\,D = cT\left(1 + \frac{D}{10}\right) = V_0\left(1 + \frac{D}{10}\right). \qquad ◂◂◂$$

## 8.8 Further applications

▸▸▸ **Application**: ***calculating limits by means of series expansion***
Determine the following limits:

$$\lim_{x\to 0}\frac{\sin x}{x},\quad \lim_{x\to 0}\frac{\frac{x^2}{2}+\cos x-1}{x^4},\quad \lim_{x\to 0}\frac{e^x-1}{x}.$$

*Solution*:
All three expressions are indeterminate at $x = 0$, namely "$\frac{0}{0}$". We use the series expansion which enables us to divide by powers of $x$ and get "reasonable values":

$$\lim_{x\to 0}\frac{\sin x}{x}=\lim_{x\to 0}\frac{x-\frac{x^3}{3!}+\dots}{x}=1-\frac{x^2}{3!}+\dots=1,$$

$$\lim_{x\to 0}\frac{\frac{x^2}{2}+\cos x-1}{x^4}=\lim_{x\to 0}\frac{\frac{x^2}{2}+\left(1-\frac{x^2}{2!}+\frac{x^4}{4!}-\frac{x^6}{6!}+\dots\right)-1}{x^4}=$$

$$=\lim_{x\to 0}\left(\frac{1}{4!}-\frac{x^2}{6!}+\dots\right)=\frac{1}{24},$$

$$\lim_{x\to 0}\frac{e^x-1}{x}=\lim_{x\to 0}\frac{\left(1+x+\frac{x^2}{2!}+\dots\right)-1}{x}=\lim_{x\to 0}\left(1+\frac{x}{2!}+\dots\right)=1.$$

◂◂◂

▸▸▸ **Application**: ***period of oscillation of the pendulum*** (Fig. 8.54)
Show that the period of oscillation $T \approx 2\pi\sqrt{L/g}$ is independent of the opening angle $\varphi$ for small $\varphi$.

*Solution*:
At point $P$, the normal acceleration is $g\sin\omega$. This reminds us of the rotation of a point $P_0$ about a fixed point – even there, we have the same acceleration distribution (Application p. 274). The oscillation duration of $P$ corresponds to the orbital period of the point $P_0$ at the auxiliary circle $c$. Its radius is $r = L\sin\varphi$. Thus, the circumference of $c$ is $U = 2\pi\,\sin\varphi\,L$.
The path velocity $v$ at $P_0$ is equal to the velocity of $P$ at the lowest point. The height difference of $P$ is $\Delta = L(1-\cos\varphi)$. At the lowest point, its potential energy is completely converted into kinetic energy:

$$m\,g\,\Delta = m\,\frac{v^2}{2}\Rightarrow v=\sqrt{2g\Delta}=\sqrt{2(1-\cos\varphi)\,L\,g}.$$

This means that we have the desired orbit time

$$T = \frac{U}{v} = \frac{2\pi\, L \sin\varphi}{\sqrt{2(1-\cos\varphi)}\,\sqrt{L\,g}} = 2\pi\, \underbrace{\frac{\sin\varphi}{\sqrt{2(1-\cos\varphi)}}}_{\approx 1} \cdot \sqrt{\frac{L}{g}}.$$

Now, we use the approximations from Application p. 383, according to which both $\sin\varphi$ and $\sqrt{2(1-\cos\varphi)}$ agree quite well with $\varphi$, so that the fraction is approximately 1. For a small $\varphi$, $T$ is independent of $\varphi$. The following table provides information on the quality of the approximation:

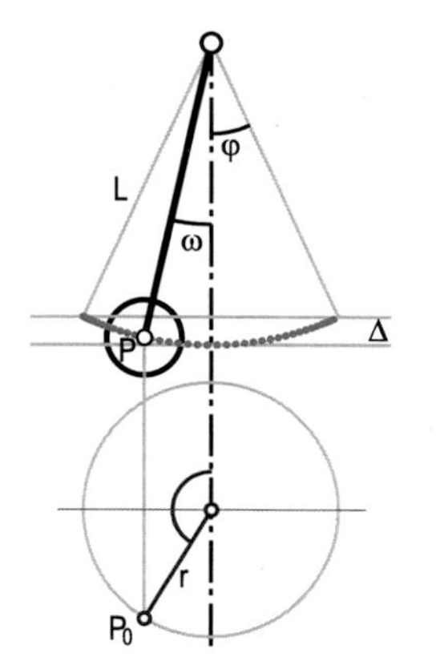


**Fig. 8.54** pendulum

| $\varphi$ | radians | $\frac{\sin\varphi}{\sqrt{2(1-\cos\varphi)}}$ | error [%] |
|---|---|---|---|
| 5° | 0.09 | 0.99905 | 0.1 |
| 10° | 0.17 | 0.99619 | 0.4 |
| 15° | 0.26 | 0.99144 | 0.9 |
| 20° | 0.35 | 0.98481 | 1.5 |
| 25° | 0.44 | 0.97630 | 2.4 |
| 30° | 0.52 | 0.96593 | 3.4 |
| 35° | 0.61 | 0.95372 | 4.6 |

A comparison with the table in Application p. 383 shows that, through the two approaches, the errors even "cancel" each other out to some extent. The error due to the approximation of the fraction is virtually negligible for $\varphi <$ 15° (long pendulum with a small deflection, see also Application p. 20). ◂◂◂

▸▸▸ **Application**: ***average brightness of the Moon*** (Fig. 8.55)
Calculate the average brightness of the Moon in the transition phase from half moon to full moon (new moon).

*Solution*:
The Moon is illuminated by the Sun. Because of the Moon's spherical shape and the nearly parallel light rays, exactly one half of the Moon is illuminated. The boundary of the shadow is a great circle of the Moon, which, viewed from the Earth, appears as an ellipse $e$ whose principal axis divides the outline of the Moon into two equal halves (the principal vertices of the ellipse are at the contour). The illuminated lunar surface, thus, consists of the area of a semicircle together with half the area of the ellipse $e$. Looking at a full moon period (it takes 29.53 days to go from one full moon to the next), the average illuminated area is equal to the area of the half moon for "symmetry reasons". This does not mean that between half moon and full moon (new moon) 3/4 (1/4) of the surface is illuminated on average, as we shall see soon:

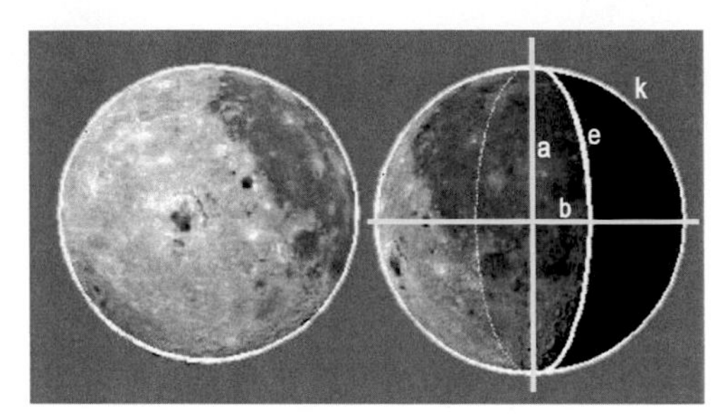


**Fig. 8.55** full moon, waning moon

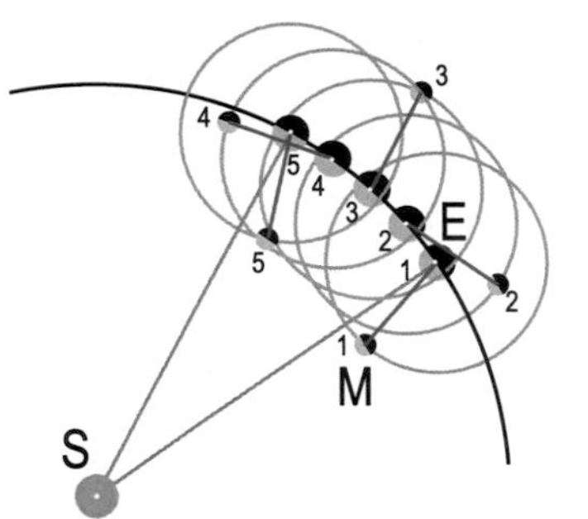


**Fig. 8.56** rotation of the Moon

Let $a$ be the radius of the contour circle and $b$ be the semi-axis of the elliptical margin of the half moon. Then, we apply Formula (8.17) to the illuminated area:

$$A = \frac{\pi}{2}a^2 \pm \frac{\pi}{2}a\,b = \frac{\pi}{2}\,a(a \pm b).$$

Since the entire lunar disc has area $A_\circ = \pi a^2$, the portion of the illuminated area $S$ (and thus, the brightness $h$) equals

$$h = \frac{A}{A_\circ} = \frac{a \pm b}{2a} = \frac{1}{2}\left(1 \pm \frac{b}{a}\right).$$

Simplifying, we assume that the lunar orbit is circular and the rotation of the Moon is uniform (which is quite close to reality). Let $T$ be the time between the half moon and full moon ($T \approx 7\,d$). In this time, the angle of rotation is about $\pi/2$ (that is, 90°). Yet, it must be remembered that the Earth and the Moon turn around the Sun during these 7 days by about 7° (in Fig. 8.57, the Moon's rotation is illustrated by 12 interpolations). Since the angle does not enter the following considerations quantitatively, nothing changes in the "average result".

After $t$ days, the angle of rotation equals $\varphi = \frac{t}{T}\frac{\pi}{2}$. The semi-axis $b$ of the ellipse is obviously proportional to $\sin\varphi$ (Fig. 8.56) and the current lunar brightness differs from the brightness of the crescent by the value $\Delta h = \frac{1}{2}\sin\varphi$. According to Formula (8.19), the average sine value is $\frac{2}{\pi}$ and $\overline{\Delta h} = \frac{1}{\pi} \approx 0.32$. In the week when the half moon waxes to the full moon, the Moon has an average of about $0.5 + 0.32 = 82\%$ of its maximum brightness. In the week when the half moon wanes to the new moon, it has only about $0.5 - 0.32 = 18\%$ of its maximum brightness.

⊕ *Remark*: In fact, from one day before the full moon until the day afterwards, there is still the subjective impression that it is a full moon. The "true" full moon is best recognized by the fact that on this day, the Moon rises from the horizon at roughly the same time as the Sun descends on the opposite side. In the preceding days, the Moon rises about 50 minutes *earlier* per day. That is why divers like to do their "night dive" a day or two before the full moon just after sunset. The diver can find a perfectly acceptable minimum amount of natural illumination.

The inclination of the half moon, which depends on the geographical latitude, is explained in a simple, geometric manner: The Sun and the Moon move relative to one another in a single plane. Both rise and descend in moderate widths at a shallower elevation than in tropical regions. In the equatorial region, the setting waxing moon is rather lit "from below", and the crescent moon "is horizontal". In the southern hemisphere, the Sun apparently travels in the opposite direction, because at noon it is generally in the North. Thus, the crescent points in the direction opposite to where it points in the northern hemisphere. ⊕

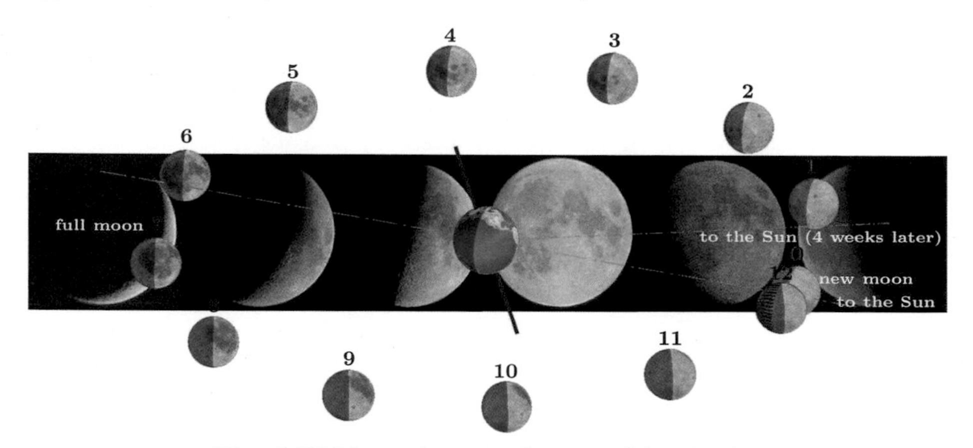


**Fig. 8.57** Moon phases in theory and in practice

⊕ *Remark*: The computer simulation shown in Fig. 8.57 shows a cycle of approximately four weeks which takes place in the northern winter (at the North Pole, there is deep polar night). It can be seen that the angle between the Earth's axis and the direction to the Sun or the Moon is approximately the same for the new moon (position 0). As a result, the Sun and the Moon migrate on nearly identical paths in the firmament (only in this phase can a solar eclipse occur). About 15 days later (between positions 6 and 7), there is a full moon. Now, the angle between the axis of the Earth and the direction to the Moon is quite different from that between the axis of the Earth and the direction to the Sun. The lunar path on the firmament will, thus, significantly deviate from the Sun's orbit (much steeper in the North). In the winter of 2005/2006, the Moon reached a particularly high elevation angle (recurring every 18.2 years) because of the extreme lunar inclination of 5.2%. ⊕

⊕ *Remark*: In the south-west of Antarctica, there is a period of several months of polar night, in which the full or almost full moon illuminates the arduous route of emperor penguins (Fig. 2.42) marching to their breeding grounds (`https://en.wikipedia.org/wiki/March_of_the_Penguins`). One more remark: How invariably beautiful must the "crescent Earth" appear from the Moon (Fig. 7.28)! Since the Moon always turns to the same side, there is no "rising Earth" or "setting Earth" on the Moon. On the side facing us, the Earth is always at the same place in the firmament because of the elliptical orbit, and there it always undergoes the "Earth phases". ⊕ ◂◂◂

### ▸▸▸ Application: *effective current*

Meters for currents measuring alternating currents do not indicate the maximum current $I_{\max}$, but the square-average current $I_{\text{eff}}$ which is also called the effective current. Calculate this for $I = I_{\max} \cos \omega t$ ($t$ is the time, $\omega$ is the angular velocity of the rotating current-generating coil).

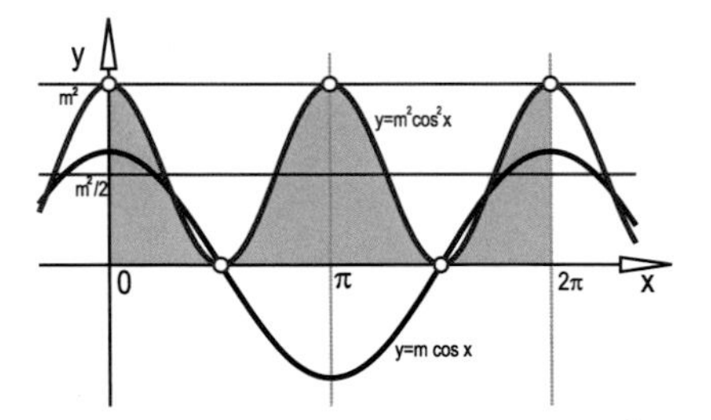


**Fig. 8.58** the average value of $y = (m \cos x)^2$ equals $m^2/2$

*Solution*:
We set $x = \omega t$ and $m = I_{\max}$. Then, we get $I^2 = m^2 \cos^2 x$. The average function value of $I^2$ in the interval $[0, \pi]$ is $\frac{1}{\pi} \int_0^{\pi} I^2 dx$. However, the average value can also be calculated immediately without using the formula $\cos^2 x = \frac{1}{2}(1 + \cos 2x)$, which can be derived from the addition theorems (see Formula (4.15)). The function $y = m^2 \cos^2 x$, therefore, has a sinusoidal graph (Fig. 8.58) which has the $x$-parallel $y = m^2/2$ as the center line. Thus, its average value in the interval in question is $m^2/2$. $I_{\text{eff}}$ is defined as the square root of this average value, and we have

$$I_{\text{eff}} = \frac{I_{\max}}{\sqrt{2}}.$$

◂◂◂

### ▸▸▸ Application: *fat deposits in the hip region*

In German, the fat belt around the hip is colloquially known as a "swim ring". Estimate its mass.

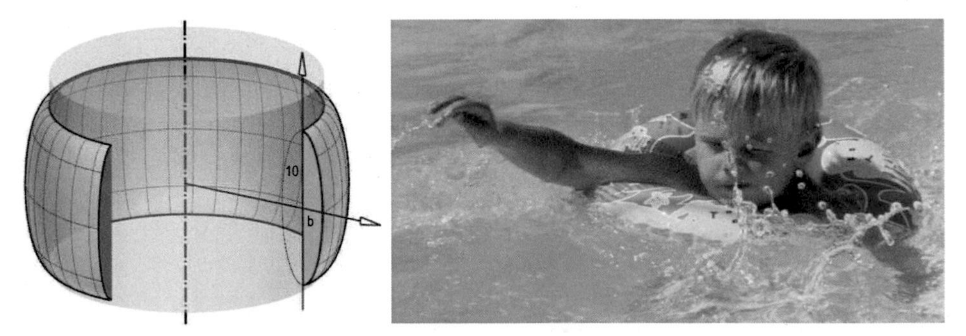


**Fig. 8.59** various swim rings (left: toroidal surface with a meridian ellipse)

*Solution*:
We shall approximate the object in question by a surface of revolution with an ellipse for its meridian curve (Fig. 8.59, left). Its meridian section is a half-ellipse with a height of 20 cm and width $b$. The area of the half-ellipse is $10\pi\, b/2\,\text{cm}^2 \approx 15\,b\,\text{cm}^2$. The ideal hip circumference (without tire) is 80 cm. Then, according to *Guldin*'s rule (Application p. 408), the tire has a volume of $1200\,b\,\text{cm}^3$. Each additional centimeter, thus, stores about 1 kg of fat (fat is lighter than water).

⊕ *Remark*: In order to lose 1 kg of fat "forever", one has to eliminate about 40,000 KJoule (or 10,000 kilocalories). People consume about a quarter of that in a day. Theoretically, one would have to drink water for four days to reduce $b$ by one centimeter (in fact, decreases or increases primarily occur in the hip area). It is best to reduce calorie intake slightly over a longer period, because it is permanent. ⊕
The maximum buoyancy of the real floating tire in Fig. 8.59, on the right, can be estimated quickly: it is a (slightly flattened) torus. At a diameter of 10 cm, the area of the meridian section is approximately $25\,\pi \approx 80\,\text{cm}^2$. With a mean diameter of 35 cm (circumference $35\,\pi \approx 110$ cm), we have a volume of about 9 liters, and thus, in order to submerge the buoyancy completely, one would need about 90 Newtons. This is easily enough to keep the head of a child completely above water. ◂◂◂

▸▸▸ **Application**: ***volume of a pyramid's frustrum*** (Fig. 8.63)
Even the ancient Egyptians used, without proof, the formula for the volume of a frustrum of a pyramid with a square base (lower edge length $a$, upper edge length $b$, height $h$):

$$V = \frac{h}{3}\,(a^2 + ab + b^2).$$

Prove the formula using *Cavalieri*'s principle.
***Proof***: The cross-section of the pyramid changes quadratically with the side length $x$: $A(x) = x^2$. The average value of the function $A(x)$ is (see Application p. 398)

$$\overline{A} = \frac{1}{a-b}\int_b^a x^2\,dx = \frac{1}{a-b}\,\frac{x^3}{3}\Big|_b^a = \frac{a^3-b^3}{3(a-b)} = \frac{a^2+ab+b^2}{3}.$$

The capacity of the volume-like prism is obtained by multiplying $\overline{A}$ by the height $h$.

⊕ *Remark*: Egyptian (and Babylonian) mathematicians did not give rigorous proofs. In this way, formulas were used which are only correct in special cases. For example, they had an incorrect formula for the calculation of the area of a general quadrangle of which the sides are known and which is not yet clearly defined. ⊕ ⊙
The advantage of this consideration is that the proof can be applied directly to any truncated pyramid or truncated cone, see also Application p. 404. ◂◂◂

### ▸▸▸ Application: *navigation problems with a GPS*

When we are in a flat terrain, there are always "more than enough" satellites that can be used for navigation using a GPS. However, when hiking in the mountains, it is often the case that satellites are "visible" only from a certain angle of elevation $\varphi$. How many satellites are "visible" from a flat terrain? What is the critical inclination angle $\varphi$?

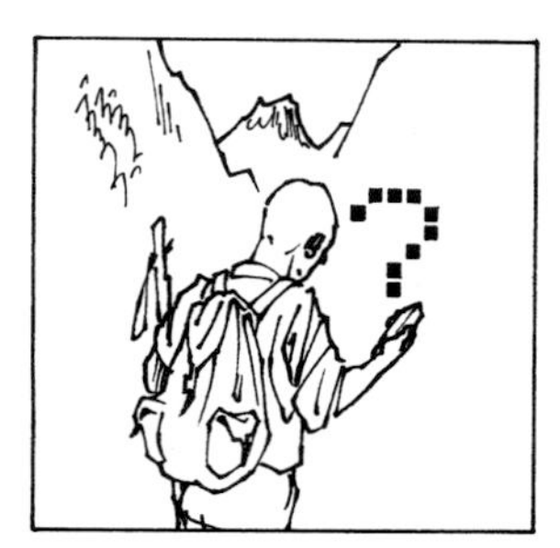

**Fig. 8.60** navigation and ...

**Fig. 8.61** ... houses as obstacles

*Solution*:

Let $R = 6,370\,\text{km}$ be the Earth's radius. The 24 satellites orbiting the Earth at an altitude of $h = 20,000\,\text{km}$ (distance $r = R + h$ from the center of the Earth) are distributed quite evenly on a spherical shell with the area $S = 4\pi r^2$. The "visible area" is nearly a spherical cap whose area $S^*$ is computed with Formula (8.26) and $\beta = \dfrac{\pi}{2}$:

$$S^* = 2\pi r^2(1 - \sin\alpha).$$

So, we have an average of

$$24 \cdot \frac{S^*}{S} = 12\,(1 - \sin\alpha)$$

"usable" satellites. The angle $\alpha$ is not the same as the inclination angle $\varphi$, but according to Fig. 8.62, we have

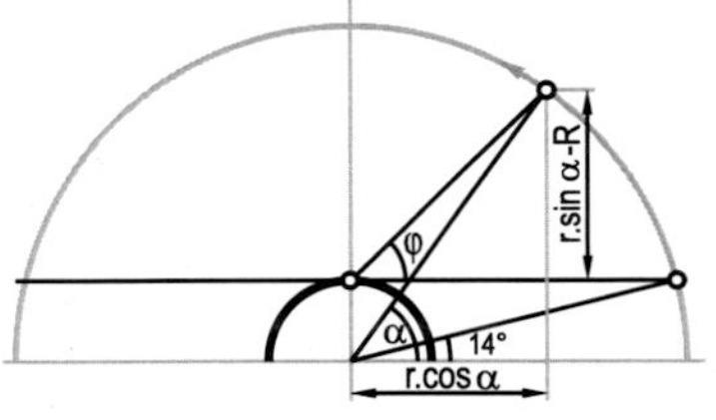


**Fig. 8.62** satellites

$$\tan\varphi = \frac{r\,\sin\alpha - R}{r\,\cos\alpha}.$$

For $\varphi = 0$, we find $\alpha = \arcsin\frac{R}{r} \approx 0.25$ ($14°$). This results in an average of eight satellites. For exact position determination, one needs "only" four satellites.

Just four satellites can be seen if

$$\alpha = \frac{2}{3}\ (\approx 42°) \Rightarrow \tan\varphi \approx 0.48 \Rightarrow \varphi \approx 26°.$$

Such a restriction occurs relatively often on a mountain hike (Fig. 8.60) and even more often in a big city (Fig. 8.61)!

⊕ *Remark*: Actual efforts have been made to increase the number of navigation satellites to at least 32. Otherwise, the navigation of cars in narrow major city lanes becomes a problem! ⊕ ◂◂◂

▸▸▸ **Application**: ***volume of a beam connection*** (Fig. 8.64)
Calculate the volume of the union of two congruent cylinders of revolution whose axes intersect at right angles (radius $r$, height $h$).

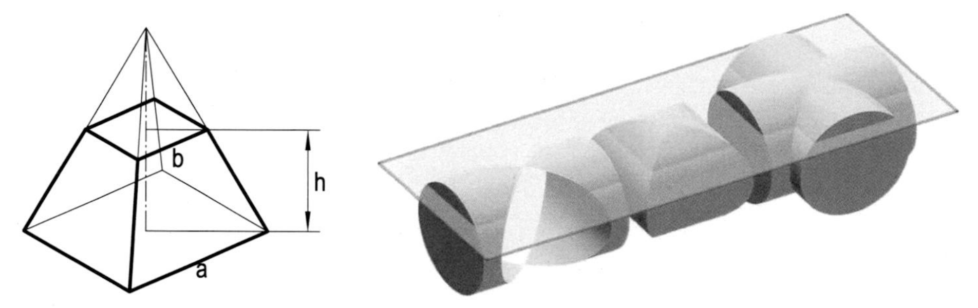


**Fig. 8.63** frustrum of a pyramid

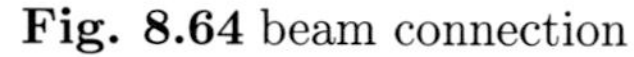

**Fig. 8.64** beam connection

*Solution*:
It is clear that the volume of the *union* of the two cylinders is equal to the total volume of the two cylindrical beams minus the (doubly counted) volume of the *average* of the two cylinders (this is a fundamental rule of *Boole*an algebra, which we will not discuss in more detail here). We first calculate the volume of the average.
As a reference plane, we select the horizontal plane of symmetry that is spanned by the cylinder's axes. The cross-section of the intersection with a plane parallel to it at a distance $z$ provides a square with the area

$$A = (2\varrho)^2 = 4\,\varrho^2 \quad (\text{with } \varrho = \sqrt{r^2 - z^2}).$$

The intersection of the cylinders is circumscribed by a sphere of radius $r$. The circle of the sphere at height $z$ has the area $\overline{A} = \pi\,\varrho^2$. The cross-sectional surfaces of the sphere and of the intersection solid differ only by the constant factor $\dfrac{4}{\pi}$. The spherical volume is $\frac{4\pi}{3}\,r^3$, which is the desired average volume

$$V_1 \cap V_2 = \frac{4}{\pi} \cdot \frac{4\pi}{3}\,r^3 = \frac{16}{3}\,r^3.$$

We also get the same value if we assume the average cross-section of the object with 2/3 of the maximum cross-section $(2r)^2$ as in Application p. 405. The squares in the horizontal intersections are always circumscribed by the circles of the sphere:

$$V_1 \cap V_2 = \frac{2}{3} \cdot (2r)^2 \cdot 2r = \frac{16}{3}\,r^3.$$

For the third time, we get the same value with *Kepler*'s rule (p. 413), although this would only be an approximation:

$$V = \frac{2r}{6}\left(0 + 4\,(2r)^2 + 0\right) = \frac{16}{3}\,r^3.$$

Finally, the volume of the beam connection is

$$V_1 \cup V_2 = V_1 + V_2 - V_1 \cap V_2 = 2\,(\pi r^2\,h) - \frac{16}{3}\,r^3.$$ ◂◂◂

▸▸▸ **Application**: ***weight of an antelope's horn*** (Fig. 8.65)
Antelopes have spirally shaped horns with a characteristic cross-section. The circumference $u$ of the cross-section of the base and the height $h$ of the horn can be easily estimated. The density of the horn substance is about $2\,\mathrm{g/cm^3}$. What is the load that the antelope has to carry?

**Fig. 8.65** kudu, oryx, nyala

*Solution*:
Horns grow according to the so-called "heli-spiral motion". If one knows a cross-section (all other parts are similar), their volume can be equated with a cone having the same cross-section according to *Cavalieri*'s principle. Therefore, it is determined by the height, but not the length of the curved center line!
If the circumference $u$ is given and the cross-section is not extremely elongated, we are content with the following approximation: We replace the cross-section by a circle of the same circumference and then reduce the area of this circle, for instance, by $2/3$, as the circle is the closed curve which encloses a maximum area for a given circumference. The corresponding volume-equivalent cone is, then, a cone of revolution with the desired volume

$$V \approx \frac{2}{3}\left(\frac{u}{2\pi}\right)^2 \pi\,\frac{h}{3} \approx \frac{h\,u^2}{60}.$$

*Numerical example*: For $u = 20\,\mathrm{cm}$ and $h = 90\,\mathrm{cm}$, we find $V \approx 600\,\mathrm{cm}^3$. Both horns together have an estimated weight of more than $2\,\mathrm{kg}$. However, parts of the horns are hollow. Therefore, we must estimate the actual weight to be lower.

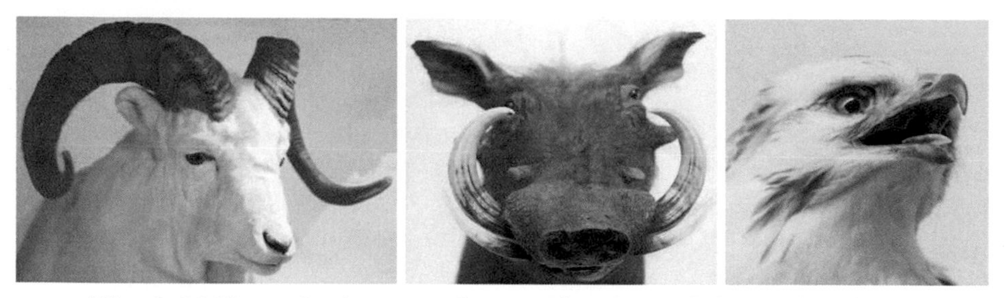

**Fig. 8.66** Horns, beaks, even claws, and tusks are heli-spiral surfaces.

⊕ *Remark*: As opposed to antler bearers, horn bearers do not discard their "headdress" every year (because they occasionally use it as a dangerous weapon, such as the oryx antelopes in the center of Fig. 8.65). Thus, their weight cannot be measured. The above estimate is, therefore, of practical value. More problematic than the weight is the great leverage that occurs with long horns. ⊕ ◂◂◂

▸▸▸ **Application**: ***marking a drinking glass*** (Fig. 8.46)
In Application p. 412, we looked at a drinking glass generated by the rotation of a sinusoidal curve. Where should the 0.25 $l$-mark be placed?

*Solution*:
This time, we will use numerical integration. For the volume of the drinking glass, we have

$$V = \pi \int_0^h \left(a - b \sin \frac{2\pi}{h} t\right)^2 dt$$

(the integration variable has been relabeled to $t$). If we consider the upper limit of the integral to be the variable $x$, then we obtain the "volume function"

$$V(x) = \pi \int_0^x \left(a - b \sin \frac{2\pi}{h} t\right)^2 dt$$

dependent on the fluid level $x$. For $x = 12$, we obtain 398 cm$^3$ from Application p. 412.
Now we look for the solution of the equation $V(x) = 250\,\text{cm}^3$, that is, the zeros of the function $V(x) - 250 = 0$. This is done using *Newton*'s method and leads to $x = 8.64$ cm: The mark must be placed at this height. ◂◂◂

▸▸▸ **Application**: ***How many people have ever been born?***
The left of Fig. 8.67 shows the development of the Earth's total population. You do not need to be a mathematician to recognize the explosiveness of the curve drawn in red. Here, we will only raise two unorthodox questions: How many people have ever been born and how many of them are still alive?

*Solution*:
The area underneath the curve can be interpreted as a measure of the number of people that have been born. The curve does not, of course, start at the year 0 but goes back a long time. It is estimated that about 5 million people lived

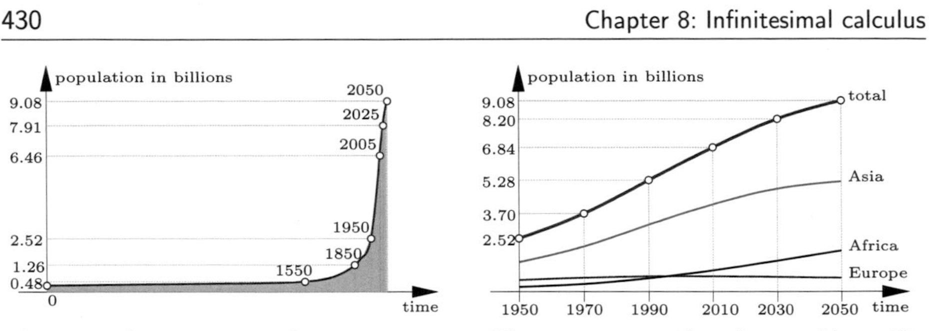


**Fig. 8.67** A steep upward movement . . . **Fig. 8.68** . . . with a foreseeable end?

10,000 years ago. Let us approximate the red curve in the left-hand region by a straight line from $(-8,000/0.005)$ to $(0/0.3)$. The contribution beyond this period can be neglected. A numerical integration shows that the number of people born in the last 2000 years is equal to the number of people born before then (in 2045, the ratio will be clearly in favour of the right half). Now we need the average age of a modern person. It is under the age of 30 (in many developing and emerging countries, it is considerably lower; in Europe and North America, it is much higher). If $y$ is the current year, then the fraction $p$ is the quotient of two definite integrals:

$$p = \int_{y-a}^{y} f(x)\,dx \Big/ \int_{-\infty}^{y} f(x)\,dx.$$

For $y = 2006$ specifically, we have $p = 0.06$. That is, 6% of all people ever born are alive. In the year $y = 2045$, it will be 9%.

⊕ *Remark*: Fig. 8.68 shows a detail of the red curve over a period of 100 years. Currently, the world's population is growing annually by about the population of Germany. The right part of the curve is, of course, an extrapolation! If this proves true, a stabilization of the world population at a high level is in sight at the end of the century. This calculation also shows how the population figures will develop on the individual continents. Obviously, everything depends on Asia. ⊕ ◂◂◂

▸▸▸ **Application**: ***differential equation of the catenary*** (Fig. 8.69)
Determine the curve that is the equilibrium position of a "heavy rope" (or, for example, of a chain).

*Solution*:
In order to calculate the equation of the curve, consider a curve point $P(x/y)$ and an infinitesimally close point $Q(x + dx/y + dy)$ (Fig. 8.69). The arc element $PQ$ has the length

$$ds = \sqrt{dx^2 + dy^2}.$$

The weight $\vec{f}$ of the rope piece from $P$ to $Q$ is proportional to the length $ds$ and acts in the $y$-direction.
The tangent $Q$ has the direction vector $\overrightarrow{t_Q}$. Those in $P$ have the direction vector $\overrightarrow{t_P}$. The two tractive forces at the rope points $P$ and $Q$ act in the

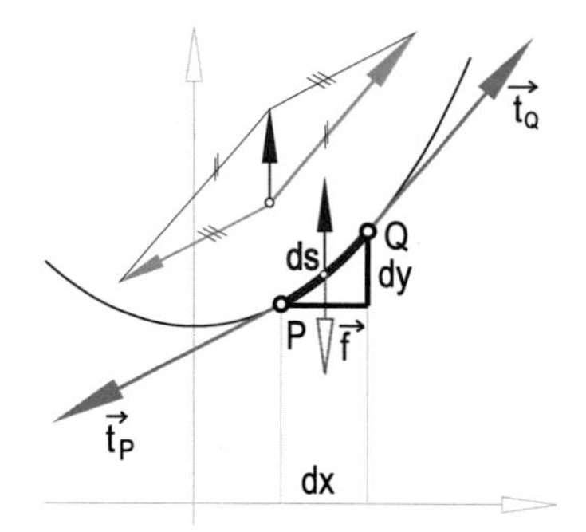


**Fig. 8.69** a differential equation

direction of the curve tangent. Thus, they are multiples of the tangent vectors. The weight and tractive forces must be balanced so that

$$A \cdot \vec{f} = B \cdot \vec{t_Q} + C \cdot \vec{t_P}.$$

The vector equation is written down in detail:

$$A \begin{pmatrix} 0 \\ \sqrt{dx^2 + dy^2} \end{pmatrix} = B \begin{pmatrix} 1 \\ y'(x + dx) \end{pmatrix} + C \begin{pmatrix} -1 \\ -y'(x) \end{pmatrix}.$$

When comparing the first row, we see that $B = C$. We divide the second row by $B$ and get

$$\frac{A}{B}\sqrt{dx^2 + dy^2} = y'(x + dx) - y'(x).$$

If we now set $A/B = a$ and divide the equation by $dx$, we get

$$a\frac{\sqrt{dx^2 + dy^2}}{dx} = \frac{y'(x + dx) - y'(x)}{dx}.$$

Now, we perform the limit $Q \to P: \; dx \to 0$, which causes $y'(x + dx) \to y''$:

$$\boxed{a\sqrt{1 + y'^2} = y''.} \tag{8.32}$$

Obviously, this differential equation is of the second order (the second derivative occurs). Since $y$ does not appear, the differential equation can be reduced to a differential equation of the first order with the substitution $z(x) = y'(x) \Rightarrow z'(x) = y''(x)$:

$$a\sqrt{1 + z^2} = z' \Rightarrow z'^2 = a^2(1 + z^2).$$

This is usually solved by separating variables: $z = ue^{p \cdot x} + ve^{q \cdot x} \Rightarrow z' = p\,u\,e^{p \cdot x} + q\,v\,e^{q \cdot x}$ is then inserted into Formula (8.32) and coefficients are compared. One can also guess the solution:

$$z = \sinh\frac{x}{a} = \frac{e^{\frac{x}{a}} - e^{-\frac{x}{a}}}{2}. \tag{8.33}$$

This results in

$$y = \int z(x)dx = a \cosh \frac{x}{a} + C.$$

A prototype of this family of curves is

$$\boxed{y = a \cosh \frac{x}{a}.} \tag{8.34}$$

The vertex $S(0/a)$ of this *catenary* is, then, on the $y$-axis. For a general vertex $S(x_0/y_0)$, the equation is

$$\boxed{y = a \cosh \frac{x - x_0}{a} + (y_0 - a).} \tag{8.35}$$

All catenaries are similar to each other (up to a scaling factor $a$) – as are all circles, all parabolas, and all equilateral hyperbolas (Application p. 242). ◂◂◂

▸▸▸ **Application**: ***arc length of the catenary***
Compute the arc length of the catenary in the interval $x_1 \le x \le x_2$.
*Solution*:

$$L = \int_{x_1}^{x_2} \sqrt{1 + y'^2} dx$$

with $y' = \sinh \frac{x-x_0}{a} \Rightarrow \sqrt{1 + y'^2} = \cosh \frac{x-x_0}{a}$. This integral can be expressed in terms of elementary functions:

$$L = a \left( \sinh \frac{x_2 - x_0}{a} - \sinh \frac{x_1 - x_0}{a} \right). \tag{8.36}$$

For $x_0 = 0$, $x_1 = 0$, and $x_2 = b$ specifically, we have $L = a \sinh \dfrac{b}{a}$. ◂◂◂

The rope pulls at the suspension point $B$ with a force $F$ that acts in the opposite direction of the tangent vector of the curve and whose magnitude is, therefore, proportional to its length $\sqrt{1 + y'^2} = a \cosh \frac{b}{a}$.

▸▸▸ **Application**: ***chain of prescribed length on two points*** (Fig. 8.70)
This task, which frequently occurs in practice, also leads to a function $f(a) = 0$, which can only be solved by means of an approximation.
Let $P_1(x_1/y_1)$ and $P_2(x_2/y_2)$ be the two end points and let $l$ be the given length of the chain. Then, we have

$$f(a) = 2a \sinh \frac{x_2 - x_1}{2a} - \sqrt{l^2 - (y_2 - y_1)^2}.$$

This leads to

$$x_0 = x_1 + x_2 - 2a \operatorname{artanh} \frac{y_2 - y_1}{l}$$

with $\operatorname{artanh} x = \frac{1}{2}\ln[(1+x)/(1-x)]$ and, subsequently,

$$b = y_1 + a - a\cosh\frac{x_2 - x_0}{a}.$$

*Numerical example*:
$P_1(2/3)$, $P_2(8/5)$, $l = 8$ ($\Rightarrow$ $x_1 = 2$, $y_1 = 3$, $x_2 = 8$, $y_2 = 5$) $\Rightarrow x_0 = 2.364$, $y_0 = 4.396$, $a = 1.678$. ◂◂◂

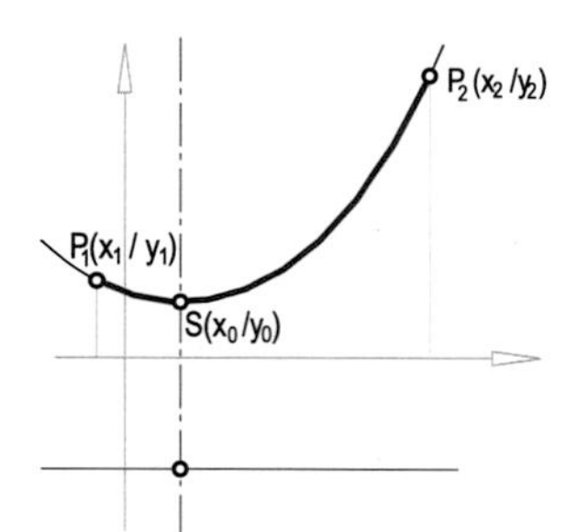


**Fig. 8.70** chain of a given length

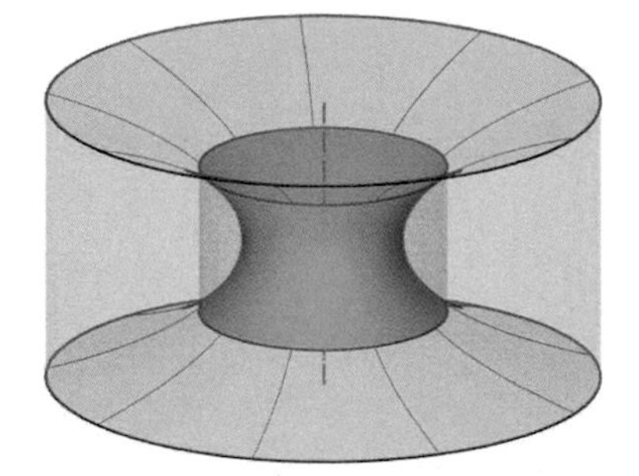

**Fig. 8.71** catenoid as a minimal surface

▸▸▸ **Application**: ***the "catenoid" as a minimal surface*** (Fig. 8.71)
If a wire rack composed of two parallel circles is immersed in a soap solution, the resulting soap film has the form of a catenoid. This surface is generated by the rotation of the catenary around the $x$-axis. Now, replace the meridian of the catenoid by a parabolic arc that has the same vertex and the same end points. This results in a new surface of revolution that is hardly distinguishable from the catenoid. Show by means of a numerical example that the catenoid has the smaller surface area.

*Solution*:
For the sake of simplicity, we choose the "standard catenary" $y = \cosh x$. The boundary circles are given by $x_1 = -1$ and $x_2 = 1$. Then, the surface area of the corresponding catenoid (the darker part in Fig. 8.71) equals

$$S = 2\pi\int_{-1}^{1}\cosh x\sqrt{1+\sinh^2 x}\,dx = 2\pi\int_{-1}^{1}\cosh^2 x\,dx = 17.677.$$

The approximation with the parabola has the equation

$$y = (\cosh 1 - 1)x^2 + 1 \quad\Rightarrow\quad y' = 2(\cosh 1 - 1)x.$$

The corresponding surface area (numerically evaluated) is 17.680, and thus, only 0.05 per mille larger.

⊕ *Remark*: The surface area of the cylinder of revolution through the two congruent boundary circles of the catenoid is larger than the surface area of the catenoid in-between these two circles provided that the circles are not too large. Fig. 8.71

illustrates that the cylindrical surface area can, however, be smaller than the surface area of the catenoid for farther away boundary circles. The "minimal surface condition" does not always provide the absolute minimum (unstable form). In the appendix on complex numbers, we will discuss the catenoid again (Application p. 548). ⊕

◂◂◂

**Fig. 8.72** Herrenkrug bridge in Magdeburg: The ropes are parabolas in non-vertical planes. The "proof": In the photograph, the images of the ropes are conics.

▸▸▸ **Application**: ***What is the shape of the ropes?*** (Fig. 8.72)

Free hanging ropes have the shape of a catenary in a vertical plane. If the ropes are strained equally, their shapes converge to parabolas. How can one verify this on a photographic image?

*Solution*:

A parabola is a conic section. We know that perspective images (such as photographs) of conics are again conics, though the type of the conic may change. A parabola only appears as parabola under specific conditions, which are usually not fulfilled in an arbitrarily taken photo. So, in most cases, the image will be either an ellipse or a hyperbola. On the other hand, a catenary will *never* appear as a conic.

Conics are defined uniquely by five points. So, if we choose five image points on the rope and there is a conic that passes through them, then the corresponding space curve is very likely to be a conic. If we are sure that the curve is planar, it *must* be a conic. Fig. 8.72 shows that for this bridge, the ropes are conics – and certainly not catenaries. It is possible to determine whether they are parabolas or not, but this would go too far here.

The "proof" is, of course, only heuristic and not strictly mathematical, but at least it provides "strong evidence". If the result is confirmed in one, or better, two additional pictures from other viewpoints, then we can be sure, and this could be accepted as a proof even by a mathematician.

⊕ *Remark*: If a rope is not strained equally, however, the result can be "basically everything". ⊕ ◀◀◀

▸▸▸ **Application**: ***differential equation for free fall with air resistance***
Find the falling speed as a function of time.

*Solution*:
In Application p. 21, we had *Newton*'s formula for the velocity in free fall, and with it, the equation $m\,a = m\,g - c_W\,A\,\varrho\frac{v^2}{2}$. If we divide by $m$, we get

$$a = \frac{dv}{dt} = g - b\,v^2 \quad \text{with} \quad b = \frac{c_W\,A\,\varrho}{2m}.$$

We can separate the variables $t$ and $v$ and get

$$dt = \frac{1}{g - b\,v^2}\,dv.$$

Now, we integrate both sides:

$$t = \int \frac{1}{g - b\,v^2}\,dv = \frac{1}{g}\int \frac{1}{1 - \frac{b}{g}v^2}\,dv = \frac{1}{g}\int \frac{1}{1 - \left(\sqrt{\frac{b}{g}}v\right)^2}\,dv.$$

With the substitution

$$x = \sqrt{\frac{b}{g}}\,v \Rightarrow dx = \sqrt{\frac{b}{g}}\,dv \Rightarrow dv = \sqrt{\frac{g}{b}}\,dx,$$

we obtain

$$t = \sqrt{\frac{1}{b\,g}}\int \frac{1}{1 - x^2}\,dx = \sqrt{\frac{1}{b\,g}}\,\text{artanh}\,x \quad \text{according to Formula (6.22).}$$

If we apply the hyperbolic tangent on both sides, we obtain $x$ (and hence $v$) as a function of $t$:

$$x = \sqrt{\frac{b}{g}}\,v = \tanh\sqrt{b\,g}\,t \quad \Rightarrow \quad v = \sqrt{\frac{g}{b}}\tanh\sqrt{b\,g}\cdot t.$$

For $t \to \infty$, the hyperbolic tangent converges to 1 and the velocity converges towards the terminal velocity $v_{\max} = \sqrt{g/b}$ (Application p. 21).

⊕ *Remark*: Of course, the dependence of the fallen distance on time is also interesting in this context. We have to integrate this again. This results (without further explanation) in

$$s(t) = 1/b\,\ln(\cosh\sqrt{b\,g}\,t).$$

Those who are not convinced only need to differentiate $s(t)$ with respect to $t$ using the chain rule which, as is well-known, is always simpler (one can still say: one has guessed the solution ...). In practice, we consult collections of formulas to check if one can find the desired integral or we work with algebra systems like Derive, Mathematica, Maple, etc. ⊕ ◂◂◂

### ▸▸▸ Application: *vector fields and flux lines*

If there is a tangent direction $y'(x, y)$ assigned to each point $P(x/y)$ in a certain domain, then there are curves that "integrate" these directions. Determine the integral curves for $y' = -\frac{y}{x}$ (Fig. 8.73, left).

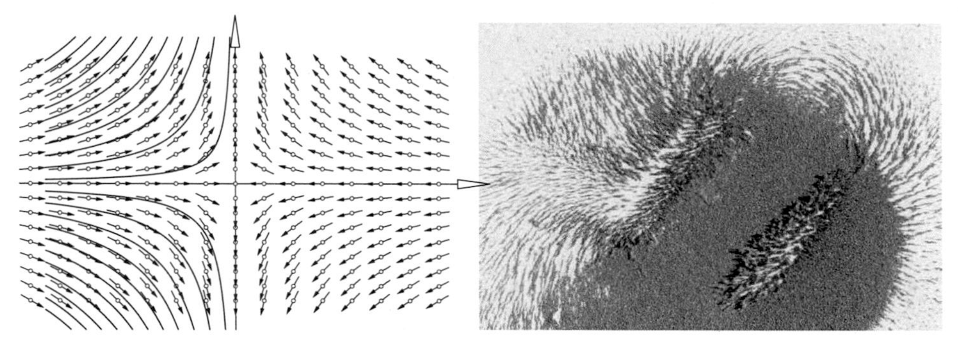

**Fig. 8.73** vector field and flux lines in theory and in practice

*Solution*:
With $y' = \frac{dy}{dx}$, we have to solve the differential equation $\frac{dy}{dx} = -\frac{y}{x}$. In this case, the variables $x$ and $y$ can easily be separated and $\frac{dy}{y} = -\frac{dx}{x}$. After integrating, the name of the variable does not matter. We can integrate this equation with Formula (8.10) and get $\ln|y| = -\ln|x| + C$ or $\ln|y| + \ln|x| = C$ (both sides provide an integration constant, which we write on the right-hand side of the equation as $C$). Using the computation rules for logarithms, we have $\ln|x\,y| = C \Rightarrow x\,y = \pm e^C = \pm a$. These are the equations of all equilateral hyperbolas with the axis of the coordinate system for their asymptotes (Application p. 242).

⊕ *Remark*: On the right of Fig. 8.73, iron chips under the influence of a magnetic field are shown. However, the corresponding differential equation is not so easy to solve. Flux in liquids and in the atmosphere can also be described by vector fields. ⊕

◂◂◂

# 9 Statistics and probability calculus

The word statistics was first employed as a synonym for "numerical descriptions of societies" in 19th century England and France. In those times, no inferences had yet been made about the properties of individual people. Only in the late 19th century did statistics experience a boom, which coincided with the rising use of mathematical methods in the natural sciences.
Today, statistics does not merely refer to population statistics, but to a branch of mathematics that investigates *mass phenomena* and attempts to develop *decision-supporting methods*. One may even put it more poetically: *Statistics is the art of learning from numbers.*
In descriptive statistics, data is sensibly ordered, grouped, and summarized through graphics, tables, or measures. This is usually done in an immediately accessible way, as it is aimed at drawing attention to the essence of the data set (a data sample).
In order to draw correct farther-reaching generalizations from a data sample, we need to employ the fundamentals of probability calculus. This will, in turn, enable us to learn about probability distributions, and in particular, about the "central limit theorem", which allows us to make predictions and draw conclusions from statistical materials.


G. Glaeser, *Math Tools*, https://doi.org/10.1007/978-3-319-66960-1_9

# 9.1 Descriptive statistics

## Bar charts, pie charts, histograms

A bar chart or histogram consists of a set of columns or bars, whose heights and surface areas are proportional to measured or calculated values.

▸▸▸ **Application**: ***number of computers worldwide***

| year | number (in thousands) |
|---|---|
| 1975 | 50 |
| 1980 | 2,100 |
| 1985 | 33,000 |
| 1990 | 100,000 |
| 1995 | 225,000 |
| 2000 | 523,000 |
| 2003 | 738,000 |
| 2005 | 896,000 |
| 2010 | 1,350,000 |

The table to the left shows statistics about the worldwide numbers of computers in the years 1975 – 2010.[a] What is particularly noticeable is that the interval differences are not always uniform (2003 is inserted). Since the table was compiled in 2006, the value for 2010 is given as a prediction.

[a] Data from `www.etforecasts.com/products/ES_pcww1203.htm`, January 2006.

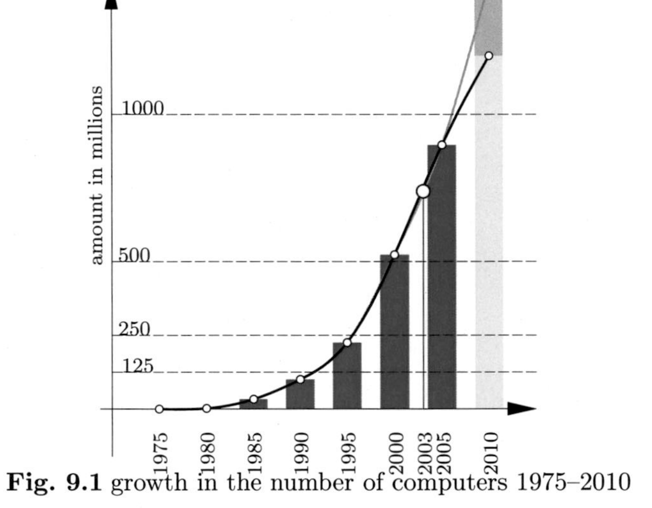


**Fig. 9.1** growth in the number of computers 1975–2010

In Fig. 9.1, the temporal range is notated on the axis of abscissas and the associated number of computers on the axis of ordinates. Two different predictions for the year 2010 can also be seen. At any given time, the associated number of computers can be estimated by interpolation. The measured data is interpolated by a curve which is as smooth as possible.

Fig. 9.1 indicates that there is a function of which only a certain number of sampling points is given. Through a smooth connection of the sampling points (spline curve), the associated function graph can be drawn (black).

Possible extrapolations towards the right (predictions) are shown in black and grey. The inserted intermediate value from 2003 is supposed to indicate the quality of the extrapolation.

⊕ *Remark*: This application stems from an earlier edition of the book and is, thus, already "dated". However, such examples may also have their advantages: It is possible to ascertain whether trends continued or predictions have actually come true. In fact, the web address given in the footnote still existed at the beginning of 2014, and we may now give the following real-world comparisons: The predicted value of 1.35 billion computers may be contrasted against the actual estimated value of 1.425 billion. For the years 2015 and 2020, numbers of the order of 2.2 billion and 2.5 billion are predicted. ⊕ ◀◀◀

Bar charts are particularly well suited for juxtaposing two or three comparable data sets. For example, related bars may be grouped together using different colors. Another way of visualizing such data sets is provided by *pie charts*, which are practical especially when dealing with percentages.

▶▶▶ **Application**: ***population profile***

| age group | 1950 | 2007 |
|---|---|---|
| 0 – 19 | 30.2% | 20.3% |
| 20 – 39 | 25.3% | 26.5% |
| 40 – 59 | 28.9% | 28.3% |
| 60 – 79 | 14.4% | 20.6% |
| 80 – 99 | 1.2% | 4.3% |

The population profile for Germany from 1950 to 2007 is to be analysed and visualized in a bar chart. The given census-based population statistics are to be used for this purpose.

*Solution*:
Here, we are dealing with comparatively few data (or rather: with strongly compressed data). Common sense provided, the results are accessible even to non-experts. One needs only the "naked eye" to conclude that the proportion of young people has decreased, while the proportion of older people has increased.

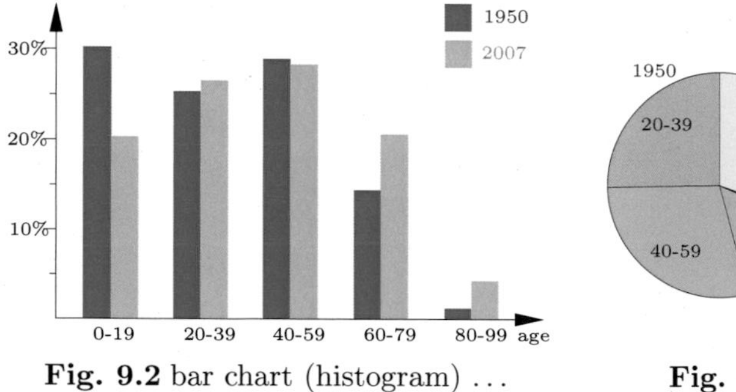


**Fig. 9.2** bar chart (histogram) ...

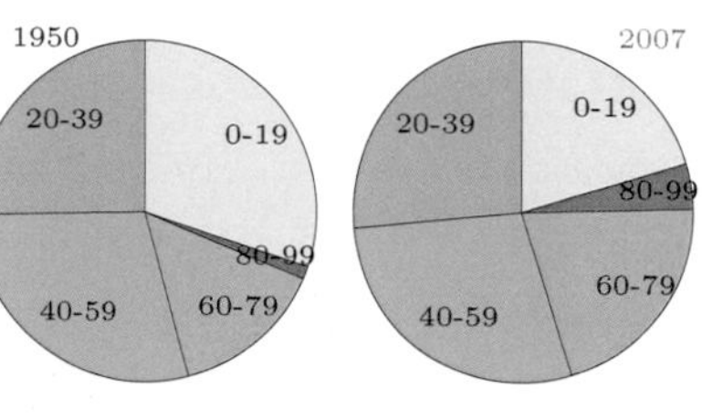


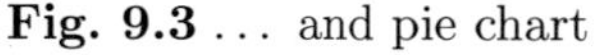

**Fig. 9.3** ... and pie chart

Concerning the *bar chart* in Fig. 9.2: On the axis of abscissas, we inscribe appropriate intervals for the five age groups. Above these, we place rectangles of appropriate width whose height is proportional to the respective percentages. Since there are only two data columns in the table, we merely need to draw two rectangles next to each other, using different colors. The chart shows nicely how what was formerly the young section of the population now constitutes today's age group of 60 – 79.

⊕ *Remark*: Such graphical representations are easier to understand for many people than decimal numbers. The ratio of the individual percentages per age group becomes particularly clear. This shows that the decline in the youngest age group is lower than the increase in the elderly or the oldest. One can see nicely that the high percentage of the age group 0 – 19 almost 60 years ago is reflected today in the high percentage of the age group 60 – 79. ⊕

Ad the *pie chart* (Fig. 9.3): The sum of the percentages is always constant, namely 100%. A sector with an opening angle of $360°/100 = 3.6°$ corresponds to one *percentage point.*
This time, we separate the two columns of the table. The color-coding refers to age groups. It is up to us where we put the starting point on the circle. If we place it in such a way that the age group 20 – 39 starts at 12 o'clock (thinking of an analog clock), one can already see the following relatively easily: If the 20 to 60-year-olds are essentially the "net payers", then they form the scarce majority of the population in the two different distributions.

⊕ *Remark*: Therefore, the education system and retirement system (still) works. The idea to raise the retirement age to compensate for the casualties of the net payers which arise due the "natural course of things" is, thus, an open secret. ⊕
Subsequently: In December 2010, the wheel had proceeded in such a way that the age groups 0 – 19, 20 – 39, etc. were given the following percentages: 18.4%, 24.2%, 31.1%, 21.0%, 5.3%.
We already see: Unless we make a statistic about the curvature of bananas or the number of brush strokes in paintings, we will always try to use statistics to arrive at results which do not appear to be listed. What is important is that we have to deal with "honest" data. Fictitious statistics are not even half as exciting! ◂◂◂

## Illustrate connections

Statistics should not only be done for their own purpose. The information in the data is often quite obvious.

▸▸▸ **Application**: ***language recognition by means of letter frequency***
How often do individual letters appear in a text? Even if we restrict ourselves to a single language, one can only work with samples. The more text is tested, the better our understanding becomes. The question is: Are there laws in the distribution of letters which are characteristic to a language?
For the present case, we have initially selected a text by *Franz Kafka* that was found on the Internet and has $n = 150,000$ characters. This amounts to

a 43-page manuscript with 3,500 characters per page. We assign the different types of umlaut, 'ä', 'ö', 'ü', which represent only a small percentage of the characters, to the letters 'a', 'o', 'u', and the more frequent letter 'ß' to 's' (counting it as one letter). If we do not distinguish upper and lower case letters, then we have 26 different classes.
If all letters appeared with the same frequency, we could expect a $(100/26)\% = 3.85\%$ share for each letter. Fig. 9.4 (vowels drawn in orange) tells us at a glance: The letter 'e' is the undisputed leader, followed by the consonant 'n'. The letters 'i' and 'r' show up nearly equally often.

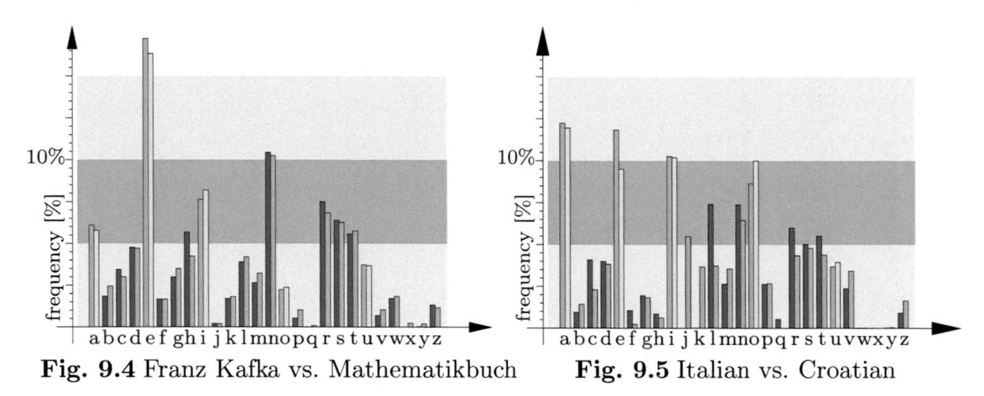


**Fig. 9.4** Franz Kafka vs. Mathematikbuch **Fig. 9.5** Italian vs. Croatian

⊕ *Remark*: It is interesting to see if the proportion of letters changes over time or from application to application. In Fig. 9.4, the "classical" distribution of the *Mathematical Toolbox* (right, brighter bars) is compared with the new spelling (again with $n = 150,000$). The distribution of letters qualitatively resembles that of the text by Franz Kafka. The amounts of 'e' and 'h' is slightly lower. We should bear in mind though that 'x' and 'y' are letters that typically appear quite often in mathematical texts. ⊕

**Fig. 9.6** different languages

Let us take a closer look at an Italian and a Croatian text (Fig. 9.5). This results in a completely different picture: In these languages, the first four places are occupied by the vowels 'a', 'e', 'i', and 'o'. In Croatian, 'j' and 'z' occur strikingly often, while ‘f’ is very rare. The letters ‘q’ and ‘w’ do not occur at all. Nevertheless, the two languages differ considerably less than, for example, German and Croatian. ◂◂◂

Similar comparisons can be made in different language texts, with relatively small differences between individual texts. Like German, English has the letter ‘e’ ranked first in the list, but ‘e’ is then followed by ‘t’ and ‘o’. The vowel ‘a’, on the other hand, is as frequent as ‘i’. Even in non-mathematical English, ‘x’ and ‘y’ are comparatively common.

⊕ *Remark*: This frequency is taken into account in the *Morse* code: Frequent letters – especially ‘e’ – have a short *Morse* character. *Morse* counted the printing press type-case in Philadelphia, where he found ‘e’ 12,000 times, ‘t’ 9,000 times, and ‘a’, ‘o’, ‘i’, ‘n’, and ‘s’ 8,000 times. This is consistent with our results. ⊕

On the basis of such diagrams, it is obvious to conjecture that the language in which a text is written can already be recognized by a frequency diagram. In the same way, one's voice can be recognized by a frequency diagram.

## Data collection and grouping

Mathematical statistics is used in a wide variety of fields: the statistic of bicycle burglaries for the insurance industry, formation of low pressure areas in meteorology, the relation between the population of a beetle species and the number of soil bacteria in biology, or the analysis of the efficacy of certain medicines.

Despite the diversity and variability of the questions one can ask, one should proceed in mathematical statistics according to the following basic steps:
First, the *problem has to be formulated.* In order to do this, clear concepts need to be created. If one is looking for the percentage of alcoholics in a country, then first clarify what makes someone an alcoholic. What if one wants to count the number of traffic fatalities on roads? Does a person who has been in critical condition for over 30 hours then fall out of the statistics? (This is how it is currently being handled.)
Secondly, the *experiment has to be planned.* Specifically, one has to ask oneself how much data will be needed to draw any conclusions from it at all and by which criteria the data is to be selected. Only then can the *practical execution* follow.
Next, the *results need to be tabulated* and described (creation of graphics). Not infrequently, data must also be grouped in a meaningful way. For better comprehension, it is important to determine certain measures of the statistics. The solution to this problem is presented in this section.
The last step is probably the most crucial and also the most difficult one: the *conclusion from the sample to the entirety.* It can only work if the previous steps are well thought-out and correctly executed. The mathematical solution is developed in the following sections.

The last step is omitted when *all* data is available. A typical example is population statistics.

▸▸▸ **Application**: ***age pyramid***

If the age distribution of the population is examined on the basis of a census, then for example, everyone born in the same calendar year can form a class. You can also distinguish between males and females. Fig. 9.7 presents the two histograms by plotting the relative frequency of a certain age group upwards or downwards depending on the sex.

**Fig. 9.7** age distribution of the Austrian population 2005, broken down by sex

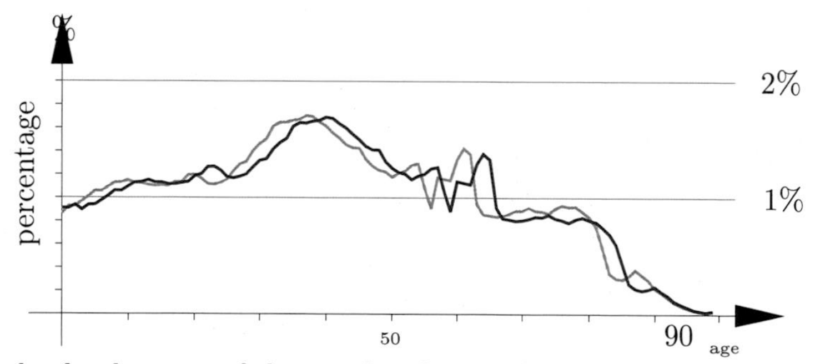


**Fig. 9.8** the development of the age distribution: Strong birth years "move to the right" (red curve: 5 years later).

Two different colors are used. To put it more precisely, both histograms are plotted on both sides, but in reverse order. This illustrates the difference between the two frequencies. In the present example, it is seen that the male proportion is greater up to the age of about 60 years. However, then the situation turns around, and the 80-year-olds and older clearly have a female surplus.

⊕ *Remark*: Such statistics are extremely important to the expert. They allow us to make fairly accurate predictions of how the age distribution will develop in the following years. Fig. 9.8 shows how the age groups move "to the right" over the years. Therefore, one can predict relatively early, for example, when there will be a surplus or a lack of teachers for certain grades. ⊕ ◂◂◂

## Mean, median, and other measures

One can link the data of a sample to a number of measures in order to assess the data better.
The *arithmetic mean* $\overline{x}$ of a measurement series $x_i,\ i = 1, \ldots, n$ is:

$$\text{arithmetic mean:}\ \ \overline{x} = \frac{x_1 + x_2 + \ldots + x_n}{n} = \frac{1}{n} \cdot \sum_{i=1}^{n} x_i.$$

Further, there is certainly at least one *minimum value* or one *maximum value* respectively, and the difference between the two values is called the *span*.
If you arrange values according to size, you can look for a value which is smaller than or equal to at least half of the values, as well as larger than or equal to the other half. This value is called the *median*.
If the amount of data is even, then there are two values which satisfy this property, so we define the median to be their arithmetic mean. Mathematically, we can formulate this roughly as follows: For a random (one-dimensional) sample $x_i\ (i = 1, \ldots, n)$, the median is the value $x_{(n+1)/2}$ if $n$ is odd and it is the value $1/2 \cdot (x_{n/2} + x_{n/2+1})$ if $n$ is even.

▸▸▸ **Application**: ***monthly wages***

We consider the monthly wages of a fictitious department supervisor of a company (4,000 €) and three other employees (2,000 € each). The amount of data ordered by size is

$$\{2{,}000,\ 2{,}000,\ 2{,}000,\ 4{,}000\}.$$

The minimum value is 2,000. The maximum value is 4,000. The mean is $\frac{1}{4}(3 \cdot 2{,}000 + 1 \cdot 4{,}000) = 2{,}500$. The median is the arithmetic mean of the second and third value, namely 2,000.
In this case, the median is already much lower than the arithmetic mean. Even more extreme is the difference when one looks at the fees of professional athletes. Here, the median reflects the reality much better than the mean. In an oil sheikdom, of course, most people have a comparatively lower – and to some extent, equal – income (the median income), while a few individuals move millions on their accounts. ◂◂◂

The arithmetic mean and the median are different in general, and neither of the two values is the arithmetic mean of the minimum and maximum value.

In contrast to the mean, the median is, at least for an odd $n$, a value occurring in the sample. On the other hand, we are accustomed to accept that, for example, a Central European owns 0.8 cars on average (the median would probably be 1).

▸▸▸ **Application**: ***The majority is often better than the average.***
If 90% of all drivers are accident-free within one calendar year and the remaining 10% have exactly one accident, each driver has an average of 0.1 accidents during this period. The median, on the other hand, is certainly 0, which insurance companies use to their advantage. ◂◂◂

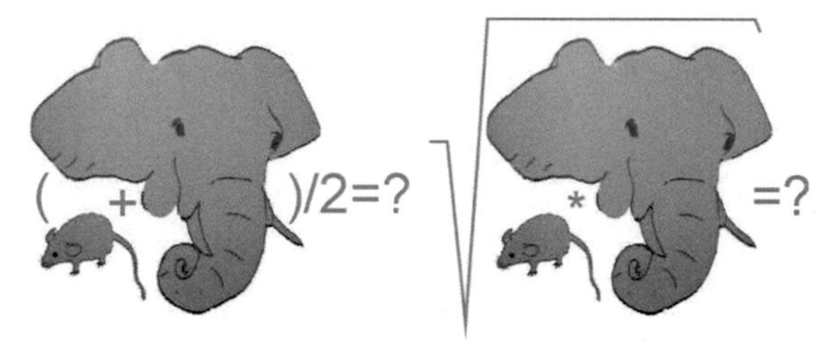


**Fig. 9.9** different mean values

▸▸▸ **Application**: ***The arithmetic mean does not always make sense.***
Consider the largest and smallest land mammal, that is, an elephant (mass 4,000 kg, body length 4 m) and a shrew (mass 4 g= 4/1,000 kg, body length 4 cm). The arithmetic mean provides an animal with a mass of 2,000 kg and a body length of 2 m. This is most likely to correspond to the mass and body length of a hippo.
The geometric mean $\tilde{x} = \sqrt{x_1 \cdot x_2}$, on the other hand, yields a living creature with a length of 40 cm and a mass of 4 kg, which comes close to a domestic cat. Among the 5,000 known mammals, there is practically no species between the hippo and the elephant, while there are certainly hundreds of species which are bigger or heavier than a cat. ◂◂◂

A variable which is important for a probability distribution (see p. 487) is the *variance*. This is the mean square deviation from the mean value $\overline{x}$. It indicates to what extent the data differs from $\overline{x}$. The less deviation there is, the more meaning the data has. Variance and *standard deviation* are defined by:

$$\text{variance } \sigma^2 = \frac{1}{n-1} \cdot \sum_{i=1}^{n} (x_i - \overline{x})^2, \quad \text{standard deviation } \sigma = \sqrt{\sigma^2}.$$

## 9.2 Probability – computing with chance

The theory of calculating probability first developed from questions of gambling. There are exact probabilities under ideal conditions. However, mathematical statistics must additionally provide methods which allow relations to be identified under non-ideal conditions (as they usually occur in practice) with sufficient certainty so that reasonable forecasts can be made.
The term "probably" is already established in colloquial language. It will probably rain tomorrow. A certain politician will probably win the election. Your partner will probably be worried if you are two hours late and you cannot be reached on the mobile phone. All of your bones will probably break if you jump down to the street from a height of 10 meters, and it would be a miracle if your bones remained unharmed. The chance of being hit by lightning is higher than that of your boss admitting to being wrong.
This is not the mathematical concept of probability, which is not necessarily contradictory to subjective perception. In the weather forecast, we are told that there will be a 60% chance of rain tomorrow. The survey results yield a 5% lead of candidate $A$ to candidate $B$. The probability of being hit by lightning is higher than that of a plane crash.

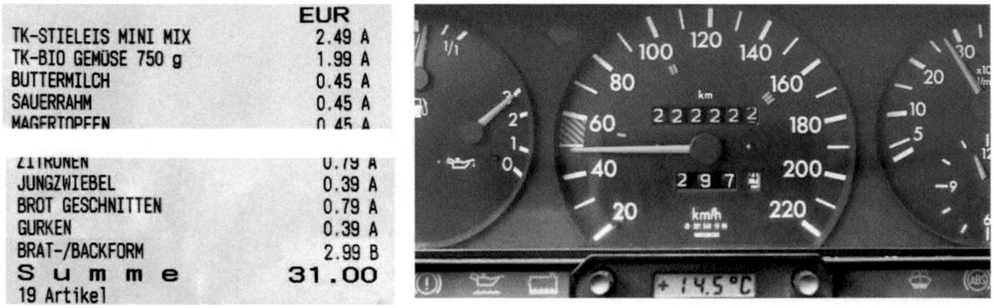

| | EUR |
|---|---|
| TK-STIELEIS MINI MIX | 2.49 A |
| TK-BIO GEMÜSE 750 g | 1.99 A |
| BUTTERMILCH | 0.45 A |
| SAUERRAHM | 0.45 A |
| JUNGZWIEBEL | 0.39 A |
| BROT GESCHNITTEN | 0.79 A |
| GURKEN | 0.39 A |
| BRAT-/BACKFORM | 2.99 B |
| Summe | 31.00 |
| 19 Artikel | |

**Fig. 9.10** Left: How often does it happen that you have to pay a round amount to the cashier and do not have to hassle with change in your hand? Right: The mileage 222, 222 in your own car is already something special.

**▸▸▸ Application**: ***the little joys of everyday life*** (Fig. 9.10)
If you pay for more than 15 or 20 items at the cashier, you will rarely get a round Euro amount (as in the image on the left). Any Cent-sum can occur, so that there is a 1 : 100 probability of a round Euro sum. However, if all items cost 0.99€, 1.99€, etc., you will need exactly 100 items.
When driving in your car and looking at the kilometer distance on your trip recorder, the probability of spotting 6 identical digits by accident (Fig. 9.10, right) is nearly zero. Even if you hope to spot the number 222, 222, there is only a moderate chance of actually seeing it in your own car.

⊕ *Remark*: According to an estimate from the USA,[1] a car drives an average of 204,000 km. Still, 10% of American cars reach 200,000 miles (≈ 322,000 km) and are, therefore, potential candidates for the 333,333 km threshold. Reaching 111,111 km, on the other hand, is very likely. ⊕ ◂◂◂

### ▸▸▸ Application: *Friday the Thirteenth*

Some people see it as a lucky day. Others dread it as a black day. Many believe that this constellation of calender day and weekday is quite unlikely to occur. However, the thirteenth can fall on any of the seven days of the week, and it occurs as often as the twelfth or fourteenth day of a month. On average, every seventh thirteenth will fall on a Friday, and we can experience the event about twice a year.

⊕ *Remark*: A more detailed study of how the thirteenth of each month is distributed among the seven days of the week (in this case, it is of no relevance when we start our calculation) will yield the ratio of 687:685:685:687:684:**688**:684 for the occurrence frequencies among weekdays beginning with Sunday. The ratio shows that the thirteenth falls more frequently on a Friday (printed here in bold) than on any other day of the week (though the difference is barely significant).[2] ⊕ ◂◂◂

**Fig. 9.11** The probability of being killed by a shark – here a white shark (*Carcharhinus carcharias*) – is almost zero. People are actually the real monsters who slaughter millions of these animals only for their fins.

### ▸▸▸ Application: *the fear of a shark attack*

Every year around 20–30 people are attacked by sharks worldwide. An ave-

---

[1] *Vehicle Survivability and Travel Mileage Schedules*. National Center for Statistics and Analysis (1/2006).

[2] `https://en.wikipedia.org/wiki/Friday_the_13th`

rage of 6 people succumb to their injuries. As a result, billions of people are afraid of swimming further out into the sea.
If we make a deliberately low estimate that perhaps every thousandth person has at least once the theoretical chance of splashing near one of the dreaded monsters during a beach holiday, then we have 6 actual and 7 million possible cases, i.e. a ratio of less than $1:1,000,000$ for this nightmare to become true. Interestingly enough, some people are comforted by the fact that about three-quarters of the attacked people survive, although this hardly changes anything about the order of the magnitude in question.
For comparison: From a total of about 100 million people in the German-speaking world, about 7,000 die in road accidents per year. Here the probability is 70 times as high $(70:1,000,000)$.
In return, unscrupulous humans slaughter 100 to 200 million sharks per year, often for the mere purpose of cutting off their the coveted fins (the meat is less tasty). The probability of a shark being killed by human hands is so high that these animals are almost extinct. ◂◂◂

## Where do we deal with probabilities?

It is comparatively easy to find examples of applications in everyday life involving probability calculations, simply because so much in life and nature has to do with chance. Here, we will discuss two examples that, at first glance, may not appear to be related to statistics or probability:

▸▸▸ **Application**: ***goals in a soccer match***

In soccer – compared to other ball games – few goals are scored. Is this good or bad for a weaker team?
Let us assume that in the last ten matches, team $A$ scored half the goals of team $B$.
If $A$ scores a maximum of one goal during the match in question and $B$ scores a maximum of two, then the six possible game outcomes are $0:0$, $0:1$, $0:2$, $1:0$, $1:1$, $1:2$. Thus, $A$ loses in three cases and wins in one case. In the remaining two cases, the game is drawn. Therefore, in half of the cases, $A$ makes at least one point $(50:50)$.
If $A$ scores a maximum of two goals and $B$ scores a maximum of four, then the outcomes $0:1$, $0:2$, $1:2$, $0:3$, $1:3$, $2:3$, $0:4$, $1:4$, and $2:4$ are a loss for $A$. The draws $0:0$, $1:1$, and $2:2$ are at least a partial success. The results $1:0$, $2:0$, and $2:1$ are a victory. This time, $A$ reaches only one point in 6 out of 15 possible cases $(40:60)$.
This shows *that if fewer goals are scored in a game, the weaker team is more likely to score a point.* This is due to the fact that the probability of a draw is higher. ◂◂◂

▸▸▸ **Application**: ***bubble formation in a water pot*** (Fig. 9.13)

When we boil water in a cooking pot, small gas bubbles develop randomly

**Fig. 9.12** few goals are thrilling ...

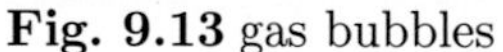

**Fig. 9.13** gas bubbles

on the ground. As the temperature increases, bubbles start rising, and the physicist speaks of Maxwell's velocity distribution – which we shall return to in the next chapter – and new and larger bubbles are formed gradually bringing the water to rest.

If the pot is centered on the hearth, the water will rise evenly despite the constantly new situation of the fire. In sum, the probabilities are the same, and the place where the bubbles are formed no longer plays a role. ◂◂◂

## Profit outlook as motivation

The beginnings of probability theory go back to *Galileo Galilei*, *Blaise Pascal*, and *Pierre de Fermat* in the 17th century. It began with the profit outlook for games of chance. At the beginning of the 18th century, two basic works on probability theory appeared: one written by *Jacob Bernoulli* (*Ars conjectandi*) and another by *Abraham de Moivre* (*The Doctrine of Chances*). *Pierre Simon de Laplace* refined the theories in his *Théorie analytique des probabilités* a hundred years later. Thus, the classic concept of probability was coined. This concept is still used today in game theory. Later, the term had to be expanded to be suitable for statistics.

## Classical experiments

**Fig. 9.14** The toss of a coin is a classical experiment with only two possible results (except when the coin lands on its edge, which is quite unlikely). Throwing a drawing pin is a random experiment, but not a Laplace experiment, because the probability of landing on the head is greater than that of balancing on the tip.

- If we throw a die, we can expect a number from one to six. Nobody can predict the exact outcome.
- A coin tossed high is, so to speak, a die with only two sides ("heads or tails").
- Also, when you draw a card in a card game, you have a well-defined number of options. Yet, if the cards are well shuffled, the result is not predictable for each attempt.
- This analogously applies to a blind person grasping for something in an urn in which there are balls that only differ by color.

## Random experiment

What we mean by a random experiment (*experiment*, for short) is an arbitrary, often repeatable process which is carried out according to a very specific set of rules, the result of which cannot be clearly predicted. In other words, the random experiment's result "depends on chance".

▸▸▸ **Application**: ***Is there a random experiment?***

- Tossing a drawing pin is a random experiment with two possible results: the tip facing up or down.
- The case of buttered bread falling from a table to the ground is, strictly speaking, not sufficiently defined to qualify as a random experiment: The result depends strongly on the height of the table: The bread does not fall on the buttered side because it is heavier. Due to the usual table height, it is, in fact, likely to perform a half-turn.
- Choosing a person for the purpose of investigating various data (e.g. cholesterol levels or telephone accounts) or checking the expiration date of a product in a supermarket *can* be a random experiment if the selection criteria is predetermined.

◂◂◂

## The result of the experiment

An experiment has different results by definition. For each attempt, a given result (*event*) $A$ may or may not come true. If we perform the experiment $n$ times and $A$ occurs $k$ times, then $k$ is the *absolute probability* of $A$. The *relative frequency* $h(A)$ is then defined by

$$h(A) = \frac{\text{number of trials with result } A}{\text{number of trials}} \quad (\Rightarrow 0 \leq h(A) \leq 1).$$

▸▸▸ **Application**: ***boy or girl (1)*** (Fig. 9.15)
In one day, 14 boys and 16 girls are born in a birth clinic. The two possible events are $A$: birth of a boy, $B$: birth of a girl. The corresponding relative frequencies are
$h(A) = 14/30 \approx 0.47$, $h(B) = 16/30 \approx 0.53$. Countless statistics show that the ratio of *boys* : *girls* at birth is not exactly 1 : 1, but that more boys are born on average around the world. See also Application p. 493.

**Fig. 9.15** Three to four babies every second worldwide ...

Smaller differences between individual countries are generally related to marriage habits, which tend to have impact on the age at which women have babies: Older mothers have a slightly higher probability of having a boy. ◂◂◂

## Can events be related to each other?

We shall define the following:

> *Sum event and product event* Considering the event in which either $A$ or $B$ occurs, one speaks of the sum event $A \cup B$. The event which occurs when both $A$ and $B$ occur is called a product event $A \cap B$.

⊕ *Remark*: In different textbooks, there are alternative notations for the sum event and the product event: Instead of $A \cup B$, one also writes $A + B$ or $A \vee B$. Instead of $A \cap B$, one also writes $AB$ or $A \wedge B$. From the context, it is usually easy to see which notation is preferred, and because there is no possibility of confusion, this is not a problem for us. ⊕

▸▸▸ **Application**: ***events when rolling dice***
$A$: rolling a 6, $B$: rolling a 3, $C$: rolling an even number. Then, by definition, $A \cup B$: rolling a number divisible by 3, $A \cup C$: rolling an even number, $B \cup C$: rolling an even number or a 3 (or alternatively phrased, rolling a number other than 1 and 5). Furthermore, $A \cap B$: is not possible, $A \cap C$: rolling a 6, $B \cap C$: is not possible. From the products one sees immediately: The events $A$ and $B$ on the one hand and $B$ and $C$ on the other hand exclude each other mutually. ◂◂◂

**▸▸▸ Application**: ***ticking-off***

A questionnaire for an examination according to the "American system" is considerably easier to fill out if only one answer is possible, that is, if the answers are mutually exclusive. A lot more knowledge is required to tick the correct answer(s) in a multiple-choice test. ◂◂◂

For the relative frequencies of two events $A$ and $B$, we have:

| | |
|---|---|
| $A$ and $B$ exclude each other: | $h(A \cup B) = h(A) + h(B),\ h(A \cap B) = 0;$ |
| in general: | $h(A \cap B) = h(A) + h(B) - h(A \cup B).$ |

***Proof***: As in the above two formulas, let $A$ and $B$ be two events that can occur in an experiment. In $n$ trials, exactly four cases can occur in each experiment:

1. $A$ and $B$ occur simultaneously ($i$ times);
2. $A$ occurs but not $B$ ($j$ times);
3. $B$ occurs but not $A$ ($k$ times);
4. Neither $A$ nor $B$ occur ($n - i - j - k$ times).

Thus, $A$ has occurred exactly $i + j$ times, $B$ occurred exactly $i + k$ times, $A \cup B$ occurred exactly $i+j+k$ times and $A \cap B$ occurs exactly $i$ times. This, by definition, gives the relative probabilities

$$h(A) = \frac{i+j}{n},\ h(B) = \frac{i+k}{n},\ h(A \cup B) = \frac{i+j+k}{n},\ h(A \cap B) = \frac{i}{n}. \qquad \odot$$

**▸▸▸ Application**: ***Fibonacci number and/or prime number***

Consider the integers $n$ up to 20 and the occurrences
A: $n$ is a Fibonacci number (p. 530) ($x \in \{2, 3, 5, 8, 13\}$);
B: $n$ is a prime number ($n \in \{2, 3, 5, 7, 11, 13, 17, 19\}$).
We have $h(A) = 5/20,\ h(B) = 8/20,\ h(A \cap B) = 4/20,\ h(A \cup B) = 9/20$. In fact, the second of the above formulas holds true. ◂◂◂

**▸▸▸ Application**: ***boy or girl (2)***

In Application p. 450 with 30 births (14 boys, 16 girls), the events $A$ and $B$ (birth of a boy or a girl) are mutually exclusive. The sum event $A \cup B$ is called the "birth of a boy or girl". Therefore, $h(A \cup B) = h(A) + h(B) = \frac{14}{30} + \frac{16}{30} = 1$. ◂◂◂

## 9.3 The probability concept

We have used the word *probability* so often that it is time to specify its definition explicitly. Let us begin with a definition that goes back to *Laplace*.

*Classical definition of mathematical probability:*
The probability $P(A)$ of an event $A$ in a random experiment is given by

$$P(A) = \frac{\text{number of events with result } A}{\text{number of equally possible cases}}.$$

Therefore, we need a number of equally possible cases. Let us examine the examples we have seen so far as to whether we have equally possible cases.

- The tossed coin provides only two cases, and both are equally possible with sufficient accuracy. Thus, the probability that a coin will land on the side $A$ is $P(A) = 1/2$.
- In the case of dice, the six possible events ideally occur equally frequently. Thus, the probability of rolling a 4 is as high as rolling a 1, namely $1/6$.
- The rolling of regular 12-sided solids (dodecahedra, Fig. p. 437) has 12 equivocal results with the probabilities $1/12$.
- In a card game (with $n$ cards), the probability of drawing a certain card from a well-shuffled deck equals $1/n$ for all cards. The same applies to the random removal of $n$ balls from an urn.
- We get in trouble with the *Laplace* definition when tossing a drawing pin, because the two probabilities are different.
- Also, human birth is, strictly speaking, not a*Laplace* experiment, because marginally more boys are born than girls.

### Probability and relative frequency

So, we see that we need to extend the original definition. Our starting point is the following observation: In most random experiments, the occurrence of events is in the long run subject to certain laws. These laws must, of course, be proven by experiments.

*Definition of probability via relative frequency:*
If the relative frequency of an event $A$ converges to a fixed value as the number of experiments increases, this value is called the probability $P(A)$ of the event $A$ for the respective random experiments.

The classical concept of probability fits seamlessly into the generalized definition. In this case, one assumes the equality of the events.

▸▸▸ **Application**: ***It will soon settle again.***

One is tempted to say: The relative frequency "oscillates". This concept, however, gives the impression that there is a certain "restoring force", but there is no such thing (see p. 476). We only have the following statement:

> A "temporary excessively high frequency" is only "broken down" by the fact that the number of events in the future will increment the total number of events (the numerator) at a much slower rate than the number of trials (the denominator).

◂◂◂

One should always be aware of this fact. Random experiments do not *remember* what happened in the past.

▸▸▸ **Application**: ***tossing a drawing-pin***

In order to ascertain if a certain relative frequency/probability occurs when tossing a drawing-pin, it should be thrown 50 times, then, for example, another 50 times, etc. One can also shorten the experiment by tossing 50 drawing-pins on a table. However, the drawing-pins can mutually influence each other, as can be seen on the right in Fig. 9.14.

The relative frequencies of the two possible events will only change in the second decimal place, and later only in the third decimal place. Even more accurate results do not make sense in this example. ◂◂◂

It can happen that empirically determined probabilities change with the passage of time. For example, with changing living habits (women being much older for their first pregnancy), the distribution of the sexes in births was found to change at least from the third decimal place onward in favor of the male sex.

The probability introduced here is, thus, the theoretical counterpart to the empirical nature of relative frequency. Therefore, it is not difficult for us to accept certain assumptions based on relative frequencies when these assumptions are not further justified. This so-called system of axioms was laid down by A.N. *Kolmogoroff* (1933).

From these axioms one can now draw conclusions. Consider, for example, two opposite (complementary) events $A$ and $\overline{A}$:

If $A$ occurs, then $\overline{A}$ does not occur, and vice versa. Thus, $A \cup \overline{A}$ is the definite event and we have:

$$P(A) + P(\overline{A}) = P(A \cup \overline{A}) = P(S) = 1.$$

This leads to the following formula, which is often used in practice for complementary probability.

Complementary probability: $P(\overline{A}) = 1 - P(A)$.

▸▸▸ **Application**: ***rolling two dice***

What is the probability of reaching a maximum of 11 total points with two dice (event $A$)?

*Solution*:

This can be solved either by writing down all the cases in which the sum of eyes is less than or equal to 11 or by looking at the complementary event $\overline{A}$: The sum of the eyes is greater than 11, i.e. equal to 12. This can only be achieved with two 6s. Among all the 36 equally probable combinations (6 possibilities for the first die and 6 possibilities for the second die), there is only one in which $\overline{A}$ is true. Thus, $P(\overline{A}) = 1/36$ and $P(A) = 1 - P(\overline{A}) = 35/36$. ◂◂◂

## Adding and multiplying probabilities

We generalize and define the sum of $m$ events $A_1 \cup A_2 \cup \ldots \cup A_m$ as the event that occurs when at least one of the events mentioned occurs. Similarly, we define the product of $m$ events as $A_1 \cap A_2 \ldots \cap A_m$. For this product, all events must occur simultaneously.

According to *Kolmogoroff*, the following arises by induction for mutually exclusive events $A_1, A_2, \ldots, A_m$:

$$P(A_1 \cup A_2 \cup \ldots \cup A_m) = P(A_1) + P(A_2) + \ldots + P(A_m).$$

**Fig. 9.16** Two dice show the same face in this throw. Apart from this, the throw is invalid because the red die does not lie flat on the board.

**Fig. 9.17** Dice was played in the 17th century and was very popular. However, if you think about the gambling boom of recent times ...

▸▸▸ **Application**: ***three different results***

How likely is it to get three different results while throwing three dice at the same time?

*Solution*:

If the first die shows 1, then the second die will only have five options, and the third will have only four, so that no numeric repetition occurs. This yields $5 \cdot 4 = 20$ possibilities. There is an equal possibility to get a 2 on the first die, etc. This yields $6 \cdot 20 = 120$ possibilities, and the probability of the sum event is $120 \cdot 1/216 = 5/9$, which is about 56% according to the addition theorem. ◂◂◂

▸▸▸ **Application**: ***two questions from the beginning of probability***
*Galileo Galilei* was able to answer the following question, which had been discussed for centuries and was asked by the *Prince of Tuscany*: Why is the sum of eyes on three dice more likely to be 10 than 9 even though both sums can occur in six ways? *Blaise Pascal* solved for the *Chevalier de Méré* (a philosopher and literary man at the court of *Ludwig XIV*) the following problem: Is it more probable to roll a 6 with one die that is rolled 4 times or to obtain two 6s with two dice after 24 rolls?
*Solution*:
First, the task of *Galilei*:
In total, we have $6^3 = 216$ equally probable cases. The sums 9 or 10 can both be achieved in 6 ways:

$$9 = 1+2+6 = 1+3+5 = 1+4+4 = 2+2+5 = 2+3+4 = 3+3+3$$

$$10 = 1+3+6 = 1+4+5 = 2+2+6 = 2+4+4 = 2+3+5 = 3+3+4$$

Triples with three different numbers of eyes, e.g. $(1,2,6)$, can occur in $3{\cdot}2{\cdot}1 = 6$ combinations. Triples with two equal numbers, e.g. $(1,4,4)$, can only occur in three ways. The triple $(3,3,3)$ can occur only once. Therefore, there are only 25 possibilities for the sum 9, but 27 possibilities for the sum 10. So, the two sums occur with the probabilities $25/216 \approx 0.116$ and $27/216 = 0.125$ respectively. On average, the sums 9 or 10 appear with almost every fourth roll.
If you consider how many attempts you have to make to show the slight difference in probabilities, you can only wonder why the fact was so well-known. In the next chapter, we will determine the significance with which we can test the hypothesis.
Let us now turn to the question of *Chevalier de Méré*:
Four rolls with a single die provide $6^4$ equally likely cases. The counter-event of "rolling at least a 6" is called "rolling no 6". For this, there are $5^4$ possibilities and, thus, the probability is $(5/6)^4 \approx 0.482$. The desired complementary probability is $1 - 0.482 = 0.518$.
A roll with two dice has 36 equally likely occurrences, including (with the probability of 1/36) the double 6. The opposite of "at least one double 6" is the event "no double 6" (probability 35/36). If this is to occur 24 times in succession, the probability is $(35/36)^{24} \approx 0.509$. The desired complementary probability is $1 - 0.509 = 0.491$. So, this event is marginally less improbable.
An additional question of *de Méré* is also worth mentioning: If the probability of rolling a double 6 with two dice is slightly less than 50%, then for how

many throws does the probability exceed 50%? Let $n$ be this number. The complementary probability must now be less than 0.5: $(35/36)^n < 0.5$. We raise the inequality to the power of $-1$, which reorients the inequality sign. Thus, we get

$$\left(\frac{36}{35}\right)^n > 2.$$

This can be solved by tests or using the logarithm:

$$n \cdot \log \frac{36}{35} > \log 2 \;\Rightarrow\; n > 24.6.$$

This means that it is favorable to bet on the double 6 from 25 rolls onward.
According to another question, which is completely different from *de Méré*'s, two players $A$ and $B$ agree that the first person who scores five points in the coin toss game will win. After seven throws, the game ends, with $A$ having scored four times and $B$ having scored three times. How should they split the winnings?
*Pascal* solved the question as follows: $A$ wins with the probability of 0.5 on the next toss. So, $A$ already has half of the money. If $A$ loses the next roll, $B$ has four points, and then both have the same chance of winning the remaining prize money. Consequently, the prize money is divided in the ratio $3:1$ for $A$. ◂◂◂

▸▸▸ **Application**: ***sweepstake winnings***

What is the probability of a sweepstake-12 or sweepstake-11? What is the probability of having no more than one correct tip?

*Solution*:

In the case of a soccer match, the first team can win (tip 1), the second team can win (tip 2), or the game ends in a draw (tip X). You have to guess the correct outcome for 12 games in succession, in order to have a 12. For each game, the probability is 1/3. According to the product rule, the probability of a 12 equals

$$\underbrace{\frac{1}{3} \cdot \frac{1}{3} \cdot \ldots \cdot \frac{1}{3}}_{12 \text{ times}} = \frac{1}{3^{12}} \approx \frac{1}{500,000}.$$

For an 11, one tip must be wrong (probability 2/3), and the remaining tips must be correct. For example, the probability of the first tip being wrong and the following tips all being correct is

$$\frac{2}{3} \cdot \underbrace{\frac{1}{3} \cdot \frac{1}{3} \cdot \ldots \cdot \frac{1}{3}}_{11 \text{ times}} = \frac{2}{3^{12}} \approx \frac{1}{250,000}.$$

However, the second or the third tip could be wrong. This results in the 12-fold value, i.e. about $1/20,000$, for guessing the outcome of 11 matches.

It is also quite difficult to get all tips wrong. The probability of this is, in fact,

$$\underbrace{\frac{2}{3} \cdot \frac{2}{3} \cdot \ldots \cdot \frac{2}{3}}_{12 \text{ times}} = \left(\frac{2}{3}\right)^{12} \approx \frac{1}{130}.$$

The probability of having exactly one right tip is calculated, in analogy to our calculations for 11 tips, as

$$12 \cdot \frac{1}{3} \cdot \underbrace{\frac{2}{3} \cdot \frac{2}{3} \cdot \ldots \cdot \frac{2}{3}}_{11 \text{ times}} = 4 \cdot \left(\frac{2}{3}\right)^{11} \approx \frac{1}{22}.$$

In sum, only every nineteenth randomly completed sweepstakes coupon will have a maximum of one correct result.

⊕ *Remark*: The result of the clash of two soccer teams is, however, not a random event. If you know the previous matches of both teams, you can increase your chances of a 12. ⊕ ◂◂◂

### ▸▸▸ **Application**: ***two are better than one***

Two different early detection devices detect a tumor with a probability of 70% or 80%. What is the probability that the tumor will be detected when both devices are used simultaneously?

We first answer the opposite event: The tumor is not detected by either the first or the second device. The event $A$ in which the first device fails to detect the tumor has a probability of $P(A) = 0.3$. The event $B$ in which the tumor is not detected by the second device has a probability of $P(B) = 0.2$. Thus, $P(A \cap B) = 0.2 \cdot 0.3 = 0.06$. There is a probability of 6% that the tumor will not be detected. Or put differently, there is a 94% probability of the tumor being detected by at least one of the two devices.

⊕ *Remark*: In practice, it may well be the case that seemingly independent devices are based on the same principle. In this case, the calculation is no longer correct. ⊕

◂◂◂

## 9.4 Conditional and independent events

So far, we have often calculated the probabilities of independent events: *New game, new chance*, as is often said in casino jargon. When dice are rolled, the same probability is obtained, no matter how many times we have already rolled. Let us now examine the probability of dependent events.

▸▸▸ **Application**: ***consider the flip-side***
On the shelf of a fruit stand (Fig. 9.18), there are 10 fruits 3 of which are rotten at the back. We pick a fruit and put it into the shopping basket because we assume the fruit to be fresh. What is the probability that the next fruit will also be good?

**Fig. 9.18** Left: The available fruits that are wonderful. Right: As beautiful as they looked in the front: one of the two apples is rotten!

The experiment consists of drawing one fruit out of ten, and all ten cases have the same probability (i.e. are equally likely to occur).
Let $A$ be the event that the first fruit is not rotten. Its probability is $7/10 = 0.7$. If this event happens, there are only nine fruits remaining on the shelf. Six of these are not rotten at the back.
Let $B$ be the event that the second fruit is not rotten when the first fruit has been returned. Its probability is the same as $A$ (if the position of the returned fruit is not memorized). We are only interested in the probability of $B$ under the condition that $A$ has already happened (symbolically $B|A$). This is

$$P(B|A) = \frac{6}{9} = \frac{2}{3} \approx 0.67.$$ ◂◂◂

### Bayes's formula

We shall now examine further dependent events in a more general way and prove a formula which is frequently applicable in practice. It is attributed to the 18th-century English mathematician *Thomas Bayes*:

*Conditional probability – Bayes's theorem*:
The probability of the event $B$ under the condition that an event $A$ has occurred equals

$$P(B|A) = P(A \cap B)/P(A). \tag{9.1}$$

⊕ *Remark*: We also speak of the conditional probability of the event $B$ under the hypothesis of $A$. ⊕

***Proof***: Let us consider an experiment in which a total of $n$ equally likely cases can be distinguished. Furthermore, two events $A$ and $B$ are to occur. In $m$ cases, the occurrence is $A \cap B$ (both $A$ and $B$). In $k$ cases, the occurrence is $A$ (and not $B$). In $l$ cases, the occurrence is $B$ (and not $A$). On the basis of the classical probability definition, we have the probabilities

$$P(A) = \frac{k}{n}, \quad P(B) = \frac{l}{n}, \quad \text{and} \quad P(A \cap B) = \frac{m}{n}.$$

Let us now determine the probability of $B$ under the additional condition that only the cases in which $A$ occurs (symbolically $P(B|A)$) are considered.
This new probability is obviously

$$P(B|A) = \frac{m}{k},$$

as this time we only consider $k$ cases where $B$ occurs $m$ times. The expression makes sense only if we exclude $k = 0$. So, we have

$$\frac{m}{k} = \frac{\frac{m}{n}}{\frac{k}{n}} = \frac{P(A \cap B)}{P(A)}.$$

⊙

By interchanging the labels, it is immediately obvious that

$$P(A|B) = \frac{P(A \cap B)}{P(B)}.$$

This yields the following theorem, which is important in practice:

*Multiplication theorem*:
If two events $A$ and $B$ have the probabilities $P(A)$ and $P(B)$ in an experiment, then the probability that $A$ and $B$ occur simultaneously is

$$P(A \cap B) = P(A) \cdot P(B|A) = P(B) \cdot P(A|B). \tag{9.2}$$

▸▸▸ **Application**: ***two at one blow***
Let us return to our previous example of the 10 fruits at the fruit stand where

three pieces are rotten at the back. What is the probability of picking two fresh fruits when exactly two pieces are taken from the fruit stand (Fig. 9.18, right)?
Let $A$ and $B$ again be the events in which the first fruit or the second fruit is not rotten. Obviously, $P(A) = P(B) = \frac{7}{10}$. Now, we have seen: $P(B|A) = \frac{6}{9} = \frac{2}{3}$.
The probability that both events occur simultaneously is given by the multiplication theorem

$$P(A \cap B) = P(A) \cdot P(B|A) = \frac{7}{10} \cdot \frac{2}{3} \approx 0.47.$$

◀◀◀

▸▸▸ **Application**: ***checking a random sample***
There are 4 smugglers among 9 people. What is the probability of getting three smugglers when inspecting a group of three persons?
The probability of catching a smuggler when picking the first person is 4/9. Now, there are still 8 people among which there are 3 smugglers. The probability of the second person being a smuggler is 3/8. Similarly, the probability that the third person is another smuggler is 2/7. According to the multiplication theorem, we have a probability of

$$\frac{4}{9} \cdot \frac{3}{8} \cdot \frac{2}{7} = \frac{1}{21},$$

that everyone is a smuggler in a group of three. This is a chance of less than 5%. An event with a probability of less than 5% is considered "unlikely" in statistics. ◀◀◀

▸▸▸ **Application**: ***the birthday problem*** (Fig. 9.19)
What is the probability that the birthday of each member in a group of 5 or 7 people will fall on a different day of the week?
The first person may still have his/her birthday on any day of the week. The second has, then, only 6 days "to select from". Thus, there is a probability of 6/7 for him/her to be born on another day, etc. In the case of 5 people, we have a probability of

$$\frac{7}{7} \cdot \frac{6}{7} \cdot \frac{5}{7} \cdot \frac{4}{7} \cdot \frac{3}{7} \approx 0.15.$$

For seven people, the probability is reduced to less than 1% by the factors 2/7 and 1/7, which is already extremely unlikely. With more than seven people, it is impossible to have only one person for each day of the week.
In the same way, only with more computational effort, one shows that 23 people in one room are enough to have more than a 50% chance that at least two people in the room have their birthday on the same day of the year.

**Fig. 9.19** This combination (7 persons whose birthdays fall on 7 different weekdays) is a very lucky event and occurs with a probability of less than 1%.

The probability that 22 people were born on different days of the year equals

$$\frac{365}{365} \cdot \frac{364}{365} \cdot \ldots \cdot \frac{365-21}{365} = 0.5243,$$

which is just barely over 50%. For 23 people, it falls below the 50% mark:

$$\frac{365}{365} \cdot \frac{364}{365} \cdot \ldots \cdot \frac{365-22}{365} = 0.4927.$$

With twice as many people, the probability of the event is

$$\frac{365}{365} \cdot \frac{364}{365} \cdot \ldots \cdot \frac{365-45}{365} = 0.05175.$$

So, in this case, the event is very probable (about 95%). With 70 people, we have

$$\frac{365}{365} \cdot \frac{364}{365} \cdot \ldots \cdot \frac{365-69}{365} = 0.00084 \approx 0.001$$

and the event is nearly certain to occur (99.9%).
We do not want to solve the problem by trial-and-error, but by a formal equation. First, according to Formula (8.5), which provides the approximation $1 + tx \approx e^{tx}$, we have

$$\prod_{i=1}^{n-1} \frac{365-i}{365} = \prod_{i=1}^{n-1} (1 - \frac{i}{365}) \approx \prod_{i=1}^{n-1} e^{-i/365} = H,$$

which is good for fairly large $n$. We can reshape using the rules for calculation with powers, and get

$$H = e^{-\sum_{i=1}^{n-1} i/365} \Rightarrow \ln H = -\frac{n(n-1)}{2}/365.$$

Now, let $H = 0.5$. Then,

$$\ln 0.5 = -\frac{n(n-1)}{730}$$

which is equivalent to

$$n^2 - n - 505.997 = 0.$$

This quadratic equation has the solutions

$$n = 0.5 \pm \sqrt{0.25 + 505.997}.$$

The positive solution is $n \approx 23$. ◂◂◂

▸▸▸ **Application**: *lottery win*

What is the chance of winning the lottery "6 from 49"?

For the lottery "6 from 49", there are 49 balls numbered from 1 to 49. Six balls are arbitrarily chosen (and the numbers are listed in order of the magnitude for the sake of clarity). What is the probability of guessing all six numbers? Suppose that we immediately guess one of the six numbers. The probability of this event is 6/49. Now, we have to guess the second number, which is already a bit harder, because only 48 numbers are available and only 5 of them are suitable for us (probability 5/48), etc. So, guessing the sixth number has a very small probability of 1/44. This gives a product probability of

$$\frac{6}{49} \cdot \frac{5}{48} \cdot \frac{4}{47} \cdot \frac{3}{46} \cdot \frac{2}{45} \cdot \frac{1}{44} \approx \frac{1}{14 \text{ million}}$$

for winning the lottery.

⊕ *Remark*: The Austrian variant of this lottery game uses 45 balls ("6 from 45"). For this variant of the game, the probability is about 1 : 8 million. So, there is no "quantum leap", and you are "just" twice as likely to guess the right combination. This tiny benefit will barely help. ⊕ ◂◂◂

▸▸▸ **Application**: *spread of an infection* (Fig. 9.20)

Three juxtaposed plants run a 50% risk of being infected with mites. They infect their neighboring plants with a probability of 20%. What is the probability of the middle plant becoming infected?

In any case, the middle plant becomes infected with a probability of 0.5. If the middle plant is not diseased, the probability that the left plant becomes infected, but not the right plant, is 1/8 (one of eight equally possible cases). However, since the left plant then infects the middle one with a probability of 1/5, the probability of the middle plant becoming infected increases by $1/8 \cdot 1/5 = 1/40$. The same applies to the case in which only the right plant is infected. Now, there still remains the case of both the left *and* right plant being diseased (each with a probability of 1/8). In this case, the probability that the middle plant will *not* become infected is $(1-1/5)^2 = (4/5)^2 = 16/25$, That is, the middle plant will become infected with a probability of $1-16/25 = 9/25$. In total, we have a probability of $1/2 + 2 \cdot 1/40 + 9/200 = 0.595$ that the middle plant will be infected with mites. ◂◂◂

**Fig. 9.20** Monocultures are particularly vulnerable.

## Reversing conclusions

As we have seen, the calculation of $P(\text{event}|\text{cause})$ is quite frequent. However, it is not infrequent for $P(\text{cause}|\text{event})$ to be sought, i.e. for the argument to be reversed.

▸▸▸ **Application**: ***testing twice before drawing conclusions ...***
Some diseases occur very rarely. Physicians use the term prevalence to describe the chance of a person carrying a disease. Let now $1 : 5,000$ be the prevalence for a particular disease. That is, on average, every five thousandth person carries the disease. Assume further that there is also a screening test for this disease which detects with a 99% reliability whether the illness is present or not. What is the probability that the test sounds the alarm even though the disease is actually not present?
We solve the example "with common sense". Every stick has two ends: If the test works with 99% reliability, it fails with a 1% probability. This means that it will sound the alarm for 1% of the 4999 non-sick persons, that is, at least 50 times. The patient is almost certainly identified. This makes 51 alarms.
If we now test the same 51 people again, the screening test looks much better: For the 50 healthy people, the test will give an alarm 0.5 times on average. The patient is discovered during the second round. It may still happen that a healthy person is mistakenly declared ill twice in a row. So, we need a third test!
If the prevalence is higher, e.g. $1 : 100$ (a typical rate for HIV-infected people in many countries), there is an average of only one false alarm during the first test, and the second test is already quite reliable.
With a low prevalence, even a 99.9% accuracy is of little use. In the above case $(1 : 5,000)$, we still have five false alarms and one correct alarm during the first test. This time, however, the second test is sufficient, because it causes the false alarms to drop to 0.005. ◂◂◂

Let us solve the above example by means of a formula. The process is also known under the name of *backward induction*. According to *Bayes*'s theorem, we have

$$P(A) \cdot P(B|A) = P(B) \cdot P(A|B). \tag{9.3}$$

Let us now eliminate $P(B)$ from the formula. To do this, we first write

$$P(B) = P(B) \cdot 1$$

and replace 1 by the probability of the definite event $A|B \cup \overline{A}|B$

$$P(\overline{A}|B) + P(A|B) = 1.$$

So, with

$$P(B) = P(B) \cdot \big(P(\overline{A}|B) + P(A|B)\big) =$$

$$= \underbrace{P(B) \cdot P(\overline{A}|B)}_{P(B|\overline{A})} + \underbrace{P(B) \cdot P(A|B)}_{P(B|A) \cdot P(A)},$$

we find

$$P(B) = P(B|\overline{A}) + P(B|A) \cdot P(A). \tag{9.4}$$

(9.3) and (9.4) together yield a new formula:

$$\boxed{P(A|B) = \frac{P(B|A) \cdot P(A)}{P(B|\overline{A}) \cdot P(\overline{A}) + P(B|A) \cdot P(A)}.} \tag{9.5}$$

▸▸▸ **Application**: ***the force of habit***

Every eighth American drinks tomato juice for breakfast (in the rest of the world only every 80th person). Let us assume that we attend a conference where 50% of the participants come from the USA and one participant is spotted drinking tomato juice at breakfast. What is the probability that he/she is an American?

*Solution*:

According to *Bayes*'s formula,

$$P(A|T) = \frac{P(A \cap T)}{P(T)} = \frac{1/16}{1/16 + 1/160} = \frac{10}{11} \quad (91\%).$$

◂◂◂

►►► **Application**: ***early detection of rare diseases***
Let us apply Formula (9.5) to our problem dealing with disease prevalence and early recognition:
Let $A$ be the event that the person being tested carries the disease and $B$ be the event that the test sounds the alarm.
Now, we have the following events:

- $A|B$: The person is actually sick when the test alerts.
- $B|A$: The test responds correctly if the person is sick.
- $B|\overline{A}$: The test mistakenly identifies a healthy person as diseased.

In the special case, $P(A) = 1/5,000 = 0.0002$, and therefore, $P(\overline{A}) = 0.9998$. Furthermore, "according to factory specification", $P(B|A) = 0.99$ and $P(B|\overline{A}) = 0.01$.
Thus, according to the above formula, we get

$$P(A|B) = \frac{0.99 \cdot 0.0002}{0.01 \cdot 0.99 + 0.99 \cdot 0.0002} \approx \frac{0.0002}{0.01} \approx \frac{1}{50}.$$

Now we could debate which of the two methods – the "common sense method" or the "formula method" – is better. When applying the formula, you must know exactly what you are doing. Once you are sure of your subject, you can of course vary more quickly. Let us put it in this way: The computer often has the formula, but the danger lies in the person entering the data erroneously! ◄◄◄

## 9.5 Combinatorics

Examples such as those of the lottery-6 or the sweepstake-12 show that probability calculations often involve having to count many (favorable or also possible) cases. It is often difficult to avoid making mistakes. Are there simple formulas for counting the number of possibilities?

### Three typical questions

1. How many combinations are there for $k$ offices to be occupied by $k$ people according to the rotation principle?
The first office can be occupied by $k$ different people. The next office can only be occupied by $k-1$ people, etc. Overall, there are $k \cdot (k-1) \cdot \ldots \cdot 2 \cdot 1$ combinations. This number already appeared in Chapter 3, and was denoted by $k!$, i.e. $k$ factorial:

$$k! = k \cdot (k-1) \cdot \ldots \cdot 2 \cdot 1.$$

▸▸▸ **Application**: ***rotation principle*** (Fig. 9.21)
Three people share an apartment and want to share the responsibility for the three most unpleasant "chores" (vacuum-cleaning, washing dishes, and cleaning the toilet) based on a weekly changing rotation principle. How many weeks will it take for the cycle to start again?

**Fig. 9.21** rotation principle

If one of the three flatmates is in charge of the vacuum cleaner, two people remain and one of the two must wash the dishes. The final remaining person must then automatically do the third chore of cleaning the toilet. Therefore, there are $3 \cdot 2 \cdot 1 = 6$ combinations, and the cycle restarts every six weeks. ◂◂◂

2. In how many combinations can $n$ people occupy $k$ different posts?
The first post can be taken by each of the $n$ people. For the second post, there are only $n-1$ people available. For the $k$-th post, only $n-k+1$ people are free. All in all, we have

$$n \cdot (n-1) \cdot (n-2) \cdot \ldots \cdot (n-k+1)$$

possibilities. With our new notation, this product is easier to write. We expand the numerator and the denominator by the factor

$$(n-k) \cdot (n-k-1) \cdot \ldots \cdot 2 \cdot 1 = (n-k)!$$

and obtain

$$\frac{n \cdot (n-1) \cdot \ldots \cdot (n-k+1) \cdot (n-k) \cdot \ldots \cdot 2 \cdot 1}{(n-k) \cdot \ldots \cdot 2 \cdot 1} = \frac{n!}{(n-k)!}.$$

▸▸▸ **Application**: ***rituals***
In one room, there are 7 people (4 women and 3 men) who will raise their glasses to each other on the New Year. What is the probability that a woman will raise her glass with a man?
In analogy to the above example, the glasses will clink $7 \cdot 6/2 = 21$ times overall. The women clink their glasses together $4 \cdot 3/2 = 6$ times and the men $3 \cdot 2/2 = 3$ times. There will be $21-6-3 = 12$ mixed pairings with a probability of $12/21 \approx 0.57$. ◂◂◂

3. In how many ways can $n$ people be grouped into equal teams of $k$ persons?
This question can be answered with the solution strategies used for the first two questions. First, we have $n!/(n-k)!$ ways to form a team where the order is still important. If the order is not important, each team is counted $k!$ times. This allows us to form

$$\frac{n!}{k! \cdot (n-k)!}$$

teams.
This is the so-called *binomial coefficient* $n$ choose $k$ (sometimes called $k$ from $n$) for integers $n \geq k$:

$$\binom{n}{k} = \frac{n}{k} \cdot \frac{n-1}{k-1} \cdot \ldots \cdot \frac{n-k+1}{1} = \frac{n!}{k! \cdot (n-k)!}.$$

⊕ *Remark*: The term binomial coefficient stems from the formula

$$(a+b)^n = \binom{n}{0} a^n b^0 + \binom{n}{1} a^{n-1} b^1 + \binom{n}{2} a^{n-2} b^2 + \ldots + \binom{n}{n-1} a^1 b^{n-1} + \binom{n}{0} a^0 b^n \quad (9.6)$$

(see also p. 265), which is easier to understand when written like this: After complete expansion of the product $(a+b)^n = (a+b) \cdot \ldots \cdot (a+b)$, the summand $a^k b^{n-k} =$

$\underbrace{a \cdot \ldots \cdot a}_{k \text{ times}} \cdot \underbrace{b \cdot \ldots \cdot b}_{n-k \text{ times}}$ shows up $\binom{n}{k}$ times if we take $a$ out of $k$ brackets and $b$ out of $n-k$ brackets. This is precisely the statement posed. ⊕

### ▸▸▸ Application: *courtesies*

Four or, for a more general consideration, $n$ people in a room shake hands to greet each other. How many handshakes occur?
In the case of the four people, it is still clear: Let $A$, $B$, $C$, and $D$ be the people. Then, we can enumerate six relevant pairings: $AB$, $AC$, $AD$, $BC$, $BD$, and finally $CD$. Thus, there are six handshakes.
One also comes to the same result as follows: One of the four people can shake hands with three others. Since this is true for every person, we get $4 \cdot 3 = 12$ pairings. However, we have counted each handshake twice so that we actually come to 6 pairings.
Now, the calculation for $n$ people is easy: Each of the $n$ people can shake hands with $n-1$ others, and in the end, this number is to be halved. The general result is $n \cdot (n-1)/2$. For $n = 4$, this is again $4 \cdot 3/2 = 6$. ◂◂◂

### ▸▸▸ Application: *alphabet soup*

In how many ways can you arrange (permute) the letters of the word *STATISTIK* (German for "statistics")?

**Fig. 9.22** alphabet soup

⊕ *Remark*: The word consists of 9 letters, but only the letters $A$ and $K$ appear once. $I$ and $S$ occur twice, and $T$ occurs three times. If we choose the position of $A$ in nine ways (9 over 1), the position of $K$ can then be selected in 8 ways (8 over 1). The two $I$s can be arranged within the remaining 7 positions in 7 choose 2 ways. Meanwhile, the two remaining $S$ have only 5 possible positions left for 5 choose 2 ways. Now, only 3 positions are free, and they necessarily have to be occupied by the three $T$ (3 choose 3 possibilities, thus, exactly one). The number of possible words is, thus,

$$s = 9 \cdot 8 \cdot \binom{7}{2} \cdot \binom{5}{2} = 72 \cdot 21 \cdot 10 = 15,120.$$

We can also write the result as

$$s = \binom{9}{1} \cdot \binom{8}{1} \cdot \binom{7}{2} \cdot \binom{5}{2} \cdot \binom{3}{3} = \tag{9.7}$$
$$= \frac{9 \cdot 8 \cdot 7 \cdot 6 \cdot 5 \cdot 4 \cdot 3 \cdot 2 \cdot 1}{1 \cdot 1 \cdot 2 \cdot 1 \cdot 2 \cdot 1 \cdot 3 \cdot 2 \cdot 1} =$$
$$= \frac{9!}{1! \cdot 1! \cdot 2! \cdot 3!}.$$

This result has a much easier interpretation. Indeed, it suggests a general formula in which the permutations appear without repetition in the numerator, while the denominator contains the products of the permutations of the individual repetitions.

⊕ ◂◂◂

**Binomial coefficients have remarkable properties**

From the definition, we immediately conclude

$$\binom{n}{k} = \binom{n}{n-k}.$$

This is useful when $k > n/2$. So, we get

$$\binom{100}{98} = \binom{100}{2} = \frac{100}{2} \cdot \frac{99}{1} = 50 \cdot 99 = 4,950.$$

The binomial coefficients $n$ choose $k$ can be calculated recursively. First, by definition

$$\binom{0}{0} = \frac{0!}{0! \cdot (0-0)!} = 1, \quad \binom{n}{n} = \binom{n}{0} = \frac{n!}{n! \cdot 0!} = 1,$$

and furthermore,

$$\binom{n+1}{k+1} = \binom{n}{k} + \binom{n}{k+1} \quad \text{(recursion formula)}.$$

▸▸▸ **Application**: ***a rule for the coefficients***

*Pascal*'s triangle can be used to determine all binomial coefficients up to the order $n$. According to the recursion formula, we can find the entries of *Pascal*'s triangle by simple additions. Each new number (starting from the third line) is the sum of the two numbers above it.

◂◂◂

**Fig. 9.23** All cans must be on the shelf. Also a kind of *Pascal*'s triangle!

$$\begin{array}{ccccccccccccccc}
 & & & & & & & 1 & & & & & & & \\
 & & & & & & 1 & & 1 & & & & & & \\
 & & & & & 1 & & 2 & & 1 & & & & & \\
 & & & & 1 & & 3 & & 3 & & 1 & & & & \\
 & & & 1 & & 4 & & 6 & & 4 & & 1 & & & \\
 & & 1 & & 5 & & 10 & & 10 & & 5 & & 1 & & \\
 & 1 & & 6 & & 15 & & 20 & & 15 & & 6 & & 1 & \\
1 & & 7 & & 21 & & 35 & & 35 & & 21 & & 7 & & 1 \\
\vdots & \vdots & \vdots & \vdots & \vdots & \vdots & \vdots & \vdots & \vdots & \vdots & \vdots & \vdots & & &
\end{array}$$

## Application to "classical" gambling games

**▸▸▸ Application**: ***lottery, the second***

On p. 463, we calculated the probability of the lottery-6. There are 6 numbers to be taken out of 49 or 45, whereby the order plays no role. In principle, this example can be handled like the third example on p. 463 with the formation of teams of size $k$ out of $n$ people. Consequently, we have

$$s = \binom{49}{6} = \frac{49 \cdot 48 \cdot 47 \cdot 46 \cdot 45 \cdot 44}{6 \cdot 5 \cdot 4 \cdot 3 \cdot 2 \cdot 1} \approx 14 \text{ million}$$

ways to draw the numbers.

Now, we can expand the example further and ask: What is the probability of guessing a lottery-5, lottery-4, etc.?

The 49 numbers are divided into two groups: one group of 6 correct numbers and a second group of 43 false numbers. In order to make an accurate guess of $r$ correct numbers, $r$ numbers must be guessed from the set of correct numbers and $6 - r$ from the set of false numbers. This can be done in

$$\binom{6}{r} \cdot \binom{43}{6-r}$$

ways. For $r = 4$, we have, for example,

$$s_4 = \binom{6}{4} \cdot \binom{43}{2} = \binom{6}{2} \cdot \binom{43}{2} = \frac{6}{2} \cdot 5 \cdot 43 \cdot \frac{42}{2} = 13,545$$

possibilities and, therefore, a probability of

$$s_4/s \approx 14,000/14,000,000 \approx 1/1,000,$$

to guess a Lottery-4.

With $r = 0$, we can calculate the probability that no single number has been guessed correctly. We have

$$s_0 = \underbrace{\binom{6}{6}}_{1} \cdot \binom{43}{6} \approx 6.1 \text{ million},$$

which results in a probability of $s_0/s \approx 0.44$. ◂◂◂

►►► **Application**: ***sweepstakes, the second***

On p. 457, we considered in how many ways a sweepstake-11 can be achieved. The result: $2 \cdot 12 = 24$ possibilities (in the wrongly guessed game, two wrong answers are possible). With the current knowledge, we would say: two times 12 choose 1 combinations. With our formula, we have much greater flexibility. The number of 10s (exactly 2 wrong tips) is $2^2$ times "12 choose 10" = 264, and the probability of guessing 10 games correctly is $264/3^{12} \approx 1 : 2,000$.

⊕ *Remark*: For an insider – in contrast to the lottery – the probability for a 10 is usually greater. For some soccer games, the result is clear anyhow, because one of the two teams is usually more likely to win. ⊕ ◄◄◄

## Combining systematically

We have already learned five different ways to group $n$ different elements of a set into a class:

• We can arrange the elements in groups of $n$. The order of the elements plays a role. You can exclude repetitions or not. This process is called *permutation*.

### 1. Permutations without repetition

The number of possible ways to arrange $n$ different elements of a set is

$$P(n) = n!.$$

►►► **Application**: ***three-letter words***

From three letters $a$, $d$, and $i$, we can build $3! = 3 \cdot 2 \cdot 1 = 6$ different words: *adi*, *aid*, *dai*, *dia*, *iad*, and *ida*. ◄◄◄

### 2. Permutations with repetition

If $k$ elements occur several times in the set, then the number is reduced to

$$P^W(n) = \frac{n!}{n_1! \cdot n_2! \cdot \ldots \cdot n_k!}.$$

►►► **Application**: ***rolling four dice***

Rolling four dice results in 1, 1, 2, and 2, giving a total of 6. You can roll the sum 6 in $4!/(2!2!) = 6$ ways with exactly the same number of eyes. ◄◄◄

• Furthermore, we can combine elements of the basic set to classes with $k \leq n$ elements, with the order of the elements either being important or not. We then speak of *variations*.

### 3. Variations without repetition

Consider the following problem: You should select $k$ elements from a set with $n$ different elements so that no element occurs more than once and the order plays a role. This can be done in

$$V(n,k) = n \cdot (n-1) \cdot \ldots \cdot (n-k+1) = \frac{n!}{(n-k)!}$$

ways.

▸▸▸ **Application**: ***two-digit numbers (1)***
How many two-digit numbers can be formed from 1, 2, and 3 if each digit can only occur once? The solution is: $\{12, 13, 21, 23, 31, 32\}$ which is

$$V(3,2) = \frac{3!}{(3-2)!} = 3 \cdot 2 = 6$$

combinations. ◂◂◂

**4. Variations with repetition**
Consider the following problem: Again $k$ elements from $n$ elements are to be selected. The order is important, but the elements may occur more than once. This can be done in

$$V^R(n,k) = \underbrace{n \cdot n \cdot \ldots \cdot n}_{k \text{ times}} = n^k$$

ways.

▸▸▸ **Application**: ***two-digit numbers (2)***
There are $3^2 = 9$ two-digit numbers that can built from 1, 2, and 3, namely 11, 12, 13, 21, 22, 23, 31, 32, and 33. ◂◂◂

• If the sequence matters, one speaks of *combinations*. We have already discussed the case when elements in the selected class cannot be repeated.

**5. Combinations without repetition**

▸▸▸ **Application**: ***mixing colors***
How many (RGB) mixtures can be composed from Red, Green, and Blue in the ratio 1:1? The mixed colors are $RG$, $RB$, and $GB$, which is $K(3,2) = 3$ possibilities. ◂◂◂

**6. Combinations with repetition**
A sixth and final case is still pending: The elements in these classes may be repeated.

▸▸▸ **Application**: ***bouquets***
How many combinations can be picked from yellow, white, and violet flowers as a gift to your sweetheart (Fig. 9.24)?
*Solution*:
First, we find all possible combinations by simple enumeration: $ggg$, $ggw$, $gww$, $www$, $wwv$, $wvv$, $vvv$, $vvg$, $vgg$, and $vgw$. So, there are 10 combinations of three flowers that differ in color.

**Fig. 9.24** Left: A sufficiently large basic amount of three different flower varieties is given. The binding of a small bouquet can be interpreted as a combination with repetition. Right: The *field violet* is the "joker" in our concept and can "transform" into each of the three flower varieties (white, yellow, violet). Formula 9.8 can thus be explained.

If there are enough flowers available, we can also bind much larger bouquets ($k$ is arbitrary). In other words, $k$ may be greater than $n$. With $n = 3$ (3 varieties), we assume that there are enough representatives of each variety. ◂◂◂

The formulas for combinations with repetition can also be easily understood: Let us imagine $n$ different elements and $k-1$ different "jokers" in the basic set (see Fig. 9.24, right). From this set, there can be $n+k-1$ over $k$ combinations without repeating the $k$ elements in each. The jokers can subsequently "turn" into any of the $n$ elements, thereby capturing all possible repetitions. More specifically, there are also $n$ cases in which one element is repeated $k$ times. We have already found the number of combinations with repetition:

$$K^R(n,k) = \binom{n+k-1}{k}. \tag{9.8}$$

We check this formula by means of the above example with the flower bouquet. We need one flower of each kind and in addition two field violets as jokers which yields a total of five flowers and allows the combinations $ggg$, $ggw$, $gww$, $www$, $wwv$, $wvv$, $vvv$, $vvg$, $vgg$, and $vgw$. This results in

$$K^R(3,3) = \binom{3+3-1}{2} = \frac{5 \cdot 4}{2} = 10.$$

## 9.6 Fallacies, traps of reasoning, and apparent contradictions

In the previous sections, we have already worked out a lot of the calculations of "classical" probabilities. Routine tasks can be solved without difficulty. Nevertheless, one should generally be aware of the possibility that there may be some ideas or supposed contradictions (paradoxes) that must be avoided.

⊕ *Remark*: Fallacies usually occur when a tried and tested mechanism of thought leads to erroneous results. The only way out is to be prepared for it – just as one, for example, can escape optical illusions by applying a ruler. Thus, for instance, the question of whether London is further North or farther South than Berlin is answered as "further North" by the majority because England "as a whole" lies further North than Germany. ⊕

▸▸▸ **Application**: ***the false queue***

Why do most people feel that they are standing in the longer queue?

**Fig. 9.25** When cycling in hilly terrain, it always seems to go uphill. No wonder, you need twice as long for the same route as when you go downhill.

*Solution*:

Queues as a whole are usually of equal length, because every newcomer wants to minimize his/her own waiting time. Nevertheless, most people feel that they have picked the wrong queue, which is, of course, paradoxical. The apparent contradiction resolves itself when one considers the following: If you were actually in the wrong queue, you would have to wait longer. Should we consider a randomly chosen situation among all experienced "queuing situations", then the possibility of standing in the longer queue is higher.

⊕ *Remark*: The "red traffic light effect" (the feeling that one constantly encounters a red traffic light) is comparable. There is also a similar effect in cycling. If the terrain is hilly, it almost always seems to go uphill. This also has to do with the choice of the time: Going downhill is much faster than going up. That is, it takes more time to cycle the same trajectory uphill. ⊕ ◂◂◂

## The gambler's fallacy

A classic illusion which has ruined many a man is known as the *gambler's fallacy*.

The fallacy occurs when someone believes that something "which has to occur in the long run" will occur the next time. One could also put it this way: An event that was expected to happen has not occurred for a long time. Therefore, it is assumed to happen soon.

We distinguish two cases: the first concerns events which are independent of each other.

▸▸▸ **Application**: *sometimes 6 comes*

In a game of dice, there has not been a six 10 times in a row. So, with every new throw of the dice, it is believed that the 6 is more likely to appear. Likewise for the coin toss: "If *heads* has appeared 10 times in a row, then the probability of *tails* has to increase, because on average, *tails* appears every other time and the long series of *heads* has to be compensated..."

However, neither dice nor coins have a "memory", and what was thrown before does not affect what the next roll will be. ◂◂◂

The second fallacy relates to events that depend on each other (or on a third event).

▸▸▸ **Application**: *statistically speaking ...*

The soccer team of a certain nation has always participated in the four-year World Championship and has been in the finals every third time for the last 36 years. The last two times, the team was not in the finals. So, the probability of entering the finals this time is believed to be extremely high. The team, however, can be in a bad shape and the series of victories can end. There is probably no mathematical law behind this. ◂◂◂

## The Saint Petersburg paradox

Even if you are well aware that a coin cannot remember whether it previously landed on *heads* or *tails*, there is still hope that the following – paradoxical – strategy might work:

▸▸▸ **Application**: *Somehow the money runs out ...*

It is understood that for a coin toss, a double bet is paid out if one guesses the right side. Now, one plays with this system and keeps insisting that the coin will land on tails. If tails does not appear, you double your bet and bet on tails again. After the $k$-th toss, if it finally lands on tails, you will always have a profit. To see this, we only need a small calculation: Let $E$ be the base bet. If tails shows up, $2 \cdot E$ is paid out giving a net-profit of $E$. Otherwise, $2 \cdot E$ is bet again. If tails appears, $4 \cdot E$ is paid out with an investment of $3 \cdot E$. So, you have again a net profit of $E$. If one has already doubled $k-1$ times, the investment is

$$(1 + 2 + \ldots + 2^{k-1}) \cdot E = (2^k - 1) \cdot E = 2^k \cdot E - E.$$

If tails shows up, $2^k \cdot E$ gets paid out and your profit is again $E$.
If you play the game all night long, again and again, it adds up to a lot. It may be seen as a form of "work", for which one gets money between his/her wages. You cannot lose in any case!
Of course, you can lose, at the latest once the stake has grown so high that you do not have enough money to pay it. If you bet $E = 1$ € per game, then, after only 20 tosses showing heads, the stake is at least already $2^{20} - 1 > 1{,}000{,}000$ €. It is no comfort that the probability that tails first appears in the 21st toss is only $1/2^{20} \approx 1 : 1{,}000{,}000$. ◂◂◂

▸▸▸ **Application**: ***infinite bets***
The above example has become famous in a modified form through *Daniel Bernoulli*'s idea from 1738. The following gamble is offered in a hypothetical casino in Saint Petersburg:
A participation fee is required. Then a coin is tossed until tails falls. This completes the game. The profit depends on the total number of coin tosses. If there was only one toss, the player would receive one ducat. For two throws, the player would receive two ducats. For three throws, the prize is four ducats, and so on. So, you get $2^{k-1}$ ducats when the coin is thrown $k$ times.
We calculate the expected value for the profit by including the associated probabilities for paying out the respective profit sum:

$$G = \sum_{k=1}^{\infty} \frac{2^{k-1}}{2^k} = \frac{1}{2} + \frac{1}{2} + \frac{1}{2} + \ldots = \infty .$$

Theoretically, the expected value is $\infty$, and the bank is not able to set the bet high enough. Otherwise, it would be bankrupt sooner or later! One way to escape the dilemma is, of course, to set a maximum number of tosses $m$. Then, the bank would have to set a basic bet of at least $m/2$ ducats per player.
This example has become famous mainly because it has sparked a series of proposals for its solution, though only *Bernoulli*'s solution shall briefly be mentioned here. He wrote: "The calculation of the value of an object need not be based on its price, but rather on the usefulness that it possesses .... Undoubtedly, a profit of 1,000 ducats is more significant for a beggar than for a wealthy man, although both receive the same amount." Therefore, *Bernoulli* suggested to distribute only $\ln 2^k$ ducats instead of $2k$ ducats. As a result, the expected value for the profit is no longer infinite

$$\overline{G} = \sum_{k=1}^{\infty} \frac{\ln(2^{k-1})}{2^k} = \sum_{k=1}^{\infty} \frac{(k-1)\ln 2}{2^k} = \ln 2 \sum_{k=1}^{\infty} \frac{(k-1)}{2^k} = \ln 2 \underbrace{(1/2^2 + 2/2^3 + 3/2^4 + \ldots)}_{=1 \text{ (without proof)}} < \infty .$$

◂◂◂

## The 50:50 error

**▸▸▸ Application**: ***as if: "no divulging of secrets"***

Suppose that three candidates $A$, $B$, and $C$ have applied for a post. $A$ is waiting alone behind the door where the committee is making its decision and thinks: "There is a 1/3 probability that I will be accepted." When a single member of the jury leaves the room, $A$ asks him or her:
"I know you cannot tell me whether I have been accepted or not. But it is clear at least that either $B$ or $C$ was not accepted. Tell me which of the two was not accepted. Then, you would not tell me a secret, and I would have no additional information about my situation. I promise to keep the information to myself only."
The jury member is convinced and says "$B$ was not accepted". Thereupon, $A$ beams and says: "Now the probability that I will be accepted has risen to 1/2." Is that correct?
The answer is not trivial: $A$ is not quite right (his probability of 1/3 has not changed), but the probability that $C$ gets the position has actually doubled! To understand this, consider the three possible decisions of the jury (+ or – in the first position indicates whether $A$ gets the position or not, etc.):

$$+--, \quad -+-, \quad --+.$$

All three are equally likely. Judging by the committee's verdict, case 2, which would have been one of the two unfavourable decisions for both $A$ and $C$, is no longer valid.
However, the difference between $A$ and $C$ is now that the question posed by $A$ did not refer to $A$, but to $B$ and $C$. Therefore, nothing changes for $A$. The probability of $C$ not being accepted is halved.
To understand this better, consider the notorious "three-door problem". ◂◂◂

**▸▸▸ Application**: ***three doors and two strategies***

Just imagine the following situation in a game show, just before the goal: One must "only" point to one of three doors hiding the main prize (a sports car). However, behind two of the doors, there are merely consolation prizes. To make things even more annoying, if you point to the wrong door, the consolation prizes hiding behind these doors are goats (blanks). That is how it worked in *Monty Hall*'s show "Let's Make a Deal". The host chimed in an interesting addendum: Once the candidate picked a door, the host, who knew exactly where the car was hidden, would make the candidate insecure by opening one of the doors hiding a goat. Then, he would offer the candidate (sighing with relief as there is now a $50:50$ chance) the possibility to change his/her initial choice. What would *you* do in such a situation?

*Solution*:

If you have the choice between a goat and a Porsche, you will probably invest some thought into figuring out a solution to this seemingly simple problem. Hundreds of people have expressed their curiosity about this problem on the

Internet and elsewhere. The answer is actually quite simple if one proceeds according to the following reasoning:

I arbitrarily choose one of the three doors. If I happen to pick the door hiding the sports car, this is good. If I pick a door hiding one of the two goats, this is bad. The probability of winning the car is 1/3. If I stick with my choice, then this probability will not change. Not even if the host shows one of the two goats to me: He will do so *in any case* – regardless of whether I have picked the car or one of the two goats.

**Fig. 9.26** Car or goat? The question has provoked much discussion!

Let us now consider the strategy of changing the initial choice "in any event". There are three cases: If my initial pick is the sports car, this is bad for me. If I pick goat 1 or goat 2, I am lucky now: In both cases, I will eventually point to the door hiding the highly anticipated car. Thus, I have doubled the chance to win! So, the information provided by the host was helpful and should not be dismissed as creating only uncertainty. Strategically, picking the third door is the much better choice!

⊕ *Remark*: At least in the original version of the game, it was possible for the host to open the door initially picked by the candidate if that door was hiding a goat. In this case, the candidate has a 50 : 50 chance of picking the correct door.
In the show "Who wants to be a Millionaire?", the "Fifty-Fifty-Joker" randomly removes 2 of the 3 incorrect answers. The decision of the computer is in no way influenced by the candidate's presumption. So, theoretically, the probability that the candidate will pick the correct answer now is exactly 50 : 50. ⊕ ◂◂◂

## Computer simulations help to review considerations

The "three-door problem" is a classic example illustrating how very smart people may proudly present different results. To determine who has found the right solution, you could spend an evening repeating the same game: A host hides a pea randomly under one of three thimbles (by secretly throwing a die, for instance). Player $A$ taps on one of the thimbles, and the host then lifts a second thimble (hiding no pea). A second player $B$ now lifts the third thimble. If $B$ has guessed the correct thimble, $B$ will keep the pea. Otherwise,

it is $A$ who will, logically, get the pea. After only a few dozen experiments, it will become clear that $B$ has obtained significantly more peas than $A$.
If we can code well and write a computer program that works with random numbers and carries out all the steps in the prescribed way, we can simulate the situation a million times and evaluate it accordingly. The point ratio will almost exactly be $1:2$ for a large number of repetitions. Here is the output of a corresponding program ($n$ is the number of iterations; the process is repeated several times so we can see that the taken numbers are chosen randomly):

```
n = 10: NO CHANGE good: 30.00%, CHANGE good: 70.00%
n = 100: NO CHANGE good: 38.00%, CHANGE good: 62.00%
n = 1000: NO CHANGE good: 33.40%, CHANGE good: 66.60%
n = 10000: NO CHANGE good: 34.44%, CHANGE good: 65.55%
n = 100000: NO CHANGE good: 33.17%, CHANGE good: 66.82%
n = 1000000: NO CHANGE good: 33.28%, CHANGE good: 66.71%
n = 10000000: NO CHANGE good: 33.34%, CHANGE good: 66.65%

n = 10: NO CHANGE good: 30.00%, CHANGE good: 70.00%
n = 100: NO CHANGE good: 22.00%, CHANGE good: 78.00%
n = 1000: NO CHANGE good: 35.00%, CHANGE good: 65.00%
n = 10000: NO CHANGE good: 33.49%, CHANGE good: 66.49%
n = 100000: NO CHANGE good: 33.57%, CHANGE good: 66.42%
n = 1000000: NO CHANGE good: 33.41%, CHANGE good: 66.58%
n = 10000000: NO CHANGE good: 33.36%, CHANGE good: 66.63%

n = 10: NO CHANGE good: 40.00%, CHANGE good: 60.00%
n = 100: NO CHANGE good: 32.00%, CHANGE good: 68.00%
n = 1000: NO CHANGE good: 34.10%, CHANGE good: 65.90%
n = 10000: NO CHANGE good: 33.16%, CHANGE good: 66.84%
n = 100000: NO CHANGE good: 33.23%, CHANGE good: 66.77%
n = 1000000: NO CHANGE good: 33.31%, CHANGE good: 66.69%
n = 10000000: NO CHANGE good: 33.33%, CHANGE good: 66.67%

n = 10: NO CHANGE good: 20.00%, CHANGE good: 80.00%
n = 100: NO CHANGE good: 31.00%, CHANGE good: 69.00%
n = 1000: NO CHANGE good: 34.20%, CHANGE good: 65.80%
n = 10000: NO CHANGE good: 33.94%, CHANGE good: 66.06%
n = 100000: NO CHANGE good: 33.33%, CHANGE good: 66.67%
n = 1000000: NO CHANGE good: 33.31%, CHANGE good: 66.68%
n = 10000000: NO CHANGE good: 33.34%, CHANGE good: 66.65%
```

## Aggregation traps: "half the truth"

On a daily basis, we are overwhelmed with statistics in newspapers and magazines that are "simplified" for the presumed sake of making them "more understandable". Yet, things are not always simple, and such statistics may – intentionally or not – end up presenting the wrong information. That is why it is often colloquially said that "everything can be proven with statistics".

*Aggregation* is the condensation of individual statistical statements to general statements. The non-specific aggregation of counted data creates connections which do not exist in the original statistics. The first of the following examples is "unproblematic". The second is tendentious:

### ▸▸▸ Application: *murders and death sentences in Florida*

| skin color of the perpetrator | death sentence yes/no | sum | share of death-sentences in % |
|---|---|---|---|
| black | 59/2,448 | 2,507 | 2.4% |
| white | 72/2,185 | 2,257 | 3.2% |

When looking at the above statistics (New York Magazine, March, 11, 1979) the answer seems clear: Being white seems to be a disadvantage.
The matter looks quite different when another variable comes into play, namely the skin color of the *victim*:

| skin color perpetrator / victim | death sentence yes/no | sum | share of death-sentences in % |
|---|---|---|---|
| black / black | 11/2,209 | 2,220 | 0.5% |
| white / black | 0/111 | 111 | 0.0% |
| black / white | 48/239 | 287 | 16.7% |
| white / white | 72/2,047 | 2,146 | 3.4% |

Now, we are suddenly faced with a completely new situation: Regardless of the victim's skin color, blacks are more likely to be condemned to death than whites, especially if the victim is white. ◂◂◂

### ▸▸▸ Application: *division of work*

Suppose work $W$ is evenly distributed among two people $A$ and $B$. On the first day, $A$ performs 60% of his daily routine, while $B$ performs 90%. The next day, $A$ performs only 10% of his work, while $B$ performs 30%. In both cases, $B$ achieves a much higher percentage – but that does not mean that $B$ has done more work.

*Numerical example*: $A$ could have split his/her share in the ratio $10 : 1$, and $B$ in the ratio $1 : 10$. So, on the first day, $A$ works $60/100 \cdot 10/11 \cdot W/2 = 600/2,200 \cdot W$, and on the second day he/she works $10/100 \cdot 1/11 \cdot W/2 = 10/2,200 \cdot W$. $B$, on the other hand, works $90/100 \cdot 1/11 \cdot W/2 = 90/2,200 \cdot W$ and $30/100 \cdot 10/11 \cdot W/2 = 300/2,200 \cdot W$. Now, the work performed by the two is $61 : 39$ in favour of $A$! ◂◂◂

# 9.7 Probability distributions

## The binomial distribution

▸▸▸ **Application**: ***vocabulary test*** (Fig. 9.27)
You have learned $p = 5/6$ of the vocabulary assigned for a spelling test and are eventually asked to spell $n$ random words from that vocabulary. What are the chances that you know $k$ of them? What is the probability that you know at least 18 out of 20 words?

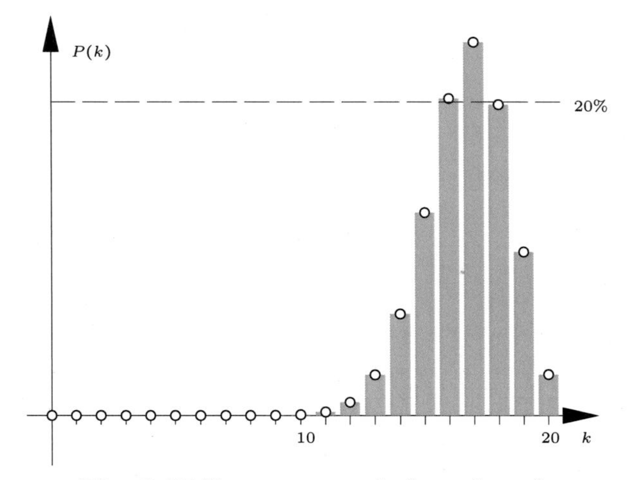


**Fig. 9.27** How many words do we know?

*Solution*:
Suppose you know the first $k$ words, and not the remaining $n - k$. For this, the probability is $p^k \cdot (1-p)^{n-k}$. You need not know the first word in order to spell a total of $k$ words correctly. There are $n$ choose $k$ combinations. Thus, the general formula is

$$P(k) = \binom{n}{k} \cdot p^k \cdot (1-p)^{n-k}.$$

Fig. 9.27 shows a diagram for $n = 20$. It shows that the probability of knowing exactly 18 words out of 20 is about 20%. There is a 10% probability of knowing 19 words, and only 3% of knowing 20. In total, the possibility of knowing at least 18 out of 20 is 1/3, i.e. 33%. ◂◂◂

▸▸▸ **Application**: ***the* Galton *board***
When balls fall from a funnel onto a board of nails as shown in Fig. 9.28, they distribute themselves in a bell-shaped manner at the lower end. This is a sequence of adjacent 50 : 50 decisions ($p = 1 - p = 0.5$).

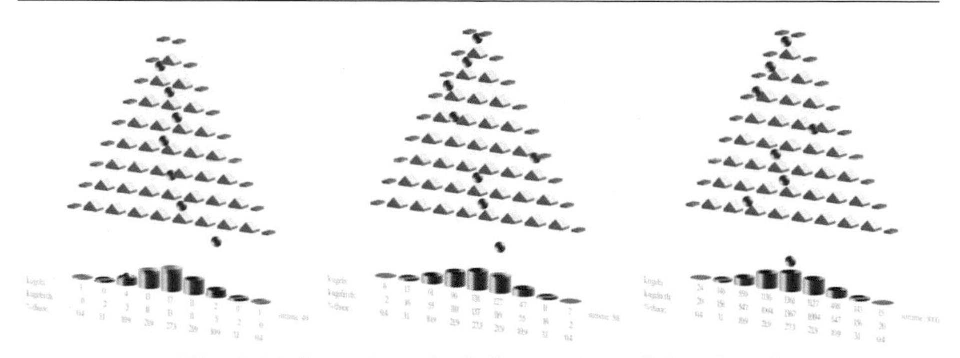

**Fig. 9.28** formation of a bell curve in a *Galton* board

The distribution of the balls is ideally like a bell-shaped "stepped polygon". Fig. 9.29 and Fig. 9.30 illustrate a *Galton* board for 10 and 20 rows of nails. Better estimates for the polygon can be made when we increase the number of dropped balls.

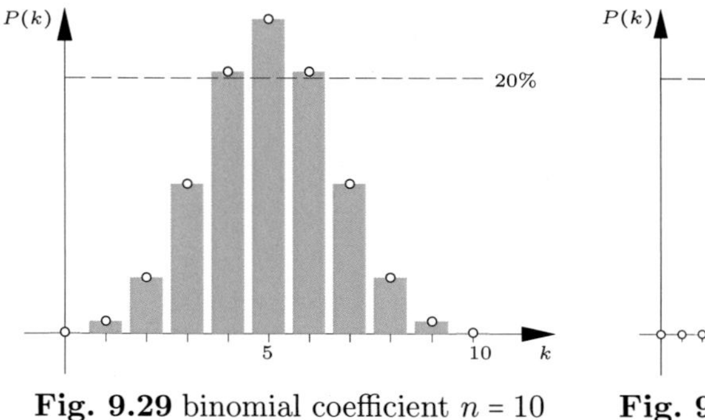


**Fig. 9.29** binomial coefficient $n = 10$

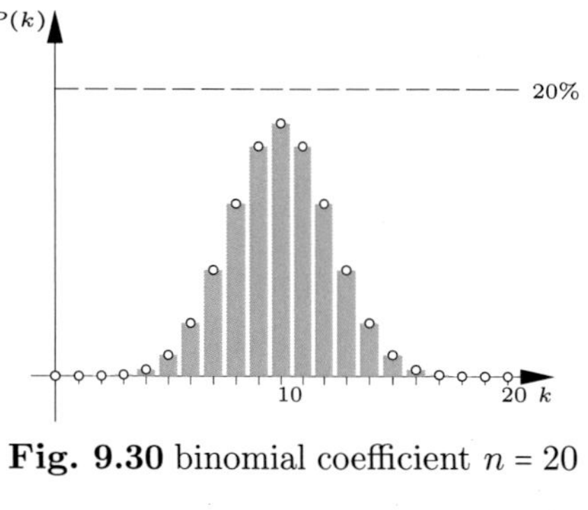


**Fig. 9.30** binomial coefficient $n = 20$

◂◂◂

## Sums of probabilities

When it comes to the probability of exceeding or falling under a certain limit, one has to sum up individual probabilities (in Application p. 482, the question was about the probability of passing a vocabulary test where at least $k$ out of $n$ words are known). If, as in Fig. 9.31 and Fig. 9.32, it is not the probabilities but the sums of preceding probabilities that are applied, it is easy to determine when such a barrier is exceeded.

▸▸▸ **Application**: ***resolute minorities***

An association of 25 members will vote on a number of topics by mail. In the case of a seemingly irrelevant proposition in which most members will more or less randomly tick *yes* or *no*, five of the members have secretly decided to vote *no*. What is the probability that the proposition will still pass based on a majority voting in favor?

*Solution*:

A total of 13 opposing votes is required for a rejection of which 5 are already

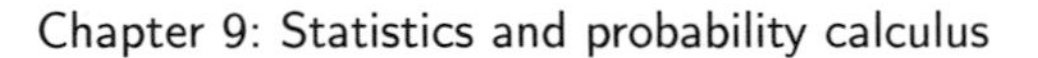
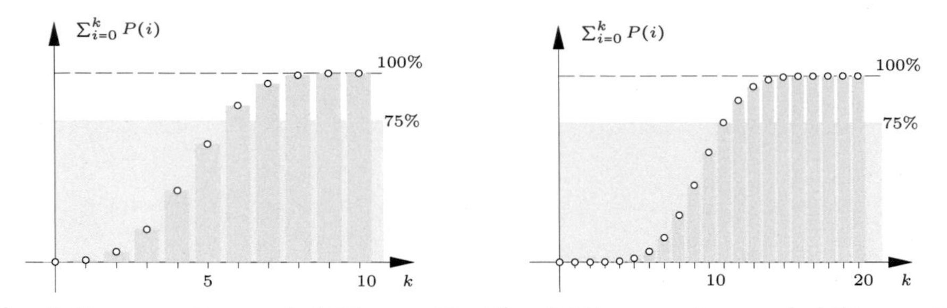


**Fig. 9.31** summation probability $n = 10$ **Fig. 9.32** summation probability $n = 20$

guaranteed. Thus, $P$ is the probability that, out of the 20 undecided members ($p = 1-p = 0.5$), a maximum of 7 will vote *no*. There is a (symmetric) binomial distribution and $P = \sum_{k=0}^{7} \binom{20}{k} \cdot 0.5^k \cdot 0.5^{20-k} \approx 0.13$. In Fig. 9.32, this value is found at the point $k = 7$. The probability that the resolute minority rooting for a *no* will win is 87%. ◂◂◂

▸▸▸ **Application**: ***crackling rain*** (Fig. 9.33, Fig. 9.34)
Raindrops strike at random. Nevertheless, there is a locally constant amount of rainfall per square meter during a strong thunderstorm. That amount may, for example, be 1 litre ($\approx 6,000$ drops) per minute. Thus, we have an average of 100 drops per second on a square meter. The impact area shown in the picture is $1/30\ \text{m}^2$. We expect there to be $100/30 \approx 3$ drops per second. What is the probability of seeing the impact of two (or three) drops simultaneously, as in the photographs below?

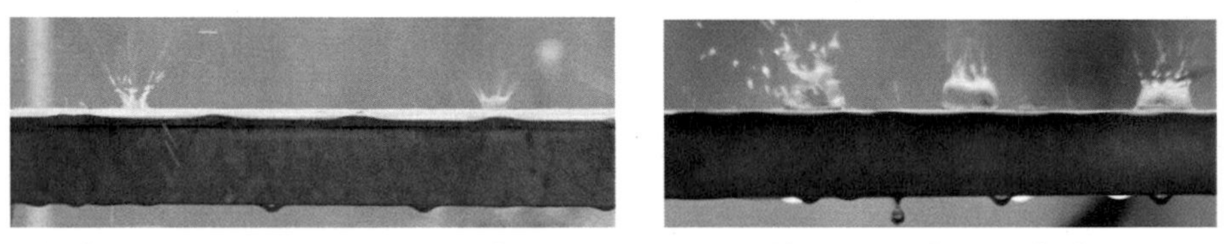

**Fig. 9.33** two simultaneous drops

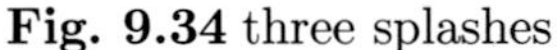

**Fig. 9.34** three splashes

Let us, first of all, arrive at a reasonable estimate by using common sense, especially since we have not made precise assumptions.
An initial assumption is that the impact of each drop will be visible for a fraction of a second in the form of an "explosion cone". If this was not the case, one would almost never see the desired impact at short exposures (in this specific case, 1/125 second). The probability that one sees the desired impact at one drop per second is assumed (as an empirical value) to be 1/5. The probability of *not* seeing it (complementary event) is $1 - 1/5 = 4/5$.
Secondly, we assume the impact of *exactly three* drops every second. The probability of *not* seeing any of these drops is $P(0) = (4/5)^3 \approx 0.51$. The probability of seeing exactly one impact is determined as follows: Suppose that we see the impact of the first drop, but not that of the other two. For this,

we have the probability $1/5 \cdot 4/5 \cdot 4/5$. The visible drop can be varied in three ways, so that exactly one drop can be seen with $P(1) = 3 \cdot 1/5 \cdot 4/5 \cdot 4/5 \approx 0.38$. The probability that the first two drops can be observed at the moment of their impact but not the third equals $(1/5)^2 \cdot (4/5) = 4/125$. Since it does not matter if we see drop 1 and drop 2 or drop 1 and drop 3 or drop 2 and drop 3, this probability is $P(2) = 3 \cdot (4/125) \approx 1/10$. On average, we see simultaneous impacts of two raindrops in every tenth picture. In fact, out of about 50 taken photographs, only five of them show what may more or less qualify as a double impact (see also Application p. 216).

⊕ *Remark*: The probability of seeing three impacts at the same time is $P(3) = (1/5)^3 = 1/125$, i.e. just under 1%. According to our highly simplified model, there would be no possibility of four or more visible impacts ($P(4) = 0$). It might occur in practice but it is extremely unlikely. This already shows: We need a more sophisticated model for probability distributions. ⊕ ◄◄◄

## The Poisson distribution

In the previous application, it was, first of all, necessary to find an average expected value $\lambda$ for the number of raindrops. We set it to be 3 drops per second. Now consider Fig. 9.35: If someone told you that he/she had analysed 1,000 attempts exactly and had thus come to this distribution, you would probably believe him/her, because the distribution looks realistic.

Let us relate this now to Fig. 9.36: If you have determined that you can photograph the impact of an average of $\lambda = 0.6$ water droplets, then the distribution shown also seems plausible. It agrees fairly well with the estimated values of 55%, instead of 51%, for zero visible impacts; 33%, instead of 38%, for one visible impact; 10% for two visible impacts; and 2%, instead of 1%, for three visible impacts.

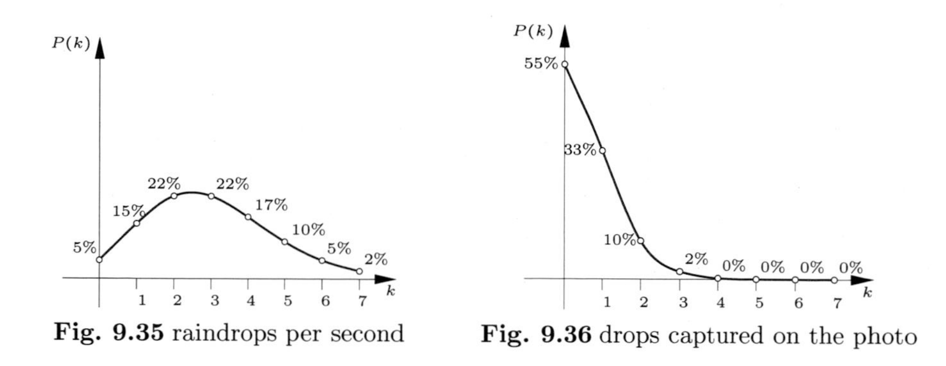


**Fig. 9.35** raindrops per second **Fig. 9.36** drops captured on the photo

Why do we compare the values so extensively? Well, the two "plausible" distributions were calculated with the following simple approximation formula by *Poisson*

$$\boxed{P(k) = \frac{\lambda^k}{k!}\,e^{-\lambda},} \quad \text{expected value } \lambda \text{ constant.} \tag{9.9}$$

*Poisson* looked at questions such as the “raindrop problem” in a more general manner, and showed that his formula approximates the binomial distribution for a large $n$, small $p$, and a constant $\lambda = np$. He has thus saved us a lot of work in the calculation of similar examples.

⊕ *Remark*: $P(0) = e^{-\lambda}$, $P(1) = \lambda\, e^{-\lambda}$, $P(2) = \frac{\lambda^2}{2}\, e^{-\lambda}$ etc. are special cases. Using Formula (8.5), it is easy to see that the sum of all probabilities $P(k)$ equals

$$\sum_{k=0}^{\infty} P(k) = \sum_{k=0}^{\infty} \frac{\lambda^k}{k!}\, e^{-\lambda} = e^{-\lambda} \sum_{k=0}^{\infty} \frac{\lambda^k}{k!} = e^{-\lambda}\, e^{\lambda} = e^0 = 1. \qquad \oplus$$

▸▸▸ **Application**: ***color-blind***

Tests have shown that every one hundredth person is color-blind. What is the probability that there are at least two color-blind people among 100 randomly selected persons? How many people have to be tested to find at least one color-blind person with a probability of no less than 95%?

*Solution*:

The expected value is $\lambda = 100\,(1/100) = 1$. The opposite of “at least 2” is “0 or 1”. We have $P(0) + P(1) = e^{-1}(1+1) \approx 0.74$, and therefore, $P(\geq 2) \approx 1 - 0.74 \approx 0.26$.

The opposite of “at least one” is “none”. The complementary probability of $P(0) < 0.05$ is $P(>0) > 0.95$. For $n$ tested people, the expected value is $n/100$ and $P(0) = e^{-n/100} < 0.05$ has to hold. Therefore, we have $-n/100 < \ln 0.05$ and $n \geq 300$. ◂◂◂

▸▸▸ **Application**: ***significant increase?***

A person is investigating a report from the Internet: “Strange accumulation of suicides in Aargau (Switzerland) since the beginning of the year. In January and February, 18 men and 5 women committed suicide in Aargau. In light of these numbers, even the Cantonal police of Aargau begin to wonder what is happening. In the year 2001, the statistics for the Canton of Aargau records 95 suicides. On the 20th of February alone, 4 people committed suicide.”

*Solution*:

For 2001, we have the expected value $\lambda = 95/6 = 15.83$ for 2 months. For $k$ suicides in 2 months, we get $p(k) = \lambda^k/k! \cdot e^{-\lambda \cdot k}$. The probability that a maximum of 22 suicides occurs per week is $\sum_{k=0}^{22} p(k) = 0.947$. Therefore, 23 or more suicides are expected with a probability of slightly more than 5%.

For a single day, $\lambda = 95/365 = 0.26$. The fact that four or more people commit suicide in one day is actually very extraordinary, because $p(<4) = \sum_{k=0}^{3} p(k) = 0.99984$. Nevertheless, the observation period extends over two months, and therefore, we have a probability of $1 - p(<4)^{60} \approx 0.001$ for such an extreme accumulation.

⊕ *Remark*: The latter was probably the trigger for the alarming report. 2001 could have been a year with fewer suicides than usual. If the number of total suicides considered to be "normal" had been, for example, 110 and only two people had taken their lives on the 20th of February, then $21 = 23 - 2$ suicides would not have been considered extraordinary.
In Germany, the number of annual suicides fell below 10,000 for the first time in the year 2006. There is no evidence that days of full moon or weekends lead to higher suicide rates. Only the seasons have an influence (fewer suicides are committed during summer and autumn). This ties in with the above report from Switzerland. The ratio of male to female is not uncommon either. In Germany, it is almost $3:1$. ⊕
◂◂◂

## The central limit theorem

**▸▸▸ Application**: ***rolling many dice***
In Application p. 456, we answered the question: "Why does the sum 10 appear more often than the sum 9 when rolling three dice?" Let us now clarify the question in a different way: If we throw a die, it is equally probable that it will show 1, 2, 3, and so on. The average number of eyes will be $(1+2+\ldots+6)/6 = 3.5$. If we throw three dice, the average number of eyes will approach 10.5 in many experiments. The number 10 is closer to this average value than 9. We will soon see that numbers closer to the expected value also occur more frequently. ◂◂◂

Let us imagine an "electronic die" that produces integers ranging from 1 and 6 (each with $p = 1/6$). If we roll $n$ dice simultaneously, then a maximum of $\mu = n \cdot p$ of 6 is to be expected.

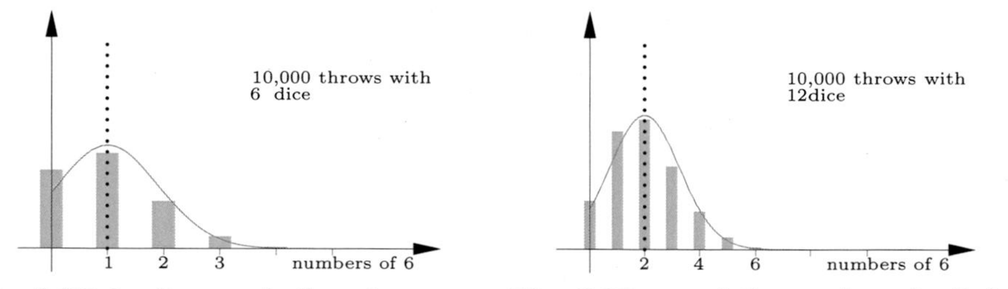


**Fig. 9.37** simultaneously throwing several "electronic dice" ... **Fig. 9.38** ... and the number of rolled 6

If the number of dice is increased, the probability distribution will begin to change into a bell-shaped curve. The curve becomes steeper the more dice are used. In other words, the "outliers" remain even closer to the expected value $\mu$ when $n$ increases. The area under the curve remains constant, because all probabilities sum up to 1. Without proving this, we formulate relatively generously:

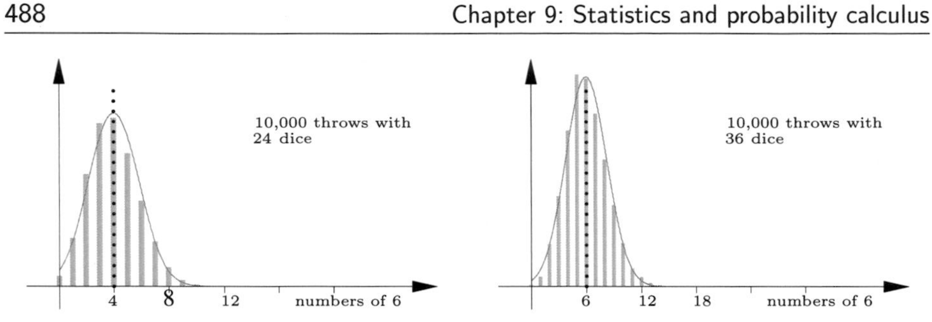


**Fig. 9.39** We raise the number of dice ... **Fig. 9.40** ... and reach a bell curve.

*Central limit theorem*: The sum of $n$ identically distributed, independent random variables $X$ obeys for a large $n$ a "normal distribution", i.e. the Gaussian bell curve (expected value $\mu = n \cdot \mu_x$ with variance $n\sigma_x^2$). This is especially true for the binomial distribution, which indicates how frequently an event with the probability $p$ occurs at an $n$-fold execution. Here, we obtain $\mu = np$; $\sigma^2 = np(1-p)$.

⊕ *Remark*: We have already discussed the prototypes of the curve ($\mu = 0$, $\sigma = 1$) in Formula (8.1). In general, the curve has its maximum at $\mu$ and the inflection points are at $\mu \pm \sigma$. The shape of the curve obviously depends on $n$ and $p$. ⊕
In practice, the following addition is important:

For a normal distribution, approximately 2/3 of all occurring values are in the interval $\mu \pm \sigma$. 95% are in the interval $\mu \pm 2\sigma$. 99.7% of all values are in the interval $\mu \pm 3\sigma$.

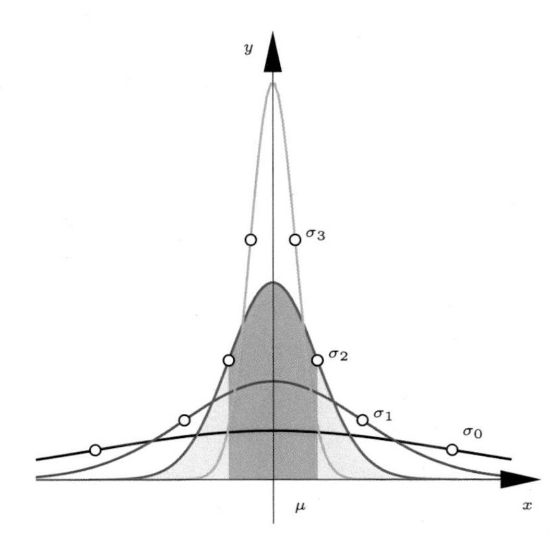


**Fig. 9.41** different variances

Fig. 9.41 shows bell curves that bulge in different manners. What they all have in common is that the area underneath the curve equals 1, namely the total sum of all probabilities. The area between $-\infty$ and $x$ is given by

$$P(x) = \int_{-\infty}^{x} \frac{1}{\sigma\sqrt{2\pi}} e^{-\frac{1}{2}(\frac{x-\mu}{\sigma})^2} \, dx.$$

This integral cannot be evaluated in terms of elementary functions, but it can be evaluated numerically, as we have seen in Application p. 417.

▸▸▸ **Application**: ***quality control*** (Fig. 9.42)
A light bulb manufacturer claims that a maximum of 2% of the delivered goods are defective. In a test of 100 bulbs, 5 were defective. Is this contradictory to the factory specification?

*Solution*:
We have $p = 0.98$ and $n = 100$. Thus, $\mu = 98$ and $\sigma^2 = \mu(1-p) \approx 2 \Rightarrow \sigma \approx 1.4$. The measured value 95 deviates by $3 > 2\sigma$ from the expected value. This means that the factory data is not correct, as we get a probability of more than 95%. ◂◂◂

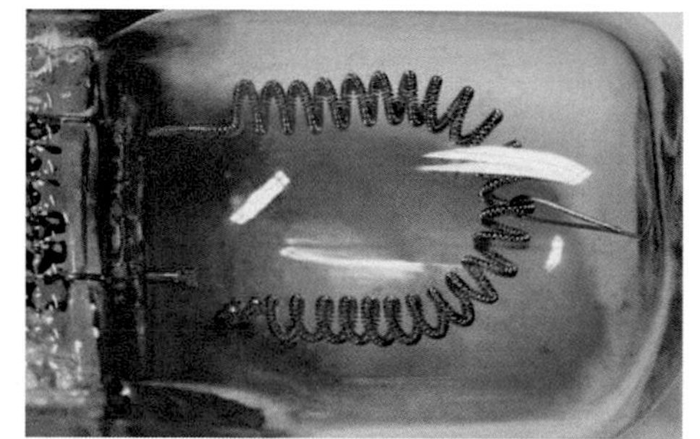

**Fig. 9.42** Another defective bulb!

**Fig. 9.43** How many will come?

▸▸▸ **Application**: ***systematic overbooking*** (Fig. 9.43)
Aircraft are usually overbooked because experience has shown that not all passengers board the plane. An airline generally overbooks a small aircraft by 1 seat and larger ones by 3 seats because 5% of all booked passengers usually do not appear. What is the probability that there will be a shortage of seats with a seating capacity of 50 or 200?

*Solution*:
On the small plane, $n = 51$ and $p = 0.95$ result in $\mu = n \cdot p = 48.45$, $\sigma^2 = 48.45 \cdot 0.05 \Rightarrow \sigma \approx 1.56$. The maximal possible value is $50 \approx \mu + \sigma$. 68% of all values lie within $\mu \pm \sigma$. 16% still lie above $\mu + \sigma$ for symmetry reasons. The probability of a shortage is, therefore, relatively high.
For larger aircraft, with $n = 203$, $\mu \approx 193$, and $\sigma \approx 3.1$, we are already outside of the $2\sigma$-range and on the safe side with 201 passengers. ◂◂◂

▸▸▸ **Application**: ***minimum sample size***

How many people must be tested so that there is a probability of 95% that the proportion of left-handed people can be determined with a precision of 1%?

*Solution*:

Let $p$ be the actual probability of left-handed people. If we examine $n$ people, the sample has the expected value $\mu = np$, and the distribution is $\sigma = \sqrt{np(1-p)}$.

Let $S_n$ be the number of left-handers and $p_n = S_n/n$ be the determined relative frequency of left-handers in the sample. Now, we require a deviation of less than one percent: $|p_n - p| \leq 0.01$.

In order to remain in the 95% confidence interval, the following occurs:

$$|S_n - \mu| \leq 2\sigma \Rightarrow |p_n - p| \leq 2\sqrt{p(1-p)/n}.$$

Therefore, we have $0.01 \geq 2\sqrt{p(1-p)/n}$. After first knowing nothing about $p$, we have to replace $p(1-p)$ by its maximum $1/4$ to be safe: $0.01 \geq 2\sqrt{1/(4n)}$. This results in $n \geq 10,000$.

Now, however, we may know from smaller preliminary investigations that $p$ is relatively small, namely about $0.1$. This shrinks $p(1-p)$, and we round to get $n = 3,600$. ◂◂◂

▸▸▸ **Application**: ***uncertain election forecasts***

For a large party, we want to make a reliable prediction of the result with a precision of 1%. What is the size of the sample?

*Solution*:

According to the results of the previous application, we need $10,000$ inquiries. For a large party, $p(1-p)$ is not much smaller than $0.25$, so that there is no hope for limitation. Neither the selection of the people nor their honesty has yet been taken into account.

If, as before, only 400–500 people are questioned by different opinion research institutes, only accuracies off by plus or minus a few percentage points can be expected. Often enough, however, it is only by a few percentage points, sometimes even by a mere few thousand votes, that a candidate gets elected. This shows that such surveys are not meaningful, and their results are strongly influenced by what political party stands closest to the research institute carrying out the survey. ◂◂◂

# 9.8 Further applications

**▸▸▸ Application**: ***It depends on the seed.***
In the computer age, many computations would be unthinkable without computer-generated random numbers. The essential task is the creation of a single random number, for example, by using the system time. Then, the remaining numbers are recursively arithmetically derived from this seed ("deterministically"). This has the advantage that it helps us to understand the scientific knowledge that can be gained with such "pseudo-random numbers".

⊕ *Remark*: For programmers: A discussion of different algorithms – partially with source code of the C functions `srand ()` and `rand ()` – can be found at `http://www.codeproject.com/KB/cpp/PRNG.aspx`. The mentioned functions produce uniformly distributed 16-bit random variables. ⊕ ◂◂◂

**▸▸▸ Application**: ***How random are the digits of $\pi$ and $e$?***
The circle number $\pi$ and Euler's number $e$ are known for the fact that the order of their decimal places is not subject to any law (the numbers can, however, through recursion be calculated arbitrarily precisely). Fig. 9.44 and Fig. 9.45 show the distribution of digits in the two numbers for 100 and 100,000 decimal places.[3]

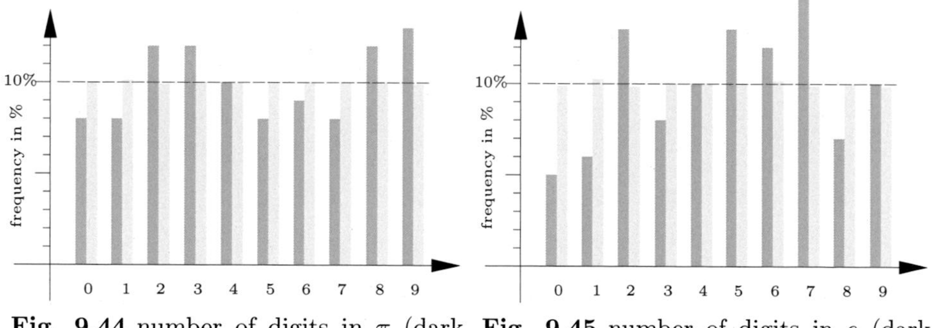


**Fig. 9.44** number of digits in $\pi$ (dark bars: $10^2$ digits, light bars: $10^6$ digits)

**Fig. 9.45** number of digits in $e$ (dark bars: $10^2$ digits, light bars: $10^6$ digits)

Even if it does not appear to be so "initially" – especially for Euler's number, the number 7 appears to be more frequent at first – the numerical frequency converges rapidly, and it is already from 100,000 digits onward that an almost uniform distribution is achieved.

⊕ *Remark*: For a random sequence of digits, it is not sufficient that all digits occur equally frequently. Another possible investigation uses the so-called *maximum test*:

[3]The files are from `http://www.arndt-bruenner.de/mathe/mathekurse.htm`

The digits are grouped into triples, and it is checked how often the middle digit is larger than its left and right neighbour. If the middle digit is, for example, 7, then there are $7^2 \cdot 7^2$ possibilities for the other two digits. Among the $10 \cdot 10 \cdot 10$ possibilities of triple digits, $1^2 + 2^2 + \ldots + 9^2 = 285$ is the maximal middle digit. It is, therefore, possible to require that such a maximum occurs with the probability $\approx 0.285$ in the case of very long random sequences. In fact, for the digits of $\pi$ and $e$, the values are 0.288 and 0.283 respectively at 100,000 digits – another indication of randomness. ⊕ ◂◂◂

▸▸▸ **Application**: ***uniform point distribution on the sphere*** (Fig. 9.46)
It sounds easier than it is: How do you "dot" a polygon or a sphere randomly but evenly?

*Solution*:
On the polygon: First, for a point $R$, one must be able to determine whether or not it is inside the polygon $P_i$ $(i = 1, \ldots, n)$. If the polygon is convex, one can, for example, use this test: Calculate the sum of all signed angles $\angle P_i R P_{i+1}$ by means of Formula (5.36). If it results in 360°, the point lies inside (for points outside, the sum is 0). If we want to reach a given point density, the number of points must be proportional to the polygon area. To calculate the area, we triangulate the polygon and calculate the area of each triangle using Formula (5.44). Now, we look at the rectangle (picture on the left) circumscribing the polygon and select random points $R(x/y)$ with random coordinates $x$ and $y$ in the rectangle area until the number of strikes in the polygon corresponding to the area is reached.

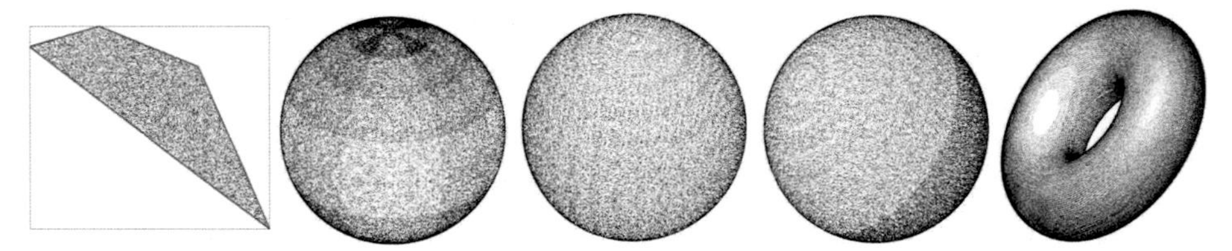

**Fig. 9.46** Distribution of random points (from left to right): polygon, sphere (bad because it is dependent on the parametrization), sphere (good because it is uniform), sphere (shaded), torus

On the sphere: The parametrization $\vec{x} = (r \cos v \cos u,\ r \cos v \sin u,\ r \sin v)$ with $0 \le u \le 2\pi$, $-\pi/2 \le v \le \pi/2$ provides an arbitrary division into quadrilaterals or triangles of different sizes. Now, we only have to fill each polygon as described (third image from the left in Fig. 9.46).
It is unfavourable to choose the parameters $u$ and $v$ randomly in their intervals, as it was done in the second picture from the left, because this results in an accumulation of points in the polar regions. If the incidence angle of the light also enters into the calculation of the number of points, then "shaded" images are obtained. This can be applied to all polygonal surfaces (rightmost image).

**Fig. 9.47** application of uniform distribution of points on the sphere in technology and in nature

⊕ *Remark*: The even distribution of points on a sphere is also known under the label *Thompson's problem*, which is concerned with minimizing the energy of $n$ equal point charges on the sphere. For more information, see the book "Geometry and its Applications". ⊕ ◂◂◂

### ▸▸▸ Application: *male or female?*

The sexual proportion is defined as the ratio of the male portion of a certain age group to the female portion of the same age group. At the time of birth, it fluctuates between 1.02 and 1.12 (Fig. 9.48). Values that go above 1.10 are reached only by gender-specific abortions. The values in Europe are between 1.05 and 1.07. We prefer mathematical notations like 1.05, because we must calculate to more decimal places. Statistics often contain information such as 1,050 (:1,000) or 105 (:100). In the following table,[4] $P(m)$ denotes the probability of a boy's birth:

| sexual-prop. | $P(m)$ | country | births / thousand | sexual-prop. | $P(m)$ | country | births / thousand |
|---|---|---|---|---|---|---|---|
| 1.02 : 1 | 0.505 | South Africa | 18.30 | 1.06 : 1 | 0.515 | Germany | 8.25 |
| | | Greenland | 15.93 | | | Denmark | 11.13 |
| | | Kenya | 39.72 | | | Sweden | 10.27 |
| 1.03 : 1 | 0.507 | Nigeria | 40.43 | | | Czech Republic | 9.02 |
| | | Mali | 49.82 | | | Greece | 9.68 |
| | | Columbia | 20.48 | | | Cuba | 11.89 |
| 1.04 : 1 | 0.510 | Finland | 10.45 | | | Bangladesh | 29.80 |
| | | New Zealand | 13.76 | 1.07 : 1 | 0.517 | Italy | 8.72 |
| | | Laos | 35.49 | | | Spain | 10.06 |
| 1.05 : 1 | 0.512 | Austria | 8.74 | | | Slovenia | 8.98 |
| | | Switzerland | 9.71 | | | Bosnia | 8.77 |
| | | Norway | 11.46 | | | Ukraine | 8.82 |
| | | France | 11.99 | | | Vietnam | 16.86 |
| | | USA | 14.14 | | | Malaysia | 22.86 |
| | | Indonesia | 20.34 | 1.08 : 1 | 0.519 | South Korea | 10.00 |
| | | Brazil | 16.56 | | | Venezuela | 18.71 |
| | | India | 22.01 | | | China until 1980 | – |
| | | Iran | 17.00 | | | Singapore | 9.34 |
| | | Japan | 9.37 | 1.10 : 1 | 0.524 | Taiwan | 12.56 |
| | | North Korea | 15.54 | 1.12 : 1 | 0.528 | China | 13.25 |
| | | Argentina | 16.73 | 1.06 : 1 | 0.515 | Worldwide | 20.05 |

**Fig. 9.48** sexual proportions of different countries ◂◂◂

[4] Data downloaded from `www.cia.gov/publications/factbooks/geos/` in 2010.

**▸▸▸ Application**: ***classification by birth order***

We want to give an invulnerable breakdown of sexual proportions for the first birth, second birth, third birth, and so on. Here, data from the USA was used (Fig. 9.49): Firstly, the US has a sex ratio that is typical for the world population. Secondly, the data is statistically secure because of their reliability and large numbers (about 2 to 4 million births per year). After all, there are 100,000 births of a 5th child per year. The data was extracted from huge files of the *National Bureau of Economic Research*.[5]

In the US, there were approximately 50 million births in the period 1992–2003. It can be seen from the table that certain circumstances are fairly stable during this period. These circumstances include the average sexual proportion (1.05), the high number of data, and the number of children per mother.

| year | 1. birth | 2. birth | 3. birth | 4. birth | 5. birth | sum 1.-5. birth | per mother |
|---|---|---|---|---|---|---|---|
| 1991 | 1.053 | 1.043 | 1.041 | 1.033 | 1.035 | 4.02 mio. | 1.96 |
| 1992 | 1.054 | 1.051 | 1.046 | 1.040 | 1.040 | 3.97 mio. | 1.96 |
| 1993 | 1.054 | 1.049 | 1.048 | 1.038 | 1.033 | 3.91 mio. | 1.95 |
| 1994 | 1.052 | 1.047 | 1.045 | 1.039 | 1.038 | 3.86 mio. | 1.94 |
| 1995 | 1.056 | 1.047 | 1.044 | 1.040 | 1.035 | 3.80 mio. | 1.93 |
| 1996 | 1.050 | 1.047 | 1.043 | 1.043 | 1.036 | 3.80 mio. | 1.94 |
| 1997 | 1.052 | 1.049 | 1.041 | 1.038 | 1.040 | 3.79 mio. | 1.95 |
| 1998 | 1.050 | 1.048 | 1.044 | 1.037 | 1.041 | 3.85 mio. | 1.96 |
| 1999 | 1.053 | 1.049 | 1.043 | 1.041 | 1.035 | 3.87 mio. | 1.96 |
| 2000 | 1.054 | 1.048 | 1.045 | 1.038 | 1.030 | 3.97 mio. | 1.96 |
| 2001 | 1.051 | 1.045 | 1.038 | 1.040 | 1.044 | 3.94 mio. | 1.97 |
| 2002 | 1.055 | 1.048 | 1.040 | 1.040 | 1.031 | 3.94 mio. | 1.97 |
| 2003 | 1.054 | 1.047 | 1.049 | 1.037 | 1.031 | 3.94 mio. | 1.97 |

**Fig. 9.49** sexual proportions in individual births (USA)

Fig. 9.50 illustrates the table graphically. In purely visual terms, one can draw quite relevant conclusions. The plotted graphs are smoothed so that all values for each single birth remain in a simple confidence interval. At the 4th birth, and even more at the 5th birth, these intervals become larger due to lower birth rates. If you allow a confidence interval of 95%, then the curves appears almost linear.

From this very reliable data, the following conclusion can be drawn, at least for the USA during the period of 1991–2003:

*The sexual proportion drops slightly but significantly from birth to birth. The average sexual proportion agrees fairly closely with the sexual proportion at the second birth.*

Of course, the latter applies only if a mother has 2 children on average. If the number of children per woman is limited by law to 1, the sexual proportion might be correspondingly higher. However, if the sexual proportion is slightly

[5] `http://www.nber.org/data/vital-statistics-natality-data.html`

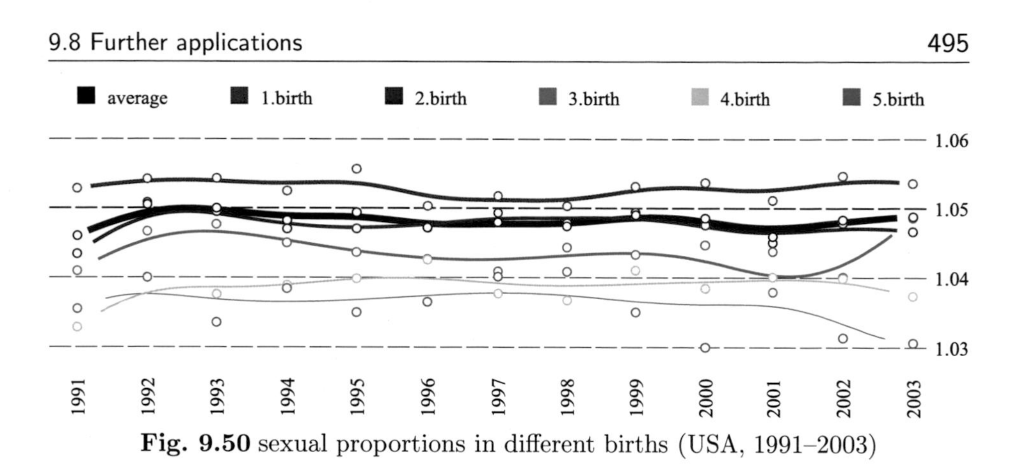


**Fig. 9.50** sexual proportions in different births (USA, 1991–2003)

dependent on the age of the mother and mothers with single children are slightly older when they give birth, then the proportion does not increase even with single children. Corresponding comparative values would have to cover huge samples and those would probably come from China. There, however, the data is falsified because more female than male foetuses are aborted. ◂◂◂

▸▸▸ **Application**: *life expectancy*

Let $s(x)$ be the proportion of the population of a country that died at the age of $x$ years, and $f(x)$ be the portion that died at the age of at least $x$ years ($s(x),\ f(x) \in [0, 1]$). Considering these functions, one shows that the average life expectancy of a newborn is

$$\int_0^m f(x)\,dx$$

years (here, $m \approx 100$ is the statistically relevant maximum age of a person, compare with Application p. 287).

*Solution:*

Let $s_i$ be the proportion of people who – for instance, in the past three years – died at the age of $i$ years. Since each person is assigned exactly one age, $\sum_{i=0}^{m} s_i = 1$. Further, according to the definition of $f(x)$,

$$\sum_{i>x}^{m} s_i = f(x).$$

Obviously, $f(0) = 1$, $f(m) = 0$, and $f(x)$ is a monotonically decreasing function converging to zero, which is, however, almost constant at first and decreases rapidly "towards the end". The functions $s(x)$ and $f(x)$ are, strictly speaking, "step functions" ("bar diagrams" with a bar width of 1). The area under these step functions can be calculated as the sum of the rectangular areas (rectangle width is 1) or, as with any other integrable function, as a definite integral:

$$A = \sum_{i=0}^{m} f(i) \cdot 1 = \int_0^m f(x)\,dx \quad (i \text{ integer},\ x \text{ real}).$$

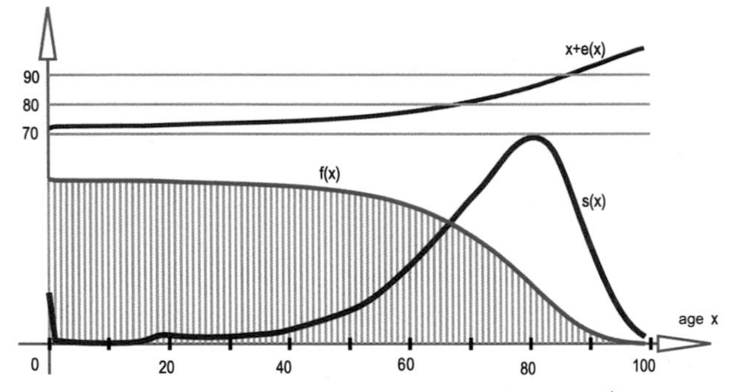


**Fig. 9.51** death statistics and life expectancy in Austria (`www.statistik.at`)

⊕ *Remark*: As is clearly shown in Fig. 9.51, mortality is comparatively high in the first year of life (especially at birth). Then, it remains constantly low until the young become victims of road accidents. From about 35 and onwards, the mortality begins to rise slowly and reaches its peak at being just over 80. ⊕

The average life expectancy $e$ of a newborn can now be calculated as follows:

$$e = 0 \cdot s_0 + 1 \cdot s_1 + 2 \cdot s_2 + \cdots =$$

$$= \underbrace{(s_1 + s_2 + s_3 + \ldots)}_{f(0)} + \underbrace{(s_2 + s_3 + \ldots)}_{f(1)} + \underbrace{(s_3 + s_4 + \ldots)}_{f(2)} + \ldots =$$

$$= f(0) + f(1) + f(2) + \ldots = \sum_{i=0}^{m} f(i) = A = \int_0^m f(x)\, dx.$$

Thus, the life expectancy of a newborn can be interpreted as the area underneath the "life step". ◂◂◂

### ▸▸▸ Application: *remaining life expectancy*

As before, let the expression $f(x)$ be the proportion of the population of a country that died at an age older than $x$ years. Show that a person who is already $a$ years old can expect an average of

$$e(a) = \frac{1}{f(a)} \int_a^m f(x)\, dx$$

remaining years to live and thus reach an average age of $a + e(a)$ years.

***Proof***: For $a = 0$ ($f(0) = 1$), we have already proven the formula. If a person has already reached a certain age $a$, the person may already have survived the early death of other people and "pushed" the average age of death. This increases the life expectancy.

This time, we restrict the group of people to those who are older than the person in question. Their proportion is defined by $f(a)$. Further, let $s_i$ be the proportion of people who died at just $i$ years old. If we now set $\hat{x} = x - a$, then our person with $\hat{x} = 0$ is, so to speak, a "newborn" with a remaining life expectancy of

**Fig. 9.52** remaining life expectancy

$$
\begin{aligned}
\hat{e}(0) &= \frac{0\,s_a + 1\,s_{a+1} + \ldots}{s_a + s_{a+1} + \ldots} = \frac{1}{f(a)} \sum_{i=0}^{m} i\,s_{a+i} = \\
&= \frac{1}{f(a)} [1\,s_{a+1} + 2\,s_{a+2} + 3\,s_{a+3} + \ldots] = \\
&= \frac{1}{f(a)} [\underbrace{(s_{a+1} + \ldots)}_{f(a)} + \underbrace{(s_{a+2} + \ldots)}_{f(a+1)} + \ldots] = \\
&= \frac{1}{f(a)} \sum_a^m f(x) = \frac{1}{f(a)} \int_a^m f(x)\,dx.
\end{aligned}
$$

Thus, the remaining life expectancy can be interpreted as the remaining area under the "life step" multiplied by the magnification factor $\frac{1}{f(a)}$.
For the individual, a calculation is, of course, of little significance. However, when it comes to pension calculations or retirement benefits, such values are of enormous importance.
*Numerical example*: Fig. 9.51 shows the graphs of the functions $s(x)$, $f(x)$, and $x + e(x)$ for Austrian males in 1992. The average life expectancy of a newborn was 73 years. A 60-year-old man could expect to become 78 – 79 years old. An 80-year-old could statistically even reach the age of 87 years. Only ten years later, all values increased by about 2 years. Remarkably, life expectancy for women is much higher (about 6 years). ⊙ ◂◂◂

▸▸▸ **Application**: ***estimation by means of point clouds*** (Fig. 9.53, Fig. 9.54)
In Application p. 292, we derived a formula for the linear fit which can be used to approximate an entire point cloud. Whether this is useful or not depends on how strongly or weakly the point cloud is "scattered". Furthermore, the meaningfulness of the linear fit depends on whether the data (that is, the coordinates of the individual points of the cloud) are actually in a linear relation.
Fig. 9.54 provides a typical example of where a linear fit is unsuitable, but a cubic fit gives remarkably good results. It concerns the (obvious) relation between body size and body mass or weight. In Chapter 2, we discussed this relation in detail (see, for example, Application p. 105). In order to prove the cubic relation (doubling the size leads to an eightfold increase in weight), 50 male and 50 female adults (students and staff at the author's university) were "measured" (Fig. 9.53).
From these 100 measurements, the linear fit and the cubic fit were calculated. The linear fit (shown in red) is practically unusable outside the range of a 150–200 cm body size.

⊕ *Remark*: The cubic parabola (shown in green) allows for a good extrapolation even far outside the interval. All the markings recorded in Fig. 9.53 that lie outside the measuring range are additional test values that were not taken into account for the calculation. The values denote the following data: average values at birth (1), daughter of a random test person (2), tallest living German woman (3), tallest living German man (4), tallest woman ever (5), and tallest human ever (6). Apparently,

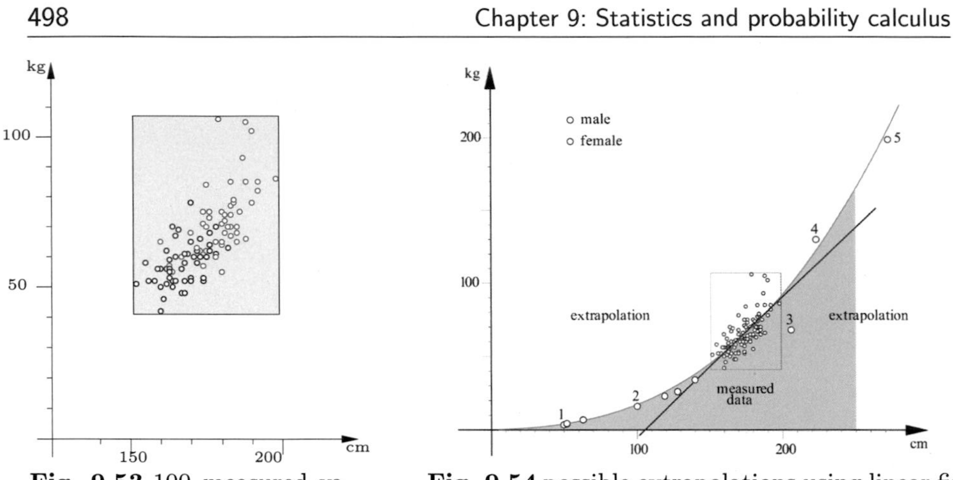


**Fig. 9.53** 100 measured values (body size, body mass)

**Fig. 9.54** possible extrapolations using linear fit or "cubic fit" (useful!)

the tallest woman ever was very overweight. The tallest German woman weighed only 68 kg at a body size of 206 cm, and was, thus, significantly underweight. ⊕ ◂◂◂

### ▸▸▸ Application: *"heart rate variability" for life prolongation*

Modern medicine has found that the intervals between individual heartbeats are not constant. The frequency of the heartbeat is determined by the interplay between the two vegetative nervous systems – the *sympathetic system* and the *parasympathetic system*. The sympathetic system keeps the heart on its toes or rushes it with stress, danger, sports, etc. The parasympathetic system, which may be regarded as the former's opponent, tries to slow the heart – for example, during the intake of food or while sleeping. This permanent interplay produces a "non-linear behaviour" of the heart, which is quite healthy for the body, because the wear of the heart is, thereby, uniform. Markus *Mooslechner* presents the seemingly chaotic structure of human heartbeats – as a strange attractor – in a remarkably visual way (`http://www.humanchaos.net/`): In the myriad of measured data (time differences between heartbeats, which are derived from commercially available hardware used for running), each of the three successive values are interpreted as the coordinates of a point in space. The points lie on the space curves which are smoothed by means of spline interpolation (Application p. 290). In order to quantify the measure of the instantaneous deviation, a sphere with the deviation of the $x$-value from the average of all deviations is placed around the point. The interpolated envelope of all these spheres results in strands that vary in thickness.

⊕ *Remark*: With appropriate knowledge, one can read out a lot about the heart rate variability of the tested person. ⊕ ◂◂◂

### ▸▸▸ Application: *unreliable extrapolation*

You have to be very cautious with extrapolations. In the marshes of Los Angeles, there are lot more carnivores than herbivores. Large mammals that

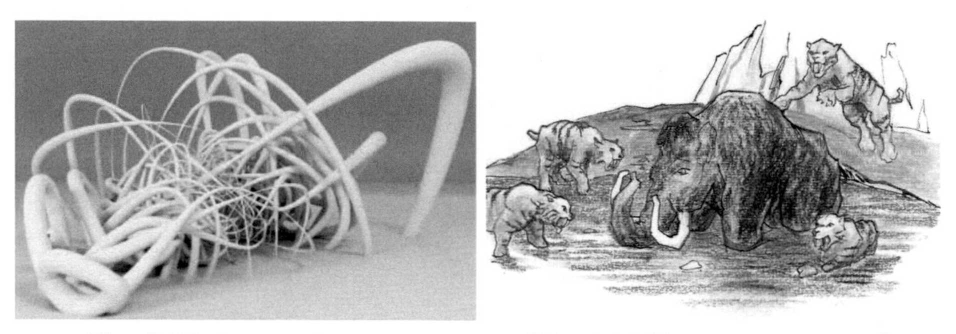

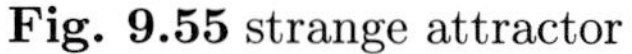

**Fig. 9.55** strange attractor

**Fig. 9.56** Too many carnivores?

would slowly sink into the swamps were clearly easy prey for a multitude of robbers, but swampy environment could just as well become fatal for the latter. ◂◂◂

▸▸▸ **Application**: ***Evolution is design by selection.***
*Albert Einstein* reportedly said: "God does not play dice". Torn from the context of quantum mechanics, this quote is often cited by "creationists" to support their belief that man as the crown of creation could not have evolved by chance. The problem is that the process of evolution is on-going and never complete: Evolution is not "design by chance" but "design by choice". Accordingly, all creatures, as they are now present on the Earth, are not mere random products, nor are they the result of an "intelligent design", but rather they are the result of a millennial – and incomplete – process based on a design principle defined by random mutations.
Among the innumerable random mutations (plants preferentially continue as they are), only those that offer a real advantage to their species will eventually assert themselves. The principle is as ingenious as it is simple: It produces living creatures – each one of its own niche – that are almost unsurpassed and extremely complex. A simple and very illustrative example is the arrangement of the seeds in a sunflower (phyllotaxis, see p. 531). ◂◂◂

**Fig. 9.57** Luckily, what is depicted here is only a firework.

**Fig. 9.58** the "God does not play dice" misunderstanding

### ▸▸▸ Application: *Murphy's law*

Murphy's law is a worldy wisdom that goes back to *Edward A. Murphy Jr.*, which makes a statement about human error or the sources of error in complicated systems. It is best known as: "If something can go wrong, then it will go wrong." Murphy's original formulation was: "If there are two or more ways to do something, and one of them can end in a catastrophe, then someone will choose that sort of thing." There are dozens of variations of this statement on the Internet, and you can spend hours searching in order to amuse yourself.

The latter is probably due to the fact that each of us has already experienced – in her/his own life or in her/his immediate surroundings – that things have gone wrong which, according to the subjective concept of probability, should never have gone wrong.

The mathematician will have less trouble understanding the matter. Let us take the example of a nuclear power station. We have already experienced a core meltdown during the Chernobyl reactor accident in 1986. Up until then, such an accident had been considered impossible. In such a complex environment, many events depend on each other. The connections are not so easy to see. One can say: The probability of all safety precautions being simultaneously suspended is zero. If, however, a person working under stress makes a mistake at one o'clock in the morning, this can cause a chain reaction in the truest sense of the word. The corresponding passage downloaded from `http://www.umweltinstitut.org/frames/all/m226.htm` in 2010 (compare Grigori Medvedev: *The Truth about Chernobyl*, Tauris Publishers, 1991) reads like a crime novel and is somehow reminiscent of the film showing the downfall of the supposedly unsinkable Titanic:

"In order to allow the experiment (the breakdown test) to take place under realistic conditions, the emergency program, the so-called "reactor protection system", was switched off .... Yet, since the start of the experiment was postponed, the unprepared staff working the night shift of April 26 had to take over the experiment, for which arrangements had been made that practically rendered the reactor unsafe. At the beginning of the experiment, the reactor performance was greatly reduced by an operating error of the inexperienced reactor operator *Leonid Toptunow*. To raise them again, the operators removed the brake rods .... Nevertheless, deputy chief engineer of the power plant, *Anatolij Djatlow*, ordered the beginning of the experiment. The operators switched on too many cooling pumps, so that the reactor, which was working with little power, could no longer evaporate the water flowing around it. The water began to boil, and the first hydraulic blows were heard. *Akimov*, the shift leader, and *Toptunov* wanted to end the test, but *Djatlow* ordered them to continue. He spoke the historical words: "Just one or two minutes, and it will be over! Somewhat more flexibility, gentlemen!" The time was 1:22:30." ◂◂◂

# A Music and mathematics

This appendix covers a topic that is underestimated in its significance by many mathematicians, who do not realize that essential aspects of music are based on applied mathematics. This short treatise is meant to show the fundamentals. It may even pique interest in the laws that govern the sensuous beauty of music. Additional literature is provided for readers who are particularly interested in this topic.
The mathematical proportions of a sounding string were first defined by Pythagoras, who made his discoveries by subdividing tense strings. This chapter describes changes in scales and tonal systems over the course of musical history. The author of this appendix is Reinhard Amon, a musician and the author of a lexicon on major-minor harmonic systems.
The chapter will conclude with exercises in "musical calculation" that show the proximity between music and mathematics.


G. Glaeser, *Math Tools*, https://doi.org/10.1007/978-3-319-66960-1

## A.1 Basic approach, fundamentals of natural science

The basic building blocks of music are rhythmic structures, certain pitches, and their distances (= intervals). Certain intervals are preferred by average listeners, such as the octave and the fifth. Others tend to be rejected or excluded. If one makes an experimentally-scientific attempt to create a system of such preferred and rejected intervals, one will find a broad correspondence between the preferences of the human ear and the overtone series, which follows from natural laws. Thus, there exists a *broad correspondence between quantity (mathematical proportion) and quality (human value judgement).*
Experiments of this kind have been conducted since the times of Hellenic philosophers – such as *Pythagoras of Samos* (≈ 570 – 480 BC), *Archytas of Tarentum* (≈ 380 BC), *Aristoxenus* (≈ 330 BC), *Eratosthenes* (3rd century BC), *Didymus* (≈ 30 BC), and *Ptolemy* (≈ 150), among others.
Pythagoras founded the religious-political community of the Pythagorean school in Kroton. He was considered to be the incarnation of *Apollo* and was, even during his lifetime, revered as a god.
During such experiments, a string was strung up on a monochord and subdivided using a measurement scale according to the strict mathematical proportions 1 : 2, 2 : 3 etc. These parts (their halves, thirds, quarters, etc.) were made to vibrate in relation to the whole. Assuming a base tone C, the following numbererd tone row is, thus, constructed.
The emphasized numbers are "ekmelic tones".
These are prime numbers greater than five. Their exact pitch cannot be represented in the Western classical notation system, whose octaves are only subdivided into twelve separate steps. They are, in fact, a little bit lower than their notation.

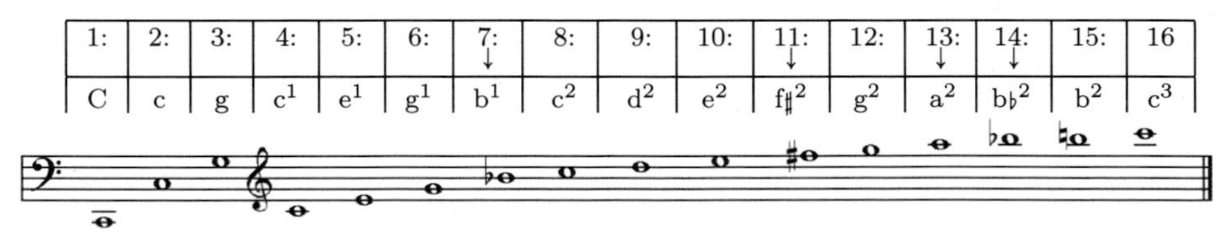

| 1: | 2: | 3: | 4: | 5: | 6: | 7: ↓ | 8: | 9: | 10: | 11: ↓ | 12: | 13: ↓ | 14: ↓ | 15: | 16 |
|---|---|---|---|---|---|---|---|---|---|---|---|---|---|---|---|
| C | c | g | $c^1$ | $e^1$ | $g^1$ | $b^1$ | $c^2$ | $d^2$ | $e^2$ | $f\sharp^2$ | $g^2$ | $a^2$ | $b\flat^2$ | $b^2$ | $c^3$ |

**Fig. A.1** tone row by the subdivision of a tense string

This row corresponds to the overtone series or harmonic series,[1] which is a physical phenomenon that accompanies every sound process. The points that appears next to the numbers transform the distances between the tones into

[1]Discovered in 1636 by *Marin Mersenne*. The associated mathematical rules were discovered in 1702 by *Joseph Sauveur*.

| 1:2 | octave | 5:8 | minor sixth | 15:16 | minor second | 2:3 | fifth |
|---|---|---|---|---|---|---|---|
| 2:3 | fifth | 5:9 | minor seventh | 8:9 | major second | 5:8 | minor sixth |
| 3:4 | fourth | 8:9 | major second | 5:6 | minor third | 3:5 | major sixth |
| 3:5 | major sixth | 8:15 | major seventh | 4:5 | major third | 5:9 | minor seventh |
| 4:5 | major third | 15:16 | minor second | 3:4 | fourth | 8:15 | major seventh |
| 5:6 | minor third | 32:45 | tritone | 32:45 | tritone | 1:2 | octave |

TABLE A.1 intervals sorted by sound consolidation and size

proportions (C - c octave = 1:2; c - g fifth = 2:3; etc.). The overtones that sound above the base tone (the lowest tone of the row – the one which is notated in the score) are not perceived by us as precisely discernible partial tones, but as "characteristic sounds". This phenomenon occurs because the strings or columns of air do not only oscillate as a whole, but also in halves, thirds, quarters, .... Thus, strings or columns of air do not only produce a single tone, but a sum of tones that produces the characteristic sound. The most important components of Western music – the intervals – can be described through mathematical proportions of the length of oscillating strings. This is also true for the oscillating columns of air in woodwind instruments. What is more, this row of pitches following from natural laws contains all important intervals for major-minor tonality.
Since modern times, numbers have been used differently in relation to tones – the basis for the description of a pitch now being the frequency, and no longer the length of a string. Here, it is important to understand that frequency is inversely proportional to the length of a sounding string. A ratio of 1:2 with respect to string lengths corresponds to a ratio of 2:1 with respect to frequency, as a string or column of air that is shortened by half oscillates at twice its original frequency. Twelve essential intervals should be given, from consonant intervals (those with a high degree of sound consolidation among both frequencies – from the octave to the minor sixth) to dissonant intervals (those with little or no sound consolidation among both frequencies).

⊕ *Remark*: The discovery of the connection between the sensual impression of music and its mathematical fundamentals is described in a story by Pythagoras (Fig. A.2):

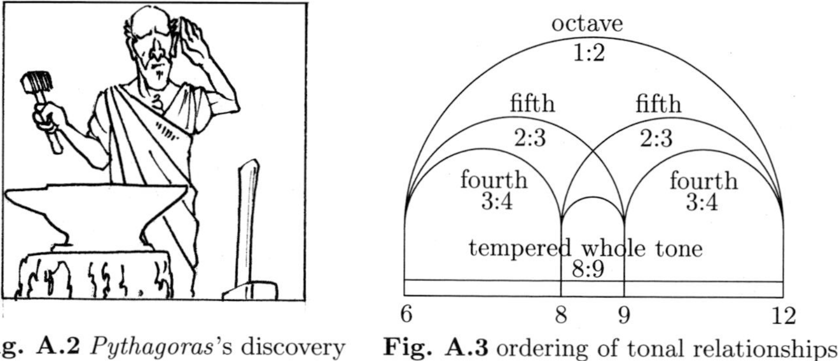


**Fig. A.2** *Pythagoras*'s discovery **Fig. A.3** ordering of tonal relationships

"One day, during a long walk, Pythagoras pondered the problem of consonance. He wondered whether he would be able to find an aid for his sense of hearing, analogous

to the yardstick or the compass as an aid for the visual sense and to the scale as an aid for the sense of touch.
Through a fortunate twist of fate, he passed by a smithy and could clearly hear that several iron hammers striking against anvil, with the exception of one pair of hammers, produced consonant sounds. Full of joy, as if a god supported his plan, he entered the smithy and realized through various experiments that it was the varying weight of the hammers, rather than the blacksmith's output of energy or the strength of the hammers, that caused the tonal difference. He carefully established the weight of the hammers and their impact, and then returned home. There, he fixed a single nail at a corner of the wall, so that no two nails, each having its specific substance, would falsify the experiment. On the nail, he hung four strings of the same substance, with the same number of wires, the same thickness, and the same rotation, and he attached a weight at the utmost end of each string. All strings had the same length, and by plucking two strings simultaneously, he managed to create the consonances he was looking for, which varied with each pair of strings. With the two weights 12 and 6, he produced an octave and thus determined that the octave is in a ratio of $2 : 1$ – something which he had already realized with the weight of the hammers hitting the anvil. With both weights 12 and 8, he received the fifth from which he derived the ratio $3 : 2$. With the weights 12 and 9, he got the fourth and thus the ratio $4 : 3$. When comparing the middle weights 9 and 8, he recognized the interval of the whole tone, which he, then, defined as $9 : 8$. Accordingly, he could define the octave as the union of fifth and fourth, that is, $2/1 = 3/2 \times 4/3$, and the whole tone as the difference of the two intervals, that is $9/8 = 3/2 \times 3/4$."[2] ⊕

The graphical representation (Fig. A.3) of this Pythagorean basic order of the tonal ratios shows that, simultaneously with the two fifths (2:3), two quarter intervals (3:4) and a whole tone interval (8:9) are produced.
These musical and mathematical rudiments discovered by *Pythagoras* and his pupils were regarded not as a "primary substance" but as a "primary law", as it was believed that the the creation of the world and the cosmos was based on invariable numerical proportions. The prevalence of this belief is reflected by the elevated status of mathematics, which is still widely considered as the first of the sciences. This is also due to the influence on a person's mental balance that is often ascribed to mathematics, as, for instance, in the following quote by the philosopher and mathematician *Wilhelm Leibniz* (1646–1716): "Music is an unconscious exercise in mathematics in which the mind is unaware that it is dealing with numbers".

[2] J. Chailley: Harmonie und Kontrapunkt, p. 7ff.

## A.2 System formation

Since ancient times, there have been attempts to derive scales and sound systems from the existing tones. To this end, people tried to fill the octave-space with intervals that were preferably of equal size and of harmonious proportions.

⊕ *Remark*: The octave – the complete fusion of the two frequencies involved (octave identity) – has a special position among the intervals, as it represents the classification framework for all music systems. It can thus be said to function analogously to the different floors of a building, with each floor having the same infrastructure. ⊕

Important for the formation of systems is the overtone series – primarily, the fourth octave from the 8th to the 16th overtone – represented by the natural wind instruments. Since the material selected for system formation is determined not by the fundamental sciences of matter, but by a person, who relies on their senses, two approaches can be formulated:

- This musical approach is based on external factors (geographical location and environment, cultural conditioning, regional listening habits, etc.) and is accepted out of habit.
- People have a predisposition to hearing (see A.5) – that is, the preference or exclusion of certain intervals is natural.

The most important sound systems (the octave space is divided into 5, 6, 7, or 12 spacings) are:

- pentatonic (5 gaps: 3 large seconds and 2 minor thirds),
- whole tone scale (6 gaps: 6 whole-tone steps = large seconds),
- pentatonics (7 gaps: 5 large and 2 small seconds (also: diatonic) → church mode major, minor),
- chromatic (12 gaps: 12 semitone steps → major / minor extension, dodecaphony).

⊕ *Remark*: The half-tone step is the smallest interval in the Western sound system and forms the "basic building block" for a definition offering a different approach to intervals (major third = 4 half-tones, pure fifth = 7 half-tones, ...). ⊕

All systems can be inserted into the above 12 chromatic-tempered spacings. The frequent occurrence of the number 12 (number of essential intervals, number of spacings in the system, ...), which is also reflected in the keyboard

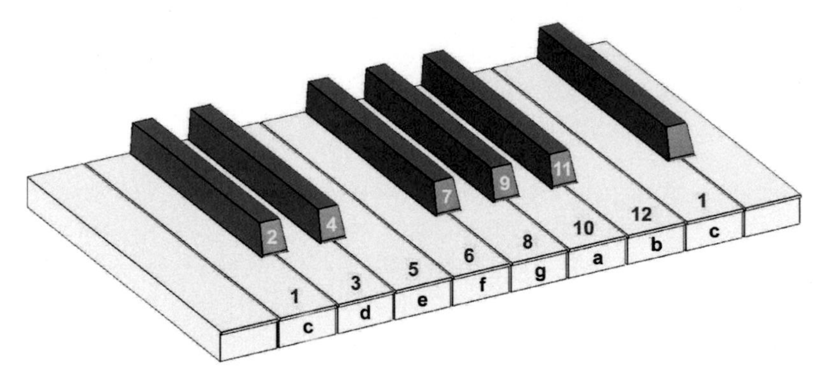


**Fig. A.4** piano keyboard

of piano instruments (as a visual illustration of our sound system), thus attains essential importance.
If we wish to start playing a major or minor scale, each of which consist of 7 tones, starting on each of the white notes, all 12 keys are required. This means that each diatonal heptatonic (major and minor) is integrated into the chromatic-enharmonic duodecatonic.

⊕ *Remark*: The term ‘enharmonic’ denotes the possibility of changing several tonal values at one place in the tonal system (e.g.: f = e♯ = g♭♭, ...). ⊕

Two different types can be distinguished:

- Enharmonic duodecatonic: 35 diatonic-chromatic *tonal values* (each of the 7 tones modified by ♭♭, ♭, ♮, ♯, ×) result in 12 places (c = h♯ = d♭♭) in the system – and thus, in the most extensive references to a key.

- Abstract tempered duodecatonic: there are only these 12 places (abstract tempered tonal values) – there can be no enharmonic confusion – without reference to a key.

Here, too, the integers provide a decisive ordering factor.

## A.3 Tuning instruments – intonation

Once the number of notes in the tuning system is fixed, the problem of tuning (i.e., the exact position of these places – also intonation) becomes relevant.

⊕ *Remark*: Intonation (lat. attunement) refers here to the meeting of the correct pitch by instruments with flexible tones (strings, singers, wind instruments). ⊕

Deviations from pure intervals, which are necessary for musical practice, are regulated through temperament control. The basic difficulty in calculating tuning systems is based on the mathematical phenomenon that integers cannot be obtained as powers of fractions. Since all pure intervals appear as mathematical fractions, a higher octave of an initial tone can never be achieved by the sequencing (= multiplication) of equidistant (pure) intervals (except the octave).

Due to these differences, it was and continues to be necessary to set certain preferences in the selection of the relevant interval sizes for the system. As a result, the respective structure of the music and the resulting sound aesthetics have produced different approaches over the centuries of Western music development.

### Pythagorean tuning (tuning in fifths)

Pythagorean tuning results from the attempt to calculate all the intervals using only the first four numbers. The system-forming interval becomes the fifth and thus the prime number 3. Twelve superimposed pure fifths in the ratio 2:3 comprise the tonal space of seven octaves. However, the 12th fifth does not agree exactly with the 7th octave. The difference (low, but clearly audible) between the 12th fifth and 7th octave is called the Pythagorean comma

$$\left(\frac{3}{2}\right)^{12} = 129.746 > \left(\frac{2}{1}\right)^{7} = 128.$$

Thus, the juxtaposition of 12 pure fifths does not result in a closed circle, but in a helix.

The bright balls are at the positions of the octaves and close the circle. The dark spheres are at the positions of the fifths and form a spiral. The 12th fifth clearly deviates from the seventh octave.

The octave 1:2 is represented by the prime number 2, the second 2:3 by the prime number 3. However, an exponentiation of 2 never coincides with an exponentiation of 3. The resulting microintervals are smaller than a semitone. Through tempering, we try to prevent this.

Calculation of the system:

| initial point C | tonal note | C | F | G | c | g | c' |
|---|---|---|---|---|---|---|---|
| | string length | 1 | $\frac{3}{4}$ | $\frac{2}{3}$ | $\frac{1}{2}$ | $\frac{1}{3}$ | $\frac{1}{4}$ |

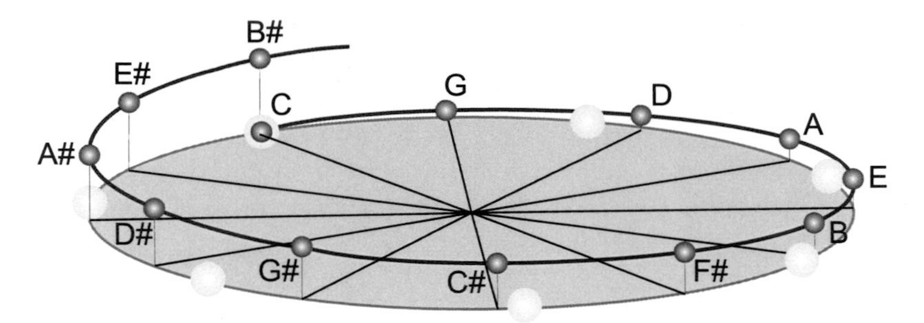


**Fig. A.5** tunes: 7 octaves, 12 fifths

D is obtained by moving two fifths ↑ from $C$: $1 \times 2/3 = 2/3$; second fifth: $2/3 \times 2/3 = 4/9$ or respectively the octave offset following ↓: $4/9 \times 2 = 8/9$.

E (4th upper fifth, 2 octaves ↓) $\Rightarrow (2/3 \times 2/3 \times 2/3 \times 2/3) \times 4 = 16/81 \times 4 = 64/81$;

A (3rd upper fifth, 1 octave ↓) $\Rightarrow (2/3 \times 2/3 \times 2/3) \times 2 = 8/27 \times 2 = 16/27$ or by the Fifth D-A $8/9 \times 2/3 = 16/27$; B (5th upper fifth, 2 octaves ↓) $\Rightarrow (2/3 \times 2/3 \times 2/3 \times 2/3 \times 2/3) \times 4 = 32/243 \times 4 = 128/243$.

Since 1884 (*A.J. Ellis*), the octave is defined as 1200 cents:

1 Cent = $\sqrt[1200]{2} = 2^{1/1200}$. Therefore, all the cent numbers occurring in the text are rounded with the exception of the number 1200. 204 cents, for example, means $\sqrt[1200]{2}^{204} = 1.125058\ldots \approx \frac{9}{8}$. The tempered semitone is then calculated as 100 cents ($2^{100/1200} = 2^{1/12} = \sqrt[12]{2}$).

This produces the following Pythagorean scale:

| tonal note | C | D | E | F | G | A | B | c |
|---|---|---|---|---|---|---|---|---|
| cent values | 0 | 204 | 408 | 498 | 702 | 906 | 1,110 | 1,200 |
| string length with respect to base tone | 1 | $\frac{8}{9}$ | $\frac{64}{81}$ | $\frac{3}{4}$ | $\frac{2}{3}$ | $\frac{16}{27}$ | $\frac{128}{243}$ | $\frac{1}{2}$ |

The series consists of five whole tone steps at a size of 204 cents and two semitones of 90 cents. This gives 5 full tones with 8/9 and 2 semitones with 243/256. Pythagorean tuning was suited for monophony and thus for the the music of the Middle Ages.

The slightly dissonant Pythagorean third 64/81 was, thus, no problem (the pure third has the proportion 64/80). As consonant intervals, only the octave, the fifth, and their complementary interval were recognized. With the advent of polyphony and the increased preference of merging (consonant) intervals, Pythagorean tuning with its dissonant major third lost its importance.

The polyphonic music that was composed up until the 17th century coped with intonation difficulties in various ways. As a rule, musicians opted for the "pure" attunement of the most common sounds of musical practice, and they avoided or concealed unpleasant harmonies (for example, by means of ornaments). Another possibility was the use of more than 12 tones within the octave. Thus, keyboard instruments with up to 31 keys were created within the octave.

## The pure tuning (natural-harmonic tuning, fifth-octave tuning)

In a sense, our ears hear pure tuning. This occurs when each interval is intoned according to its tone ratio in the series of natural tones. The obvious conclusion is to apply pure tuning to both theory and music, because pure thirds and pure fifths exclude each other. The difficulty that is encountered here is that such a tuning requires far more than the maximum of 12 positions in the system. Depending on the key type, the modulations to be performed or the need for transposing, up to 171 positions are required.
*Didymos* of Alexandria (≈30 BC) divided the monochord strings into 5 equal parts and replaced the third 64/81 by 4/5. The deviation in the tones E, A, and B as the major thirds of C, F, and G is thus less:

$$64:81 = 0.79012345678 \neq 4:5 = 0.8$$

The following scale is formed (the proportions are directly from the harmonic series):

| tonal note | C | D | E | F | G | A | B | c |
|---|---|---|---|---|---|---|---|---|
| cent values | 0 | 204 | 386 | 498 | 702 | 884 | 1,088 | 1,200 |
| string length with respect to base tone | 1 | $\frac{8}{9}$ | $\frac{4}{5}$ | $\frac{3}{4}$ | $\frac{2}{3}$ | $\frac{3}{5}$ | $\frac{8}{15}$ | $\frac{1}{2}$ |

In pure tuning, two different whole-tone steps appear: the large whole tone 8:9 between c-d with 204 cents and the small whole tone 9:10 between d-e with 182 cents. This difference causes, for example, an impure 5th between d and a.
$e : h = 5/4 \times 3/2 = 15/8 = 1.875$ (r.5)
$d : a = 9/8 \times 3/2 = 27/16 = 1.687$ (no r.5)
The "difference" 80/81 (mathematically, it is a quotient) is called the syntonic comma. It is by the same value that the fourth a-d is too large. The difference between the large whole tone 9/8 and the small whole tone 10/9 ("difference" $= \frac{8}{9} : \frac{9}{10} = 81/80$, the syntonic comma) is evident at many points in the system. For instance, at the duplicate tone e, we get the following as the octave third of c and as the compound twelfth (i.e. fifth) of c:
The 5th overtone of C = E 5=E 10=E 20=E 40=E 80=E.
The compound twelfth 1:3 of C → the third overtone G 3=G 9=D 27=A 81=E.
Duplicate tone 80/81.
The increasing use of instruments with fixed pitches (organs, piano instruments, etc.) and the inclusion of keys with more and more signatures gave impetus to a solution that would eliminate these differences. The resulting tempered tunings finally made it possible to play music in all keys.

## Mid-tone temperament

This term was coined by *Arnold Schlick* in 1511 and by *Pietro Aron* in 1523/29. The term "mid-tone" is derived from the fact that the mean value between 8/9 (large whole tone) and 9/10 (small whole tone) determines the new whole tone. The natural major third – which takes precedence over the fifth – is thus not divided as in the overtone series 8 : 9 : 10, but as a geometric mean. That is, it is divided into two equal whole-tone steps.

| tones | c | d | **b** | d | e |
|---|---|---|---|---|---|
| proportion | $\frac{1}{1}$ | $\frac{10}{9}$ | **MT** | $\frac{9}{8}$ | $\frac{5}{4}$ |
| cent | 0 | 182 | **193** | 204 | 386 |

The mid-tone scale is made in the following way (according to *Arnold Schlick*), with K standing for the Pythagorean comma:

| tonal note | C | D | E | F | G | A | B | c |
|---|---|---|---|---|---|---|---|---|
| cent values | 0 | 193 | 386 | 503 | 696.5 | 890 | 1,083 | 1,200 |
| string length with respect to base tone | 1 | $\frac{8}{9} : \frac{K}{2}$ | $\frac{4}{5}$ | $\frac{3}{4} \times \frac{K}{4}$ | $\frac{2}{3} : \frac{K}{4}$ | $\frac{3}{5} \times \frac{K}{4}$ | $\frac{8}{15} : \frac{K}{4}$ | $\frac{1}{2}$ |

Mid-tinted tones were used during the 19th century, when transposing was limited and there was no enharmony yet. This resulted in a clear *tonality characteristic.*

⊕ *Remark*: The term 'tonality characteristic' means that each tonality has a more or less clearly distinct mood or expressive content due to its interval disposition (sharp, mild, serious, soft, ...). ⊕

In order to make the "wolf tone" (an extremely impure fifth) between g♯ and d♯ more bearable, one had to increase either the sound reserve (one key for g♯ and a♭ or for d♯ and e♭) or the fifths at the expense of improving the pure thirds (⇒ "non-equilibrating temperament")

## Tempered tones

There are two basic possibilities with only 12 places in the system.

On the one hand, one can unevenly distribute the Pythagorean comma over thirds and fifths with 23.5 cents – these are uneven tones. On the other hand, one can evenly distributed over all the fifths – this is the equilibrium or equal temperament. The dissimilar tones were all calculated by *Andreas Werckmeister* (1645–1706), who was for a long time credited with the 'invention' of constant temperament control. *Johann Phillipp Kirnberger* (1721–1783), a composer and music historian who had studied as a pupil of *J.S. Bach* from 1739–41, developed three temperaments (K1, K2, and K3). Kirnberger's temperaments became not only the most widespread, but they also asserted themselves among all irregular temperaments for the longest time.

The C major scale with Kirnberger tones:

| tonal note | C | D | E | F | G | A | B | c |
|---|---|---|---|---|---|---|---|---|
| cent values | 0 | 204 | 386 | 498 | 702 | 895 | 1,088 | 1,200 |
| string length with respect to base tone | 1 | $\frac{8}{9}$ | $\frac{4}{5}$ | $\frac{3}{4}$ | $\frac{2}{3}$ | $\frac{96}{161}$ | $\frac{8}{15}$ | $\frac{1}{2}$ |

The most primitive possibility, "equal temperament" (calculated by the Dutch mathematician *Simon Stevin* in 1585 and also in China around the same time), prevailed in the 18th century, showing the various possibilities of balance (medium-temperament, uneven-floating temperament). The 12th fifth exceeds the 7th octave by 23.46 cents. In order to make a circle from the spiral of fifths, the "Pythagorean comma" is distributed uniformly (equidistantly) over all twelve fifths. The octave is, thus, divided into twelve equal half-tone intervals $1 : \sqrt[12]{2}$. In this equilibrium, no interval aside from the octave is really pure. However, the differences are so small that the ear can tolerate it very well. This eliminates the possibility of tonal characteristics (= differences in intonation), (which was the contemporaries' main objection against this new temperament).

⊕ *Remark*: The desire to maintain tonal characteristics prevented the "new" tempered equilibrium from being accepted in musical practice until well into the middle of the 19th century. ⊕

Since no fifth is pure (i.e. C - G sounds just as good as C♯ - G♯ or C♭ - G♭), no interval has to be avoided for performance reasons. A certain "instability" that may arise from this can be countered in singing or when playing string instruments by spontaneously adjusting the intonation of the respective tonal situation in the piece.

Thus, we have the following equidistantly tempered (= equilibrium) scale, where $\lambda$ stands for $\sqrt[12]{2}$:

| tonal note | C | D | E | F | G | A | H | c |
|---|---|---|---|---|---|---|---|---|
| cent values | 0 | 200 | 400 | 500 | 700 | 900 | 1,100 | 1,200 |
| string length with respect to base tone | $1/\lambda^0$ | $1/\lambda^2$ | $1/\lambda^4$ | $1/\lambda^5$ | $1/\lambda^7$ | $1/\lambda^9$ | $1/\lambda^{11}$ | $1/2$ |

## Summary

The tuning with the Pythagorean numbers 1, 2, 3, and 4 was, indeed, restricted (consider the third!), but unproblematic. With the addition of the number 5 and thus the pure third, problems arise: (a) One needs more than 12 pitches if one wants to have pure fifths and pure thirds in the system. (b) If a certain selection of tone places is determined, one must put up with impure thirds and/or fifths. As long as the third note was handled in early multi-voice music and considered as a dissonance, tuning problems were of little significance. The acknowledgement of the third as a consonance and its use in polyphony result in the above-described problems and different approaches to solving them. In spite of the lost tonal characteristics, the resulting equal temperament is a system which offers a satisfactory and understandable performance practice for the listener with a relatively small number of fixed pitches.

## A.4 Numerical symbolism

There is an almost universal view of the "symbolic meaning" of numbers that can even be found in philosophical speculation and according to which numbers and their relations constitute the essence of reality. Especially in the field of the arts, it is common to include meta-levels and go beyond sensory perception.
Some examples of this shall be given here:

- The fundamental note (see natural tone series) as reference point, i.e. the number 1 of the tone series, refers to God, the origin, the unitas.

- Exceeding the "god-willed" order brings misfortune: The 5 (after the divine 3 and cosmic all-embracing 4) in the overtone series – the number of thirds and sixths – was avoided for a long time. The fifth day of the week brings misery. At the 5th trumpet sound, a star will fall from the sky .... Similar beliefs can be found in relation to the number 11 (number of transgressions, sins, ...).

For Pythagoreans, the number is raised to the dimension of a general principle in the appearance of beings. The harmony of the world, the cosmos, and human order was seen and valued from the validity of numerical relations. Therefore, the personal union of the mathematician, astronomer, musician, and philosopher was a self-image. The most important number was 10, being the sum of the four first numbers 1+2+3+4. This is a typical starting point for Western thinking and its philosophy. What is fundamental and in accordance with the striving for perfection is the metaphysical, ontological, and axiological primacy of that which is finite, orderly and harmonious, as opposed to chaos, disorder, infinity, and irregularity. The extent of these speculations is reflected by the fact that the pupils of Pythagoras introduced a division into mathematicians (those dealing with quantitative realities) and so-called acousmatics (those who consider qualitative aspects, i.e. those areas that give emphasis to sound and its cosmic and above-all mental dimensions). Some also included a third category in this division – the "esoteric".
Especially from the Baroque period onwards, references based on numerical symbolism, biographical aspects of the composer, text-related cross-connections, monograms in the form of numbers, etc. were incorporated in musical works. This was accomplished through the selection of the time meter, number of bars, number of parts, interval structures (second stands for 2, third stands for 3, ...), number of tones of a melody, and so on. Especially in *J.S. Bach*'s works, we find countless examples of this. The cross-sum of his name based on the the numerical alphabet (this was such common practice that one could assume it would be generally understood [I = J, U = V !])

resulted in the number 14, which, as well as the reversal number 41, would appear in melodies, measures, and other references in the work.
The number 29 based on the initials J.S.B., with which *Bach* signed many of his works, has a second meaning: S.D.G. (Soli Deo Gloria = Only to the glory of the Lord). The number 158 is also quite frequent: *Johann Sebastian Bach*. When he decided to join Mizler's society,[3] he waited for the membership number 14. Obviously, the waistcoat that he wears in his portraits has 14 buttons. Moreover, the canon (BWV 1076), which he wrote for this occasion, references the year 1747 when he joined Mizler's society. It also cites a bass line by *G.F. Handel*, consists of 11 notes – *Handel* was the 11th member of the society – and the number 14 is constantly present.

### Timing – rhythm

In addition to tones and intervals, the temporal ordering process is probably the most important component of music. Music is regarded as a period of time, with the regular recurrence of rhythmic sequences in the continuous flow of metrics forming the basis of perception. Sounds are not randomly ordered, but instead their temporal sequence exhibits a structure, a certain order. The time of music is the pulse. The pulse consists of two-part or three-part units of this pulse. The continuous values are produced by simple multiples or fractions of the largest unit of measurement. Here as well, with this temporal division of the material, we are faced with the principle of counting and thus with that of mathematics.

## A.5 Harmonics (basic fundamental research)

### Basics

This science, based on the integral proportions of the overtone series, is essentially anthropological. The fundamental idea is that the order that appears in the numerical law of overtone series is universal and cosmic as it governs both animate and inanimate nature, as well as all human beings. Simple proportions have an effect on the mind and the mood of human beings. In music, these are the intervals. In the fine arts and architecture, there is, for example, the "golden section" or the "Fibonacci series".
By means of analogies, proportionality is sought in natural sciences such as chemistry, physics, astronomy, biology, medicine, as well as in art and architecture.

[3] *Lorenz Christoph Mizler* – a former *Bach* scholar – later became a lecturer in mathematics, physics, and music. He founded this "Societät der musikalischen Wissenschaften" (Society of Musical Sciences) in 1738.

| number of intersections | accordance in percentages | interval | proportion |
|---|---|---|---|
| 72 | 100 | octave | 1:2 |
| 42 | 58 | fifth | 2:3 |
| 28 | 39 | fourth | 3:4 |
| 24 | 33 | major sixth | 3:5 |
| 18 | 25 | major third | 4:5 |
| 14 | 19 | minor third | 5:6 |
| 10 | 14 | minor sixth | 5:8 |
| 8 | 11 | minor seventh | 5:9 |
| 2 | 3 | major second | 8:9 |
| 2 | 3 | major seventh | 8:15 |
| 0 | 0 | minor second | 15:16 |
| 0 | 0 | tritone | 32:45 |

TABLE A.2 from *R. Haase* - "harmonical synthesis"

## Hearing disposition

The preference of intervals when forming systems from integer proportions is fundamental in the nature of man. In the ear itself, in addition to the combination of sounds (sum and/or difference of two fundamental frequencies), "subjective overtones" and "ear overtones" also arise. The series of harmonics superimposed in interval formation have more or less common "intersections" depending on the proportion. These result in an order on which the qualitative division of the intervals into consonances and dissonances has been based for centuries (Table A.2).

## Harmonious laws in other areas

Analogous references to integer proportions of the musical intervals can be made in many fields of science. Astronomy: The discovery of harmonious laws was already a matter of concern for the ancient philosopher *Johannes Kepler*. He was a harmonist in the best sense of the word and a modern-day natural scientist who tried to prove the spherical harmony by calculating the planets in their orbits around the Sun, the day arcs, and the angles that could be measured from the sun to the extreme points lying on the planet's elliptical orbits within 24 hours. Proportional orders are detected in other areas, such as crystallography, chemistry (where proportional orders appear in the periodic system of the elements), physics, biology, and, last but not least, anthropology, where numerous "musical" proportions can be found in the human form.

Further reading:

Amon, R.: *Lexikon der Harmonielehre*, 2nd edtion, Doblinger/Metzler, Wien-München 2015
Amon, R.: *Lexikon der musikalischen Form*, in coop. with Gerold Gruber, Metzler, Vienna 2011
Amon, R.: *Piano Essentials, Eine Anthologie von kurzen Stücken in Theorie und Praxis*, Doblinger, Vienna 2008
Gauldin, R.: *Harmonic practice in tonal music*, New York 1997
Levine, M.: *Das Jazz Theorie Buch*, Advanced Music 1996
Piston, W.: *Counterpoint, New York 1947*
Piston, W.: *Harmony*, Fifth edition, New York 1987
Ulehla, L.: *Contemporary Harmony*, 1994

## A.6 Numerical examples

▸▸▸ **Application**: ***Pythagorean scale***

The scale that is created through Pythagorean tuning consists, like our modern scale, of seven intervals (five whole tones and two semitones): Take a vibrating string that produces the fundamental C. The notes F and G are then produced by reducing the string to 3/4 (fourth) and 2/3 (fifth) of the total length. When reducing the string to 1/2 (octave), the first tone c of the next scale is obtained. Now the remaining tones between C and c, namely D, E, A, B can be calculated from the octave and the fifth as a proportion $(2/3)^n : (1/2)^m$. For which powers $n$ and $m$ can the smallest possible fundamental tones be obtained? What are the corresponding reductions? What gradations result in this scale?

*Solution*:

We have
$$t = \left(\frac{2}{3}\right)^n : \left(\frac{1}{2}\right)^m = \left(\frac{2}{3}\right)^n \cdot 2^m \frac{2^{n+m}}{3^n}.$$

We only search for integers $m$ and $n$ such that
$$\frac{1}{2} \le \frac{2^{n+m}}{3^n} \le 1 \tag{A.1}$$

holds. We solve the problem through testing:

| $m$ | $n$ | $2^{n+m}$ | $3^n$ | (A.1) satisfied? | base tone | $t$ |
|---|---|---|---|---|---|---|
| −1 | 0 | 1/2 | 1 | yes | c | 1 : 2 |
| 0 | 0 | 1 | 1 | yes | C | 1 |
| 0 | 1 | 2 | 3 | yes | G | 2 : 3 |
| 0 | 2 | 4 | 9 | no | | |
| 1 | 0 | 2 | 1 | no | | |
| 1 | 1 | 4 | 3 | no | | |
| 1 | 2 | 8 | 9 | yes | D | 8 : 9 |
| 1 | 3 | 16 | 27 | yes | A | 16 : 27 |
| 1 | 4 | 32 | 81 | no | | |
| 2 | 4 | 64 | 81 | yes | E | 64 : 81 |
| 2 | 5 | 128 | 243 | yes | B | 128 : 243 |

The gradations are now the quotients from the reductions for the two successive notes. This quotient is either 8 : 9 ("full tone") or 243 : 256 ("semitone"). Semitones are available between E and F $\left(\frac{3}{4} : \frac{64}{81} = \frac{243}{256}\right)$ or B and c respectively.

The same applies to the vibrating air columns in wind instruments. ◂◂◂

▸▸▸ **Application**: ***equally tempered scale***

In the so-called "equally tempered scale", all tones are produced by dividing the string length corresponding to the previous tone by a constant ratio of $\lambda$, once for a semitone, and twice for a whole tone. In addition to the initial tone C, the Pythagorean values (Application p. 515) are somewhat corrected.

What is the value of $\lambda$ and how large are the deviations from the Pythagorean scale?

*Solution*:

In the scale (Application p. 515), we first have two whole tones, then a half-tone, then three whole tones, and finally a last half-tone. The shortening to 1/2, thus, occurs as follows:

$$\lambda^2 \cdot \lambda^2 \cdot \lambda \cdot \lambda^2 \cdot \lambda^2 \cdot \lambda^2 \cdot \lambda = \frac{1}{2} \Rightarrow \lambda^{12} = \frac{1}{2} \Rightarrow \lambda = 1 : \sqrt[12]{2}$$

Now we create a table:

| base tone | Pythagoras | equally tempered | deviation |
|---|---|---|---|
| C | 1 | 1 | 0 |
| D | $8:9 = 0.88889$ | $1:\lambda^2 = 0.89090$ | −0.00201 |
| E | $64:81 = 0,79012$ | $1:\lambda^4 = 0.79370$ | −0.00358 |
| F | $3:4 = 0.75000$ | $1:\lambda^5 = 0.74915$ | 0.00085 |
| G | $2:3 = 0.66667$ | $1:\lambda^7 = 0.66742$ | −0.00075 |
| A | $16:27 = 0.59259$ | $1:\lambda^9 = 0.59460$ | −0.00201 |
| H | $128:243 = 0.52675$ | $1:\lambda^{11} = 0.52973$ | −0.00298 |

The error in the individual tones seems to be quite small, but it is quite audible to the trained ear. The new classification has the advantage that there are now no inconsistencies in transposing. For the construction of the string lengths, see Application p. 516. ◂◂◂

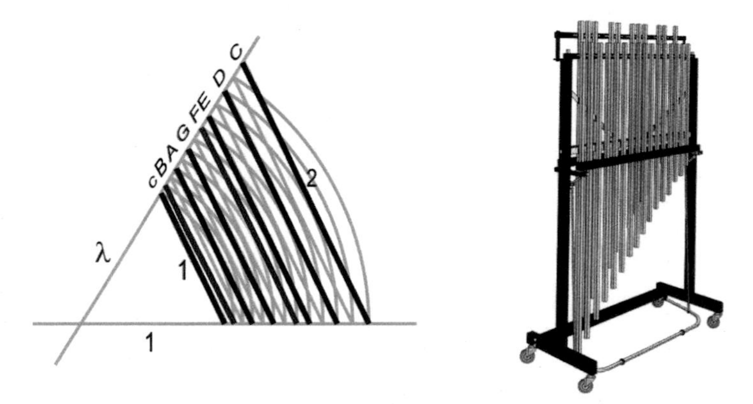


**Fig. A.6** string lengths

▸▸▸ **Application**: ***construction of the string lengths of the tempered scale***
For the equally tempered scale (Application p. 515), the powers of $\lambda = \sqrt[12]{2}$ have to be determined. Show the correctness of the construction of the string lengths in Fig. A.6.

*Solution*:

The triangle with side lengths 1, 1, $\lambda$ is isosceles (Fig. A.6). The ratio of the shorter side to the longer side is $1:\lambda$. Next, we draw the circular arc centered

at the left vertex of the triangle and passing through the top vertex. This arc intersects the base line at distance $\lambda$ from the left vertex. By drawing the parallels to the opposite short side, the long side is also multiplied by $\lambda$ due to the intercept theorem, and it thus has the length $\lambda^2$.
For twelve repetitions, the string should be twice as long. (The twelfth root can also be constructed geometrically.) The black strings in the diagram correspond to the C major pentatonic scale and the two red strings complete it to a C major scale. ◂◂◂

▸▸▸ **Application**: ***length of the air columns in wind instruments*** (Fig. A.7)
We have discussed the "equally tempered scale". In this scale, all tones are produced by shortening the length of the string or the length of the air column corresponding to the respective preceding tone in the constant ratio $\lambda = 1 : \sqrt[12]{2}$. What is the arrangement of the pipes when a) pipes with the same diameter are always used, b) the diameter of the pipes that are used is proportional to their length (as is the case with the organ).

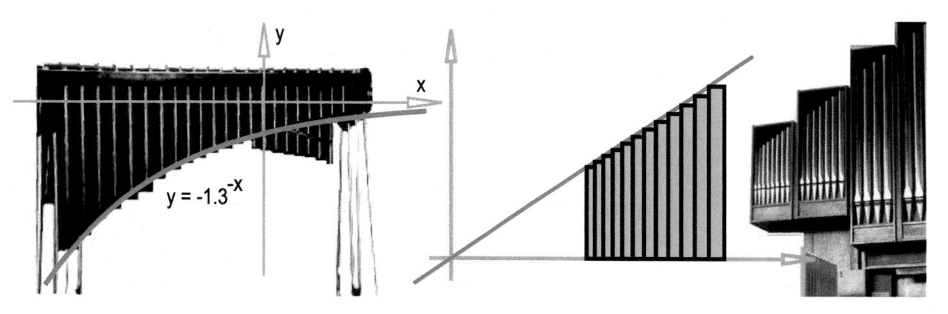


**Fig. A.7** The *width* of the air columns can grow linearly or exponentially.

*Solution*:
First, suppose the pipe diameter $d$ is constant, e.g. $d = 1$. The longest pipe has length $L$. The next whistle has length $\lambda L$, then $\lambda^2 L$, and so on. The pipe of the $x$-th tone, thus, has the length

$$y = \lambda^x L \quad (\text{with } \lambda = 1 : \sqrt[12]{2} \approx 0.9439)$$

or

$$y = c^{-x} L \quad (\text{with } c = 1/\lambda = \sqrt[12]{2} \approx 1.0595)$$

respectively. Fig. A.7, left (tubular chimes) shows an exponential function placed over a photograph (similar to a xylophone). It can be seen that the pipe length is longer than the functional value by a constant distance: In this area, the air cannot vibrate.
Now suppose the pipe diameter $d$ is not constant, but proportional to the length of the vibrating air column (Fig. A.7, right). Comparable points are on a straight line: The rectangular cross sections are similar to one another

(always with the same similarity factor) and can be mapped into each other by dilation.

**Fig. A.8** organ pipes ...

An organ has hundreds of different pipes. It is by no means necessary to arrange these pipes in order of height (even though there is a German saying – "standing like organ pipes" meaning to stand in a row from tallest to shortest – that seems to suggest so). Some organ builders arrange their pipes in such a way as to create interesting patterns, which are more difficult to classify mathematially (Fig. A.8, part of the organ in Altnagelberg, Lower Austria, during reassembly). ◂◂◂

# B Numbers

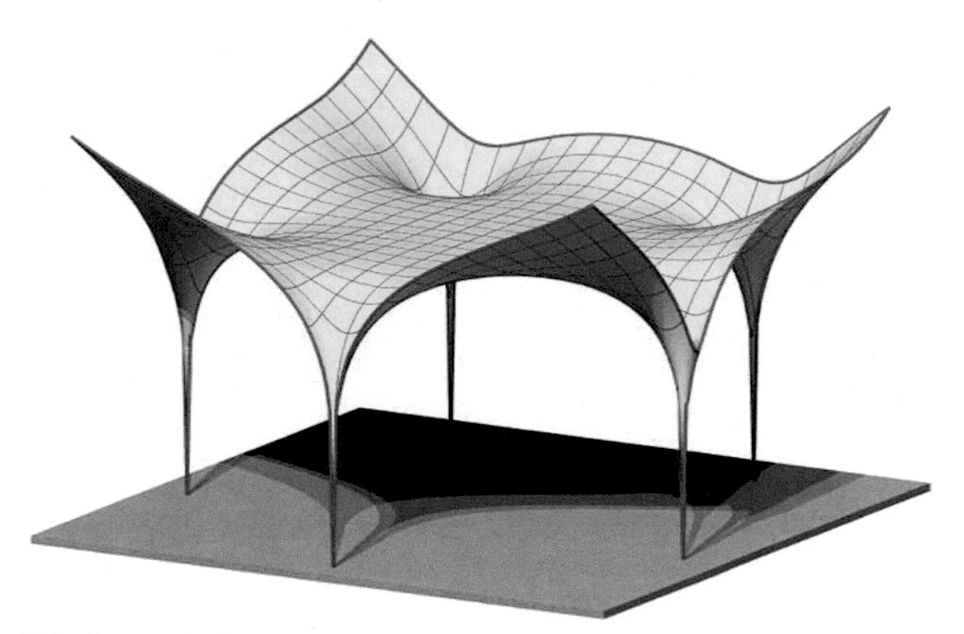

This chapter deals with special numbers that have had a significant influence on the history of mathematics. First, we will discuss two simple examples from the world of "numerology". One of these is *Goethe*'s famous "magic formula", with which we can describe a magic square.

Then, we will discuss the number $\pi = 3.14159\ldots$ and Euler's number $e = 2.71828\ldots$ in greater detail. These numbers play an exceptional role in mathematics, and can be found in many diverse applications. Both numbers are transcendental – in other words, they cannot be found as the solutions of algebraic equations with integer coefficients.

We will then turn our attention to the golden mean and to the closely related *Fibonacci* numbers, which are often found in nature – for instance, in the structures of flower blossoms.

Finally, we will discuss complex numbers, which are greatly fascinating despite their strange definition (for which it is necessary to introduce an "imaginary unit" $i$ with $i^2 = -1$). Complex numbers allow us to solve many problems that seem perplexing "in the world of real numbers", and they explain the fundamental theorem of algebra, according to which every algebraic equation of degree $n$ possesses exactly $n$ solutions (if multiple solutions are counted with their multiplicity).

An almost magical relation between $e$, $\pi$, and $i$ becomes manifest in the famous equation $0 = 1 + e^{i\pi}$. By performing calculations in the *complex plane*, we can use complex numbers to accomplish elegant solutions to problems in kinematics and other technical disciplines.


G. Glaeser, *Math Tools*, https://doi.org/10.1007/978-3-319-66960-1

## B.1 Numerology

Since the dawn of human culture, numbers have exerted a magical fascination on humans. Many situations prompt us to say: "This can be no accident." The "witch's one-time-one" by *Goethe* from his magnum opus "Faust" is almost certainly no accident.

### Witch's one-time-one and magic squares of order 3

Magic squares have fascinated humans for millennia. These are square matrices of numbers ($n$ rows, $n$ columns) where the first $n^2$ integers are arranged such that the sum of each row, column, and both diagonals is equal. This sum $s$ is, thus, given:

$$s = \frac{1}{n}\sum_{k=1}^{n^2} k, \text{ for } n = 3\text{: } s = \frac{1}{3}\sum_{k=1}^{9} k = \frac{1+2+\ldots+9}{3} = \frac{45}{3} = 15.$$

There are many known examples of magic squares of higher order, but only one for the order $n = 3$ (of course, it is still possible to flip and mirror this solution in several ways). A magic square of order 4 was found by *A. Dürer*.[1]

▸▸▸ **Application**: ***How can one memorize a magic square of order 3?***
The answer is found in *J. W. von. Goethe*'s "Faust" in a section called "witch's one-time-one":[2]

*This must thou ken:*
*Of one make ten,*
*Pass two, and then,*
*Make square the three,*
*So rich thou'lt be.*
*Drop out the four!*
*From five and six,*
*Thus says the witch,*
*Make seven and eight.*
*So all is straight!*
*And nine is one,*
*And ten is none,*
*This is the witch's one-time-one!*

Faust then ends the verse with a fairly terse comment:
*The hag doth as in fever rave.*

[1] http://mathworld.wolfram.com/DuerersMagicSquare.html
[2] `http://www.fullbooks.com/Faust-Part-13.html`

*Goethe* gave obfuscated advice on how to create a magic square of order 3. Several interpretations exist, the following being by the author:

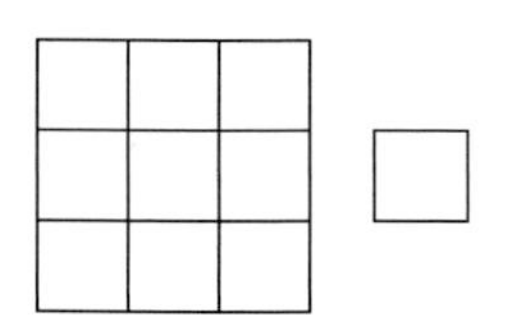

**Fig. B.1** *Of one make ten,*

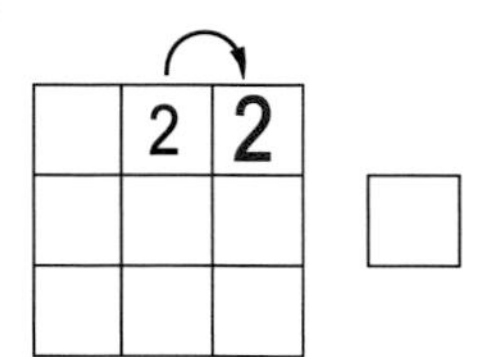


**Fig. B.2** *Pass two, and then,*

1. "*This must thou ken: Of one make ten.*" does not refer to the numbers themselves, but to the number of squares! Turn one square into 10, by adding 9 more (in the form of a matrix) (Fig. B.1).

2. "*Pass two, and then,*" is to be taken literally: The 2 moves from the second to the third position (Fig. B.2).

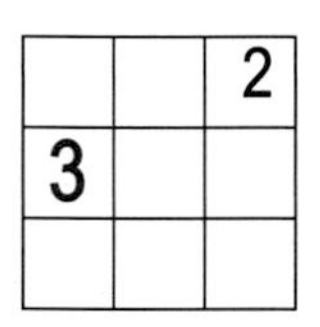


**Fig. B.3** *Make square the three,*

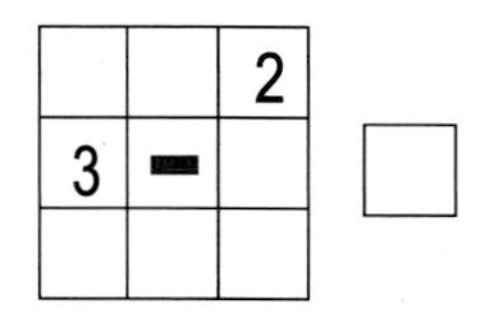


**Fig. B.4** *Drop out the four!*

3. "*Make square the three, so rich thou'lt be.*": The 3 is inserted (at position four, Fig. B.3). With this step, much is already accomplished, as we will soon realize.

4. "*Drop out the four!*": Do not insert the 4 and proceed to the next square (Fig. B.4).

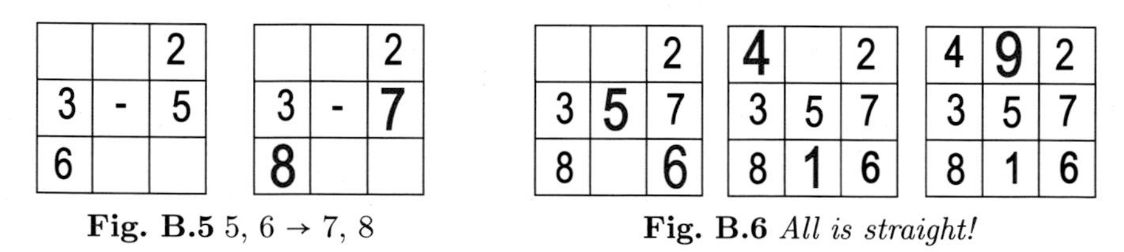


**Fig. B.5** 5, 6 → 7, 8

**Fig. B.6** *All is straight!*

5. "*From five and six, thus says the witch, make seven and eight.*": Now, it is the turn of 5 and 6. In their place, we insert 7 and 8 (Fig. B.5).

6. "*So all is straight!*": This is the key sentence: All is truly accomplished. We can now supplement the sums towards 15 (Fig. B.6)!

7. "*And nine is one, and ten is none, this is the witch's one-time-one!*" We must take this literally: Nine squares are combined into one, omitting

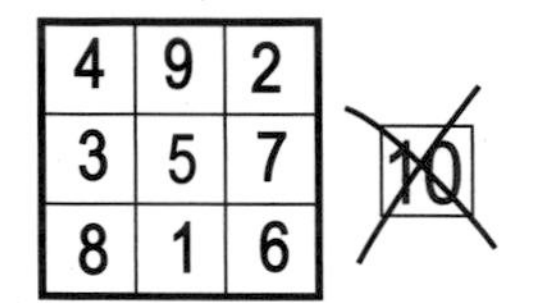


**Fig. B.7** *And nine is one, And ten is none, ...*

square number ten (Fig. B.7). Done! Now, try to repeat the procedure from *Goethe*'s original poem on your own!

◂◂◂

▸▸▸ **Application**: ***Magic squares of order four*** (Fig. B.8, Fig. B.9)
Let us now consider magic squares of order four (total sum $1 + \ldots + 16 = 136$, therefore, the sum of each row/column/diagonal is $136/4 = 34$).

**Fig. B.8** Dürer's square

| 7 | 12 | 1 | 14 |
|---|---|---|---|
| 2 | 13 | 8 | 11 |
| 16 | 3 | 10 | 5 |
| 9 | 6 | 15 | 4 |

| ७ | १२ | १ | १४ |
|---|---|---|---|
| २ | १३ | ८ | ११ |
| १६ | ३ | १० | ५ |
| ९ | ६ | १५ | ४ |

**Fig. B.9** ancient Indian magic square

In the year 1514, Albrecht Dürer showed his discovery in an engraving entitled *Melancholia I*. Remarkably, the bottom row of the engraving reads 4 15 14 1 Not only do the numbers 15 14 appear, but also his initials *D. A.* → 4 1. Another famous magic square of order four can be found in the Parshvanath Jain temple in Khajuraho, India. Fig. B.9 not only shows this square, but it also provides a "translation table" for the ancient Indian numbers engraved in the square. ◂◂◂

▸▸▸ **Application**: ***ancient temples and tables*** (Fig. B.10)
In Fig. B.9, we have an example of a "table" engraved with ancient Indian numbers. The famous Rosetta Stone, found in 1799, is engraved with three versions of the same text translated into ancient Egyptian hieroglyphs, Demotic script, and ancient Greek. It provided the key to the modern understanding of Egyptian hieroglyphs. The example given in Fig. B.10 on the right-hand side shows how numbers were displayed.

*Solution*:
The image shows a list collecting the celebrations donated by Tutmose (Tutmosis) III (18. dynasty, 15th century BC) at the temple of Karnak (rows on

**Fig. B.10** Luxor: temples and tables with numbers

the image's left-hand side) including the sacrifice ratios (small boxes above the squares). The numbers in this table can clearly be recognized: A bar stands for 1, an upside-down U stands for 10, and a curl stands for 100. The donations denoted by these numbers consisted of bread and pastries, or the equivalent amount of cereals required for the baking of these goods. The unit is one *bushel* and the necessary amount of bushels are given in the corresponding boxes. ◂◂◂

▸▸▸ **Application**: ***dates in Denmark***

Time and again, we are confronted with simple number games that yield astonishing results. The mathematical background underlying these games is usually quite simple. We will discuss only one example of such a game here, but this example is sure to astonish the reader:

*Think of a number from 1 to 10. Multiply it by 9. From the result of the digit sum, subtract 5. The result corresponds to a letter in the alphabet. Think of a European country with this initial letter that does not border on Switzerland. Now think of a fruit that begins with the same initial letter. The final question is: What do dates have to do with Denmark?*

*Solution*:

The mathematics behind this is extremely simple: One will realize quickly that the number $9n$ for $n = 1, 2, \cdots, 10$ always has the digit sum 9. So, if you subtract 5, you will get 4 in any case, and the corresponding letter in the alphabet is $D$. In the English language, there is only one European country with the intial letter $D$, namely Denmark. (In German, there are two - Germany and Denmark - but Germany is ruled out because it shares a border with Switzerland. So, here as well, the only possible answer is Denmark). Since there are hardly any fruits with the initial letter $D$ (dragon fruits and dewberries are not that common), it is very likely that the person playing the game will pick the date ... ◂◂◂

## B.2 Rational and irrational numbers

A *rational number* is a fraction $\frac{p}{q}$ where the numerator $p$ and the denominator $q$ are integers. For $q = 1$, one obtains integers. Conversely, a number is *irrational* if it cannot be represented as a fraction.

▸▸▸ **Application**: ***weekday and date***

The following formula goes back to the young *Gauß* and is used for the determination of the weekday $w$ for a given date (century $c$, year $y$ $(0 \le y \le 99)$, day $d$. The month $m$ is given according to the Julian calender: So, March means $m = 1$, February corresponds to $m = 12$:

$$w = (d + [2.6 \cdot m - 0.2] + y + [y/4] - 2c + [c/4]) \bmod 7.$$

Prove the formula.

***Proof***: In contrast to the "Danish date example", this task is not easy. The orbital period of the Earth lasts 365.242198... days, and not 52 weeks or 365 days or $365\frac{1}{4}$ days (as in the Julian year). In order to account for this irrational length of an Earth year, the year must be approximated as closely as possible by means of a fraction. In 1582, the approximation of 365.2425 days (Gregorian year) was introduced, and on October 4th, 1582, 10 days were skipped as a measure to correct the Julian calender. This Gregorian year can now be represented as follows:

$$365.2425 = 52 \cdot 7 + 1 + \frac{1}{4} - \frac{1}{100} + \frac{1}{400}.$$

The expression mod 7 ("modulo 7") refers to the remainder after dividing by 7, i.e. a number from 0 to 6. The corresponding day of the week is then assigned to each modulus (Sunday, Monday, ...). When $d$ is increased by one, $w$ jumps cyclically by one.

The expression $[2.6 \cdot m - 0.2]$ means the following: Take the smallest integer equal or greater than the expression $2.6 \cdot m - 0.2$. How did *Gauß* come to this expression? Consider the $x$-th of March and the $x$-th of April. Since March has 31 days, the corresponding day of the week jumps forward by 3, e.g. from Sunday (=0) to Wednesday (=3). The $x$-th of May jumps by 2 days. In this case, the day falls on a Friday (=5) because April has 30 days. The $x$-th of June is a Monday (=8=1) because May has 31 days, and so on. Until February, the "number of jumps" amounts to 29, which is an average (at 11 jumps) of about 2.6 days per month. Now, you have to retain the somewhat irregular alternation between 30 and 31 days a month (historically evolved), and *Gauß* discovered by trial-and-error that this works with a subtraction of 0.2. However, the result of this always exceeds the "number of jumps" by 2, which is actually quite favorable, because it will yield the correct weekday (otherwise all days would be cyclically shifted).

The year $Y$ is divided by *Gauß* into $c$ hundreds and the remainder $y < 100$ (for example, 2007 is divided into 20 and 0 = 7), and $Y = 100 \cdot c + y$. Let us approximate the year through $52 \cdot 7 + 1 + \frac{1}{4} - \frac{1}{100} + \frac{1}{400}$ days. If $Y$ years have elapsed, the weekday is advanced by the $Y \cdot 365.2425$-fold value. The expression $Y \cdot 52 \cdot 7$ is certainly divisible by 7, and is, therefore, not relevant for the weekday. The following number is important:

$$(100 \cdot c + y) \cdot \left(1 + \frac{1}{4} - \frac{1}{100} + \frac{1}{400}\right) = 100 \cdot c + 25 \cdot c - c + \frac{c}{4} + y + \frac{y}{4} - \frac{y}{100} + \frac{y}{400}.$$

So, we are almost done. The number $124 \cdot c$ has the same remainder when divided by 7 as $-2 \cdot c$ (because 126 is divisible by 7). The summand $-\frac{y}{100} + \frac{y}{400}$ does not provide an integer contribution. ⊙

*Numerical example*: February 16th, 1988 is equated to December 16th, 1987 ($d = 16$, $m = 12$, $c = 19$, $y = 87$) in the Julian calendar. This results in $w = 121 \bmod 7 = 2$ and is, thus, a Tuesday.
September, October, November, and December are, as the Latin names suggest, the seventh, eighth, ninth, and tenth months. ◂◂◂

## The diagonal of the square

As we have seen in this book, the strange irrational numbers $\pi$ (the "circle number"), $e$ (Euler's number), and $\varphi$ (the golden ratio) play a fundamental role in mathematics. It is not so easy to explain why they are fundamental. It has been shown that not only natural numbers and fractions are of essential importance.

▸▸▸ **Application**: *a proof with fatal consequences*

A pupil of *Pythagoras* was able to prove that the ratio of a side $a$ and the diagonal $d = \sqrt{2}\,a$ of a square are not rational. Prove that $\sqrt{2}$ cannot be represented by a fraction.

***Proof***: We prove this indirectly by showing how the contrary statement results in a contradiction.
Suppose there is a fraction

$$\frac{p}{q} = \sqrt{2}.$$

The fraction is "coprime", i.e. already "reduced", so that the greatest common divisor of $p$ and $q$ equals 1. Therefore, it is not possible for both $p$ and $q$ to be even (divisible by 2), because then one could reduce by 2.
If we square the equation, then we have

$$\frac{p^2}{q^2} = 2 \Rightarrow p^2 = 2\,q^2.$$

This means that $p^2$ is divisible by 2, and so is $p$ (the square of an odd number is also odd). Therefore, $q$, and thus also $q^2$, is odd. From this we conclude that $2\,q^2$ is divisible by 2, but not by 4. On the other hand, the square of an even number $p$ is always divisible by 4, and therefore, we have a contradiction to the assumption.
In a completely analogous way, one can prove that every square root of a number which is not a full square must be irrational. ⊙

⊕ *Remark*: This famous proof can be said to be really "classically Greek": a guided verbal series of logical conclusions. *Pythagoras* was apparently very angry about the arrogance of his pupil, because such irrational numbers seemed to have no place in a "rational world-order" – let us remember his scale (Application p. 515), which abounds with rational conditions. However, we want to believe that the alleged murder of his student was not instigated by him. This proof was first laid down by the Greek mathematician and "geometer" *Euclid* (ca. 325–270 BC). *Euclid*'s 13-volume work *The Elements* served as a basic textbook of geometry for more than 2, 000 years! ⊕ ◂◂◂

▸▸▸ **Application**: ***the regular polyhedra and Greek natural philosophy***
A Platonic solid is a regular, convex polyhedron consisting of congruent regular polygonal faces with the same number of faces meeting at each vertex (see also Application p. 231). In Greek natural philosophy, the first four of these polyhedra (tetrahedron, cube, octahedron, and icosahedron) stood for earth, fire, air, and water. The fifth – the later discovered dodecahedron (Fig. B.11) – represented the entire universe (called quintessence = fifth essence).

**Fig. B.11** rolling dodecahedra ... **Fig. B.12** non-convex regular polyhedra

If we do not insist on convexity, there are some other star-shaped polyhedra that fulfil the other conditions, such as the Kepler-Poinsot polyhedra, which are named after their discoverers and are shown in Fig. B.12: The top row shows the great stellated dodecahedron and the great icosahedron. Below, there are the great dodecahedron and the small stellated dodecahedron (the latter had already been discovered earlier by *Paolo Ucello*). These solids were also interpreted mystically. ◂◂◂

## B.3 Famous irrational numbers

Some numbers have enormous significance in mathematics, not only from a historical perspective. Examples of such numbers are $\pi$, Euler's number $e$, and the "golden ratio" $\varphi$. None of these numbers can be represented as a fraction. The first two are even transcendental numbers: They cannot be found as the solution of an algebraic equation.

### The number $\pi$

Since antiquity, it has been known that there is a strange number which can be used to calculate the circumference of the circle or the circular area: the "circle number"

$$\pi = 3.1415926\ldots.$$

More or less good approximations to this number date as far back as antiquity. The ancient Egyptians approximated the number by $(16/9)^2 = 3.1605$ and the Babylonians by $3 + \frac{1}{8} = 3.125$. *Archimedes* proved that the number must lie between $3\frac{1}{7} \approx 3.1428$ and $3\frac{10}{71} \approx 3.1408$.
In 1659, *John Wallis* proved

$$\frac{4}{\pi} = \frac{3\cdot 3\cdot 5\cdot 5\cdot 7\cdot\ldots}{2\cdot 4\cdot 4\cdot 6\cdot 6\cdot 8\cdot\ldots}.$$

In the year 1736, *Euler* proved the formula

$$\frac{\pi^2}{6} = \sum_{k=1}^{\infty}\frac{1}{k^2}.$$

In 1761, *Lambert* showed that $\pi$ is not a fraction, but it was not until 1882 that *Lindemann* succeeded in presenting the sophisticated proof to demonstrate that $\pi$ could never be the solution of an algebraic equation, i.e. $\pi$ is transcendental.

### Euler's number

The transcendental number $e$ appears in calculations of interest rates as follows:
If one has 1 monetary unit (MU) in a year ($1\,a$) with 100% charge in interest, then one obtains 2 MU. At a 50% interest rate per half-year, one receives $\left(1 + \frac{1}{2}\right)^2$ MU. With $\frac{100}{n}$ percent in the period $\frac{1a}{n}$, the value is then $\left(1 + \frac{1}{n}\right)^n$ at the end of the year.
For $n = 1, 2, 3, 4, 5, \ldots$ the obtained values are

$$2,\ 2.25,\ 2.37,\ 2.44,\ 2.49,\ \ldots.$$

If you refine this "infinitely accurately" ($n \to \infty$), you get the strange limit

$$e = \lim_{n\to\infty}\left(1+\frac{1}{n}\right)^n = 2.71828\ldots.$$

The function $e^x$ is known to agree with its derivative. It is, with the exception of a multiplicative factor, the only function that behaves like this. This can be meaningfully interpreted as: The increase in the function $e^x$ is as large as the corresponding function value at each point, and this has to do with natural growth. The offspring of a population is, indeed, proportional to its current size.

Finally, the function $e^x$ has the simple *Taylor* expansion

$$e^x = \sum_{k=0}^{\infty}\frac{x^k}{k!} = 1 + \frac{x}{1} + \frac{x^2}{1\cdot 2} + \frac{x^3}{1\cdot 2\cdot 3} + \ldots,$$

which makes the base $e$, in the age of the computer, an ideal reference number for all exponential functions.

## The strange golden ratio

The two solutions of the quadratic equation

$$\frac{1}{a} = \frac{a}{a+1}$$

are the numbers

$$a_1 = \frac{1+\sqrt{5}}{2} \quad \text{and} \quad a_2 = \frac{1-\sqrt{5}}{2} \tag{B.1}$$

(Application p. 37). In writing, the two numbers are often denoted by $\Phi$ and $\varphi$. They possess a wealth of interesting properties.

**▸▸▸ Application**: ***a "maximally irrational" number***

Since the two numbers $\Phi$ and $\varphi$ (Formula (B.1)) contain the irrational number $\sqrt{5}$, they are irrational (Application p. 525). Therefore,

$$\Phi = a_1 = 1.6180339887\ldots \quad \text{and} \quad \varphi = a_2 = -0.6180339887\ldots$$

have no periodicity, even if you extend the number of decimal places arbitrarily.[3]

Show the simple relations

$$(1)\quad \frac{1}{\Phi} = \Phi - 1, \qquad (2)\quad \frac{1}{\Phi} = -\varphi,$$

and thus,

[3] There are websites on which $\Phi$, $e$, and $\pi$ are listed with an accuracy of several thousands of decimal digits.

$$(3) \quad \Phi = 1 + \cfrac{1}{1 + \cfrac{1}{1 + \cfrac{1}{1 + \cfrac{1}{1 + \ldots}}}}, \qquad (4) \quad \varphi = -\cfrac{1}{1 + \cfrac{1}{1 + \cfrac{1}{1 + \cfrac{1}{1 + \ldots}}}}.$$

The two numbers can be represented by "infinite continued fractions".[4]

***Proof***: Ad (1):

$$\frac{1}{\Phi} = \frac{2}{\sqrt{5}+1} = \frac{2(\sqrt{5}-1)}{(\sqrt{5}+1)(\sqrt{5}-1)} = \frac{2(\sqrt{5}-1)}{5-1} =$$

$$= \frac{\sqrt{5}-1}{2} = \frac{\sqrt{5}+1-2}{2} = \frac{\sqrt{5}+1}{2} - 1 = \Phi - 1.$$

Ad (2):

$$\frac{1}{\Phi} = \frac{2}{\sqrt{5}+1} = \frac{2(\sqrt{5}-1)}{(\sqrt{5}+1)(\sqrt{5}-1)} = \frac{2(\sqrt{5}-1)}{5-1} = \frac{\sqrt{5}-1}{2} = -\varphi.$$

Ad (3) and (4): The number $\Phi$ is a solution of the equation

$$\frac{1}{\Phi} = \frac{\Phi}{\Phi+1}.$$

On the right-hand side, we divide by $\Phi$ and get the relation

$$\frac{1}{\Phi} = \frac{1}{1+\frac{1}{\Phi}}.$$

The left-hand side is inserted into the right-hand side. Thus, we replace $\frac{1}{\Phi}$ in the right-hand side with $\frac{1}{1+\frac{1}{\Phi}}$ and obtain

$$\frac{1}{\Phi} = \frac{1}{1+\frac{1}{1+\frac{1}{\Phi}}}.$$

Repeated insertion results in a "continuous fraction". The equations (3) and (4), therefore, follow from (1) and (2). ⊙ ◀◀◀

[4]It can be shown that the golden number is irrational in the "maximal" sense so that the worst way to approximate it with a given denominator is by means of a rational number. This is due to the fact that the elements of the continued fraction (the numbers that appear first in the denominator) are all one.

## B.4 The Fibonacci numbers

"A rabbit couple gives birth to a young couple every month from the second month onward. The young couple, in turn, gives birth to another baby couple on a monthly basis from the second month onward. How many rabbits are there after $n$ months if we had one young couple at the beginning?"
This was the question posed by *Leonardo da Pisa* (ca. 1170 – 1240), also known as *Fibonacci*, who gave a first mathematical model for natural growth. Centuries later, *Euler* gave recursion formulas for these rabbit numbers:

$$k_n = k_{n-1} + k_{n-2} \quad \text{with} \quad k_1 = k_2 = 1. \tag{B.2}$$

**Fig. B.13** flowers with 3, usually 3, and 13 petals

The sequence is, thus, $1, 1, 2, 3, 5, 8, 13, 21, 34, 55, \ldots$.
Although this model is very simple and ignores many additional factors, it is astonishing how often we encounter *Fibonacci* numbers, especially in the world of plants: The number of petals is very often a *Fibonacci* number (Fig. B.13).
Roses often have 34 or 55 petals (Fig. B.14).

**Fig. B.14** Only verifiable after dismantling: 55 petals!

In the case of sunflowers, daisies, and pine cones, the seeds are arranged spirally in such a way that when the seeds on each spiral are counted, we will always get a Fibonacci number (Fig. B.15).

**Fig. B.15** Fibonacci numbers in the seed arrangement

It is easy to prove that the quotient of two successive Fibonacci numbers converges to the golden proportion $\Phi$ (Formula (B.1)):

$$q = \lim_{n\to\infty} \frac{k_n}{k_{n-1}} = \lim_{n\to\infty} \frac{k_{n-1} + k_{n-2}}{k_{n-1}} = 1 + \lim_{n\to\infty} \frac{k_{n-2}}{k_{n-1}} = 1 + \frac{1}{q} \Rightarrow q = \Phi.$$

In this respect, the golden proportion also holds in botany ...

## A mathematical model

*Phyllotaxis* is now understood as a mathematical model for the spiral-shaped systems seen in pine trees, sun-flowers, and other plants.

A plant's seedbed can grow in the following ways: In order to make room for a new seed, the previous seed structure is simply rotated. To avoid collision with the first seed structure after a full rotation, the radial distance between the two is increased. A smaller magnification gives a more efficient space distribution.

Let us now take the following approach: We set a point $P_0(d, 0)$ near the origin of the coordinate system. Each additional point $P_n$ arises from its predecessor by being rotated around the origin by a fixed multiple $\delta$ of the full angle. At the same time, its distance from the origin is increased by the factor $(1 + \frac{1}{n})^c$. Depending on the efficiency of the algorithm, the constant $c$ can be kept as small as possible (in Fig. B.16, $c = 0.58$). Therefore, $P_n$ has the polar coordinates $\big(d(1 + \frac{1}{n})^c; n\delta \cdot 360°\big)$.

The seeds naturally have a certain size, which we want to illustrate with a small circular disk. If we test the given rule with a computer, we soon realize that the angle of rotation should not be a simple fractional amount of the full angle, because then, the seeds are concentrated on radial lines and the plenum remains free. The factor $\delta$ should, if possible, not be a fraction: So, we take the "maximal irrational number", the golden proportion $\delta = \Phi$.

We divide the full angle in the golden ratio and get the angle

$$\gamma = \Phi \cdot 360° \approx 582.4922°.$$

For $c$, a value around 0.6 is found to be meaningful, depending on the seed size. Minor changes of $\gamma$ result in entirely different arrangements (Fig. B.16). This simulation is very close to reality.

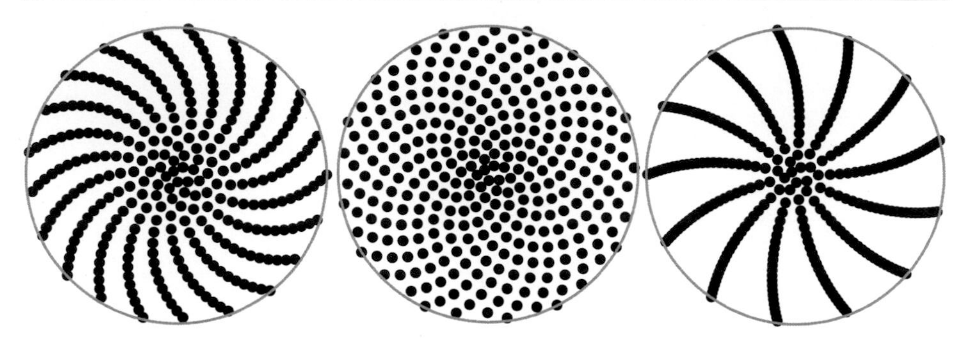

**Fig. B.16** arrangement of 300 elements at $\gamma - 0.25°$, $\gamma$, and $\gamma + 1°$

**Fig. B.17** "Real" sunflowers turn with their "back" to the Sun!

This does not mean that nature has genetically prescribed the abovementioned algorithm for plants and "knows" the golden proportion to 10 decimal places. As already stated in the introduction, mathematics is a self-contained microcosm which, however, has the strong ability to reflect and model arbitrary processes.

The quality of mathematical models depends, on the one hand, on the creativity of the person who has made it and, on the other hand, on the fine-tuning of the model's individual components. In any case, models should be tested on as many examples as possible.

Furthermore, we should always keep in mind that nature produces *inaccurate copies.* So, it can never be predicted to the last detail. This is the *secret of evolution*, without which there would be no adaptation to change (see also Application p. 499).

## B.5 Imaginary and complex numbers

In Section 1.5, we discussed the fundamental theorem of algebra (p. 41) which states that any algebraic equation of degree $n$ with complex coefficients has exactly $n$ complex solutions, if we take into account their multiplicities. The complex numbers play a very important role in many areas of graduate mathematics (and also in theoretical physics and kinematics).
What are "complex numbers" really?
To answer the question, we need the concept of an "imaginary unit" $i$. It is defined by

$$i^2 = -1. \tag{B.3}$$

⊕ *Remark*: The conclusion

$$i = \sqrt{-1} \tag{B.4}$$

drawn from this is to be used with some caution, as one can see from the following calculation: With

$$i^2 = i \cdot i = \sqrt{-1}\,\sqrt{-1} = \sqrt{(-1)\,(-1)} = \sqrt{1} = 1,$$

one immediately has a contradiction to the usual arithmetic rules in the real number field. ⊕

Now, a new set of numbers – the *imaginary numbers* – is defined. These numbers emerge from the real numbers by multiplying each element by $i$ (analogously to the definition of negative integers where the natural numbers are multiplied by $-1$). With real and imaginary numbers, one can solve all purely quadratic equations, e.g.

$$x^2 = -16 = 16 \cdot (-1) \Rightarrow x = \pm\sqrt{16}\,i = \pm 4\,i.$$

By definition, there are as many real numbers as there are imaginary numbers. Now, the real and imaginary numbers can be combined to a new type of number:

$$z = x + i\,y. \tag{B.5}$$

Here, $x$ and $y$ are real, and hence, $i\,y$ is imaginary. A complex number is defined as the *sum of a real number and an imaginary number*. The values $x$ and $y$ are respectively called the "real part" and the "imaginary part" of the complex number $z$. Symbolically, we write $z \in \mathbb{C}$ if a number $z$ is complex. The validity of this new definition can be shown, for example, by solving the following quadratic equation with Formula (2.14):

$$z^2 + 2z + 2 = 0 \Rightarrow z_{1,2} = \frac{-2 \pm \sqrt{2^2 - 4 \cdot 1 \cdot 2}}{2} = \frac{-2 \pm \sqrt{4}\,i}{2} = -1 \pm i. \tag{B.6}$$

## Conjugating complex numbers

The solutions $z_1 = -1 - i$ and $z_2 = -1 + i$ in (B.6) differ only in the sign of the imaginary part. In this case, one speaks of "complex conjugate numbers". Symbolically, one writes $z_2 = \overline{z}_1$. Note that $\overline{z_2} = \overline{\overline{z_1}} = z_1$. In general, for complex conjugate numbers $z$, $\overline{z}$, we have

$$z = x + i\,y \Rightarrow \overline{z} = x - i\,y.$$

**Fig. B.18** elliptic, hyperbolic, and parabolic pencils of circles

▸▸▸ **Application**: ***pencil of circles through complex conjugate points***
The solutions sought in (B.6) are the zeros of the function $y = x^2 + 2x + 2$ (parabola), or more generally, the zeros of all parabolas from the pencil of parabolas with the equations $y = a(x^2+2x+2)$ with $a \in \mathbb{R}$ (Fig. B.18, left). It is easy to imagine that all the circles with the equation $(x+1)^2+(y-q)^2 = q^2-1$ ($q \in \mathbb{R}$, $q \geq 1$) also have the same zeros.
No parabolas or circles of these pencils have any real points in common, but they intersect at a pair of conjugate imaginary points. One can compute with such points in the same way as with real points.
Fig. B.18 (right) illustrates a *hyperbolic pencil of circles* (where the circles share the *zero circles* of the elliptic pencil) together with the elliptic pencil. The two pencils constitute an orthogonal grid of circles: Each circle of either family intersects each circle from the complimentary pencil at right angles. Fig. B.19 shows how a hyperbolic pencil of circles can give rise to a hyperbolic pencil of spheres in 3-space. ◂◂◂

▸▸▸ **Application**: ***the absolute circle points***
With the help of complex numbers, one can prove the following curious statement: All circles in the plane have two points in common. These are complex conjugate points on the line "at infinity".
***Proof***: An arbitrary circle (center $(p/q)$, radius $r$) has the equation

$$(x-p)^2 + (y-q)^2 = r^2 \Rightarrow y = q \pm \sqrt{r^2 - (x-p)^2}.$$

**Fig. B.19** A hyperbolic pencil of circles gives rise to a hyperbolic pencil of spheres by simply revolving it about its axis.

For us, the quotient

$$\frac{y}{x} = \frac{q}{x} \pm \sqrt{\left(\frac{r}{x}\right)^2 - \left(1 - \left(\frac{p}{x}\right)\right)^2}$$

is of interest. It converges to the absolute values $\pm i$ as $x \to \infty$ because the expressions $\frac{p}{x}$, $\frac{q}{x}$, and $\frac{r}{x}$ converge to zero independently of $p$, $q$, and $r$. This means that *all* circles containing those points are defined by the complex conjugate directions $y = \pm i\,x$. ⊙

⊕ *Remark*: By means of the absolute circle points, one can explain why two circles have a maximum of two real intersections in common, although – as with all other curves of degree 2 – a maximum of four can be expected: The two absolute circle points are always intersections, albeit complex conjugates and "at infinity". In an elliptic pencil of circles, the remaining two intersections are real points whereas in a hyperbolic pencil the common points of all circles are a pair of complex conjugate points. If we calculate the distance $d$ of the points $(x/\pm i\,x)$ from the origin, the value $d = \sqrt{x^2 + (\pm i\,x)^2} = x\sqrt{1-1}$, i.e. zero, is obtained. Therefore, the two directions $y : x = \pm i$ are also called *minimal lines*. Even the term "at infinity" should be handled with care, because the limit value for $x \to \infty$ is undetermined. ⊕ ◂◂◂

## Basic arithmetic operations with complex numbers

Addition and subtraction of complex numbers result "informally" by adding or subtracting the real and imaginary parts. However, multiplication uses the distributive law and the definition $i^2 = -1$ of the imaginary unit. The arithmetic mean of two complex conjugate numbers equals

$$\frac{z+\overline{z}}{2} = \frac{(x+i\,y)+(x-i\,y)}{2} = x,$$

which is the real part of $z$. The difference between two complex conjugate numbers is purely imaginary; the product is real:

$$z\,\overline{z} = (x+i\,y)\,(x-i\,y) = x^2 - i^2\,y^2 = x^2 + y^2. \tag{B.7}$$

With Formula (B.7), you can calculate the reciprocal value of a complex number:

$$\frac{1}{z} = \frac{\overline{z}}{z\,\overline{z}} = \frac{1}{x^2+y^2}\,\overline{z}. \tag{B.8}$$

The division of complex numbers is treated in a similar way: Expand numerator and denominator by the complex conjugate of the denominator.

## The complex plane

Obviously, there are substantially more complex numbers than real numbers (or imaginary numbers):[5] You can choose the real part $x$ and the imaginary part $y$ freely. The best way to imagine the situation is geometrically. If one interprets $x$ and $y$ as the coordinates of a point $P(x/y)$ in the plane:
*There are as many complex numbers as there are points in the plane.*
In this regard, one speaks of the *complex plane*. Conjugate numbers are symmetric with respect to the "real axis" (abscissa). The distance of the point $(x/y)$ (which corresponds to the complex number $z = x + i\,y$) from the origin is given by $\sqrt{x^2+y^2}$, and this is equal to the expression

$$|z| = \sqrt{z\,\overline{z}}. \tag{B.9}$$

Analogous to vector calculus, the real number $|z|$ is called the *absolute value* of the complex number $z$. Complex numbers – interpreted in the complex plane – and two-dimensional vectors generally have something in common. The sum and difference of two complex numbers can geometrically be interpreted as vector addition or vector subtraction. That is to say, one can also perform elegant translations with complex numbers in the plane.
According to Formula (B.8), the reciprocal of a complex number $z$ is the reflection of $z$ in the real axis combined with a dilation from the origin with the factor $1/(z\cdot\overline{z})$.

[5] In the sense that complex numbers form a two-dimensional continuum. In set-theoretic terms, the sets of complex numbers and real numbers contain the same (infinite) number of elements.

**▸▸▸ Application**: ***straight lines in the complex plane***
Show that

$$\overline{m}\,z + m\,\overline{z} = n \quad (m \in \mathbb{C},\ n \in \mathbb{R}) \tag{B.10}$$

is the general equation of a straight line in the complex plane, where $m$ is the normal direction.
***Proof***: We set $m = p + i\,q$ and $z = x + i\,y$. Then, we have

$$(p - i\,q)(x + i\,y) + (p + i\,q)(x - i\,y) = n.$$

When multiplying, only

$$2p\,x + 2q\,y = n$$

remains, i.e. a linear equation in $x$ and $y$. The normal vector of the straight line is a multiple of $(p, q)$. ⨀ ◂◂◂

**▸▸▸ Application**: ***circles in the complex plane***
Show that

$$z\,\overline{z} - \overline{m}\,z - m\,\overline{z} + n = 0 \quad (m \in \mathbb{C},\ n = m\,\overline{m} - r^2 \in \mathbb{R}) \tag{B.11}$$

is the general equation of a circle (center $m$) in the complex plane.
***Proof***: An arbitrary circle with center $m$ and radius $r$ can be described by

$$|z - m| = r \Rightarrow (z - m)(\overline{z} - \overline{m}) = r^2 \Rightarrow z\overline{z} - \overline{m}z - m\overline{z} + \underbrace{m\overline{m} - r^2}_{n\in\mathbb{R}} = 0.$$

For $m\overline{m} = r^2$, we get the circle through the origin $z = 0$. Conversely, each curve of the type

$$z\overline{z} - \overline{m}z - m\overline{z} + n = 0 \tag{B.12}$$

is a circle with center $m$ and radius $r = \sqrt{m\,\overline{m} - n}$. ⨀ ◂◂◂

**▸▸▸ Application**: ***"reflection in the circle" (inversion)***
Consider the image $w = \frac{1}{\overline{z}}$ (reflection in the unit circle) and show that it "preserves circles", i.e. circles are transformed into circles. Especially, circles through the point $z = 0$ are transformed into "infinitely large circles", i.e. straight lines.
***Proof***: We consider an arbitrary circle given by Formula (B.11)

$$z\,\overline{z} - \overline{m}\,z - m\,\overline{z} + n = 0 \quad (m \in \mathbb{C},\ n = m\,\overline{m} - r^2 \in \mathbb{R}).$$

According to the rules

$$w = \frac{1}{\overline{z}} \Rightarrow \overline{z} = \frac{1}{w} \Rightarrow z = \frac{1}{\overline{w}},$$

we obtain

$$\frac{1}{\overline{w}}\frac{1}{w} - \overline{m}\frac{1}{\overline{w}} - m\frac{1}{w} + n = 0.$$

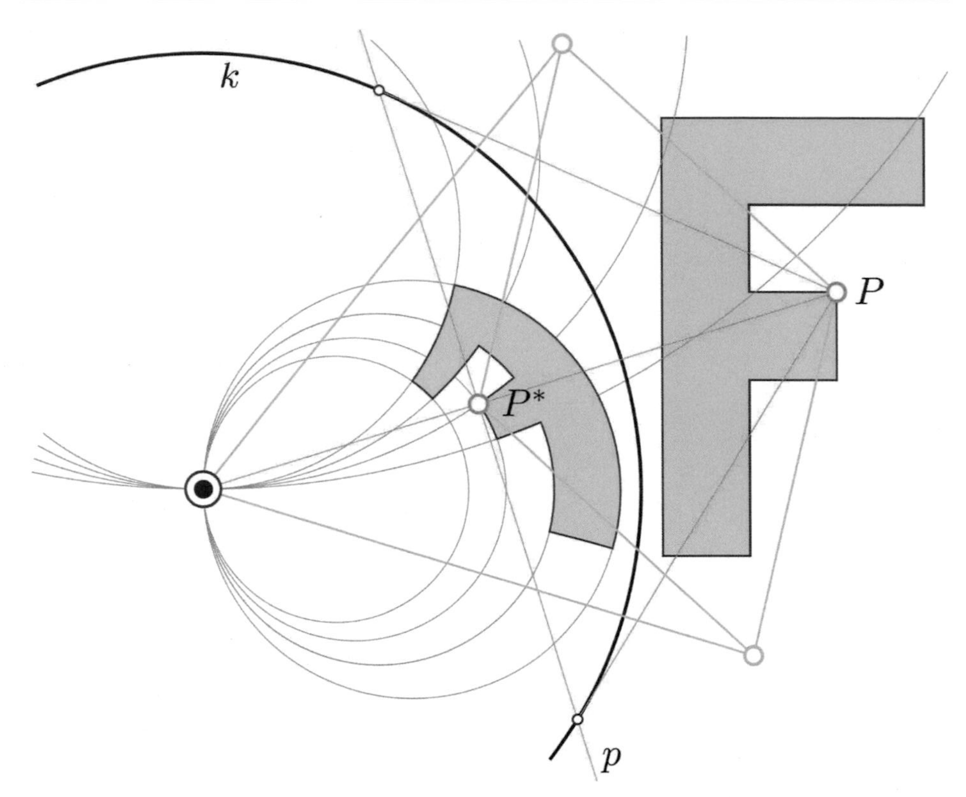


**Fig. B.20** reflection in the circle (inversion)

We multiply by $w\overline{w}$, and with

$$1 - \overline{m}\,w - m\,\overline{w} + cn\,w\overline{w} = 0,$$

we obtain once again an equation of the same form as shown in Formula (B.12). Therefore, the transformed circle is again a circle. The exception is $n = 0$ (then, the circle passes through $z = 0$): Here the quadratic part vanishes, and according to Application p. 537, the result is a straight line with the equation

$$\overline{m}\,w + m\,\overline{w} = 1$$

perpendicular to the connection of the circle center $m$ and the origin of the coordinate system. ⊙

⊕ *Remark*: This special case is utilized in the theory of wheel gears for so-called "inversors" with which, for example, a technically easy-to-implement rotation can be converted into exactly a straight line without the need for a straight directrix. ⊕

◂◂◂

▸▸▸ **Application**: ***inversion in space*** (Fig. B.21)

The inversion can be performed in three-dimensional space equally well. The inversion circle is replaced by an inversion sphere $(M; r)$. The points $P$ and $P^*$ are, once again, linked via the relation $\overline{MP} \cdot \overline{MP^*} = r^2$.

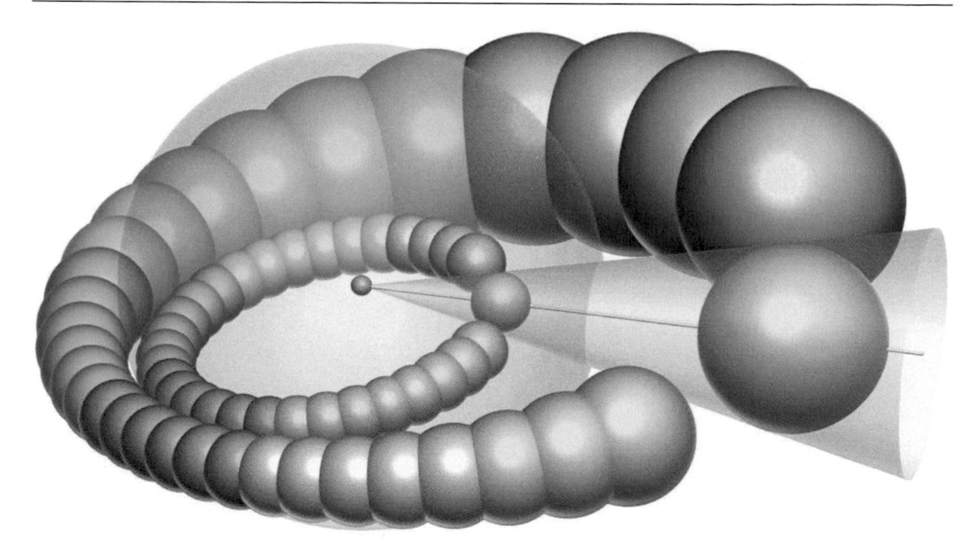

**Fig. B.21** The inversion in a sphere "is sphere preserving". Here, a torus (as an envelope of spheres) is transformed into a so-called Dupin cyclide (also of degree four). It inherits a multitude of beautiful properties from the torus.

The transformation $P \mapsto P^*$ is quadratic: Planes are, thus, mapped to spheres (in general). One can prove this by rotating the planar situated straight line (→ circle) about the normal of the straight line through the center $M$. Surfaces of $n$-th order are in general transformed into surfaces of $2n$-th order. Planes are also considered as spheres with "infinitely large radius" and "center at infinity". Spheres and planes are transformed into spheres or planes, i.e., "inversions preserve spheres"). The example in Fig. B.21 shows that surfaces created by spheres are transformed into surfaces of the same order. By analogy to circle reflection, we may, at this point, speak of reflection in a sphere. ◂◂◂

The following example and also Application p. 543 show that under certain circumstances, sophisticated formulas can be represented in a compact form by means of complex numbers. Using appropriate software, computers can calculate with complex numbers as well as real numbers. The effort for the user of such software is, thereby, enormously reduced.

▸▸▸ **Application**: ***equipotential surface***

If $z_1$ and $z_2$ are two arbitrary complex numbers, then they can be used to define with the equation

$$w(z) = \left|\frac{z - z_1}{z - z_2}\right| \quad (w \in \mathbb{R})$$

the surface depicted in Fig. B.22 (the surface has been trimmed close to the poles). Show that for $z_1 = -1 + i$ and $z_2 = -1 - i$ above the real axis, a horizontal line lies on the surface.

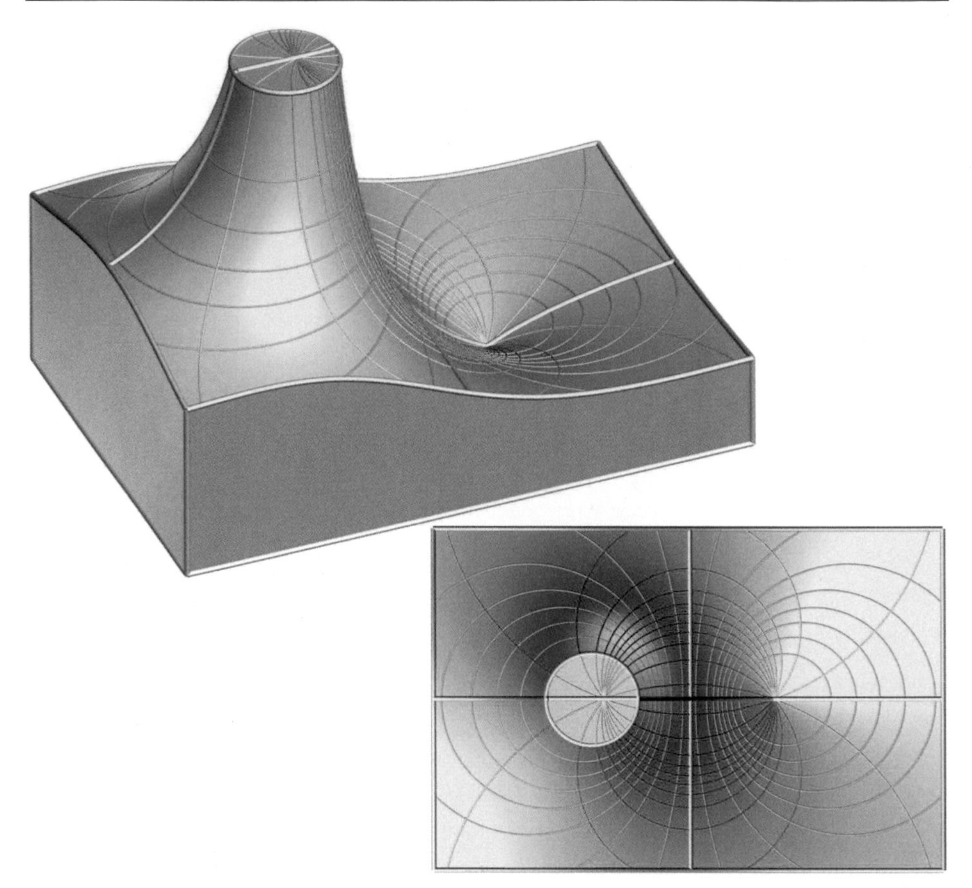

**Fig. B.22** potential surface with circular contour lines

*Solution*:
The real axis is described by $z = t$ $(t \in \mathbb{R})$. Thus,

$$w(t) = \left|\frac{t - z_1}{t - z_2}\right| = \left|\frac{t - (-1 + i)}{t - (-1 - i)}\right| = \left|\frac{t + 1 - i}{t + 1 + i}\right|.$$

If we let $u = t + 1$, we further have

$$w(t) = \left|\frac{u - i}{u + i}\right| = \left|\frac{(u - i)(u - i)}{(u + i)(u - i)}\right| =$$

$$= \frac{|u^2 - 1 - 2u\,i|}{u^2 + 1} = \frac{\sqrt{(u^2 - 1)^2 + 4u^2}}{u^2 + 1} = \frac{\sqrt{(u^2 + 1)^2}}{u^2 + 1} = 1.$$

Above the real axis, the surface points have a constant height of 1 so that the corresponding curve on the surface is actually a straight line.

⊕ *Remark*: It is a little bit more tedious to show that the horizontal curves on the surface are mapped to a hyperbolic pencil of circles in the top view (Fig. B.22, left). These correspond exactly to the hyperbolic pencil shown in (Fig. B.18, left). The corresponding elliptic pencil (Fig. B.18, left) is the image of the lines of greatest slope on the surface. In physics, $w$ can be interpreted as the potential of two electric fields. The horizontal curves are, then, called the equipotential curves, and the curves of greatest slope are called field lines. ⊕ ◂◂◂

## The function $e^z$ with complex numbers

If we extend the real function $e^x$ to complex numbers by means of its power series expansion in Formula (8.5), then we obtain

$$e^z = 1 + z + \frac{z^2}{2!} + \frac{z^3}{3!} + \frac{z^4}{4!} + \ldots \quad (z \in \mathbb{C}). \tag{B.13}$$

### ▸▸▸ Application: *Euler's formula*

The following famous relation between real trigonometric functions and the exponential function is due to *Euler*. Show that

$$e^{i\varphi} = \cos\varphi + i\,\sin\varphi \tag{B.14}$$

or, equivalently,

$$\cos\varphi = \frac{e^{i\varphi} + e^{-i\varphi}}{2}, \quad \sin\varphi = \frac{e^{i\varphi} - e^{-i\varphi}}{2i} \qquad (\varphi \in \mathbb{R}).$$

***Proof***: According to our formulas developed for real arguments, we set $\varphi \in \mathbb{R}$ (Application p. 382 ff.) and get

$$e^{\varphi} = 1 + \varphi + \frac{\varphi^2}{2!} + \frac{\varphi^3}{3!} + \frac{\varphi^4}{4!} + \ldots,$$

$$\sin\varphi = \varphi - \frac{\varphi^3}{3!} + \frac{\varphi^5}{5!} - \frac{\varphi^7}{7!} + \ldots,$$

$$\cos\varphi = 1 - \frac{\varphi^2}{2!} + \frac{\varphi^4}{4!} - \frac{\varphi^6}{6!} + \ldots.$$

The extension to complex numbers holds for pure imaginary numbers $i\,\varphi$ $(\varphi \in \mathbb{R})$:

$$e^{i\,\varphi} = 1 + i\,\varphi + i^2\frac{\varphi^2}{2!} + i^3\frac{\varphi^3}{3!} + i^4\frac{\varphi^4}{4!} + \ldots = 1 + i\,\varphi - \frac{\varphi^2}{2!} - i\frac{\varphi^3}{3!} + \frac{\varphi^4}{4!} + \ldots.$$

The above assertions can now be verified directly with the use of series expansion. ⊙

◂◂◂

## Geometric interpretation of the multiplication

The relation between the complex function $e^z$ and the real functions $\sin\varphi$ and $\cos\varphi$ leads to a very elegant means of calculation in the complex plane. For this, we first change from Cartesian coordinates to polar coordinates. It is obvious that

$$r = |z| = \sqrt{x^2 + y^2}, \ \tan\varphi = \frac{y}{x}. \tag{B.15}$$

The real and imaginary part can be computed from polar coordinates by

$$z = r\,(\cos\varphi + i\,\sin\varphi). \tag{B.16}$$

Together with Formula (B.14), we have the elegant representation

$$z = r\,e^{i\,\varphi}. \tag{B.17}$$

For $r = 1$ and $\varphi = \pi/2$, we get the unit $i$; hence, for $r = 1$ and $\varphi = \pi$, we get Euler's identity:

$$\boxed{e^{i\,\pi/2} = i, \quad e^{i\,\pi} = -1.} \tag{B.18}$$

If we apply the calculation rules for power functions to the multiplication of two complex numbers $z_1$ and $z_2$, we get

$$z_1 \cdot z_2 = r_1\,e^{i\,\varphi_1} \cdot r_2\,e^{i\,\varphi_2} = r_1 r_2\,e^{i(\varphi_1+\varphi_2)}. \tag{B.19}$$

We see that the argument (polar angle $\varphi_1 + \varphi_2$) of the product $z_1 \cdot z_2$ of the two complex numbers $z_1$ and $z_2$ is the sum of the arguments (polar angles $\varphi_1$ and $\varphi_2$) of the two complex numbers.
Geometrically, multiplication by a complex number $z = r\,e^{i\varphi}$ can be interpreted as the combination of a *rotation* (roation angle $\varphi$) with a dilation (scaling factor $r$). In fact, multiplication by the number $z = e^{i\varphi}$ is a pure rotation. This is the reason why complex numbers are extensively used in planar kinematics, in which, besides translations, rotations play an important role.

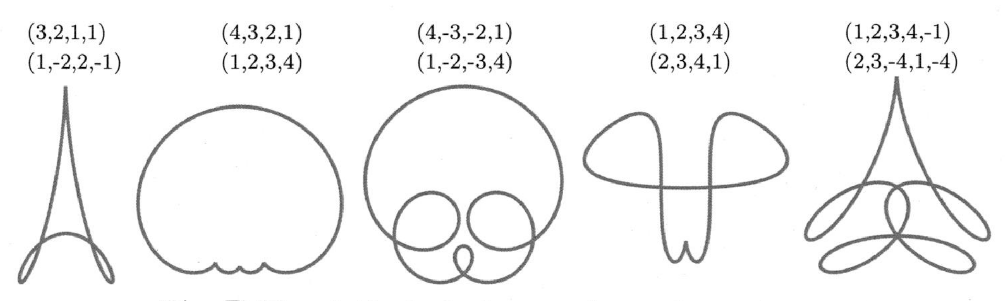


**Fig. B.23** trochoids of order 4: upper values: $r_k$, lower values: $\omega_k$

▸▸▸ **Application**: ***trochoids*** (Application p. 300)
A trochoid is generated by rotating a first rod (length $r_1$) at a constant angular velocity $\omega_1$ while a second rod (length $r_2$) rotates about the end point of the first at the proportional angular velocity $\omega_2$. What is the equation of such a curve in the complex plane?

*Solution*:
Assume that $t$ is the elapsed time. Then, the center of the second rotation has the position $z_1 = r_1\, e^{i\omega_1 t}$. The end point of the second rod has the position

$$z = z_1 + r_2\, e^{i\omega_2 t} = r_1\, e^{i\omega_1 t} + r_2\, e^{i\omega_2 t}.$$

The superposition of rotations can easily be generalized, and this will result in so-called "trochoids of order $n$", which can be represented in the complex plane as

$$z = \sum_{k=1}^{n} r_k\, e^{i\,\omega_k\, t}.$$

It can be shown that any planar curve can be approximated in parts to an arbitrary precision by a trochoid of sufficiently high order. ◂◂◂

## Powers and roots of complex numbers

The representation of complex numbers by Formula (B.17) also allows us to compute powers and roots of complex numbers: If one multiplies a number $z$ by itself, the radius must be squared and the polar angle must be doubled. If one takes the number to the power $n$ (i.e. multiplying it $n$ times with itself), one has

$$z^n = (r(\cos\varphi + i\sin\varphi))^n = r^n(\cos n\varphi + i\sin n\varphi).$$

The exponent $n$ may also take non-integer values, and the calculation rules for powers (and for roots) can be sensibly carried over to complex numbers. The coefficients of a quadratic equation, for example, can also be complex. Then, the complex solutions following Formula (2.14) are generally no longer complex conjugate.

### ▸▸▸ Application: *"Fermat surfaces"*

A famous theorem of *Fermat*, which has only recently been proved in a sophisticated way, states that the expression $a^n + b^n = c^n$ has positive integer solutions $(a, b, c)$ only for $n \le 2$ (Application p. 121). Thus, for $n > 2$, we can safely divide by $c^n$ (without omitting integer solutions) and obtain $w^n + z^n = 1 \Rightarrow w(z) = \sqrt[n]{1 - z^n}$ (with $w = a/c$ and $z = b/c$, ignoring the sign). If we also allow complex numbers $z$, then of course, the result $w$ is also complex. By forming the absolute value, we make the result real (as we did before, in the case of the equipotential surface) and consider the surface

$$w(z) = |\sqrt[n]{1 - z^n}|$$

over the complex plane. Fig. B.24 shows two extreme views of the surface for $n = 5$ (for a "common view" see p. 519). Calculate its zeros.

*Solution*:
The equation $w = |\sqrt[5]{1 - z^5}| = 0$ is to be solved. From $|w| = 0$, it follows unambiguously that $w = 0 + 0\,i = 0$, so that we obtain $\sqrt[5]{1 - z^5} = 0 \Rightarrow 1 - z^5 =$

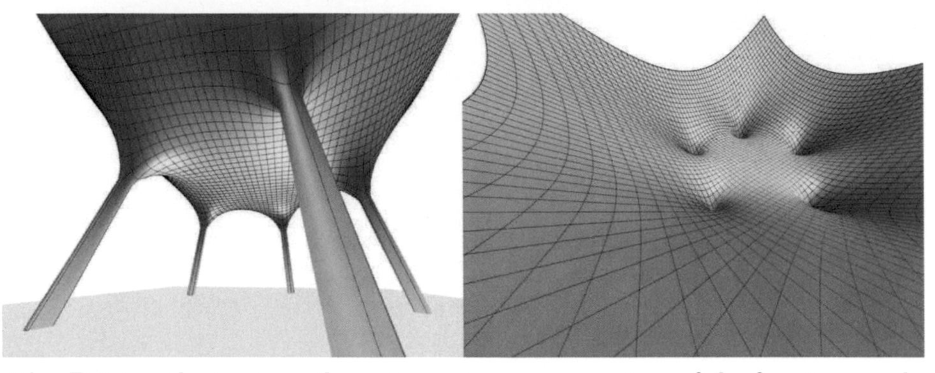

**Fig. B.24** aesthetics in mathematics: extreme perspectives of the function graph

$0 \Rightarrow z^5 = 1$. Transformations of equations in complex numbers have to be done very carefully, especially when absolute values are involved.
In polar coordinates, we have $1 = (1; 2\pi k)$ with an arbitrary integer $k$. The fifth roots are, thus, $(\sqrt[5]{1}; \frac{2\pi}{5} k)$ with $k = 1, \ldots, 5$, and they form the vertices of a regular pentagon (Fig. B.24) with the coordinates $(1/0)$, $(\cos 72°/\pm \sin 72°)$, $(\cos 144°/\pm\sin 144°)$. ◂◂◂

▸▸▸ **Application**: ***formulas for*** $\sin z$ ***and*** $\cos z$ ***for complex numbers***
Let us define the complex sine function by allowing complex arguments in Formula (B.14):

$$\sin z = \frac{e^{iz} - e^{-iz}}{2i}.$$

With $z = x + i\,y$ $(x, y \in \mathbb{R})$, we have $i\,z = -y + i\,x$, $-i\,z = y - i\,x$, and then,

$$e^{iz} = e^{-y+i\,x} = e^{-y}e^{i\,x} = e^{-y}(\cos x + i \sin x),$$

$$e^{-iz} = e^{y-i\,x} = e^{y}e^{-i\,x} = e^{y}(\cos x - i \sin x).$$

Therefore, we have

$$\sin z = \frac{e^{-y}(\cos x + i\sin x) - e^{y}(\cos x - i \sin x)}{2i} =$$

$$= \underbrace{\frac{1}{i}}_{-i} \frac{e^{-y} - e^{y}}{2} \cos x + \frac{e^{-y} + e^{y}}{2} \sin x,$$

and with Formula (6.27), the relation

$$\sin z = \sin x \cosh y + i \cos x \sinh y. \tag{B.20}$$

Analogously, for the complex cosine function, we have

$$\cos z = \frac{e^{-y}(\cos x + i\sin x) + e^{y}(\cos x - i \sin x)}{2} = \frac{e^{-y} + e^{y}}{2}\cos x - i\frac{e^{y} - e^{-y}}{2}\sin x$$

$$\Rightarrow \cos z = \cos x \cosh y - i \sin x \sinh y. \tag{B.21}$$

◂◂◂

## Conformal mappings

A conformal mapping in the plane is an angle preserving transformation of a domain to another domain. Under this mapping, orientations of planar figures and the angle of intersection of two arbitrary curves are also preserved. Such mappings occur, for example, in electrical engineering, geodesy, hydrodynamics, and aerodynamics.

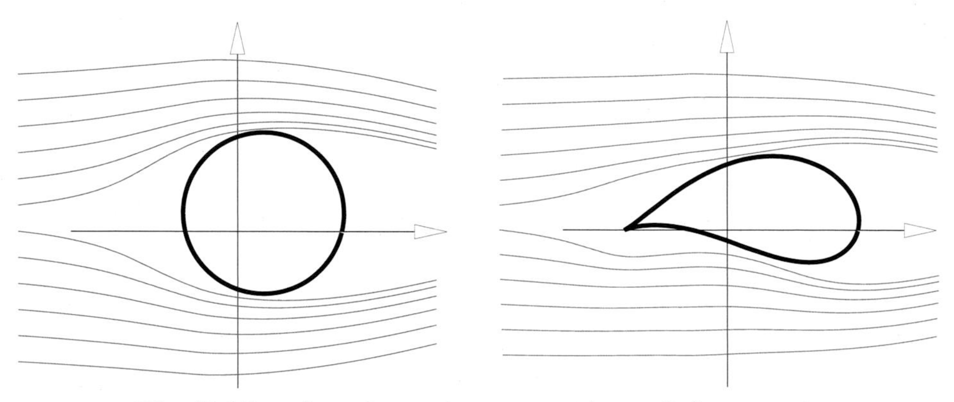

**Fig. B.25** conformal mapping $w = z + 1/z$, applied to a circle

▸▸▸ **Application**: *flow profiles*

In flow theory, the conformal mapping $w = z + 1/z$ is important. Fig. B.25 shows how a circle (center $z = 0.5 + 0.35\,i$, radius 2) is "transformed" by this mapping into a wing profile. Flows around circular profiles are easier to handle and the results can be carried over to wing profiles. ◂◂◂

▸▸▸ **Application**: *conformal patterns*

Describe the images of $x$-parallel and $y$-parallel lines as well as the images of arbitrary straight lines under the conformal mapping $w = e^z$. Show further that this mapping is angle preserving.

*Solution*:

We have $w = e^{x+iy} = e^x{\cdot}e^{iy}$. For constant $y$ ($x$-parallel), $w$ is a complex number with radius $e^x$ and a fixed polar angle $y$. Thus, all these complex numbers correspond to points on straight lines through $w = 0$. The distance from the origin increases exponentially with $x$. For constant $x$ ($y$-parallels), all these complex numbers correspond to points with constant (equal) absolute value and varying polar angle $y$.

For a generic straight line $y = kx + d$ ($k$, $d \in \mathbb{R}$), the mapping results in

$$w = e^x \cdot e^{i(kx+d)} = e^{id} \cdot e^x \cdot e^{ikx}.$$

With variable $x$, the number $e^{ikx}$ travels along the unit circle with the velocity $kx$ proportional to $x$. By multiplying by the real value $e^x$, the radial distance $x$ is scaled exponentially. A rotation through the angle 1 corresponds to a

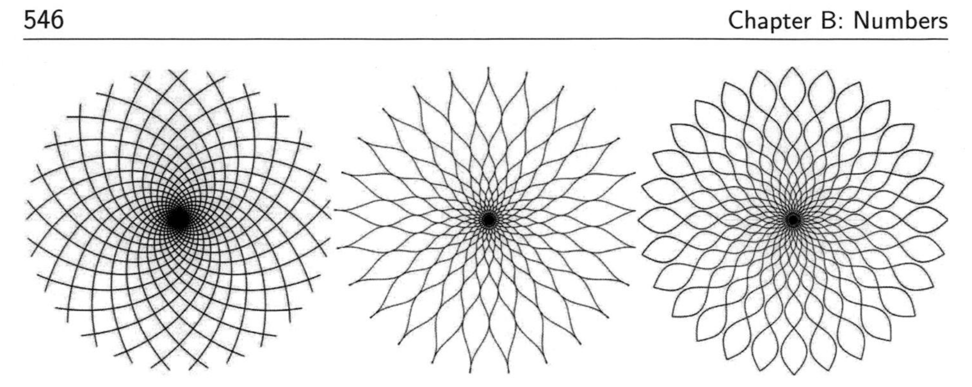

**Fig. B.26** flower-like patterns using $w = e^z$

scaling with the factor $1/k$. Therefore, the complex number (point in the complex plane) moves along a logarithmic spiral with the constant course angle $\varphi = \arctan k$ (Application p. 327), which corresponds to the slope of the initial line. Multiplication by the constant value $e^{id}$ causes an additional rotation of the spiral through the constant angle $d$. Thus, parallel lines are mapped to a *family of congruent logarithmic spirals.* The fact that the slope of the lines and the course angles of the spirals agree confirms that this mapping is conformal. In fact, the angle of intersection of two curves is measured between the tangents at the common point. These are two arbitrary straight lines (see Application p. 264).

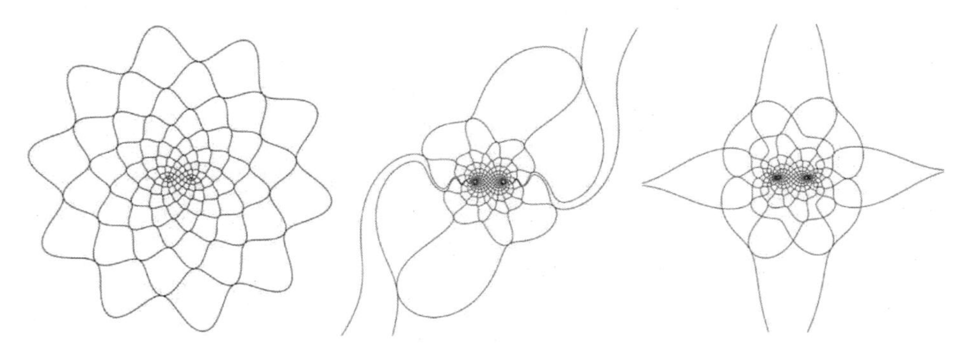

**Fig. B.27** conformal mappings of groups of sinusoidal curves (Franz *Gruber*)

Fig. B.26 on the left illustrates this for $k = \pm 1$, i.e. straight lines inclined at $\pm 45°$. The spirals intersect each other like the straight lines at right angles. In order to obtain the image in Fig. B.26 (center) or the one on the right, congruent sinusoids whose central lines are parallel to the $x$-axis are subjected to the mapping. In this way, you can easily obtain pretty patterns that are reminiscent of flowers and are difficult to describe without conformal transformations.

In Fig. B.27 on the left, the complex sine function was applied to a family of sinusoidal curves. The middle axes of the sinusoid is $x$-parallel. The two other figures show images of the sinusoids after applying the complex tangent function $w = \tan z = \sin z / \cos z$. Read more on the website that accompanies this book. ◂◂◂

**▸▸▸ Application:** ***a practical nautical chart***

In the 16th century, *Kremer* (lat. Mercator) designed a nautical chart on which straight lines should give the best lines of navigation between the continents. If $P(\lambda, \varphi)$ are the geographic coordinates (length and width) of a point $P$ on the globe, then the point $P^\circ$ in the map has the coordinates $(\lambda, \ln \varphi)$ (Fig. B.28).

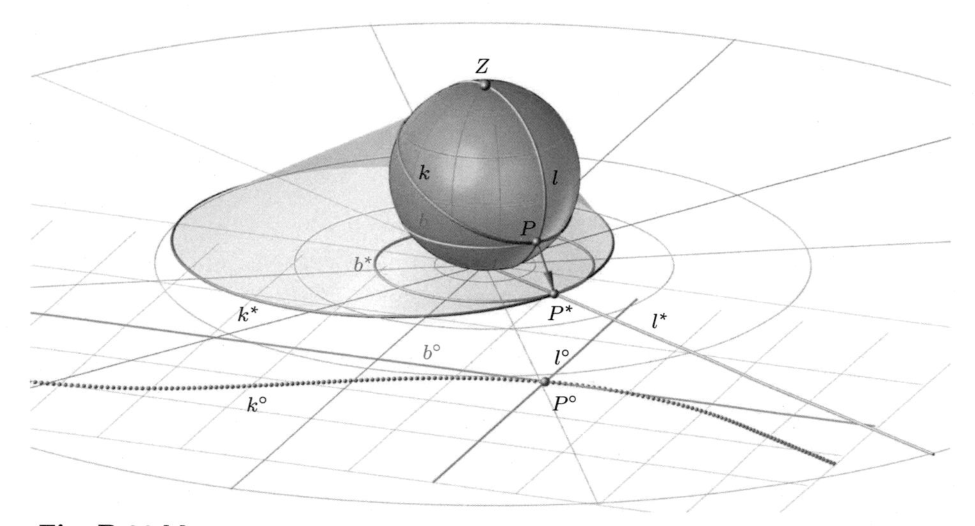


**Fig. B.28** Mercator projection = stereographic projection + conformal mapping

The people did not trust the matter until, a century later, *Kauffmann* (lat. also Mercator) verified the case. Not surprisingly, the projection is called "Mercator projection". With today's knowledge, the matter can quickly be explained: The points of the Earth are stereographically projected (Application p. 331) from the North Pole to the tangential plane in the South Pole ($P \mapsto P^*$). The meridian $l$ and the circle of latitude $k$ are mapped to a straight line $l^*$ and the circle $k^*$. The mapping is conformal and circle-preserving.
The pattern that we have obtained consists of radial lines and concentric circles. We achieved this by mapping a square grid with the conformal mapping $w = e^z$ (Application p. 545). If we apply to the above-mentioned pattern the inverse of the exponential function, that is, the equally conformal (angle-preserving) complex natural logarithm $w = \ln z$, then we can expect to obtain a rectangular grid: This results in points $P^\circ$, straight meridian circles $l^\circ$, and straight circles of latitudes $b^\circ$. The shortest routes for shipping would be great circles $k$. Unfortunately, these are mapped – viewed globally – to sinusoidal waved curves $k^\circ$. At that time and in the ideal case, a captain was steering a course with constant angle (against the meridians), and therefore, not too far from a great circle on a *loxodrome* on the sphere (Application p. 330).

This loxodrome is mapped to a straight line on the nautical chart because of the angle preserving property. ◂◂◂

### ▸▸▸ Application: *minimal curves*

Assume that $R$ = constant; $R, T, X, Y, Z \in \mathbb{C}$ and show that the complex helix

$$X = R \sin T, \ Y = R \cos T, \ Z = i\, RT \tag{B.22}$$

has arc length zero, and thus, deserves the name *minimal curve*. With real numbers, this is impossible.

***Proof***: We have to show that the distance between two neighboring points is always zero, that is, the "arc element" vanishes: $\sqrt{\dot{X}^2 + \dot{Y}^2 + \dot{Z}^2} \equiv 0$. However, with $\dot{X} = R \cos T$, $\dot{Y} = -R \sin T$ ($\Rightarrow \dot{X}^2 + \dot{Y}^2 = R^2$), $\dot{Z} = i\, R$ ($\Rightarrow \dot{Z}^2 = -R^2$), this is actually always fulfilled. ⊙ ◂◂◂

### ▸▸▸ Application: *bending of minimal surfaces*

*Weierstrass* proved that the real part of a minimal curve is a minimal surface (see Application p. 433). Calculate the real part of the complex helix from Formula (B.22) for $R = 1$ and $R = i$.

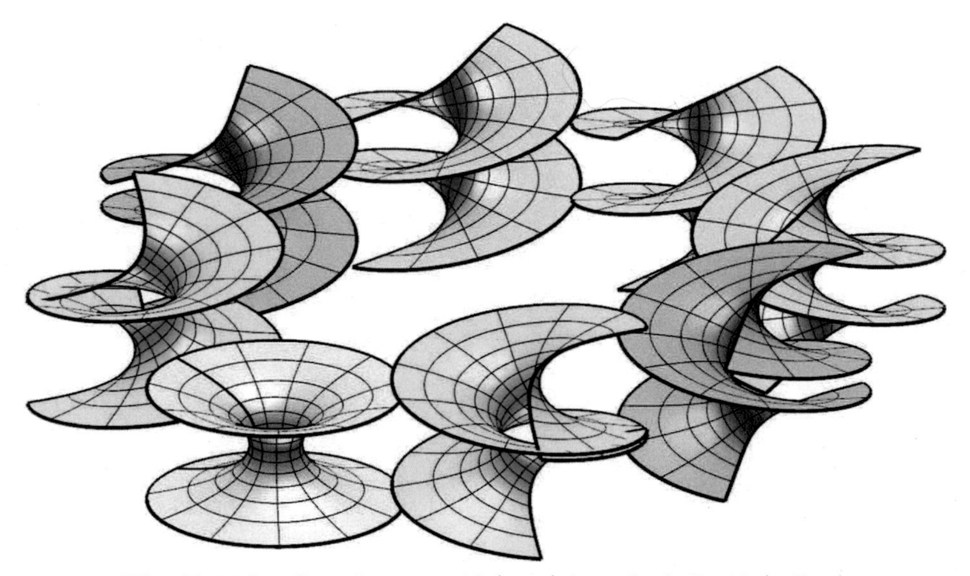

**Fig. B.29** bending the catenoid (gray) into the helicoid (yellow)

*Solution*:

In this example, for the sake of simplicity, real numbers are written in small characters and complex numbers (except the imaginary unit $i$) in capitals.

For $R = 1$, Formula (B.22) gives $X = \sin T$, $Y = \cos T$, $Z = i\, T$.

We set $T = u + i\, v$. Then, according to Formula (B.20) and Formula (B.21), the real parts thereof equal

$$x = \sin u \cosh v,\ y = \cos u\ \cosh v,\ z = -v. \tag{B.23}$$

If $u$ and $v$ are interpreted as surface parameters, then we can see in these equations the parameter representation of a catenoid (see Application p. 433 or Fig. B.29 on the left).
For $R = i$, Formula (B.22) gives $X = i \sin T,\ Y = i \cos T,\ Z = -T$.
The real parts thereof are $x = -\cos u \sinh v,\ y = \sin u\ \sinh v,\ z = -u$ or, with the substitution $r = \sinh v$

$$x = -r \cos u,\ y = r \sin u,\ z = -u. \tag{B.24}$$

With the surface parameters $r$ and $u$, this is a parameter representation of a helicoid (see Application p. 309 and Fig. B.29, right).

**Fig. B.30** "dance of minimal helical surfaces"

For the remaining complex unit numbers $R = e^{i\varphi}$, there are surfaces which are "intermediate stages" between the catenoid and the helicoid (again, a parameter representation is available). They are all minimal surfaces. Their calculation and representation is left to the computer. On the website that accompanies this book you will find a program that shows you the continuous bending from any desired view (`minimalflaechen.exe`). Since the program can calculate with complex numbers as well as with real numbers, the programming effort was very low.

If the minimal helix from Application p. 548 is replaced by another minimal curve, a variety of minimal surfaces is obtained which can be continuously bent into one another. The surfaces in Fig. B.30 are found when spiral lines are selected instead of helices.

⊕ *Remark*: The last examples should have made clear in what an incredibly efficient manner certain facts can be described using complex numbers. "From a higher point of view", or "in a higher dimension", complicated connections sometimes appear quite obvious. Likewise, when we move from the plane into space, we may see many similarities of conic sections in a different light. ⊕ ◂◂◂

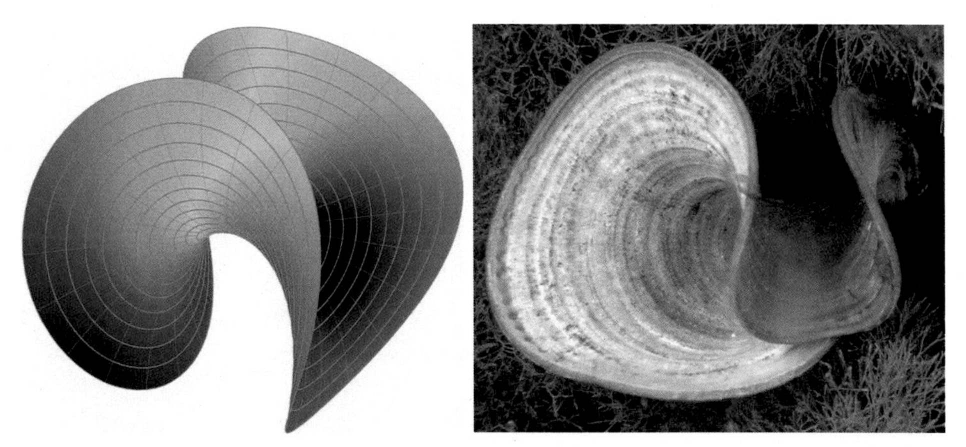

**Fig. B.31** Two fairly similar minimal surfaces: The one on the left is generated by formulas, the one on the right by "biological necessities".

▸▸▸ **Application**: ***minimal surfaces in nature*** (Fig. B.31)

If we look at the computer generated surface (it is, by the way, the famous Enneper surface, a rational minimal surface of degree nine), and compare it with the brown algae to the right, the similarity is stunning. Why do surfaces in nature tend to be minimal?

*Solution*:

A minimal surface can be bent into an infinite series of associated minimal surfaces, whereby the metric on the surface does not change. Points on the surface (skin, etc.), thus, have a constant distance (measured by means of "geodesic curves", i.e. shortest curves between the points). Even angles do not change. Therefore, a right triangle, for instance, would remain a right triangle. For animals or plants with surfaces that come close to minimal surfaces, this allows fast changes of shape without major impact to the surface. ◂◂◂

▸▸▸ **Application**: ***Mandelbrot sets*** (Fig. B.32, Fig. B.33)

Since *Benoît Mandelbrot* published his classic "Fractal Geometry of Nature" in 1977, fractals have become very popular. They play an important role in science. Let us consider the complex function $f(z) = z^2 + c$. Then, we define a sequence of complex numbers $z_0, z_1, z_2, \ldots$ as follows: $z_0 = 0, z_1 = f(z_0) = c, \ldots, z_{n+1} = f(z_n) = z_n^2 + c$. If the sequence $z_n$ remains bounded, then the number c is considered to be "good", and we mark it with a black point in the Gaussian plane. The result is a strange and very popular set, named after *Benoît Mandelbrot*. How is it possible to determine whether $z_n$ is bounded?

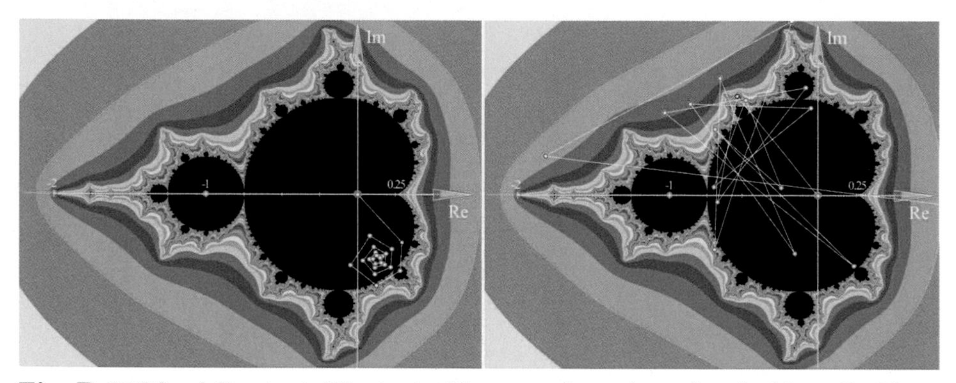


**Fig. B.32** Mandelbrot set: We start with a complex value $c$ (marked in red). Then, we apply the function $f(z) = z^2 + c$ several times (green). If the results stay inside a circle of radius 2 for, say, 100 times (left-hand side), we mark $c$ in black. If they do not (right-hand side), we mark the point in another color, depending on the number of iterations needed to determine that $z_n$ does not remain bounded.

*Solution*:
Without proof, $z_n$ will "blow up" if its radius of convergence is greater than 2. We let the computer do the work: Apply $f(z)$ until this happens. If it does not happen for a reasonable number of tries (e.g., 100), we consider $c$ as forming part of the set.

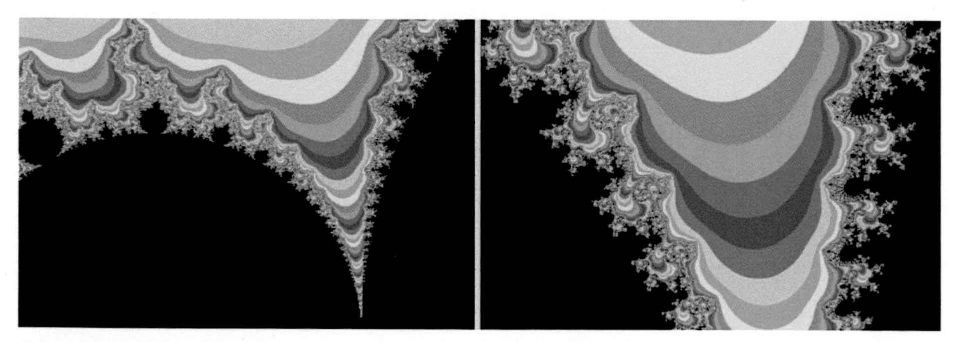

**Fig. B.33** The Mandelbrot set is a classical fractal: You can zoom in as much as you want. The resulting borderlines will not change.

Yet, in very few cases, this might be too inaccurate and we have to consider that the set is so extremely complicated that it is simply not worth checking further: Even if we zoom into the set again and again, no distinct borderline will appear.

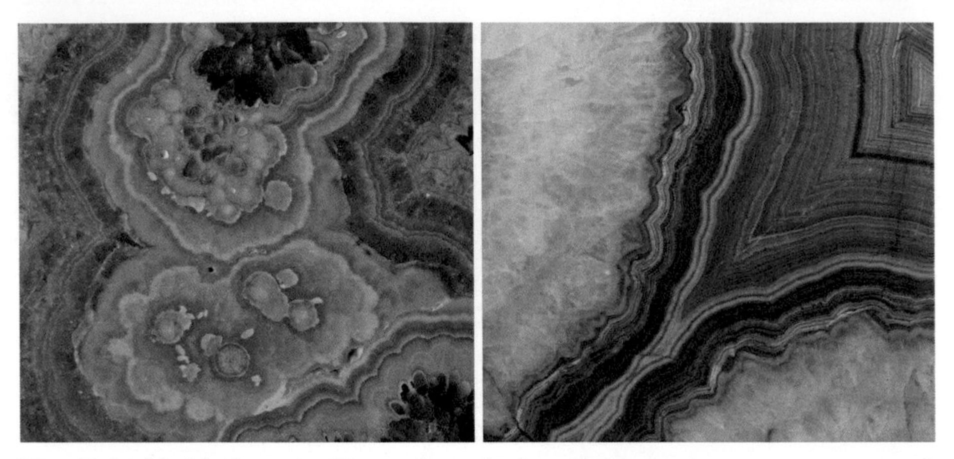

**Fig. B.34** Marble fractals: These photos look as if they were computer-generated.

Strictly mathematically, a fractal is a set that exhibits a repeating pattern displayed at every scale. In nature, this will not happen with the same accuracy. Nevertheless, when cutting stone, especially metamorphic rock composed of recrystallized carbonate minerals (marble), the cross-sections look quite similar to mathematically generated fractals (Fig. B.34). ◂◂◂

# Index




G. Glaeser, *Math Tools*, https://doi.org/10.1007/978-3-319-66960-1

## C

## J



## K



## L



## M

## N



## O

## P



## Q

Math is beautiful!

Printed in the United States
By Bookmasters